WITHDRAWN
FAIRFIELD UNIVERSITY
LIBRARY

Methods in Enzymology

Volume 198
PEPTIDE GROWTH FACTORS
Part C

METHODS IN ENZYMOLOGY

EDITORS-IN-CHIEF

John N. Abelson   Melvin I. Simon

DIVISION OF BIOLOGY
CALIFORNIA INSTITUTE OF TECHNOLOGY
PASADENA, CALIFORNIA

FOUNDING EDITORS

Sidney P. Colowick and Nathan O. Kaplan

Methods in Enzymology

Volume 198

Peptide Growth Factors

Part C

EDITED BY

David Barnes

DEPARTMENT OF BIOCHEMISTRY AND BIOPHYSICS
OREGON STATE UNIVERSITY
CORVALLIS, OREGON

J. P. Mather

DEPARTMENT OF CELL CULTURE
RESEARCH AND DEVELOPMENT
GENENTECH, INC.
SOUTH SAN FRANCISCO, CALIFORNIA

Gordon H. Sato

W. ALTON JONES CELL SCIENCE CENTER, INC.
LAKE PLACID, NEW YORK

ACADEMIC PRESS, INC.
Harcourt Brace Jovanovich, Publishers
San Diego   New York   Boston
London   Sydney   Tokyo   Toronto

This book is printed on acid-free paper. ∞

COPYRIGHT © 1991 BY ACADEMIC PRESS, INC.
All Rights Reserved.
No part of this publication may be reproduced or transmitted in any form or by any means, electronic or mechanical, including photocopy, recording, or any information storage and retrieval system, without permission in writing from the publisher.

ACADEMIC PRESS, INC.
San Diego, California 92101

United Kingdom Edition published by
ACADEMIC PRESS LIMITED
24-28 Oval Road, London NW1 7DX

LIBRARY OF CONGRESS CATALOG CARD NUMBER: 54-9110

ISBN 0-12-182099-8 (alk. paper)

PRINTED IN THE UNITED STATES OF AMERICA
91 92 93 94   10 9 8 7 6 5 4 3 2 1

Table of Contents

CONTRIBUTORS TO VOLUME 198 xi
PREFACE xvii
VOLUMES IN SERIES xix

Section I. Insulin-like Growth Factor, Nerve Growth Factor, and Platelet-Derived Growth Factor

1. Expression and Purification of Recombinant Insulin-like Growth Factors from *Escherichia coli* — BJÖRN NILSSON, GÖRAN FORSBERG, AND MARIS HARTMANIS   3

2. Insulin-like Growth Factor I Receptor cDNA Cloning — AXEL ULLRICH   17

3. Autoradiographic Localization of Insulin-like Growth Factor I Receptors in Rat Brain and Chick Embryo — M. A. LESNIAK, L. BASSAS, J. ROTH, AND J. M. HILL   26

4. *In Vivo* Assay of Neuron-Specific Effects of Nerve Growth Factor — MARK H. TUSZYNSKI AND FRED H. GAGE   35

5. Detection and Assay of Nerve Growth Factor mRNA — MARGARET FAHNESTOCK   48

6. Detection of Nerve Growth Factor Receptors after Gene Transfer — MOSES V. CHAO   61

7. Cloning and Expression of Human Platelet-Derived Growth Factor α and β Receptors — LENA CLAESSON-WELSH, ANDERS ERIKSSON, BENGT WESTERMARK, AND CARL-HENRIK HELDIN   72

8. Identification and Quantification of Polyphosphoinositides Produced in Response to Platelet-Derived Growth Factor Stimulation — LESLIE A. SERUNIAN, KURT R. AUGER, AND LEWIS C. CANTLEY   78

Section II. Fibroblast Growth Factor

9. Biaffinity Chromatography of Fibroblast Growth Factors — YUEN SHING   91

10. Cloning, Recombinant Expression, and Characterization of Basic Fibroblast Growth Factor	STEWART A. THOMPSON, ANDREW A. PROTTER, LOUISE BITTING, JOHN C. FIDDES, AND JUDITH A. ABRAHAM	96
11. Construction and Expression of Transforming Gene Resulting from Fusion of Basic Fibroblast Growth Factor Gene with Signal Peptide Sequence	SNEZNA ROGELJ, DAVID F. STERN, AND MICHAEL KLAGSBRUN	117
12. Identification and Characterization of Fibroblast Growth Factor-Related Transforming Gene *hst-1*	TERUHIKO YOSHIDA, KIYOSHI MIYAGAWA, HIROMI SAKAMOTO, TAKASHI SUGIMURA, AND MASAAKI TERADA	124
13. Phosphorylation and Identification of Phosphorylated Forms of Basic Fibroblast Growth Factor	JEAN-JACQUES FEIGE AND ANDREW BAIRD	138
14. Derivation of Monoclonal Antibody to Basic Fibroblast Growth Factor and Its Application	YOSHINO YOSHITAKE, KOUICHI MATSUZAKI, AND KATSUZO NISHIKAWA	148
15. Identification and Assay of Fibroblast Growth Factor Receptors	MIKIO KAN, ER-GANG SHI, AND WALLACE L. MCKEEHAN	158

Section III. Epidermal Growth Factor, Transforming Growth Factor α, and Related Factors

16. Expression of Epidermal Growth Factor Precursor cDNA in Animal Cells	BARBARA MROCZKOWSKI	175
17. Generation of Antibodies and Assays for Transforming Growth Factor α	CATHERINE LUCAS, TIMOTHY S. BRINGMAN, AND RIK DERYNCK	185
18. Molecular and Biochemical Approaches to Structure–Function Analysis of Transforming Growth Factor α	DEBORAH DEFEO-JONES, JOSEPH Y. TAI, AND ALLEN OLIFF	191
19. Assessment of Biological Activity of Synthetic Fragments of Transforming Growth Factor α	GREGORY SCHULTZ AND DANIEL TWARDZIK	200
20. Purification of Amphiregulin from Serum-Free Conditioned Medium of 12-*O*-Tetradecanoylphorbol 13-Acetate-Treated Cell Lines	MOHAMMED SHOYAB AND GREGORY D. PLOWMAN	213

Section IV. Epidermal Growth Factor Receptor and Related Receptors

21. Construction and Expression of Chimeric Cell Surface Receptors	REINER LAMMERS AND AXEL ULLRICH	225

22. Identification of Phosphorylation Sites: Use of the Epidermal Growth Factor Receptor	GARY J. HEISERMANN AND GORDON N. GILL	233
23. *In Vitro* Transcription of Epidermal Growth Factor Receptor Gene	RYOICHIRO KAGEYAMA AND GLENN T. MERLINO	242
24. Isolation of Cell Membrane for Epidermal Growth Factor Receptor Studies	PETER H. LIN, RICHARD SELINFREUND, AND WALKER WHARTON	251
25. Phosphorylation of Lipocortin-1 by the Epidermal Growth Factor Receptor	R. BLAKE PEPINSKY	260
26. Cloning, Expression, and Biological Effects of *erbB-2/neu* Gene in Mammalian Cells	PIER PAOLO DI FIORE, ORESTE SEGATTO, AND STUART A. AARONSON	272
27. Biological Effects of Monoclonal Antireceptor Antibodies Reactive with *neu* Oncogene Product, p185^{neu}	JEFFREY N. MYERS, JEFFREY A. DREBIN, TAKURO WADA, AND MARK I. GREENE	277
28. Quantification of *erbB-2/neu* Levels in Tissue	SOONMYOUNG PAIK, C. RICHTER KING, SUSAN SIMPSON, AND MARC E. LIPPMAN	290

Section V. Transforming Growth Factor β and Related Factors

29. Generation of Antibodies and Assays for Transforming Growth Factor β	CATHERINE LUCAS, BRIAN M. FENDLY, VENKAT R. MUKKU, WAI LEE WONG, AND MICHAEL A. PALLADINO	303
30. Purification of Transforming Growth Factors β1 and β2 from Bovine Bone and Cell Culture Assays	YASUSHI OGAWA AND SAEID M. SEYEDIN	317
31. Identification and Activation of Latent Transforming Growth Factor β	DAVID A. LAWRENCE	327
32. Assay of Astrocyte Differentiation-Inducing Activity of Serum and Transforming Growth Factor β	YOSHIO SAKAI AND DAVID BARNES	337
33. Erythroid Differentiation Bioassays for Activin	RALPH H. SCHWALL AND CORA LAI	340
34. Labeling Inhibin and Identifying Inhibin Binding to Cell Surface Receptors	TERESA K. WOODRUFF, JANE BATTAGLIA, JAMES BORREE, GLENN C. RICE, AND JENNIE P. MATHER	347

35. Bioassay, Purification, Cloning, and Expression of Müllerian Inhibiting Substance	DAVID T. MACLAUGHLIN, JAMES EPSTEIN, AND PATRICIA K. DONAHOE	358

Section VI. Other Growth Factors and Growth Inhibitors

36. Purification and Characterization of Recombinant Melanoma Growth Stimulating Activity	H. GREG THOMAS, JIN HEE HAN, EDDY BALENTIEN, RIK DERYNCK, RODOLFO BORDONI, AND ANN RICHMOND	373
37. Purification, Cloning, and Expression of Platelet-Derived Endothelial Cell Growth Factor	CARL-HENRIK HELDIN, ULF HELLMAN, FUYUKI ISHIKAWA, AND KOHEI MIYAZONO	383
38. Purification and Cloning of Vascular Endothelial Growth Factor Secreted by Pituitary Folliculostellate Cells	NAPOLEONE FERRARA, DAVID W. LEUNG, GEORGE CACHIANES, JANE WINER, AND WILLIAM J. HENZEL	391
39. Preparation and Bioassay of Connective Tissue Activating Peptide III and Its Isoforms	C. W. CASTOR, E. M. SMITH, M. C. BIGNALL, P. A. HOSSLER, AND T. H. SISSON	405
40. Purification of Growth Factors from Cartilage	YUKIO KATO, KAZUHISA NAKASHIMA, KATSUHIKO SATO, WEIQUN YAN, MASAHIRO IWAMOTO, AND FUJIO SUZUKI	416
41. Purification, Biological Assay, and Immunoassay of Mammary-Derived Growth Inhibitor	R. GROSSE, F.-D. BOEHMER, P. LANGEN, A. KURTZ, W. LEHMANN, M. MIETH, AND G. WALLUKAT	425
42. Assay and Purification of Naturally Occurring Inhibitor of Angiogenesis	PETER J. POLVERINI, NOEL P. BOUCK, AND FARZAN RASTINEJAD	440
43. Derivation of Monoclonal Antibody Directed against Fibroblast Growth Regulator	JOHN L. WANG	451

Section VII. Techniques for Study of Growth Factor Activity

44. Iodination of Peptide Growth Factors: Platelet-Derived Growth Factor and Fibroblast Growth Factor	ANGIE RIZZINO AND PETER KAZAKOFF	467

45. Localization of Peptide Growth Factors in the Nucleus	Bruno Gabriel, Véronique Baldin, Anna Maria Roman, Isabelle Bosc-Bierne, Jacqueline Noaillac-Depeyre, Hervé Prats, Justin Teissié, Gérard Bouche, and François Amalric	480
46. Antiphosphotyrosine Antibodies in Oncogene and Receptor Research	David F. Stern	494
47. Assays for Bone Resorption and Bone Formation	G. R. Mundy, G. D. Roodman, L. F. Bonewald, R. O. C. Oreffo, and B. F. Boyce	502
48. 3-[(3-Cholamidopropyl)dimethylammonio]-1-propane Sulfonate as Noncytotoxic Stabilizing Agent for Growth Factors	Yuhsi Matuo, Nozomu Nishi, Kunio Matsumoto, Kaoru Miyazaki, Keishi Matsumoto, Fujio Suzuki, and Katsuzo Nishikawa	511
49. Transgenic Mouse Models for Growth Factor Studies	Nora Sarvetnick	519

Section VIII. Cross-Index to Prior Volumes

50. Previously Published Articles from *Methods in Enzymology* Related to Peptide Growth Factors	529

AUTHOR INDEX 533

SUBJECT INDEX 563

Contributors to Volume 198

Article numbers are in parentheses following the names of contributors.
Affiliations listed are current.

STUART A. AARONSON (26), *Laboratory of Cellular and Molecular Biology, National Cancer Institute, National Institutes of Health, Bethesda, Maryland 20892*

JUDITH A. ABRAHAM (10), *California Biotechnology, Inc., Mountain View, California 94043*

FRANÇOIS AMALRIC (45), *Centre de Recherche de Biochimie et de Génétique Cellulaires, Centre National de la Recherche Scientifique, 31062 Toulouse Cédex, France*

KURT R. AUGER (8), *Department of Physiology, Tufts University School of Medicine, Boston, Massachusetts 02111*

ANDREW BAIRD (13), *Department of Molecular and Cellular Growth Biology, The Whittier Institute for Diabetes and Endocrinology, La Jolla, California 92037*

VÉRONIQUE BALDIN (45), *Centre de Recherche de Biochimie et de Génétique Cellulaires, Centre National de la Recherche Scientifique, 31062 Toulouse Cédex, France*

EDDY BALENTIEN (36), *Clinical Immunopathology, Presbyterian University Hospital, Pittsburgh, Pennsylvania 15213*

DAVID BARNES (32), *Department of Biochemistry and Biophysics, Oregon State University, Corvallis, Oregon 97331*

L. BASSAS (3), *Foundacíon Puitvert, Barcelona 08025, Spain*

JANE BATTAGLIA (34), *Department of Cell Culture, Research and Development, Genentech, Inc., South San Francisco, California 94080*

M. C. BIGNALL (39), *Department of Internal Medicine, Rackham Arthritis Research Unit and Rheumatology Division, University of Michigan Medical School, Ann Arbor, Michigan 48109*

LOUISE BITTING (10), *Sleep Research Center, Stanford University School of Medicine, Stanford, California 94305*

F.-D. BOEHMER (41), *Central Institute of Molecular Biology, Academy of Sciences of the German Democratic Republic, D-1115 Berlin-Buch, Germany*

L. F. BONEWALD (47), *Department of Medicine, Division of Endocrinology and Metabolism, University of Texas Health Science Center, San Antonio, Texas 78284*

RODOLFO BORDONI (36), *Medical Research Service, VA Medical Center (Atlanta), Decatur, Georgia 30033*

JAMES BORREE (34), *Department of Immunology Research Technology, Genentech, Inc., South San Francisco, California 94080*

ISABELLE BOSC-BIERNE (45), *Centre de Recherche de Biochimie et de Génétique Cellulaires, Centre National de la Recherche Scientifique, 31062 Toulouse Cédex, France*

GÉRARD BOUCHE (45), *Centre de Recherche de Biochimie et de Génétique Cellulaires, Centre National de la Recherche Scientifique, 31062 Toulouse Cédex, France*

NOEL P. BOUCK (42), *Department of Microbiology/Immunology and Cancer Center, Northwestern University Medical and Dental Schools, Chicago, Illinois 60611*

B. F. BOYCE (47), *Department of Pathology, University of Texas Health Science Center, San Antonio, Texas 78284*

TIMOTHY S. BRINGMAN (17), *Protein Design Labs, Mountain View, California 94043*

GEORGE CACHIANES (38), *Department of Molecular Biology, Genentech, Inc., South San Francisco, California 94080*

LEWIS C. CANTLEY (8), *Department of Physiology, Tufts University School of Medicine, Boston, Massachusetts 02111*

C. W. CASTOR (39), *Department of Internal Medicine, Rackham Arthritis Research Unit and Rheumatology Division, University of Michigan Medical School, Ann Arbor, Michigan 48109*

MOSES V. CHAO (6), *Department of Cell Biology and Anatomy, Cornell University Medical College, New York, New York 10021*

LENA CLAESSON-WELSH (7), *Ludwig Institute for Cancer Research, S-751 24 Uppsala, Sweden*

DEBORAH DEFEO-JONES (18), *Department of Cancer Research, Merck Sharp and Dohme Research Laboratories, West Point, Pennsylvania 19486*

RIK DERYNCK (17, 36), *Department of Developmental Biology, Genentech, Inc., South San Francisco, California 94080*

PIER PAOLO DI FIORE (26), *Laboratory of Cellular and Molecular Biology, National Cancer Institute, National Institutes of Health, Bethesda, Maryland 20892*

PATRICIA K. DONAHOE (35), *Pediatric Surgical Research Laboratory, Massachusetts General Hospital, Boston, Massachusetts 02114*

JEFFREY A. DREBIN (27), *Department of Surgery, The Johns Hopkins University Hospital, Baltimore, Maryland 21205*

JAMES EPSTEIN (35), *Pediatric Surgical Research Laboratory, Massachusetts General Hospital, Boston, Massachusetts 02114*

ANDERS ERIKSSON (7), *Ludwig Institute for Cancer Research, S-751 24 Uppsala, Sweden*

MARGARET FAHNESTOCK (5), *Department of Molecular Biology, SRI International, Menlo Park, California 94025*

JEAN-JACQUES FEIGE (13), *Unité INSERM 244, Fédération des Laboratoires de Biologie, Centre d'Etudes Nucléaires de Grenoble, 38041 Grenoble Cédex, France*

BRIAN M. FENDLY (29), *Department of Medicinal and Analytical Chemistry, Genentech, Inc., South San Francisco, California 94080*

NAPOLEONE FERRARA (38), *Department of Developmental Biology, Genentech, Inc., South San Francisco, California 94080*

JOHN C. FIDDES (10), *California Biotechnology, Inc., Mountain View, California 94043*

GÖRAN FORSBERG (1), *Department of Biochemistry, KabiGen AB, S-112 87 Stockholm, Sweden*

BRUNO GABRIEL (45), *Centre de Recherche de Biochimie et de Génétique Cellulaires, Centre National de la Recherche Scientifique, 31062 Toulouse Cédex, France*

FRED H. GAGE (4), *Department of Neurosciences, University of California, San Diego, La Jolla, California 92093*

GORDON N. GILL (22), *Department of Medicine, University of California, San Diego, La Jolla, California 92093*

MARK I. GREENE (27), *Department of Pathology and Laboratory Medicine, University of Pennsylvania School of Medicine, Philadelphia, Pennsylvania 19104*

R. GROSSE (41), *Central Institute of Molecular Biology, Academy of Sciences of the German Democratic Republic, D-1115 Berlin-Buch, Germany*

JIN HEE HAN (36), *Department of Cell Biology, Vanderbilt University School of Medicine, Nashville, Tennessee 37235*

MARIS HARTMANIS (1), *Department of Biochemistry, KabiGen AB, S-112 87 Stockholm, Sweden*

GARY J. HEISERMANN (22), *Cold Spring Harbor Laboratory, Cold Spring Harbor, New York 11724*

CARL-HENRIK HELDIN (7, 37), *Ludwig Institute for Cancer Research, S-751 24 Uppsala, Sweden*

ULF HELLMAN (37), *Ludwig Institute for Cancer Research, S-751 24 Uppsala, Sweden*

WILLIAM J. HENZEL (38), *Department of Protein Chemistry, Genentech, Inc., South San Francisco, California 94080*

J. M. HILL (3), *Unit on Neurochemistry, National Institute of Child Health and Human Development, National Institutes of Health, Bethesda, Maryland 20892*

P. A. HOSSLER (39), *Department of Internal Medicine, Rackham Arthritis Research Unit and Rheumatology Division, University of Michigan Medical School, Ann Arbor, Michigan 48109*

FUYUKI ISHIKAWA (37), *Third Department of Internal Medicine, Faculty of Medicine, University of Tokyo, Hongo, Tokyo 113, Japan*

MASAHIRO IWAMOTO (40), *Department of Anatomy and Histology, School of Dental Medicine, University of Pennsylvania, Philadelphia, Pennsylvania 19104*

RYOICHIRO KAGEYAMA (23), *Institute for Immunology, Kyoto University Faculty of Medicine, Kyoto 606, Japan*

MIKIO KAN (15), *W. Alton Jones Cell Science Center, Inc., Lake Placid, New York 12946*

YUKIO KATO (40), *Department of Biochemistry, Faculty of Dentistry, Osaka University, Osaka 565, Japan*

PETER KAZAKOFF (44), *Eppley Institute for Research in Cancer and Allied Diseases, University of Nebraska Medical Center, Omaha, Nebraska 68198*

C. RICHTER KING (28), *Laboratory of Oncogene Research, Molecular Oncology, Inc., Gaithersburg, Maryland 20878*

MICHAEL KLAGSBRUN (11), *Departments of Surgery and Biological Chemistry, The Children's Hospital and Harvard Medical School, Boston, Massachusetts 02115*

A. KURTZ (41), *Central Institute of Molecular Biology, Academy of Sciences of the German Democratic Republic, D-1115 Berlin-Buch, Germany*

CORA LAI (33), *Department of Developmental Biology, Genentech, Inc., South San Francisco, California 94080*

REINER LAMMERS (21), *Department of Molecular Biology, Max-Planck-Institute for Biochemistry, D-8033 Martinsried, Germany*

P. LANGEN (41), *Central Institute of Molecular Biology, Academy of Sciences of the German Democratic Republic, D-1115 Berlin-Buch, Germany*

DAVID A. LAWRENCE (31), *Unité 532 du Centre National de la Recherche Scientifique (CNRS), Institut Curie-Biologie, Centre Universitaire, F-91405 Orsay Cédex, France*

W. LEHMANN (41), *Central Institute of Molecular Biology, Academy of Sciences of the German Democratic Republic, D-1115 Berlin-Buch, Germany*

M. A. LESNIAK (3), *Diabetes Branch, National Institute of Diabetes and Digestive and Kidney Diseases, National Institutes of Health, Bethesda, Maryland 20892*

DAVID W. LEUNG (38), *Department of Molecular Biology, Genentech, Inc., South San Francisco, California 94080*

PETER H. LIN (24), *Roche Diagnostic Systems, Hoffmann-La Roche, Inc., Belleville, New Jersey 07109*

MARC E. LIPPMAN (28), *Vincent T. Lombardi Cancer Center, Georgetown University, Washington, D.C. 20007*

CATHERINE LUCAS (17, 29), *Department of Medicinal and Analytical Chemistry, Genentech, Inc., South San Francisco, California 94080*

DAVID T. MACLAUGHLIN (35), *Pediatric Surgical Research Laboratory, Massachusetts General Hospital, Boston, Massachusetts 02114*

JENNIE P. MATHER (34), *Department of Cell Culture, Research and Development, Genentech, Inc., South San Francisco, California 94080*

KEISHI MATSUMOTO (48), *Department of Pathology, Osaka University, Osaka 530, Japan*

KUNIO MATSUMOTO (48), *Department of Dermatology, Osaka University, Osaka 530, Japan*

KOUICHI MATSUZAKI (14), *Third Department of Internal Medicine, Toyama Medical and Pharmaceutical University, Toyama 930-01, Japan*

YUHSI MATUO (48), *Upstate Biotechnology, Inc., Lake Placid, New York 12946*

WALLACE L. MCKEEHAN (15), *W. Alton Jones Cell Science Center, Inc., Lake Placid, New York 12946*

GLENN T. MERLINO (23), *Laboratory of Molecular Biology, National Cancer Institute, National Institutes of Health, Bethesda, Maryland 20892*

M. MIETH (41), *Central Institute of Molecular Biology, Academy of Sciences of the German Democratic Republic, D-1115 Berlin-Buch, Germany*

KIYOSHI MIYAGAWA (12), *Genetics Division, National Cancer Center Research Institute, Tokyo 104, Japan*

KAORU MIYAZAKI (48), *Kihara Biological Institute, Yokohama 232, Japan*

KOHEI MIYAZONO (37), *Ludwig Institute for Cancer Research, S-751 24 Uppsala, Sweden*

BARBARA MROCZKOWSKI (16), *Department of Biochemistry, Vanderbilt University Medical School, Nashville, Tennessee 37232*

VENKAT R. MUKKU (29), *Department of Medicinal and Analytical Chemistry, Genentech, Inc., South San Francisco, California 94080*

G. R. MUNDY (47), *Department of Medicine, Division of Endocrinology and Metabolism, University of Texas Health Science Center, San Antonio, Texas 78284*

JEFFREY N. MYERS (27), *Department of Pathology and Laboratory Medicine, University of Pennsylvania School of Medicine, Philadelphia, Pennsylvania 19104*

KAZUHISA NAKASHIMA (40), *Department of Biochemistry, Faculty of Dentistry, Osaka University, Osaka 565, Japan*

BJÖRN NILSSON (1), *Department of Biochemistry, KabiGen AB, S-112 87 Stockholm, Sweden*

NOZOMU NISHI (48), *Department of Endocrinology, Kagawa Medical School, Kagawa 761-07, Japan*

KATSUZO NISHIKAWA (14, 48), *Department of Biochemistry, Kanazawa Medical University, Kanazawa 920-02, Japan*

JACQUELINE NOAILLAC-DEPEYRE (45), *Centre de Recherche de Biochimie et de Génétique Cellulaires, Centre National de la Recherche Scientifique, 31062 Toulouse Cédex, France*

YASUSHI OGAWA (30), *Department of Protein Chemistry, Celtrix Laboratories, Palo Alto, California 94303*

ALLEN OLIFF (18), *Department of Cancer Research, Merck Sharp and Dohme Research Laboratories, West Point, Pennsylvania 19486*

R. O. C. OREFFO (47), *ICI Pharmaceuticals, Cheshire, England*

SOONMYOUNG PAIK (28), *Department of Pathology, Vincent T. Lombardi Cancer Center, Georgetown University, Washington, D.C. 20007*

MICHAEL A. PALLADINO (29), *Immunology Research and Assay Technologies, Genentech, Inc., South San Francisco, California 94080*

R. BLAKE PEPINSKY (25), *Department of Protein Chemistry, Biogen, Inc., Cambridge, Massachusetts 02142*

GREGORY D. PLOWMAN (20), *Oncogen, Seattle, Washington 98121*

PETER J. POLVERINI (42), *Department of Pathology, Northwestern University Medical and Dental Schools, Chicago, Illinois 60611*

HERVÉ PRATS (45), *Centre de Recherche de Biochimie et de Génétique Cellulaires, Centre National de la Recherche Scientifique, 31062 Toulouse Cédex, France*

ANDREW A. PROTTER (10), *California Biotechnology, Inc., Mountain View, California 94043*

FARZAN RASTINEJAD (42), *Department of Pharmacology, Stanford University Medical School, Stanford, California 94305*

GLENN C. RICE (34), *Department of Immunology Research Technology, Genentech, Inc., South San Francisco, California 94080*

ANN RICHMOND (36), *Departments of Cell Biology and Medicine, Vanderbilt University, Nashville, Tennessee 37235 and, Veterans Administration Medical Center, Nashville, Tennessee 37232*

ANGIE RIZZINO (44), *Eppley Institute for Research in Cancer and Allied Diseases, University of Nebraska Medical Center, Omaha, Nebraska 68198*

SNEZNA ROGELJ (11), *Astra Research Centre India, Bangalore 560 003, India*

ANNA MARIA ROMAN (45), *Centre de Recherche de Biochimie et de Génétique Cellulaires, Centre National de la Recherche Scientifique, 31062 Toulouse Cédex, France*

G. D. ROODMAN (47), *Department of Medicine, Division of Hematology, University of Texas Health Science Center, San Antonio, Texas 78284*

J. ROTH (3), *National Institute of Diabetes and Digestive and Kidney Diseases, National Institutes of Health, Bethesda, Maryland 20892*

YOSHIO SAKAI (32), *Department of Anatomy, Mie University School of Medicine, Tsu, Mie 514, Japan*

HIROMI SAKAMOTO (12), *Genetics Division, National Cancer Center Research Institute, Tokyo 104, Japan*

NORA SARVETNICK (49), *Department of Neuropharmacology, Scripps Clinic and Research Foundation, La Jolla, California 92037*

KATSUHIKO SATO (40), *Department of Biochemistry, Faculty of Dentistry, Osaka University, Osaka 565, Japan*

GREGORY SCHULTZ (19), *Department of Obstetrics and Gynecology, University of Florida, Gainesville, Florida 32610*

RALPH H. SCHWALL (33), *Department of Developmental Biology, Genentech, Inc., South San Francisco, California 94080*

ORESTE SEGATTO (26), *Department of Molecular Biology, Max-Planck Institute for Biochemistry, D-8033 Martinsried, Germany*

RICHARD SELINFREUND (24), *Department of Pharmacology, Yale University School of Medicine, New Haven, Connecticut 06510*

LESLIE A. SERUNIAN (8), *Morgan and Finnegan, New York, New York 10022*

SAEID M. SEYEDIN (30), *Matrix Biosystems, Palo Alto, California 94304*

ER-GANG SHI (15), *W. Alton Jones Cell Science Center, Inc., Lake Placid, New York 12946*

YUEN SHING (9), *Departments of Surgery and Biological Chemistry, The Children's Hospital and Harvard Medical School, Boston, Massachusetts 02115*

MOHAMMED SHOYAB (20), *Oncogen, Seattle, Washington 98121*

SUSAN SIMPSON (28), *Vincent T. Lombardi Cancer Center, Georgetown University, Washington, D.C. 20007*

T. H. SISSON (39), *Department of Internal Medicine, Rackham Arthritis Research Unit and Rheumatology Division, University of Michigan Medical School, Ann Arbor, Michigan 48109*

E. M. SMITH (39), *Department of Internal Medicine, Rackham Arthritis Research Unit and Rheumatology Division, University of Michigan Medical School, Ann Arbor, Michigan 48109*

DAVID F. STERN (11, 46), *Department of Pathology, Yale University School of Medicine, New Haven, Connecticut 06510*

TAKASHI SUGIMURA (12), *Genetics Division, National Cancer Center Research Institute, Tokyo 104, Japan*

FUJIO SUZUKI (40, 48), *Department of Biochemistry, School of Dentistry, Osaka University, Osaka 565, Japan*

JOSEPH Y. TAI (18), *Department of Cancer Research, Merck Sharp and Dohme Research Laboratories, West Point, Pennsylvania 19486*

JUSTIN TEISSIÉ (45), *Centre de Recherche de Biochimie et de Génétique Cellulaires, Centre National de la Recherche Scientifique, 31062 Toulouse Cédex, France*

MASAAKI TERADA (12), *Genetics Division, National Cancer Center Research Institute, Tokyo 104, Japan*

H. GREG THOMAS (36), *Pharmaceutical Analysis, Reid-Powell, Marietta, Georgia 30062*

STEWART A. THOMPSON (10), *California Biotechnology, Inc., Mountain View, California 94043*

MARK H. TUSZYNSKI (4), *Department of Neurosciences, University of California, San Diego, La Jolla, California 92093*

DANIEL TWARDZIK (19), *Oncogen, Seattle, Washington 98121*

AXEL ULLRICH (2, 21), *Department of Molecular Biology, Max-Planck-Institute for Biochemistry, D-8033 Martinsried, Germany*

TAKURO WADA (27), *Department of Orthopedic Surgery, Sapporo Medical Clinic, Sapporo, Japan*

G. WALLUKAT (41), *Central Institute of Molecular Biology, Academy of Sciences of the German Democratic Republic, D-1115 Berlin-Buch, Germany*

JOHN L. WANG (43), *Department of Biochemistry, Michigan State University, East Lansing, Michigan 48824*

BENGT WESTERMARK (7), *Department of Pathology, University Hospital, S-751 85 Uppsala, Sweden*

WALKER WHARTON (24), *Life Sciences Division, Los Alamos National Laboratory, Los Alamos, New Mexico 87545*

JANE WINER (38), *Department of Molecular Biology, Genentech, Inc., South San Francisco, California 94080*

WAI LEE WONG (29), *Department of Medicinal and Analytical Chemistry, Genentech, Inc., South San Francisco, California 94080*

TERESA K. WOODRUFF (34), *Department of Cell Culture, Research and Development, Genentech, Inc., South San Francisco, California 94080*

WEIQUN YAN (40), *Department of Biochemistry, Faculty of Dentistry, Osaka University, Osaka 565, Japan*

TERUHIKO YOSHIDA (12), *Genetics Division, National Cancer Center Research Institute, Tokyo 104, Japan*

YOSHINO YOSHITAKE (14), *Department of Biochemistry, Kanazawa Medical University, Uchinada, Ishikawa 920-02, Japan*

Preface

Peptide Growth Factors, Part C, is a supplement to and update of Volumes 146 and 147 of *Methods in Enzymology*. As such, the aims and general organization remain the same, although there is a greater emphasis in this volume on molecular biological techniques and, to some extent, on techniques for *in vivo* studies. Major additions that reflect the direction of recent advances in the field include a separate section dealing with the EGF receptor and a number of chapters on growth inhibitors. Contributions are also included on the TGF beta-related peptides, activin, inhibin, and Müllerian inhibiting substance, serving to point out that the field of peptide growth factors really represents an extension of classic endocrinology.

The volume is divided into sections dealing with specific growth factors in which new, updated, or alternative procedures are presented for purification, bioassay, radiolabeling, and radioreceptor assay, immunoassay, receptor identification, and quantitation. These are followed by a section dealing with general techniques for the study of growth factors. In a few cases, such as situations in which oncogene products are treated as growth factors, we have included orientation chapters as introductions to the area for those unfamiliar with the oncogene–growth factor relationship.

As in the previous volumes, we have not addressed the specific growth factors of lymphoid cells, since these are covered in the Immunochemical Techniques volumes of *Methods in Enzymology*. Recently, the distinction between these factors and the growth factors covered in our volumes has become less marked, and investigators interested in the subject may wish to consult the other volumes for helpful procedures. Likewise, we have not attempted to cover completely the growth factor-related aspects of phospholipases, protein phosphorylation, G proteins, neuroendocrine peptides, calcium regulation or platelets, because volumes of *Methods in Enzymology* dealing with these topics have appeared since publication of Volumes 146 and 147. In article [50] a list of previous articles of *Methods in Enzymology* related to peptide growth factors is given.

We thank the authors who submitted chapters for this volume, particularly those who did so on time, and the authors who pleasantly tolerated our merciless attempts to condense chapters in order to reduce the size of the volume. We also thank the Editors-in-Chief, the staff of Academic

Press, who helped greatly with the details of preparation, and Emily Amonett–Wood, who kept it all organized. Finally, we wish to acknowledge our continuing debt to the late Sidney Colowick and Nate Kaplan. Many times in the preparation of this volume we were reminded of the countless ways in which Sidney and Nate helped bring out the best in those around them.

DAVID BARNES
J. P. MATHER
GORDON H. SATO

METHODS IN ENZYMOLOGY

VOLUME I. Preparation and Assay of Enzymes
Edited by SIDNEY P. COLOWICK AND NATHAN O. KAPLAN

VOLUME II. Preparation and Assay of Enzymes
Edited by SIDNEY P. COLOWICK AND NATHAN O. KAPLAN

VOLUME III. Preparation and Assay of Substrates
Edited by SIDNEY P. COLOWICK AND NATHAN O. KAPLAN

VOLUME IV. Special Techniques for the Enzymologist
Edited by SIDNEY P. COLOWICK AND NATHAN O. KAPLAN

VOLUME V. Preparation and Assay of Enzymes
Edited by SIDNEY P. COLOWICK AND NATHAN O. KAPLAN

VOLUME VI. Preparation and Assay of Enzymes (*Continued*)
Preparation and Assay of Substrates
Special Techniques
Edited by SIDNEY P. COLOWICK AND NATHAN O. KAPLAN

VOLUME VII. Cumulative Subject Index
Edited by SIDNEY P. COLOWICK AND NATHAN O. KAPLAN

VOLUME VIII. Complex Carbohydrates
Edited by ELIZABETH F. NEUFELD AND VICTOR GINSBURG

VOLUME IX. Carbohydrate Metabolism
Edited by WILLIS A. WOOD

VOLUME X. Oxidation and Phosphorylation
Edited by RONALD W. ESTABROOK AND MAYNARD E. PULLMAN

VOLUME XI. Enzyme Structure
Edited by C. H. W. HIRS

VOLUME XII. Nucleic Acids (Parts A and B)
Edited by LAWRENCE GROSSMAN AND KIVIE MOLDAVE

VOLUME XIII. Citric Acid Cycle
Edited by J. M. LOWENSTEIN

VOLUME XIV. Lipids
Edited by J. M. LOWENSTEIN

VOLUME XV. Steroids and Terpenoids
Edited by RAYMOND B. CLAYTON

VOLUME XVI. Fast Reactions
Edited by KENNETH KUSTIN

VOLUME XVII. Metabolism of Amino Acids and Amines (Parts A and B)
Edited by HERBERT TABOR AND CELIA WHITE TABOR

VOLUME XVIII. Vitamins and Coenzymes (Parts A, B, and C)
Edited by DONALD B. MCCORMICK AND LEMUEL D. WRIGHT

VOLUME XIX. Proteolytic Enzymes
Edited by GERTRUDE E. PERLMANN AND LASZLO LORAND

VOLUME XX. Nucleic Acids and Protein Synthesis (Part C)
Edited by KIVIE MOLDAVE AND LAWRENCE GROSSMAN

VOLUME XXI. Nucleic Acids (Part D)
Edited by LAWRENCE GROSSMAN AND KIVIE MOLDAVE

VOLUME XXII. Enzyme Purification and Related Techniques
Edited by WILLIAM B. JAKOBY

VOLUME XXIII. Photosynthesis (Part A)
Edited by ANTHONY SAN PIETRO

VOLUME XXIV. Photosynthesis and Nitrogen Fixation (Part B)
Edited by ANTHONY SAN PIETRO

VOLUME XXV. Enzyme Structure (Part B)
Edited by C. H. W. HIRS AND SERGE N. TIMASHEFF

VOLUME XXVI. Enzyme Structure (Part C)
Edited by C. H. W. HIRS AND SERGE N. TIMASHEFF

VOLUME XXVII. Enzyme Structure (Part D)
Edited by C. H. W. HIRS AND SERGE N. TIMASHEFF

VOLUME XXVIII. Complex Carbohydrates (Part B)
Edited by VICTOR GINSBURG

VOLUME XXIX. Nucleic Acids and Protein Synthesis (Part E)
Edited by LAWRENCE GROSSMAN AND KIVIE MOLDAVE

VOLUME XXX. Nucleic Acids and Protein Synthesis (Part F)
Edited by KIVIE MOLDAVE AND LAWRENCE GROSSMAN

VOLUME XXXI. Biomembranes (Part A)
Edited by SIDNEY FLEISCHER AND LESTER PACKER

VOLUME XXXII. Biomembranes (Part B)
Edited by SIDNEY FLEISCHER AND LESTER PACKER

VOLUME XXXIII. Cumulative Subject Index Volumes I–XXX
Edited by MARTHA G. DENNIS AND EDWARD A. DENNIS

VOLUME XXXIV. Affinity Techniques (Enzyme Purification: Part B)
Edited by WILLIAM B. JAKOBY AND MEIR WILCHEK

VOLUME XXXV. Lipids (Part B)
Edited by JOHN M. LOWENSTEIN

VOLUME XXXVI. Hormone Action (Part A: Steroid Hormones)
Edited by BERT W. O'MALLEY AND JOEL G. HARDMAN

VOLUME XXXVII. Hormone Action (Part B: Peptide Hormones)
Edited by BERT W. O'MALLEY AND JOEL G. HARDMAN

VOLUME XXXVIII. Hormone Action (Part C: Cyclic Nucleotides)
Edited by JOEL G. HARDMAN AND BERT W. O'MALLEY

VOLUME XXXIX. Hormone Action (Part D: Isolated Cells, Tissues, and Organ Systems)
Edited by JOEL G. HARDMAN AND BERT W. O'MALLEY

VOLUME XL. Hormone Action (Part E: Nuclear Structure and Function)
Edited by BERT W. O'MALLEY AND JOEL G. HARDMAN

VOLUME XLI. Carbohydrate Metabolism (Part B)
Edited by W. A. WOOD

VOLUME XLII. Carbohydrate Metabolism (Part C)
Edited by W. A. WOOD

VOLUME XLIII. Antibiotics
Edited by JOHN H. HASH

VOLUME XLIV. Immobilized Enzymes
Edited by KLAUS MOSBACH

VOLUME XLV. Proteolytic Enzymes (Part B)
Edited by LASZLO LORAND

VOLUME XLVI. Affinity Labeling
Edited by WILLIAM B. JAKOBY AND MEIR WILCHEK

VOLUME XLVII. Enzyme Structure (Part E)
Edited by C. H. W. HIRS AND SERGE N. TIMASHEFF

VOLUME XLVIII. Enzyme Structure (Part F)
Edited by C. H. W. HIRS AND SERGE N. TIMASHEFF

VOLUME XLIX. Enzyme Structure (Part G)
Edited by C. H. W. HIRS AND SERGE N. TIMASHEFF

VOLUME L. Complex Carbohydrates (Part C)
Edited by VICTOR GINSBURG

VOLUME LI. Purine and Pyrimidine Nucleotide Metabolism
Edited by PATRICIA A. HOFFEE AND MARY ELLEN JONES

VOLUME LII. Biomembranes (Part C: Biological Oxidations)
Edited by SIDNEY FLEISCHER AND LESTER PACKER

VOLUME LIII. Biomembranes (Part D: Biological Oxidations)
Edited by SIDNEY FLEISCHER AND LESTER PACKER

VOLUME LIV. Biomembranes (Part E: Biological Oxidations)
Edited by SIDNEY FLEISCHER AND LESTER PACKER

VOLUME LV. Biomembranes (Part F: Bioenergetics)
Edited by SIDNEY FLEISCHER AND LESTER PACKER

VOLUME LVI. Biomembranes (Part G: Bioenergetics)
Edited by SIDNEY FLEISCHER AND LESTER PACKER

VOLUME LVII. Bioluminescence and Chemiluminescence
Edited by MARLENE A. DELUCA

VOLUME LVIII. Cell Culture
Edited by WILLIAM B. JAKOBY AND IRA PASTAN

VOLUME LIX. Nucleic Acids and Protein Synthesis (Part G)
Edited by KIVIE MOLDAVE AND LAWRENCE GROSSMAN

VOLUME LX. Nucleic Acids and Protein Synthesis (Part H)
Edited by KIVIE MOLDAVE AND LAWRENCE GROSSMAN

VOLUME 61. Enzyme Structure (Part H)
Edited by C. H. W. HIRS AND SERGE N. TIMASHEFF

VOLUME 62. Vitamins and Coenzymes (Part D)
Edited by DONALD B. MCCORMICK AND LEMUEL D. WRIGHT

VOLUME 63. Enzyme Kinetics and Mechanism (Part A: Initial Rate and Inhibitor Methods)
Edited by DANIEL L. PURICH

VOLUME 64. Enzyme Kinetics and Mechanism (Part B: Isotopic Probes and Complex Enzyme Systems)
Edited by DANIEL L. PURICH

VOLUME 65. Nucleic Acids (Part I)
Edited by LAWRENCE GROSSMAN AND KIVIE MOLDAVE

VOLUME 66. Vitamins and Coenzymes (Part E)
Edited by DONALD B. MCCORMICK AND LEMUEL D. WRIGHT

VOLUME 67. Vitamins and Coenzymes (Part F)
Edited by DONALD B. MCCORMICK AND LEMUEL D. WRIGHT

VOLUME 68. Recombinant DNA
Edited by RAY WU

VOLUME 69. Photosynthesis and Nitrogen Fixation (Part C)
Edited by ANTHONY SAN PIETRO

VOLUME 70. Immunochemical Techniques (Part A)
Edited by HELEN VAN VUNAKIS AND JOHN J. LANGONE

VOLUME 71. Lipids (Part C)
Edited by JOHN M. LOWENSTEIN

VOLUME 72. Lipids (Part D)
Edited by JOHN M. LOWENSTEIN

VOLUME 73. Immunochemical Techniques (Part B)
Edited by JOHN J. LANGONE AND HELEN VAN VUNAKIS

VOLUME 74. Immunochemical Techniques (Part C)
Edited by JOHN J. LANGONE AND HELEN VAN VUNAKIS

VOLUME 75. Cumulative Subject Index Volumes XXXI, XXXII, XXXIV–LX
Edited by EDWARD A. DENNIS AND MARTHA G. DENNIS

VOLUME 76. Hemoglobins
Edited by ERALDO ANTONINI, LUIGI ROSSI-BERNARDI, AND EMILIA CHIANCONE

VOLUME 77. Detoxication and Drug Metabolism
Edited by WILLIAM B. JAKOBY

VOLUME 78. Interferons (Part A)
Edited by SIDNEY PESTKA

VOLUME 79. Interferons (Part B)
Edited by SIDNEY PESTKA

VOLUME 80. Proteolytic Enzymes (Part C)
Edited by LASZLO LORAND

VOLUME 81. Biomembranes (Part H: Visual Pigments and Purple Membranes, I)
Edited by LESTER PACKER

VOLUME 82. Structural and Contractile Proteins (Part A: Extracellular Matrix)
Edited by LEON W. CUNNINGHAM AND DIXIE W. FREDERIKSEN

VOLUME 83. Complex Carbohydrates (Part D)
Edited by VICTOR GINSBURG

VOLUME 84. Immunochemical Techniques (Part D: Selected Immunoassays)
Edited by JOHN J. LANGONE AND HELEN VAN VUNAKIS

VOLUME 85. Structural and Contractile Proteins (Part B: The Contractile Apparatus and the Cytoskeleton)
Edited by DIXIE W. FREDERIKSEN AND LEON W. CUNNINGHAM

VOLUME 86. Prostaglandins and Arachidonate Metabolites
Edited by WILLIAM E. M. LANDS AND WILLIAM L. SMITH

VOLUME 87. Enzyme Kinetics and Mechanism (Part C: Intermediates, Stereochemistry, and Rate Studies)
Edited by DANIEL L. PURICH

VOLUME 88. Biomembranes (Part I: Visual Pigments and Purple Membranes, II)
Edited by LESTER PACKER

VOLUME 89. Carbohydrate Metabolism (Part D)
Edited by WILLIS A. WOOD

VOLUME 90. Carbohydrate Metabolism (Part E)
Edited by WILLIS A. WOOD

VOLUME 91. Enzyme Structure (Part I)
Edited by C. H. W. HIRS AND SERGE N. TIMASHEFF

VOLUME 92. Immunochemical Techniques (Part E: Monoclonal Antibodies and General Immunoassay Methods)
Edited by JOHN J. LANGONE AND HELEN VAN VUNAKIS

VOLUME 93. Immunochemical Techniques (Part F: Conventional Antibodies, Fc Receptors, and Cytotoxicity)
Edited by JOHN J. LANGONE AND HELEN VAN VUNAKIS

VOLUME 94. Polyamines
Edited by HERBERT TABOR AND CELIA WHITE TABOR

VOLUME 95. Cumulative Subject Index Volumes 61–74, 76–80
Edited by EDWARD A. DENNIS AND MARTHA G. DENNIS

VOLUME 96. Biomembranes [Part J: Membrane Biogenesis: Assembly and Targeting (General Methods; Eukaryotes)]
Edited by SIDNEY FLEISCHER AND BECCA FLEISCHER

VOLUME 97. Biomembranes [Part K: Membrane Biogenesis: Assembly and Targeting (Prokaryotes, Mitochondria, and Chloroplasts)]
Edited by SIDNEY FLEISCHER AND BECCA FLEISCHER

VOLUME 98. Biomembranes (Part L: Membrane Biogenesis: Processing and Recycling)
Edited by SIDNEY FLEISCHER AND BECCA FLEISCHER

VOLUME 99. Hormone Action (Part F: Protein Kinases)
Edited by JACKIE D. CORBIN AND JOEL G. HARDMAN

VOLUME 100. Recombinant DNA (Part B)
Edited by RAY WU, LAWRENCE GROSSMAN, AND KIVIE MOLDAVE

VOLUME 101. Recombinant DNA (Part C)
Edited by RAY WU, LAWRENCE GROSSMAN, AND KIVIE MOLDAVE

VOLUME 102. Hormone Action (Part G: Calmodulin and Calcium-Binding Proteins)
Edited by ANTHONY R. MEANS AND BERT W. O'MALLEY

VOLUME 103. Hormone Action (Part H: Neuroendocrine Peptides)
Edited by P. MICHAEL CONN

VOLUME 104. Enzyme Purification and Related Techniques (Part C)
Edited by WILLIAM B. JAKOBY

VOLUME 105. Oxygen Radicals in Biological Systems
Edited by LESTER PACKER

VOLUME 106. Posttranslational Modifications (Part A)
Edited by FINN WOLD AND KIVIE MOLDAVE

VOLUME 107. Posttranslational Modifications (Part B)
Edited by FINN WOLD AND KIVIE MOLDAVE

VOLUME 108. Immunochemical Techniques (Part G: Separation and Characterization of Lymphoid Cells)
Edited by GIOVANNI DI SABATO, JOHN J. LANGONE, AND HELEN VAN VUNAKIS

VOLUME 109. Hormone Action (Part I: Peptide Hormones)
Edited by LUTZ BIRNBAUMER AND BERT W. O'MALLEY

VOLUME 110. Steroids and Isoprenoids (Part A)
Edited by JOHN H. LAW AND HANS C. RILLING

VOLUME 111. Steroids and Isoprenoids (Part B)
Edited by JOHN H. LAW AND HANS C. RILLING

VOLUME 112. Drug and Enzyme Targeting (Part A)
Edited by KENNETH J. WIDDER AND RALPH GREEN

VOLUME 113. Glutamate, Glutamine, Glutathione, and Related Compounds
Edited by ALTON MEISTER

VOLUME 114. Diffraction Methods for Biological Macromolecules (Part A)
Edited by HAROLD W. WYCKOFF, C. H. W. HIRS, AND SERGE N. TIMASHEFF

VOLUME 115. Diffraction Methods for Biological Macromolecules (Part B)
Edited by HAROLD W. WYCKOFF, C. H. W. HIRS, AND SERGE N. TIMASHEFF

VOLUME 116. Immunochemical Techniques (Part H: Effectors and Mediators of Lymphoid Cell Functions)
Edited by GIOVANNI DI SABATO, JOHN J. LANGONE, AND HELEN VAN VUNAKIS

VOLUME 117. Enzyme Structure (Part J)
Edited by C. H. W. HIRS AND SERGE N. TIMASHEFF

VOLUME 118. Plant Molecular Biology
Edited by ARTHUR WEISSBACH AND HERBERT WEISSBACH

VOLUME 119. Interferons (Part C)
Edited by SIDNEY PESTKA

VOLUME 120. Cumulative Subject Index Volumes 81–94, 96–101

VOLUME 121. Immunochemical Techniques (Part I: Hybridoma Technology and Monoclonal Antibodies)
Edited by JOHN J. LANGONE AND HELEN VAN VUNAKIS

VOLUME 122. Vitamins and Coenzymes (Part G)
Edited by FRANK CHYTIL AND DONALD B. MCCORMICK

VOLUME 123. Vitamins and Coenzymes (Part H)
Edited by FRANK CHYTIL AND DONALD B. MCCORMICK

VOLUME 124. Hormone Action (Part J: Neuroendocrine Peptides)
Edited by P. MICHAEL CONN

VOLUME 125. Biomembranes (Part M: Transport in Bacteria, Mitochondria, and Chloroplasts: General Approaches and Transport Systems)
Edited by SIDNEY FLEISCHER AND BECCA FLEISCHER

VOLUME 126. Biomembranes (Part N: Transport in Bacteria, Mitochondria, and Chloroplasts: Protonmotive Force)
Edited by SIDNEY FLEISCHER AND BECCA FLEISCHER

VOLUME 127. Biomembranes (Part O: Protons and Water: Structure and Translocation)
Edited by LESTER PACKER

VOLUME 128. Plasma Lipoproteins (Part A: Preparation, Structure, and Molecular Biology)
Edited by JERE P. SEGREST AND JOHN J. ALBERS

VOLUME 129. Plasma Lipoproteins (Part B: Characterization, Cell Biology, and Metabolism)
Edited by JOHN J. ALBERS AND JERE P. SEGREST

VOLUME 130. Enzyme Structure (Part K)
Edited by C. H. W. HIRS AND SERGE N. TIMASHEFF

VOLUME 131. Enzyme Structure (Part L)
Edited by C. H. W. HIRS AND SERGE N. TIMASHEFF

VOLUME 132. Immunochemical Techniques (Part J: Phagocytosis and Cell-Mediated Cytotoxicity)
Edited by GIOVANNI DI SABATO AND JOHANNES EVERSE

VOLUME 133. Bioluminescence and Chemiluminescence (Part B)
Edited by MARLENE DELUCA AND WILLIAM D. MCELROY

VOLUME 134. Structural and Contractile Proteins (Part C: The Contractile Apparatus and the Cytoskeleton)
Edited by RICHARD B. VALLEE

VOLUME 135. Immobilized Enzymes and Cells (Part B)
Edited by KLAUS MOSBACH

VOLUME 136. Immobilized Enzymes and Cells (Part C)
Edited by KLAUS MOSBACH

VOLUME 137. Immobilized Enzymes and Cells (Part D)
Edited by KLAUS MOSBACH

VOLUME 138. Complex Carbohydrates (Part E)
Edited by VICTOR GINSBURG

VOLUME 139. Cellular Regulators (Part A: Calcium- and Calmodulin-Binding Proteins)
Edited by ANTHONY R. MEANS AND P. MICHAEL CONN

VOLUME 140. Cumulative Subject Index Volumes 102–119, 121–134

VOLUME 141. Cellular Regulators (Part B: Calcium and Lipids)
Edited by P. MICHAEL CONN AND ANTHONY R. MEANS

VOLUME 142. Metabolism of Aromatic Amino Acids and Amines
Edited by SEYMOUR KAUFMAN

VOLUME 143. Sulfur and Sulfur Amino Acids
Edited by WILLIAM B. JAKOBY AND OWEN GRIFFITH

VOLUME 144. Structural and Contractile Proteins (Part D: Extracellular Matrix)
Edited by LEON W. CUNNINGHAM

VOLUME 145. Structural and Contractile Proteins (Part E: Extracellular Matrix)
Edited by LEON W. CUNNINGHAM

VOLUME 146. Peptide Growth Factors (Part A)
Edited by DAVID BARNES AND DAVID A. SIRBASKU

VOLUME 147. Peptide Growth Factors (Part B)
Edited by DAVID BARNES AND DAVID A. SIRBASKU

VOLUME 148. Plant Cell Membranes
Edited by LESTER PACKER AND ROLAND DOUCE

VOLUME 149. Drug and Enzyme Targeting (Part B)
Edited by RALPH GREEN AND KENNETH J. WIDDER

VOLUME 150. Immunochemical Techniques (Part K: *In Vitro* Models of B and T Cell Functions and Lymphoid Cell Receptors)
Edited by GIOVANNI DI SABATO

VOLUME 151. Molecular Genetics of Mammalian Cells
Edited by MICHAEL M. GOTTESMAN

VOLUME 152. Guide to Molecular Cloning Techniques
Edited by SHELBY L. BERGER AND ALAN R. KIMMEL

VOLUME 153. Recombinant DNA (Part D)
Edited by RAY WU AND LAWRENCE GROSSMAN

VOLUME 154. Recombinant DNA (Part E)
Edited by RAY WU AND LAWRENCE GROSSMAN

VOLUME 155. Recombinant DNA (Part F)
Edited by RAY WU

VOLUME 156. Biomembranes (Part P: ATP-Driven Pumps and Related Transport: The Na,K-Pump)
Edited by SIDNEY FLEISCHER AND BECCA FLEISCHER

VOLUME 157. Biomembranes (Part Q: ATP-Driven Pumps and Related Transport: Calcium, Proton, and Potassium Pumps)
Edited by SIDNEY FLEISCHER AND BECCA FLEISCHER

VOLUME 158. Metalloproteins (Part A)
Edited by JAMES F. RIORDAN AND BERT L. VALLEE

VOLUME 159. Initiation and Termination of Cyclic Nucleotide Action
Edited by JACKIE D. CORBIN AND ROGER A. JOHNSON

VOLUME 160. Biomass (Part A: Cellulose and Hemicellulose)
Edited by WILLIS A. WOOD AND SCOTT T. KELLOGG

VOLUME 161. Biomass (Part B: Lignin, Pectin, and Chitin)
Edited by WILLIS A. WOOD AND SCOTT T. KELLOGG

VOLUME 162. Immunochemical Techniques (Part L: Chemotaxis and Inflammation)
Edited by GIOVANNI DI SABATO

VOLUME 163. Immunochemical Techniques (Part M: Chemotaxis and Inflammation)
Edited by GIOVANNI DI SABATO

VOLUME 164. Ribosomes
Edited by HARRY F. NOLLER, JR., AND KIVIE MOLDAVE

VOLUME 165. Microbial Toxins: Tools for Enzymology
Edited by SIDNEY HARSHMAN

VOLUME 166. Branched-Chain Amino Acids
Edited by ROBERT HARRIS AND JOHN R. SOKATCH

VOLUME 167. Cyanobacteria
Edited by LESTER PACKER AND ALEXANDER N. GLAZER

VOLUME 168. Hormone Action (Part K: Neuroendocrine Peptides)
Edited by P. MICHAEL CONN

VOLUME 169. Platelets: Receptors, Adhesion, Secretion (Part A)
Edited by JACEK HAWIGER

VOLUME 170. Nucleosomes
Edited by PAUL M. WASSARMAN AND ROGER D. KORNBERG

VOLUME 171. Biomembranes (Part R: Transport Theory: Cells and Model Membranes)
Edited by SIDNEY FLEISCHER AND BECCA FLEISCHER

VOLUME 172. Biomembranes (Part S: Transport: Membrane Isolation and Characterization)
Edited by SIDNEY FLEISCHER AND BECCA FLEISCHER

VOLUME 173. Biomembranes [Part T: Cellular and Subcellular Transport: Eukaryotic (Nonepithelial) Cells]
Edited by SIDNEY FLEISCHER AND BECCA FLEISCHER

VOLUME 174. Biomembranes [Part U: Cellular and Subcellular Transport: Eukaryotic (Nonepithelial) Cells]
Edited by SIDNEY FLEISCHER AND BECCA FLEISCHER

VOLUME 175. Cumulative Subject Index Volumes 135–139, 141–167

VOLUME 176. Nuclear Magnetic Resonance (Part A: Spectral Techniques and Dynamics)
Edited by NORMAN J. OPPENHEIMER AND THOMAS L. JAMES

VOLUME 177. Nuclear Magnetic Resonance (Part B: Structure and Mechanism)
Edited by NORMAN J. OPPENHEIMER AND THOMAS L. JAMES

VOLUME 178. Antibodies, Antigens, and Molecular Mimicry
Edited by JOHN J. LANGONE

VOLUME 179. Complex Carbohydrates (Part F)
Edited by VICTOR GINSBURG

VOLUME 180. RNA Processing (Part A: General Methods)
Edited by JAMES E. DAHLBERG AND JOHN N. ABELSON

VOLUME 181. RNA Processing (Part B: Specific Methods)
Edited by JAMES E. DAHLBERG AND JOHN N. ABELSON

VOLUME 182. Guide to Protein Purification
Edited by MURRAY P. DEUTSCHER

VOLUME 183. Molecular Evolution: Computer Analysis of Protein and Nucleic Acid Sequences
Edited by RUSSELL F. DOOLITTLE

VOLUME 184. Avidin–Biotin Technology
Edited by MEIR WILCHEK AND EDWARD A. BAYER

VOLUME 185. Gene Expression Technology
Edited by DAVID V. GOEDDEL

VOLUME 186. Oxygen Radicals in Biological Systems (Part B: Oxygen Radicals and Antioxidants)
Edited by LESTER PACKER AND ALEXANDER N. GLAZER

VOLUME 187. Arachidonate Related Lipid Mediators
Edited by ROBERT C. MURPHY AND FRANK A. FITZPATRICK

VOLUME 188. Hydrocarbons and Methylotrophy
Edited by MARY E. LIDSTROM

VOLUME 189. Retinoids (Part A: Molecular and Metabolic Aspects)
Edited by LESTER PACKER

VOLUME 190. Retinoids (Part B: Cell Differentiation and Clinical Applications)
Edited by LESTER PACKER

VOLUME 191. Biomembranes (Part V: Cellular and Subcellular Transport: Epithelial Cells)
Edited by SIDNEY FLEISCHER AND BECCA FLEISCHER

VOLUME 192. Biomembranes (Part W: Cellular and Subcellular Transport: Epithelial Cells)
Edited by SIDNEY FLEISCHER AND BECCA FLEISCHER

VOLUME 193. Mass Spectrometry
Edited by JAMES A. MCCLOSKEY

VOLUME 194. Guide to Yeast Genetics and Molecular Biology
Edited by CHRISTINE GUTHRIE AND GERALD R. FINK

VOLUME 195. Adenylyl Cyclase, G Proteins, and Guanylyl Cyclase
Edited by ROGER A. JOHNSON AND JACKIE D. CORBIN

VOLUME 196. Molecular Motors and the Cytoskeleton
Edited by RICHARD B. VALLEE

VOLUME 197. Phospholipases
Edited by EDWARD A. DENNIS

VOLUME 198. Peptide Growth Factors (Part C)
Edited by DAVID BARNES, J. P. MATHER, AND GORDON H. SATO

VOLUME 199. Cumulative Subject Index Volumes 168–174, 176–194 (in preparation)

VOLUME 200. Protein Phosphorylation (Part A: Protein Kinases: Assays, Purification, Antibodies, Functional Analysis, Cloning, and Expression) (in preparation)
Edited by TONY HUNTER AND BARTHOLOMEW M. SEFTON

VOLUME 201. Protein Phosphorylation (Part B: Analysis of Protein Phosphorylation, Protein Kinase Inhibitors, and Protein Phosphatases) (in preparation)
Edited by TONY HUNTER AND BARTHOLOMEW M. SEFTON

VOLUME 202. Molecular Design and Modeling: Concepts and Applications (Part A: Proteins, Peptides, and Enzymes) (in preparation)
Edited by JOHN J. LANGONE

VOLUME 203. Molecular Design and Modeling: Concepts and Applications (Part B: Antibodies and Antigens, Nucleic Acids, Polysaccharides, and Drugs) (in preparation)
Edited by JOHN J. LANGONE

Section I

Insulin-like Growth Factor, Nerve Growth Factor, and Platelet-Derived Growth Factor

[1] Expression and Purification of Recombinant Insulin-like Growth Factors from *Escherichia coli*

By BJÖRN NILSSON, GÖRAN FORSBERG, and MARIS HARTMANIS

Introduction

In this chapter we describe the use of a general system for the expression of peptide hormones in a soluble form in bacteria. The peptides of interest are expressed at very high levels of fusion proteins secreted to the growth medium of *Escherichia coli*. The products are fused to an *in vitro* designed immunoglobulin G (IgG)-binding domain based on staphylococcal protein A (SpA), which allows the fusion proteins to be purified by IgG affinity chromatography. Recently, the general uses of SpA fusions were described in a volume of this series.[1] Here, the use of SpA fusions for the expression of peptide hormones is presented by describing a set of optimized expression vectors, cloning strategies, affinity purification, as well as cleavages of fusion proteins. This SpA fusion concept, also known as the EcoSec expression system, is demonstrated by the expression and purification of active and native insulin-like growth factor I (IGF-I), truncated IGF-I (tIGF-I), and insulin-like growth factor II (IGF-II).

Bacterial expression of mammalian peptides is an obvious application of recombinant DNA technology, which started with the expression of human somatostatin in *E. coli* in 1977.[2] Naturally, the production of eukaryotic proteins in bacteria is of considerable commercial interest for the pharmaceutical industry, but the techniques have also opened new possibilities for the scientific community to obtain large quantities of proteins and peptides that are hard to purify from natural sources and too large to be synthesized chemically. It is also possible to express modified proteins, constructed by *in vitro* mutagenesis, by bacterial expression systems. To date, a large number of recombinant mammalian proteins have been expressed in bacteria by the use of various bacterial expression systems (for a review, see Ref. 3). The major reason for the development of a number of similar bacterial expression systems is the large number of problems observed when expressing heterologous proteins in bacteria,

---

[1] B. Nilsson and L. Abrahmsén, this series, Vol. 185, p. 144.
[2] K. Itakura, T. Hirose, R. Crea, A. D. Riggs, H. L. Heynecker, F. Bolivar, and H. W. Boyer, *Science* **198**, 1056 (1977).
[3] F. A. O. Marston, *Biochem. J.* **240**, 1 (1986).

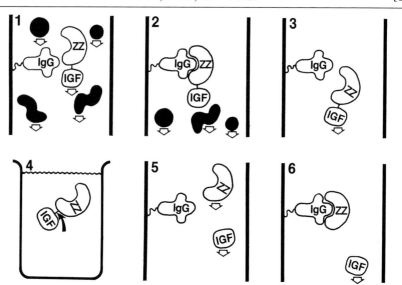

FIG. 1. Concept of affinity purification of IGFs using the IgG-binding ZZ fusion protein concept. (1) Application of filtered culture growth medium, or a crude extract from an osmotic shock procedure, to the IgG-Sepharose column. (2) Binding of the ZZ–IGF fusion protein to the IgG column while other proteins flow through. (3) Elution of the fusion protein using 0.2 M acetic acid, pH 3.2. (4) Cleavage of the fusion protein by chemical or enzymatic methods to release the IGF moiety from the ZZ affinity handle. (5) Application of the cleavage mixture to the IgG affinity column. (6) Uncleaved fusion protein and the ZZ affinity handle bind to the IgG while the liberated product flows through.

for example, proteolytic degradation, low biological activity of purified proteins and peptides, and precipitation of the product *in vivo*.[3,4]

The difficulties involved in recovering active peptides from bacteria have resulted in the development of "second generation" expression systems, in which purification of the heterologous proteins is simplified by expressing the product peptide as a fusion protein. The scheme to purify IGFs by such a fusion strategy is shown in Fig. 1. The fusion partner should be proteolytically stable and have a strong affinity to a ligand so that fusion proteins can be purified by ligand affinity chromatography. To date, several proteins have been utilized as such partners in fusion proteins (for a review, see Uhlén and Moks[4]). These fusion techniques have greatly facilitated the recovery of heterologous gene products expressed in bacteria, even though difficulties in finding general methods to cleave fusion proteins have been a serious bottleneck.

[4] M. Uhlén and T. Moks, this series, Vol. 185, p. 129.

Fusions to Staphylococcal Protein A

Staphylococcal protein A is to date one of the most widely used fusion partners in second-generation expression systems. SpA fusions have been utilized for high level expression of peptide hormones,[5-8] for immobilization of enzymes,[9,10] and for production of specific antibodies against gene products.[11-13] SpA has its biological function as a pathogenicity factor on the cell surface of the bacterium *Staphylococcus aureus*.[14] Its pathogenic effect is correlated to its ability to bind to the Fc portion of most mammalian class G immunoglobulins,[14,15] but the detailed biological significance of this activity is not yet established.[16] Extensive *in vitro* studies of the IgG-binding activity have been performed through the years (for a review, see Langone[14]). The SpA protein, as concluded from protein analysis[17] and from the deduced amino acid sequence of the *spa* structural gene,[18] consists of three structurally and functionally distinct regions (Fig. 2A): (i) From the amino terminus, the first region is the signal peptide, consisting of 36 amino acid residues, which is cleaved off during translocation through the cytoplasmic membrane.[19] (ii) The second region spans five highly homologous IgG-binding domains, E, D, A, B, and C, each with approximately 58 amino acid residues.[20] These domains can be cleaved apart by

---

[5] T. Moks, L. Abrahmsén, E. Holmgren, M. Bilich, A. Olsson, G. Pohl, C. Sterky, H. Hultberg, S. Josephson, A. Holmgren, H. Jörnvall, M. Uhlén, and B. Nilsson, *Biochemistry* **26**, 5239 (1987).

[6] T. Moks, L. Abrahmsén, B. Österlöf, S. Josephson, M. Östling, S.-O. Enfors, I. Persson, B. Nilsson, and M. Uhlén, *Bio/Technology* **5**, 379 (1987).

[7] L. Abrahmsén, T. Moks, B. Nilsson, and M. Uhlén, *Nucleic Acids Res.* **14**, 7487 (1986).

[8] B. Nilsson, E. Holmgren, S. Josephson, S. Gatenbeck, and M. Uhlén, *Nucleic Acids Res.* **13**, 1151 (1985).

[9] B. Nilsson, L. Abrahmsén, and M. Uhlén, *EMBO J.* **4**, 1075 (1985).

[10] M. Uhlén, B. Nilsson, B. Guss, M. Lindberg, S. Gatenbeck, and L. Philipson, *Gene* **23**, 419 (1983).

[11] B. Löwenadler, B. Nilsson, L. Abrahmsén, T. Moks, L. Ljungqvist, E. Holmgren, S. Paleus, S. Josephson, L. Philipson, and M. Uhlén, *EMBO J.* **5**, 2393 (1986).

[12] B. Löwenadler, B. Jansson, S. Paleus, E. Holmgren, B. Nilsson, T. Moks, G. Palm, S. Josephson, L. Philipson, and M. Uhlén, *Gene* **58**, 87 (1987).

[13] B. Guss, M. Uhlén, B. Nilsson, M. Lindberg, J. Sjöquist, and J. Sjödahl, *Eur. J. Biochem.* **138**, 413 (1984).

[14] J. J. Langone, *Adv. Immunol.* **32**, 157 (1982).

[15] R. Lindmark, K. Thorén-Tolling, and J. Sjöquist, *J. Immunol. Methods* **62**, 1 (1983).

[16] A. H. Patel, P. Nowlan, E. D. Weavers, and T. Foster, *Infect. Immun.* **55**, 3103 (1987).

[17] J. Sjödahl, *Eur. J. Biochem.* **78**, 471 (1977).

[18] M. Uhlén, B. Guss, B. Nilsson, S. Gatenbeck, L. Philipson, and M. Lindberg, *J. Biol. Chem.* **259**, 1695 (1984).

[19] L. Abrahmsén, T. Moks, B. Nilsson, U. Hellman, and M. Uhlén, *EMBO J.* **4**, 3901 (1985).

[20] T. Moks, L. Abrahmsén, B. Nilsson, U. Hellman, J. Sjoquist, and M. Uhlén. *Eur. J. Biochem.* **14**, 7487 (1986).

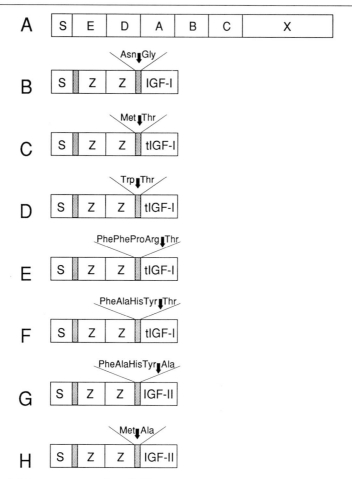

FIG. 2. Linear representation of different proteins. S is the signal peptide from SpA, which is cleaved off from the fusion protein during translocation through the cytoplasmic membrane. E, D, A, B, and C are the different IgG-binding domains of SpA. X is the portion of the SpA which is localized in the cell wall of *Staphylococcus aureus*. Z is an IgG-binding domain as described in the text and discussed by Nilsson *et al.* [B. Nilsson, T. Moks, B. Jansson, L. Abrahmsén, A. Elmblad, A. Holmgren, C. Henrichson, T. A. Jones, and M. Uhlén, *Protein Eng.* **1**, 107 (1987)]. (A) Structure of natural SpA from *S. aureus*.[18] (B) ZZ–IGF-I designed for release of IGF-I with hydroxylamine. (C) ZZ–tIGF-I designed for release of tIGF-I with partial cyanogen bromide treatment. (D) ZZ–tIGF-I designed for release of tIGF-I with N-chlorosuccinimide treatment. (E) ZZ–tIGF-I designed for proteolytic release of tIGF-I using thrombin. (F) ZZ–tIGF-I designed for release of tIGF-I with H64A subtilisin. (G) ZZ–IGF-II designed for release of IGF-II with H64A subtilisin. (H) ZZ–IGF-II designed for release of IGF-II with cyanogen bromide treatment.

trypsin and are individually IgG binding. (iii) The third region is the C-terminal region X which is partly anchored in the cytoplasmic membrane, with the remaining very hydrophilic portion located in the cell wall.[13]

The IgG-binding region of SpA has many properties which make it an ideal partner in fusion proteins for the expression of heterologous proteins. Some of the most important of these features are the following: (1) SpA has a strong affinity for the constant region (Fc) of IgG from most mammals, including man.[15] This specific affinity permits the purification of fusion proteins in a single step by IgG affinity chromatography.[6] (2) The three-dimensional structure of domain B of SpA, in a complex with Fc, has been solved, from X-ray crystallographic diffraction data, to a resolution of 2.8 Å.[21,22] The structure suggests that the IgG-binding motif consists of two antiparallel α helices and that the C-terminal end of the domain is flexible. Fusions to the flexible part of the domain may facilitate enzymatic cleavages of fusion proteins in engineered cleavage sequences and allow independent folding of a fused product and the SpA moiety. (3) SpA is proteolytically stable, and recombinant *spa* DNA constructions can be expressed in cells to high levels, not only in the homologous host *S. aureus*,[8,23] but also in the cytoplasm of *E. coli*,[9,13,24] or secreted through the cytoplasmic membrane.[5,6,9,19] (4) The proteolytic stability of gene products is often increased *in vivo* when fused to SpA.[8,25] Even though fusion to SpA does not abolish proteolytic activity, an increase in half-life of the fusion protein is a general observation where comparative studies have been undertaken.[25] A possible explanation is stabilization by the fusion to the proteolytically stable SpA, but as the effect is most apparent for secreted basic peptide hormones, a contribution to stability may also come from electrostatic interactions between the basic product and the acidic portion of SpA (p*I* 4) due to their opposite charges at neutral pH. Even though not fully explained, the increase in half-life *in vivo* is an important advantage of the SpA fusion strategy for the expression of soluble heterologous gene products in bacteria. (5) The expression of SpA in *E. coli* induces a leaky phenotype of the gram-negative bacterial cell, and periplasmic, but not intracellular, proteins will quantitatively leak out into the growth medium.[7] This unexplained phenomenon is technically very convenient, as the recombinant product can be recovered directly from the growth medium.[6]

---

[21] J. Deisenhofer, T. A. Jones, R. Huber, J. Sjödahl, and J. Sjöquist, *Hoppe-Seyler's Z. Physiol. Chem.* **359**, 975 (1978).
[22] J. Deisenhofer, *Biochemistry* **20**, 2361 (1981).
[23] M. Uhlén, B. Guss, B. Nilsson, F. Götz, and M. Lindberg, *J. Bacteriol.* **159**, 713 (1984).
[24] L. Monaco, H. M. Bond, K. E. Howell, and R. Cortese, *EMBO J.* **6**, 3253 (1987).
[25] B. Nilsson, I. D. Kuntz, and S. Anderson, in "Protein Folding" (J. King and L. Gierasch, eds.), p. 117. American Association for the Advancement of Science, Washington, D.C., 1990.

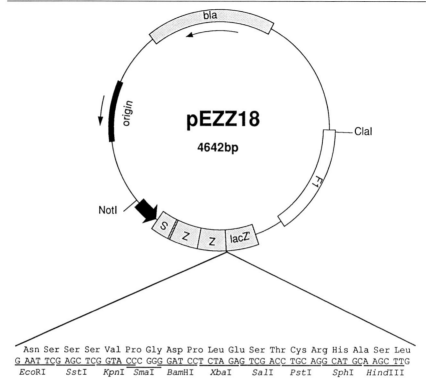

FIG. 3. The plasmid vector pEZZ18. Product genes are fused to the ZZ gene by cloning the product gene into the multirestriction enzyme cloning linker. The amino acid sequence of the protein, as deduced from the nucleotide sequence, is shown spanning the linker region. The box designated bla shows the position of the *bla* gene encoding TEM-β-lactamase, and the arrow under the box indicates the direction of transcription. F1 is the recognition sequence for packing the DNA into single-stranded DNA phage particles after infection with M13 or f1, S shows the SpA signal sequence, Z is the Z gene as described in the text, and lacZ' shows the position of the *lacZ'* structural gene. The black arrow indicates the position of the double promoter motif 5' of the *SZZ* gene. The origin of replication from pEMBL8[+ 27] is also indicated (see text for details).

Construction of pEZZ Expression Vector

The SpA fusion system has been developed sequentially, and here we present the pEZZ18 plasmid vector (Fig. 3) for the expression of peptide hormones and immunogenic peptides.[5,12] The design and construction of this expression plasmid have previously been described in detail.[26] The

---

[26] B. Nilsson, T. Moks, B. Jansson, L. Abrahmsén, A. Elmblad, E. Holmgren, C. Henrichson, T. A. Jones, and M. Uhlén, *Protein Eng.* **1,** 107 (1987).

plasmid, based on pEMBL8+,[27] harbors the *bla* structural gene encoding β-lactamase. The *spa* gene fragment encodes a two-domain variant of SpA, designated ZZ. The ZZ protein is secreted by the SpA signal sequence. Transcription is initiated from two promoters in tandem, the *lacUV5* promoter[28] and the *spa* promoter.[18] The *in vitro* designed Z domain is based on domain B from SpA with the following significant differences and characteristics: (1) The codon for the glycine residue in the asparagine-glycine dipeptide sequence, present in the second α helix of all five IgG-binding domains, was substituted for a codon for alanine. Thus, the expressed peptide is resistant to hydroxylamine, since treatment with this nucleophile selectively cleaves the peptide bond between asparagine and glycine residues.[6] (2) The codon for the methionine residue, normally present in SpA domains, was substituted for a leucine codon. (3) The codon for amino acid 2, which in all domains encodes an alanine residue, was changed to a valine codon to furnish a nonpalindromic *Acc*I site at the DNA level (G↓TAGAC), by which it is possible to polymerize Z gene fragments in an obligatory head-to-tail fashion.[12] (4) Downstream from the ZZ gene fragment, the mp18 multirestriction enzyme linker and the *lacZ'* gene, originating from pUC18,[28] were inserted to facilitate the cloning of product gene fragments into the expression system by a number of restriction enzymes. Thus, in spite of the fact that expressed ZZ fusions will be translocated through the cytoplasmic membrane, recombinant plasmids may be identified by their white phenotype in the *lacZ* complementation screen widely used in the pUC plasmids and M13 vector series of Messing.[28]

Procedures

Method 1: Selection of Different Cleavage Methods to Release IGF-I, tIGF-I, and IGF-II from ZZ Fusion Proteins

Chemical synthesis of the structural genes encoding IGF-I and IGF-II has previously been described.[29,30] Unsuccessful attempts were made to express the synthetic gene fragments in *E. coli* using different promoter fragments.[29] In fact, there are a number of problems associated with the

---

[27] L. Dente, G. Cesarini, and R. Cortese, *Nucleic Acids Res.* **11**, 1645 (1983).
[28] C. Yanisch-Perron, J. Viera, and J. Messing, *Gene* **33**, 103 (1985).
[29] A. Elmblad, L. Fryklund, L.-O. Hedén, E. Holmgren, S. Josephson, M. Lake, B. Löwenadler, G. Palm, and A. Skottner-Lundin, in "Third European Congress on Biotechnology III," p. 287. Verlag Chemie, Weinheim, 1984.
[30] B. Hammarberg, T. Moks, M. Tally, A. Elmblad, E. Holmgren, M. Murby, B. Nilsson, S. Josephson, and M. Uhlén, *J. Biotechnol.* **14**, 423 (1990).

expression of IGFs in bacteria, owing to proteolysis in the cytoplasm and in the periplasm.[8,29] In this study, each synthetic IGF gene was cloned into the pEZZ18 expression plasmid as in-frame fusions to the ZZ structural gene. Thereafter, the nucleotide sequences at the junction of the ZZ gene and each IGF gene were mutagenized *in vitro* to introduce codons for amino acid residues furnishing cleavage sites to release mature IGF-I, tIGF-I, and IGF-II, respectively, from the ZZ–IGF fusion protein. The different cleavage methods that were considered for each product are discussed and described below. The expression and purification of ZZ fusion proteins are described in Method 2, the cleavages of the fusion proteins are described in Method 3, and the purification of the growth factors is described in Method 4.

1. For the release of IGF-I, hydroxylamine (HA) cleavage is used.[6] HA selectively cleaves the peptide bond between asparagine (Asn) and glycine (Gly) residues. IGF-I has Gly as the native N-terminal residue and lacks internal Asn-Gly dipeptide sequences. Therefore, HA may be used to release native IGF-I from a ZZ–IGF-I fusion protein. Codons for an Asn-Gly dipeptide sequence are introduced into the ZZ–IGF-I gene fusion by *in vitro* mutagenesis[5] (Fig. 2B).

2. Four different methods are used for the release of tIGF-I from the ZZ–tIGF-I fusion protein. Truncated IGF-I is a slightly modified IGF-I found in brain,[31] colostrum,[32] and uterus.[33] It has been shown that tIGF-I is more potent than either IGF-I or IGF-II in radioreceptor assays using either brain or placenta membranes and in stimulation of brain cell DNA synthesis.[34] The primary structure of tIGF-I is identical to that of IGF-I, except for the deletion of the three amino acid residues glycine-proline-glutamic acid from the amino terminus. Thus, the growth factor has an N-terminal threonine (Thr) residue, which excludes the use of hydroxylamine to release mature tIGF-I. Instead, four other cleavage methods are used.

Cleavage after a Methionine (Met) Residue with Cyanogen Bromide (CNBr).[35] A codon for Met is introduced into the ZZ–tIGF-I gene construction 5' of the codon for the N-terminal Thr residue of tIGF-I (Fig. 2C). CNBr cleaves the peptide bond on the C-terminal side of methionine

---

[31] V. R. Sara, C. Carlsson-Skwirut, C. Andersson, K. Hall, B. Sjögren, A. Holmgren, and H. Jörnvall, *Proc. Natl. Acad. Sci. U.S.A.* **83**, 4904 (1986).

[32] G. L. Francis, L. C. Read, F. J. Ballard, C. J. Bagley, F. M. Upton, P. M. Gravestock, and J. C. Wallace, *Biochem. J.* **233**, 207 (1986).

[33] M. Ogasawara, K. P. Karey, H. Marquardt, and D. A. Sirbasku, *Biochemistry* **28**, 2710 (1989).

[34] C. Carlsson-Skwirut, M. Lake, M. Hartmanis, K. Hall, and V. R. Sara, *Biochim. Biophys. Acta* **1011**, 192 (1989).

[35] E. Gross and B. Witkop, *J. Biol. Chem.* **237**, 1856 (1962).

residues, converting the Met to a homoserine or a homoserine lactone.[35] However, as tIGF-I has an internal Met, a peptide bond in the product will also be cleaved by this method. Therefore, the cleavage must be performed partially.[36]

Cleavage on the C-Terminal Side of Tryptophan (Trp) with N-Chlorosuccinimide (NCS).[36,37] There are no Trp residues present in ZZ or tIGF-I. A codon for Trp is introduced into the tIGF-I gene 5' of the codon for the N-terminal Thr residue (Fig. 2D).

Cleavage with Thrombin C-Terminal to the Tetrapeptide Sequence Phenylalanine-Phenylalanine-Proline-Arginine (PhePheProArg).[38] Codons for the tetrapeptide are engineered into the ZZ–tIGF-I gene construction 5' of the codon for the N-terminal Thr residue of tIGF-I (Fig. 2E).

Cleavage with H64A Subtilisin on the C-Terminal Side of the Tetrapeptide Phenylalanine-Alanine-Histidine-Tyrosine (PheAlaHisTyr).[39] H64A subtilisin is a variant subtilisin BPN' which has been engineered for site-specific proteolysis by Carter and Wells.[40] In this study, an enhanced activity pentamutant variant of the H64A subtilisin is used.[39] A DNA linker encoding the H64A subtilisin recognition sequence (PheAlaHisTyr) is introduced into the ZZ–tIGF-I gene fragment (Fig. 2F) so that the codon for the N-terminal Thr residue in tIGF-I will follow the Tyr codon in the cleavage linker. It should be mentioned that the preferred target sequence for H64A subtilisin cleavage, as defined by the k_{cat}/K_m ratio, has recently been found to be AlaAlaHisTyr. On peptide substrates, this target sequence has a 4-fold greater k_{cat}/K_m value compared to the PheAlaHisTyr target sequence.[39,40]

3. Two separate methods are used for the release of IGF-II from the ZZ–IGF-II fusion protein.

Cleavage on the C-terminal side of Met with CNBr,[35] *as Described Above for the Release of tIGF-I.* IGF-II has no internal Met in its primary structure. Thus, CNBr cleavage may be utilized to release mature IGF-II from the ZZ–IGF-II fusion protein. A codon for Met is introduced into the ZZ–IGF-II gene fusion, 5' of the codon for the N-terminal alanine residue of IGF-II (Fig. 2H).

Cleavage with H64A Subtilisin on the C-Terminal Side of the Tetrapep-

---

[36] G. Forsberg, B. Baastrup, M. Brobjer, M. Lake, H. Jörnvall, and M. Hartmanis, *BioFactors* **2**, 105 (1989).
[37] Y. Schechter, A. Patchornik, and Y. Burstein, *Biochemistry* **15**, 5071 (1976).
[38] B. Blombäck, B. Hessel, D. Hogg, and G. Claesson, in "Chemistry and Biology of Thrombin" (R. L. Lundblad, J. W. Fenton II, and K. G. Mann, eds.), p. 275. Ann Arbor Science Publ., Ann Arbor, Michigan, 1977.
[39] P. Carter, B. Nilsson, J. P. Burnier, D. Burdick, and J. A. Wells, *Proteins* **6**, 240 (1989).
[40] P. Carter and J. A. Wells, *Science* **237**, 394 (1987).

tide PheAlaHisTyr,[39] *as Described Above for the Release of tIGF-I.* Codons for the H64A subtilisin recognition peptide are introduced at the junction of the ZZ and the IGF-II moieties in the ZZ–IGF-II gene fusion (Fig. 2G).

Method 2: Expression and Purification of Fusion Proteins

The ZZ–IGF fusion proteins (Fig. 2) are expressed and purified as described in this section and in Ref. 1. *Escherichia coli* 514(λ) is used as the host strain,[41] but most *E. coli* strains may be used successfully, even though differences in terms of expression levels and efficiency of product leakage to the growth medium can be observed. Expression in shaker flasks may yield sufficient material for many applications as the expression level is rather high, but the ZZ–IGF process has also been scaled up to a culture volume of 1000[6] and 3000 liters (M. Hartmanis, unpublished). In shaker flasks, the leakage to the growth medium is not so efficient as in the fermentor. In such cases, an osmotic shock procedure can be added to the recovery procedure of the fusion protein.[1] The expression levels of fusion proteins vary from 20 to 200 mg/liter in shaker flasks[1] and up to 2000 mg/liter in a fermentor (M. Hartmanis, unpublished). Shaker flask cultures are typically grown overnight at 37° in a medium based on LB (10 g/liter tryptone, 5 g/liter yeast extract, and 10 g/liter NaCl) supplemented with 0.1% glucose and 250 mg/liter ampicillin.

After cell growth, all fusion proteins are purified from microfiltered (pore size 0.5 μm) growth media by affinity chromatography on IgG-Sepharose equilibrated with 50 mM Tris-HCl, pH 7.6, containing 150 mM NaCl and 0.05% Tween 20. After sample application, the column is washed with the equilibration buffer followed by 5 mM ammonium acetate, pH 4.8. The fusion protein is eluted with 0.2 M acetic acid, pH 3.2, and lyophilized. Thereafter, each IGF product is released from the ZZ–IGF fusion protein by the procedures described in Method 3.

Method 3: Cleavages of ZZ–IGF-I, ZZ–tIGF-I, and ZZ–IGF-II

We have explored and optimized the different chemical and enzymatic methods described under Method 1 for cleavage of ZZ–IGF fusion proteins. All cleavage methods described are carried out as batch reactions only. The enzymatic methods, however, may also be performed in continuous systems utilizing immobilized enzymes.

Cleavage of ZZ–IGF-I with Hydroxylamine.[6] The ZZ–IGF-I fusion protein (Fig. 2B) is dissolved to a protein concentration of 20 mg/ml in a

---

[41] M. Zabeau and K. Stanley, *EMBO J.* **1,** 1217 (1982).

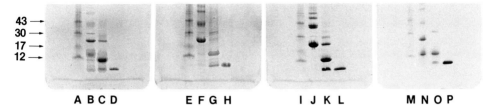

FIG. 4. Sodium dodecyl sulfate–polyacrylamide gel electrophoresis of IGF-1, IGF-II, and tIGF-I before and after cleavage of the fusion proteins. Lanes A, E, I, and M: Molecular weight markers (LKB Pharmacia Biotechnology, Uppsala, Sweden) of 12.3K, 17.2K, 30.0K, and 43.0K, respectively. Lane B: ZZ–IGF-I (~22.5K) with an Asn-Gly linker (Fig. 2B) before cleavage. Lane C: ZZ–IGF-I after chemical cleavage with hydroxylamine. Lane D: IGF-I (~7.5K) after purification. Lane F: ZZ–tIGF-I (~22.5K) with a subtilisin linker (Fig. 2F) before cleavage. Lane G: ZZ–tIGF-I after enzymatic cleavage with subtilisin. Lane H: tIGF-I (~7.5K) after purification. Lane J: ZZ–IGF-II (~22.5K) with a methionine linker (Fig. 2H) before cleavage. Lane K: ZZ–IGF-II after chemical cleavage with cyanogen bromide. Lane L: IGF-II (~7.5K) after purification. Lane N: ZZ–IGF-II (~22.5K) with a subtilisin linker (Fig. 2G) before cleavage. Lane O: ZZ–IGF-II after enzymatic cleavage with subtilisin. Lane P: IGF-II (~7.5K) after purification.

solution containing 2 M hydroxylamine, 1 mM EDTA, and 0.2 M Tris-HCl, pH 9.2. Cleavage is carried out for 6 hr at 45° with a cleavage yield of approximately 80%. After termination of the reaction by lowering the pH to 6 with acetic acid, the hydroxylamine is removed by gel filtration on Sephadex G-25 in 50 mM ammonium acetate, pH 6.0. The result of the cleavage reaction is shown in Fig. 4.

Partial Cleavage of ZZ–tIGF-I with CNBr.[35,36] The fusion protein (Fig. 2C) is dissolved in 0.13 M HCl to a protein concentration of 20 mg/ml. Chemical cleavage is initiated by the addition of 92 mg CNBr per gram of fusion protein. The reaction is stopped after 3 hr at 25° by evaporation under reduced pressure. The sample is subsequently lyophilized to complete dryness. A cleavage yield of approximately 30% is obtained (data not shown).

Cleavage of ZZ–tIGF-I with N-Chlorosuccinimide (NCS).[36,37] The fusion protein (Fig. 2D) is dissolved in 30% acetic acid containing 4 M urea to a protein concentration of 10 mg/ml. Addition of 24 mg NCS per gram of fusion protein initiates the reaction. After 45 min at 25°, the reaction is terminated by addition of methionine in excess. The cleavage yield is approximately 30% (data not shown).

Cleavage of ZZ–tIGF-I and ZZ–IGF-II with H64A Subtilisin.[39,40] Each fusion protein (Figs. 2F and G) is dissolved to a protein concentration of 1 mg/ml in 100 mM Tris-HCl, pH 8.6, containing 5 mM CaCl$_2$. H64A subtilisin is added to a concentration of 53 mg enzyme per gram of fusion protein, and the reaction is allowed to proceed for 4 hr at 37°. The reaction is terminated by injecting the sample onto a reversed-phase HPLC column. The components are separated using a gradient of aqueous acetonitrile containing 0.1% trifluoroacetic acid. An almost quantitative cleavage (>95%) is obtained with both fusion proteins (Fig. 4).

Cleavage of ZZ–tIGF-I with Thrombin.[38] The fusion protein (Fig. 2E) is dissolved in 50 mM ammonium carbonate, pH 8.0, to a protein concentration of 5 mg/ml. Enzymatic cleavage is initiated by addition of 40 mg thrombin per gram of fusion protein. The reaction is allowed to proceed for 18 hr at 37°, and cleavage is terminated by lowering the pH to approximately 4.5 with acetic acid. This method gives a cleavage yield of approximately 30% (data not shown).

Cleavage of ZZ–IGF-II with CNBr.[35] The fusion protein (Fig. 2H) is dissolved in 70% formic acid to a protein concentration of 10 mg/ml. Cleavage is initiated by the addition of 1 g CNBr per gram of fusion protein. After 6 hr at 25°, the reaction is terminated by evaporation under reduced pressure. The sample is subsequently lyophilized to complete dryness. A cleavage yield of 65% is obtained by this procedure (Fig. 4).

SDS–PAGE analysis, illustrating results from cleavage experiments involving the above procedures, is shown in Fig. 4. All cleavage methods, except that with NCS, yield a native growth factor exhibiting full biological activity (see Method 4). The oxidative cleavage of the Trp linker by NCS causes conversion of the internal Met in tIGF-I to methionine sulfoxide.[36] However, the oxidized growth factor still exhibits full biological activity in a radioreceptor assay.[36,42]

Method 4: Purification and Characterization of IGF-I, tIGF-I, and IGF-II

In this section we describe optimized procedures to purify the IGF molecules after cleavage of the ZZ–IGF fusion proteins. The purification scheme is rather extensive and will result in greater than 99% purity of the IGF molecules. It should be mentioned that this procedure has been developed to obtain an acceptable purity of the growth factors for pharmaceutical applications.

In general, it is possible to use a second affinity chromatography step on IgG-Sepharose after the chemical or enzymatic cleavage procedure in order to separate the growth factor from ZZ and uncleaved fusion protein as outlined in Fig. 1. On a preparative scale, however, this affinity step is usually replaced by other chromatographic procedures. For the purifica-

---

[42] K. Hall, K. Takano, and L. Fryklund, *J. Clin. Endocrinol. Metab.* **39,** 973 (1974).

tion of the IGFs, the purification procedure involves three subsequent chromatographic steps. Each cleavage reaction mixture is first purified by cation-exchange chromatography using S-Sepharose, equilibrated and eluted with ammonium acetate buffers, followed by gel filtration on Sephadex G-50 in 0.1 M ammonium acetate, pH 6.0. Ammonium acetate is selected since it is desirable to use a volatile buffer for lyophilization purposes. The ion-exchange step is an efficient purification procedure, owing to the large differences in isoelectric point between the three major components, the growth factor (IGF), the fusion partner (ZZ), and the fusion protein (ZZ-IGF). Gel filtration is utilized for removal of misfolded (dimeric and multimeric) forms of the growth factors, held together with intermolecular disulfide bridges.

The eluted material from the gel-filtration step consists of native, monomeric growth factor and some contaminants, consisting of modified forms of the growth factor itself. Several such modifications have been identified and characterized,[36,43,44] such as variants in which the internal methionine residue has been oxidized to a methionine sulfoxide (IGF-I, tIGF-I), variants in which misincorporation of norleucine for methionine has occurred during fermentation (IGF-I, tIGF-I), forms with incorrectly paired disulfide bonds, and proteolytically degraded forms (IGF-I, tIGF-I, and IGF-II). The modified forms are all removed in the last purification step on reversed-phase HPLC, using shallow gradients of aqueous acetonitrile containing trifluoroacetic acid or pentafluoropropionic acid.

All three highly purified growth factors have been extensively characterized by HPLC analysis, amino acid analysis, N- and C-terminal sequencuing, plasma desorption mass spectrometry, electrophoresis, and isoelectric focusing. Analysis of the purified IGFs by reversed-phase HPLC indicates that the purified recombinant human growth factors are identical to the naturally occurring peptides. In addition, the growth factors have been assayed biologically by radioreceptor assays using human placenta membranes[42] and a cell proliferation assay,[45] showing equal or even higher potencies than the products isolated from human sources.

Concluding Remarks

In this chapter, we have described experimental procedures to express IGF-I, tIGF-I, and IGF-II using a fusion protein strategy. The fusion

[43] M. Hartmanis and Å. Engström, in "Techniques in Protein Chemistry" (T. E. Hugli, ed.), p. 327. Academic Press, San Diego, California, 1989.
[44] G. Forsberg, G. Palm, A. Ekebacke, S. Josephson, and M. Hartmanis, *Biochem. J.* **271**, 357 (1990).
[45] M. Tally, C. H. Li, and K. Hall, *Biochem. Biophys. Res. Commun.* **148**, 811 (1987).

proteins were expressed in *E. coli* in soluble form to high levels and could easily be recovered from *E. coli* fermentation broths by IgG affinity chromatography. The general purification procedure is rather simple, following the scheme in Fig. 1. This concept could potentially be used to express any gene product in *E. coli,* and no product assay would be needed during the course of the purification. Thus, this method of expressing peptides in bacteria is well suited for cloned products of unknown properties or modified peptides and proteins constructed by *in vitro* mutagenesis.

Site-specific cleavage is often considered to be the bottleneck in expression strategies based on fusion proteins.[4] In this chapter, chemical as well as enzymatic cleavage methods were described that have been successfully utilized to release IGF molecules from the ZZ affinity handle. Chemical cleavage methods are preferred in process scale, because of ease of scale-up and for economical reasons. However, chemical methods have restrictions in their poor specificities, and they would not meet demands as being general for any fused product. In our hands, the most promising general cleavage method for ZZ fusion proteins is treatment with H64A subtilisin. This enzyme specifically and quantitatively cleaves engineered cleavage sites in a number of ZZ fusion proteins.[39,46] As this enzyme is (i) specific, (ii) produced in large quantities by recombinant methods, (iii) thermodynamically rather stable, (iv) amenable to immobilization by an engineered cysteine residue, and (v) maintains some activity in the presence of denaturants and detergents,[39] H64A subtilisin may be close to ideal for release of proteins from recombinant fusion proteins, both on a laboratory scale and on a process scale.

Acknowledgments

The ZZ fusion system was developed at the Department of Biochemistry and Biotechnology at the Royal Institute of Technology, Stockholm, and KabiGen AB, and we wish to thank all collaborators participating in the project, especially, Dr. Mathias Uhlén, Dr. Erik Holmgren, Dr. Staffan Josephson, Dr. Tomas Moks, Dr. Lars Abrahmsén, and Dr. Lennart Philipson. We also thank Dr. Paul Carter and Dr. James A Wells, Genentech, Inc., for the kind gift of purified H64A subtilisin and for communicating results prior to publication. The authors are grateful to Dr. Paul Carter, Dr. Lars Abrahmsén, and Dr. Roger Bishop for critical comments on the manuscript. Finally, we thank KabiGen AB, the Swedish National Board for Technical Development, and the Swedish National Science Research Council for financial support of several projects which made the development of the ZZ gene fusion concept possible.

---

[46] P. Carter and J. A. Wells, personal communication.

[2] Insulin-like Growth Factor I Receptor cDNA Cloning

By AXEL ULLRICH

Introduction

The mitogenic activity of a number of polypeptide growth factors is mediated by cell surface receptors that possess an intrinsic, ligand-controlled, protein tyrosine kinase activity. These receptor tyrosine kinases (RTKs) carry out a variety of activities which are specified by sequences within a multidomain structure, including an extracellular ligand-binding domain and a cytoplasmic catalytic domain which are connected by a hydrophobic membrane anchor region.

Among the peptides that interact with such RTKs are insulin and the insulin-like growth factor I (IGF-I). The IGF-I receptor is similar to but distinct from the insulin receptor.[1,2] Like the insulin receptor, the IGF-I receptor is a membrane glycoprotein of M_r 300,000–350,000, consisting of two α subunits (M_r ~135,000) and two β subunits (M_r ~90,000) that are connected by disulfide bonds to form the functional β-α-α-β heterotetrameric receptor complex.[2-6] In analogy with the insulin receptor,[7] IGF-I receptor α and β subunits are thought to be encoded within a single 180,000 molecular weight receptor precursor[8] that is glycosylated, dimerized, and proteolytically processed to yield the mature $\alpha_2\beta_2$ form of the receptor. On binding to the extracellular domain, IGF-I stimulates an intracellular, tyrosine-specific protein kinase activity which leads to β subunit autophosphorylation[9-11] and presumably phosphorylation of cytoplasmic components of an IGF-I receptor-specific signal transfer cascade.

[1] E. R. Froesch, C. Schmid, J. Schwander, and J. Zapf, *Annu. Rev. Physiol.* **47**, 443 (1985).
[2] M. M. Rechler and S. P. Nissley, *Annu. Rev. Physiol.* **47**, 425 (1985).
[3] S. D. Chernausek, S. Jacobs, and J. J. van Wyk, *Biochemistry* **20**, 7345 (1981).
[4] B. Bhaumick, R. M. Bala, and M. D. Hollenberg, *Proc. Natl. Acad. Sci. U.S.A.* **78**, 4279 (1981).
[5] J. Massague and M. P. Czech, *J. Biol. Chem.* **257**, 5038 (1982).
[6] F. C. Kull, Jr., S. Jacobs, F.-Y. Su, M. E. Svoboda, J. J. van Wyk, and P. Cuatrecasas, *J. Biol. Chem.* **258**, 6561 (1983).
[7] G. V. Ronnett, V. P. Knutson, R. A. Kohanski, T. L. Simpson, and M. D. Lane, *J. Biol. Chem.* **259**, 4566 (1984).
[8] S. Jacobs, F. C. Kull, and P. Cuatrecasas, *Proc. Natl. Acad. Sci. U.S.A.* **80**, 1228 (1983).
[9] S. Jacobs, F. C. Kull, H. S. Earp, M. E. Svoboda, J. J. van Wyk, and P. Cuatrecasas, *J. Biol. Chem.* **258**, 9581 (1983).
[10] J. B. Rubin, M. A. Shia, and P. F. Pilch, *Nature (London)* **305**, 438 (1983).
[11] Y. Zick, N. Sasaki, R. W. Rees-Jones, G. Grunberger, S. P. Nissley, and M. M. Rechler, *Biochem. Biophys. Res. Commun.* **119**, 6 (1984).

Despite functional and structural similarities, the receptors for IGF-I and insulin are thought to play different biological roles during mammalian development and mature life. Whereas insulin plays a key role in regulation of a variety of metabolic processes, IGF-I appears to be more potent in promoting cell growth. To enhance our understanding of the molecular mechanisms that evoke specific hormonal effects in cellular systems, and to attempt to identify structural features underlying these processes, cDNA clones encoding the human IGF-I receptor precursor were isolated and characterized.

Purification and Partial Amino Acid Sequence Determination

The IGF-I receptor is isolated from human placental membranes that are solubilized by Triton X-100, which prevents aggregation of the amphiphilic molecule in aqueous solution. Immunoaffinity chromatography using antibody αIR3[6,12,13] is then used for purification. The flow-through fraction of an insulin-Sepharose column representing about 20 placentas is applied to a 2×6.4 cm αIR3 column (~0.5 mg antibody/ml of packed gel), which has been equilibrated with 50 mM Tris-HCl buffer, pH 7.4. The column is washed extensively with this buffer and eluted first with 50 mM acetate buffer, pH 5, and then with 1 M glycine buffer, pH 2.2. All buffers contain 1 M NaCl, 0.1% Triton X-100, 0.1 mM phenylmethylsulfonyl fluoride, and 2 mM N-α-benzoyl-L-arginine ethyl ester. Fractions (2 ml) are collected in tubes containing 3 ml of 0.5 M Tris to neutralize the eluate.

αIR-3 Sepharose column fractions eluted with 1 M glycine, pH 2.2, are pooled, and IGF-I receptor is isolated by precipitation with methanol–chloroform.[14] The precipitate is redissolved in 50 mM Tris-HCl, pH 6.8, containing 2% (w/v) sodium dodecyl sulfate (SDS) and 20 mM dithiothreitol and heated at 60° for 30 min. This material is subjected to SDS–PAGE in 7% polyacrylamide gels in the presence of 0.1 mM thioglycolate. After electrophoresis, a guide strip of the gel is stained with Coomassie blue, and the bands corresponding to M_r 120,000 and about 90,000 are electroeluted.[15] The purified α and β subunits obtained by electroelution are precipitated by the methanol–chloroform procedure. A

---

[12] J. S. Flier, P. Usher, and A. C. Moses, *Proc. Natl. Acad. Sci. U.S.A.* **83**, 664 (1986).

[13] T. R. Le Bon, S. Jacobs, P. Cuatrecasas, S. Kathuria, and Y. Fujita-Yamaguchi, *J. Biol. Chem.* **261**, 7685 (1986).

[14] D. Wessel and U. I. Flügge, *Anal. Biochem.* **138**, 141 (1984).

[15] M. W. Hunkapiller, E. Lujan, F. Ostrander, and L. E. Hood, this series, Vol. 91. p. 227.

TABLE I
INSULIN-LIKE GROWTH FACTOR I RECEPTOR PROTEIN SEQUENCE ANALYSIS[a]

Peptide 1:	Glu Ile XXX Gly Pro Gly Ile Asp Ile Arg Asn Asp Tyr Gln Gln Leu Lys Arg Leu Gln XXX XXX Thr Val Ile Glu
Peptide 2:	Met Tyr Phe Ala Phe Asn Pro Lys
Peptide 3:	Asn (Arg) Ile (Ile) Ile Thr Trp His (Arg) Tyr (Arg) Pro Pro Asp Tyr Arg Asp Leu Ile (Ser) Phe Thr
Peptide 4:	Asp Val Gln Pro Gly (Ile) Leu (Leu) His Gly Leu Lys Pro (Trp) Thr Gln Tyr Ala Val
Peptide 5:	Ile Pro Ile Arg Lys
Peptide 6:	(Val) Phe Glu Asn Phe Leu His Asn Ser

[a] The amino-terminal amino acid sequence (peptide 1) and peptide sequences from purified IGF-I receptor α-subunit fragments are given. Parentheses indicate uncertainty in assignment. X's indicate unknown amino acids. Reprinted from A. Ullrich, A. Gray, A. W. Tam, T. Yang-Feng, M. Tsubokawa, C. Collins, W. Henzel, T. Le Bon, S. Kathuria, E. Chen, S. Jacobs, U. Francke, J. Ramachandran, and Y. Fujita-Yamaguchi, *EMBO J.* **5**, 2503 (1986).

portion of the α subunit preparation (130 pmol) is subjected to amino acid sequence analysis, and the rest (~300 pmol) is digested by lysyl peptidase. The digests are separated by reversed-phase HPLC on a 2mm × 10 cm C_4 column by elution with a linear gradient of 0.1% trifluoroacetic acid–70% 2-propanol. The peptide peaks which seem to be pure are collected, concentrated, and subjected to amino acid sequence determination on the Applied Biosystems (Foster City, CA) Protein Sequencer 470B equipped with PTH Analyzer 120A (Table I).

Design of Insulin-like Growth Factor I Receptor-Specific Oligonucleotide Probes

Previous cloning strategies had utilized stretches of amino acid sequences that contained mostly amino acids with low genetic degeneracy to synthesize pools of short (11–20 nucleotides) probes that contained all possible nucleotide sequences encoding the particular amino acid sequence. The approach employed for the IGF-I receptor followed the "long probe" or "guessmer" strategy[16-18] in which statistical codon usage frequency data are used to design unique probes of 40–80 nucleotides in length. Peptide sequences 1 and 4 were chosen for the design of probes of

[16] M. Jaye, H. de la Salle, F. Schamber, A. Balland, V. Kohli, A. Findeli, P. Tolstoshev, and J-P. Lecocq, *Nucleic Acids Res.* **11**, 2325 (1983).
[17] S. Anderson and I. B. Kingston, *Proc. Natl. Acad. Sci. U.S.A.* **80**, 6838 (1983).
[18] A. Ullrich, C. H. Berman, T. J. Dull, A. Gray, and J. Lee, *EMBO J.* **3**, 361 (1984).

```
                  1   2   3   4   5   6   7   8   9   10  11  12  13  14  15  16  17  18  19
PEPTIDE 1:       GLU ILE XXX GLY PRO GLY ILE ASP ILE ARG ASN ASP TYR GLN GLN LEU LYS ARG LEU
           5'-GAG ATC TGT GGC CCC GGC ATC GAT ATC CGG AAC GAC TAC CAG CAG TTG AAG AGG CTG
           3'-CTC TAG ACA CCG GGG CCG TAG CTA TAG GCC TTG CTG ATG GTC GTC AAC TTC TCC GAC
                  *       *   *   *           *           *   *                   *   * *

                  20  21  22  23  24  25  26
                 GLU XXX XXX THR VAL ILE GLU
                 GAG AAT TGC ACT GTC ATC GAA-3'
                 CTC TTA ACG TGA CAG TAG CTT-5'
                          *       *   *   *
```

```
                  531 532 533 534 535 536 537 538 539 540 541 542 543 544 545 546 547 548 549
PEPTIDE 4:       ASP VAL GLU PRO GLY XXX LEU XXX HIS GLY LEU LYS PRO XXX THR GLN TYR ALA VAL
                     3'-GGG CCG GAC GAC ACC GTG CCG GAC TTC GGG ACC TGG GTC ATG CGG-5'
                         *   *   *   *  ***     *   *                   *
```

FIG. 1. IGF-I receptor α-subunit peptides and nucleotide sequences of synthetic oligonucleotide probes deduced from the peptide sequences. Asterisks indicate differences between predicted and actual nucleotide sequences. [Reprinted from A. Ullrich, A. Gray, A. W. Tam, T. Yang-Feng, M. Tsubokawa, C. Collins, W. Henzel, T. Le Bon. S. Kathuria, E. Chen, S. Jacobs, U. Francke, J. Ramachandran, and Y. Fujita-Yamaguchi, *EMBO J.* **5,** 2503 (1986).]

78 and 45 nucleotides, respectively, for both coding and noncoding strands (Fig. 1). Since both peptides 1 and 4 contained sequence uncertainties, homologous regions of the human insulin receptor nucleotide sequence[19,20] were used to fill the gaps.

Success in using these probes is dependent on the presence of stretches of correctly guessed, uninterrupted sequences of eight or more nucleotides. The shorter the probe, the stronger the destabilizing impact of a mismatch in a cDNA–probe hybrid. On the other hand, the hybridization stringency used should be as high as possible to prevent isolation of false positives.

Synthetic oligonucleotide probes (0.02 OD_{260}) are radiolabeled by phosphorylation with T4 polynucleotide kinase (PL Biochemicals, Madison WI, 1 U/10 μl) and [γ-^{32}P]ATP (Amersham, Arlington Heights, IL, >5000 Ci/mmol) yielding probes with a specific activity of $0.5-1 \times 10^9$ counts/min(cpm) per microgram DNA. Unincorporated [γ-^{32}P]ATP is not removed from the probe-labeling reaction.

---

[19] A. Ullrich, J. R. Bell, E. Y. Chen, R. Herrera, L. M. Petruzzelli, T. J. Dull, A. Gray, L. Coussens, Y.-C. Liao, M. Tsubokawa, A. Mason, P. H. Seeburg, C. Grunfeld, O. M. Rosen, and J. Ramachandran, *Nature (London)* **313,** 756 (1985).

[20] Y. Ebina, L. Ellis, K. Jarnagin, M. Edery, L. Graf, E. Clauser, J. Ou, F. Masiarz, Y. W. Kan, I. D. Goldfine, R. A. Roth, and W. J. Rutter, *Cell (Cambridge, Mass.)* **40,** 747 (1985).

cDNA Cloning and Library Screening

Preparation of human term placenta poly(A)$^+$ mRNA,[21] cDNA cloning in λgt10,[22] and screening of a cDNA library of 2×10^6 recombinant phages are carried out as previously described.[23–25] In order to minimize background hybridization, 0.7% agarose (Seakem, FMC Bio Products, Rockland, ME) is used rather than agar to plate the phage library. About 5×10^5 recombinant λ phage are mixed with 300 μl *Escherichia coli* C600-HFl culture (~2 OD/ml) and 300 μl MgCa phage adsorption buffer (10 mM CaCl$_2$, 10 mM MgCl$_2$), and the mixture is incubated for 15 min at 37°. Then 12 ml of liquid (55°) 0.7% agarose (w/v) in NZYDT medium (GIBCO, Grand Island, NY) is added, mixed, and poured onto 13-cm diameter NZYDT/agar plates. Subsequently the plates are incubated for 12–16 hr at 37°. To avoid the formation of puddles on the bacterial lawn, plates are first incubated for 1 hr with the lid partially open to allow evaporation of excess water.

After incubation plates are cooled for 1 hr at 4°. Nitrocellulose filters are then placed on the plate lysate. Two filter replicas are generated for each plate. Marking of plate and filters is essential to allow subsequent alignment of hybridization signals and plaques. To process the filters for hybridization, each filter is placed phage-side-up for 3 min on filter paper soaked with denaturation solution (0.5 M NaOH, 1.5 M NaCl) followed by 3 min on neutralizing solution (0.5 M Tris-HCl, pH 7.5, 1.5 M NaCl) and a wash in 2× SSC (0.3 M NaCl, 0.3 M sodium citrate). Filters are air-dried and then backed at 70° in a vacuum oven for 2 hr. Nitrocellulose filters carrying copies of gene library phage plaques are pretreated (12–16 hr) and hybridized (12–16 hr) in a solution containing 5× Denhardt's solution,[26] 5× SSC (1× SSC is 0.15 M NaCl, 0.15 M trisodium citrate), 50 μg/ml sonicated salmon sperm DNA, 50 mM sodium phosphate buffer, pH 6.8, 1 mM sodium pyrophosphate (Na$_2$P$_4$O$_7$ · 10H$_2$O), 100 μM ATP, and 30% formamide (v/v) at 42°. Approximately 5×10^5 cpm/ml of radioactive DNA probe is used during hybridization.

---

[21] G. Cathala, J.-F. Savouret, B. Mendez, B. L. West, M. Karin, J. A. Martial, and J. D. Baxter, *DNA* **2**, 329 (1983).

[22] T. Huynh, R. Young, and R. Davis in "DNA Cloning: A Practical Approach" (D. Glover, ed.), Vol. 1, p. 49. IRL, Oxford, 1985.

[23] A. Ullrich, L. Coussens, J. S. Hayflick, T. J. Dull, A. Gray, A. W. Tam, J. Lee, Y. Yarden, T. A. Libermann, J. Schlessinger, J. Downward, J. Bye, N. Whittle, M. D. Waterfield, and P. H. Seeburg, *Nature (London)* **309**, 418 (1984).

[24] W. D. Benton and R. W. Davis, *Science* **196**, 180 (1977).

[25] T. Maniatis, R. C. Hardison, E. Lacy, J. Lauer, C. O'Connell, D. Quon, G. K. Sim, and A. Efstratiadis, *Cell (Cambridge, Mass.)* **15**, 687 (1978).

[26] D. T. Denhardt, *Biochem. Biophys. Res. Commun.* **23**, 641 (1966).

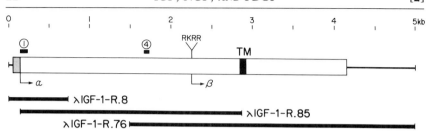

FIG. 2. Schematic of the cDNA structure of the IGF-I receptor. Overlapping cDNA clones used in sequence determinations are shown below. Translated sequences are boxed, with the signal sequence shaded. Also indicated are the putative precursor processing site (RKRR), the transmembrane domain (TM), and the positions of the synthetic oligonucleotide probes derived from peptides 1 and 4. [Reprinted from A. Ullrich, A. Gray, A. W. Tam, T. Yang-Feng, M. Tsubokawa, C. Collins, W. Henzel, T. Le Bon, S. Kathuria, E. Chen, S. Jacobs, U. Francke, J. Ramachandran, and Y. Fujita-Yamaguchi, *EMBO J.* **5**, 2503 (1986).]

Hybridized filters are washed 4 times for 20 min in 0.2× SSC, 0.1% SDS at 40° with agitation. Parameters that should be adjusted when using probes of different length and base composition are the wash temperature and/or the formamide concentration.

Filters are then air-dried and wrapped in plastic (Saran Wrap) in sets of six and placed on X-ray film (35 × 43 cm X-OMAT AR, Kodak Cat. No. 165 1512, Rochester, NY) for exposure at −70° (12–24 hr) using Cronex Lightning Plus (Du Pont, Wilmington, DE) intensifying screens. Developed films are aligned with markings on corresponding exposures of duplicate filters and then with the original plates for isolation of a small section of the phage plate lysate containing the hybridization-positive recombinant λ phage. The identified area on the plate is isolated using the wide end of a Pasteur pipette. The agar plug is suspended in phage suspension buffer (PSB: 10 mM Tris, pH 7.5, 100 mM NaCl, 10 mM $MgCl_2$ with 0.5 g gelatin per liter) containing chloroform for sterility. In subsequent steps, the IGF-I receptor cDNA-containing λgt10 phage are purified by repeated dilution, plating, and hybridization as described above.

Characterization of Insulin-like Growth Factor I Receptor cDNA

Initial screening yielded clone λIGF-1-R.8 (Fig. 2), which contains sequences complementary to probe 1. The 730-base pair (bp) *Eco*RI insert of λIGF-1-R.8 and the synthetic probe derived from peptide 4 (Fig. 1) were used for subsequet screening of the same placental library to yield overlapping clones λIGF-1-R.85 and λIGF-1-R.76, of 2.8 and 3.5 kilobases (kb), respectively.

Complete nucleotide sequence analysis of the cloned cDNA resulted in deduction of the complete primary sequence of the human IGF-I receptor precursor (Fig. 3). The 4989-nucleotide sequence contains an open reading frame of 4101 nucleotides, which begins with an ATG codon that is flanked by sequences meeting the requirements for an initiation codon as defined by Kozak.[27] The 45-nucleotide sequence preceding this potential initiation codon includes a purine-rich portion and a 5'-terminal oligo(T) sequence, and it is likely to be part of a 5'-untranslated sequence, although no in-frame stop codon can be identified. At the 3' end, the open reading frame is flanked by a TGA signal for translation termination and 840 nucleotides of 3'-untranslated sequence.

The initiation methionine is the first residue of a 30-residue sequence, preceding the chemically identified amino-terminal glutamic acid residue, that displays structural features characteristic of signal peptides, necessary for transfer of the nascent polypeptide chain into the membrane of the endoplasmic reticulum. The IGF-I receptor signal sequence is unusually rich in polar residues such as threonine and serine (30%). Cleavage by signal peptidase occurs after a glycine residue to expose the amino-terminal IGF-I receptor α-subunit glutamic acid residue; this establishes the preproreceptor organization to be NH_2–signal peptide–α subunit–β subunit–COOH. The α-subunit region contains a single cysteine-rich (24 Cys) region between residues 148 and 302 and 11 potential N-linked glycosylation sites. An Arg-Lys-Arg-Arg tetrapeptide after residue 706 marks the putative cleavage site of the $\alpha\beta$ proreceptor polypeptide, suggesting that the β subunit begins at Asp-711.

The β-subunit portion is characterized by the presence of a single, 24-amino acid hydrophobic sequence (residues 906–929) that is likely to be the transmembrane domain, and it is flanked at its carboxy terminus by a stretch of basic amino acids that may facilitate plasma membrane anchoring. Five potential N-linked glycosylation sites are found between residues 711 and 905 upstream from this hydrophobic sequence. This region probably represents β-subunit extracellular sequences. Structural features characteristic of tyrosine kinase enzymatic domains are found downstream from the transmembrane sequence; thus, the cytoplasmic domain is generated by the carboxy-terminal 407 amino acids of the β subunit.

Tyrosine kinase features are located within a 257-residue region between amino acids 973 and 1229 and include a potential ATP binding site

---

[27] M. Kozak, *Nucleic Acids Res.* **12**, 857 (1984).

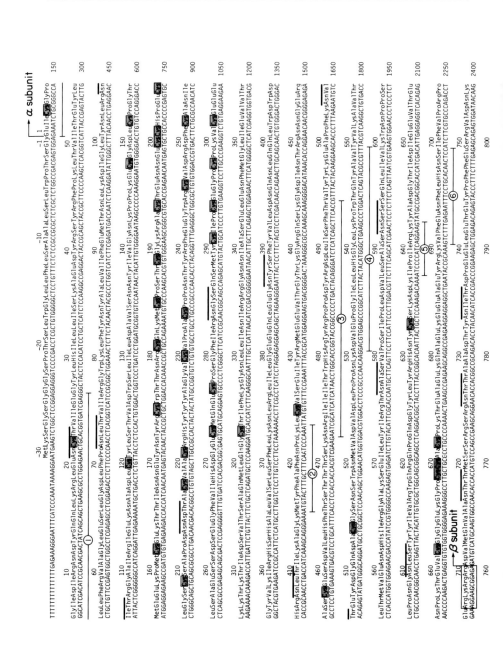

FIG. 3. Nucleotide and predicted amino acid sequences of the IGF-I receptor. Amino acids of the proreceptor are numbered above, starting at Glu-1, and are preceded by a 30-residue signal sequence; nucleotides are numbered to the right. Experimentally determined peptide sequences are underlined and numbered; potential N-linked glycosylation sites are overlined; cysteine residues are shaded; and the transmembrane domain is heavily underlined. The potential ATP binding site is indicated by asterisks over Gly-976, -978, and -981, and by an arrow over Lys-1003. The putative precursor processing site is boxed. [Reprinted from A. Ullrich, A. Gray, A. W. Tam, T. Yang-Feng, M. Tsubokawa, C. Collins, W. Henzel, T. Le Bon, S. Kathuria, E. Chen, S. Jacobs, U. Francke, J. Ramachandran, and Y. Fujita-Yamaguchi, *EMBO J.* **5**, 2503 (1986).]

(Gly-976 to Gly-981 and Lys-1003). The entire unmodified IGF-I proreceptor polypeptide chain having a predicted M_r of 151,869 can thus be subdivided into the 80,423 α subunit that lacks sequences with membrane-spanning characteristics, and a 70,866 β subunit that contains tyrosine kinase-homologous sequences and appears to anchor the intact receptor complex in the plasma membrane.

[3] Autoradiographic Localization of Insulin-like Growth Factor I Receptors in Rat Brain and Chick Embryo

By M. A. LESNIAK, L. BASSAS, J. ROTH, and J. M. HILL

Introduction

Insulin-like growth factors I and II (IGF-I, IGF-II), as well as their specific receptors, have been identified and characterized in many tissues, including brain, by methods using whole cells and membrane preparations. However, the functions of these peptides and their receptors in brain are not yet known. We have used autoradiographic techniques to determine the anatomical distribution of the receptors in rat and chicken brain and in whole chick embryo.[1-3] The techniques for locating the IGF-I receptors have been adapted from the methods developed by Herkenham and Pert for the localization of drug and neurotransmitter receptors in fresh tissue[4] and were used by us previously to study brain insulin. (Readers should also refer to Refs. 5 and 6 for additional practical details.)

Preparation of Tissues and Tissue Sections

Animals

Rats, Sprague-Dawley, male, 200 g (Harlan Farms, Indianapolis, IN)
Chicken embryos, White Leghorn

---

[1] M. A. Lesniak, J. M. Hill, W. Kiess, M. Rojeski, C. B. Pert, and J. Roth, *Endocrinology* **23,** 2089 (1989).
[2] L. Bassas, M. Girbau, M. A. Lesniak, J. Roth, and F. De Pablo, *Endocrinology* **125,** 2320 (1989).
[3] J. M. Hill, M. A. Lesniak, C. B. Pert, and J. Roth, *Neuroscience* **17,** 1127 (1986).
[4] M. Herkenham and C. B. Pert, *J. Neurosci.* **2,** 1129 (1982).
[5] M. Herkenham, *in* "Brain Receptor Methodologies" (P. J. Marangos, I. C. Campbell, and R. M. Cohen, eds.), p. 127. Academic Press, New York, 1984.
[6] M. Herkenham, *in* "Molecular Neuroanatomy" (S. W. Van Leeuwen, R. M. Buijs, C. W. Pool, and O. Pach, eds.), p. 111. Elsevier, Amsterdam, 1988.

Materials for Tissue Preparation

Rongeurs (Biomedical Research Instruments, Inc., Rockville, MD)
Guillotine, scalpel, scissors
Isopentane (2-methylbutane) at $-30°$; this temperature is achieved by placing the solution in a glass beaker on dry ice
Dry ice (powdered)

Materials for Tissue Sections

Cryostat
Embedding matrix (Lipshaw M-1; Detroit, MI)
Glass slides (frosted at one end), Kimax
Slides, gelatin coated[7]
Brush, fine camel's hair, to transfer tissue
Slide box, 25 positions
Desiccator
Slide-warming table (optional)

Methods

Rats are decapitated with a guillotine without anesthesia. The skin over the skull is cut along the midline and retracted before the skull is carefully opened with rongeurs. The meninges that remain are gently teased away before removing the brain from the skull. Often, the olfactory bulbs and pituitary can be removed intact with the whole brain. To maintain morphology, immediately immerse the brain in $-30°$ isopentane for about 20 sec, that is, until just hard to the touch with forceps (longer immersion can result in dehydration and cracking of tissue). Remove the tissue and bury it in powdered dry ice until it is thoroughly frozen. Chick embryos are handled the same as brains except that they are thoroughly washed in iced isotonic saline before freezing by immersion in isopentane. When tissues are not processed on the day of preparation, they can be stored at $-20°$ or colder for up to 1 week.

The frozen brain or other tissues are mounted on cryostat pedestals with embedding matrix and allowed to equilibrate to the temperature of the cryostat, typically $-14°$ for 25-μm sections and $-18°$ for thinner sections, for about 20–30 min. Cut sections can be transferred with a fine camel's hair brush to gelatin-coated slides that have been kept cold in the cryostat, and they are made to adhere to the slide by melting; this is accomplished by pressing a finger on the back of the slide or by placing the slide on a warming table.

---

[7] A. Hendrickson and S. B. Edwards, *in* "Neuroanatomical Research Techniques" (R. T. Robertson, ed), p. 2141. Academic Press, New York, 1978.

To obtain an accurate determination of receptor distribution, brains are often cut in each of the three planes (coronal, horizontal, and sagittal). When small enough, more than one section can be accommodated on a slide (e.g., two or more pairs of olfactory bulbs, or two coronal sections of brain). In addition, it is preferable to position tissue near, but not touching, the edges of the slide. Usually adjacent sections (4–8 per area) from about 12 distinct areas of brain are cut for binding studies. Collect the slides in a small slide box that is kept on ice. Cut sections should be maintained cold, but not frozen.

Tissues must be dehydrated at 4° for 18–24 hr before being used for binding studies. This is accomplished by placing the uncovered sections in a desiccator under vacuum. When dried under the described conditions, sections have a transparent, glasslike appearance. Once dried, sections can be stored at $-20°$ or colder indefinitely. However, tissue quality sometimes suffers with long-term storage. Best results are usually obtained by keeping sections desiccated at $-4°$ for 3 or fewer days before use. (For additional details on tissue handling, see Refs. 5 and 6.)

Binding Studies

Materials

Staining dishes
Slide racks, 30 positions, stainless steel
Forceps
Slide mailers (Labtek Products, Miles, Naperville, IL).

Reagents

Insulin-like growth factor I (AMGEN Biologicals, Thousand Oaks, CA)
^{125}I-Labeled insulin-like growth factor I, receptor grade (Amersham Corporation, Arlington Heights, IL)
Normal saline
Potassium phosphate, 30 mM in normal saline, pH 7.4
Assay buffer: HEPES (100 mM) NaCl (120 mM), MgSO$_4$ (1.2 mM), KCl (2.5 mM), sodium acetate (15 mM), glucose (10 mM), EDTA (1 mM), and bovine serum albumin (fraction V, 1.0 mg/ml), pH 7.8
Developer D-19 (Kodak, Rochester, NY)
Rapid fix, prepared without hardener (Kodak)

Protocol

Preincubation. Slides are brought to room temperature in the desiccator to prevent condensation when exposed to room air and are then transferred to a slide rack for rinsing in staining dishes. The frosted end of the slide is used for handling and for labeling the experimental protocol. Adjacent sections are used to compare experimental conditions. Prior to incubation, the sections are rinsed twice in a chilled (1° to 4°) normal saline with 30 mM KCl, pH 7.4, for 10 min per rinse. Purportedly this is to dissociate any endogenous peptide from tissue. Excess liquid is allowed to drain onto absorbent paper for a few minutes before the sections are immersed in the assay buffer.

Incubation. Incubation is carried out in plastic slide mailers. The mailers are placed in a vertical position. Binding buffer is added either with [^{125}I]IGF-I alone or in the presence of unlabeled IGF-I.[1] Four test slides are placed in each mailer (two slides are positioned back-to-back in the mailer). If necessary, use blank slides to fill the mailer. We routinely use 4.0 ml per mailer, but a minimum of 2.5 ml can be used if care is taken to place the tissue on the lower half of the glass slide.

The conditions for incubation have been adapted from the protocol used for studying the binding of [^{125}I]IGF-I to cell membranes. The tissue sections are incubated at 15° for 2.5–3 hr. A simple 15° water bath can be easily assembled and maintained with ice-filled beakers set in the bath container. The binding conditions for chick embryo are the same except that bacitracin (1 mg/ml), a protease inhibitor, is added to the binding buffer and the incubation is carried out at 4° for 6–8 hr.[2]

To determine nonspecific binding, a few sections are incubated with labeled IGF-I in the presence of an excess of unlabeled IGF-I; however, this can be expensive. In place of an excess of unlabeled peptide, a concentration of 250–500 ng/ml can be used, which inhibits binding about 60%. Specific binding can be determined by subtracting this "nonspecific" from total binding.

Rinsing. At the end of the incubation period the binding buffer is decanted and transferred to containers reserved for radioactive waste. Cold (1°) phosphate-buffered saline is immediately added to the incubation containers. We believe that lowering of the solution temperature slows the dissociation reaction of labeled ligand and as well as keeps the tissue moist. The slides are then removed and returned to the slide rack, and the unbound radioactive ligand is removed by rinsing the slides at least 5 times in cold phosphate-buffered saline, pH 7.4, approximately 1 min/rinse. The slides are air-dried quickly with a hair dryer set on cool. The slide rack can be tipped at an angle to accelerate draining. When two adjacent sections have been placed on a single slide, one section can be used for counting and the other for autoradiography. Tissue that is to be counted is scraped from the slide with a razor blade. Laboratory wipes, such as Kimwipes, can be cut or torn into small squares to handle the tissue, which then is placed into a counting vial (e.g., 12 × 75 mm tube) for counting in a γ counter. The remaining tissue section is used for autoradiographic analysis.

Tissue Fixation. After the tissue has dried, the slides are placed in a glass desiccator. Positioned in the desiccator is an open container of paraformaldehyde, which has been exposed to room humidity for 24 hr. The slides are placed over, not touching, the fixative. Under reduced pressure, the desiccator is placed in an 80° oven for a minimum of 2 hr. After the fixation process, the vacuum is released in a chemical hood and the slides immediately removed from the warm fumes. Do not allow condensation of fumes to occur on the slides. Before being positioned with X-ray film, the slides are usually left in the open air for a few hours to allow for the evaporation of residual formaldehyde. Although fixation is not necessary to obtain good autoradiographs, it sometimes improves autoradiograph quality and allows staining of the tissue sections after exposure to film.

Slides are placed in X-ray cassettes and exposed to autoradiographic film. [^{3}H]Ultrofilm (LKB, Uppsala, Sweden) provided the best results for sensitivity as well as low background for ^{125}I. After 5–7 days at room temperature, the slides are developed in D-19 developer at 20° for 4 min, fixed in rapid fix for 2 min, and washed in running water at least 20 min before being air dried.

The developed film is analyzed using computerized densitometry. To identify the brain or embryo structures, reference atlases[8,9] are used. To identify neuroanatomical sites or specific organs, sections can be stained with routine histological staining techniques, for example, thionin for brain tissue and hematoxylin and eosin for chick embryo tissues.

Additional Steps. The IGF-I receptor on preparations was characterized further by competition studies (Figs. 1–3). Competition studies are best done on sections through brain regions which are as similar as possible. For example, the olfactory bulbs and cerebellum are composed of several millimeters of tissue of very similar cell composition and anatomical arrangement. Thus, even though all experimental treatments might not occur on a section adjacent to the control they can be performed on very similar tissue. [^{125}I]IGF-I is competed for by unlabeled peptides, IGF-I > IGF-II > insulin, as described for the IGF-I receptor in other cell and membrane preparations.

In the competition binding studies with low concentrations of IGF-I,

---

[8] G. Paxinos and C. Watson, "The Rat Brain in Stereotaxic Coordinates." Academic Press, New York, 1982.

[9] W. J. Kuenzel, M. Masson, "A Stereotaxic Atlas of the Brain of the Chick (*Gallus domesticus*)." Johns Hopkins Press, Baltimore, Maryland, 1988.

  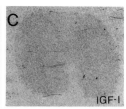

FIG. 1. Autoradiographs of labeled IGF-I binding to olfactory bulb. Coronal sections (36 μm) of adult rat olfactory bulbs were incubated as described in the text and apposed to film for 7 days. The images are as follows: tissue was incubated (a) with [^{125}I]IGF-I alone, (b) in the presence of 25 ng/ml unlabeled IGF-I, and (c) in the presence of 1000 ng/ml IGF-I. (From Ref. 1.)

there appears to be an increase in binding of [^{125}I]IGF-I. This is a reproducible observation, for which we have no explanation. By image analysis, the increase in radioactivity is widely distributed and not localized to any one structure in the brain.

Quantitation of the density of receptor sites is possible with the use of ^{125}I-labeled polymer-based standards (Amersham), which are included with the sections during exposure to X-ray film. For additional practical information regarding selection of isotope, see Refs. 10 and 11.

We routinely test the integrity of the labeled peptide by measuring the trichloroacetic acid [TCA, 5% (v/v) final concentration] precipitation of radioligand that has been incubated in the presence and absence of tissue. Note that intact labeled IGF-I is not so TCA-precipitable as is insulin and other similar peptides. Further, the tracer that had been incubated with tissue can be incubated with a liver membrane preparation in a membrane radioreceptor assay. The binding values can be compared to the results obtained during incubation with tracer that has not been previously incubated with tissue. Our results with this type of experiment indirectly suggest that growth factor binding proteins have minimal, if any, interference with the IGF-I binding pattern. The decrease in TCA precipitability is about 10% in our experiments.

Other Ligands

Insulin-like Growth Factor II. We have determined the distribution of insulin-like growth factor II (IGF-II) receptors in rat brain using the above protocol. IGF-II receptors were not found in chick embryo by us, but others report IGF-II receptors in chick tissue. In addition to using

[10] M. J. Kuhar and Unnerstall, *Trends Neurosci.* **8,** 49 (1985).
[11] D. G. Baskin, B. Brewitt, D. A. Davidson, E. Corp, T. Paquette, S. P. Figlewicz, T. K. Lewellen, M. K. Graham, S. G. Woods, and D. M. Dorsa, *Diabetes* **35,** 246 (1986).

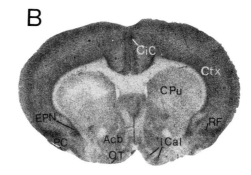

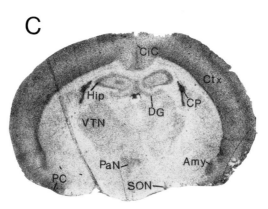

Fig. 2. Autoradiographic distribution of receptors for IGF-I in rat brain. Tissue sections (36 μm) were incubated with [^{125}I]IGF-I (0.2 ng/ml) and processed as described. The sections had been apposed to film for 5 days. Coronal sections of olfactory bulb (A) and brain, anterior to posterior (B → F), are illustrated. Acb, Nucleus accumbens; Amy, amygdala; CiC, cingulate cortex; Ctx, cerebral cortex; CC, cerebellar cortex (molecular layer); CN, cochlear nucleus; CP, choroid plexus; CPu, caudate putamen; DG, dentate gyrus; EC, entorhinal cortex, EPL, external plexiform layer; EPN, endopyriform nucleus; FN, facial nucleus; GL,

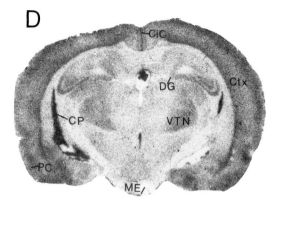

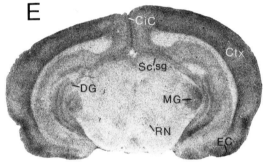

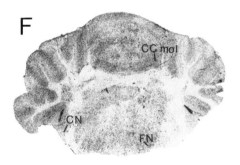

Fig. 2 (*continued*)

glomerular layer; Hip, hippocampus; ICal, islands of Calleja; ME, median eminence; MG, medial geniculate; ON, olfactory nerve; OT, olfactory tubercle; PaN, paraventricular nucleus; PC, pyriform cortex; RF, rhinal fissure; RN, red nucleus; SC, sg, superior colliculus, superficial gray; SON, supraoptic nucleus; VTN, ventral thalamic nuclei.[1]

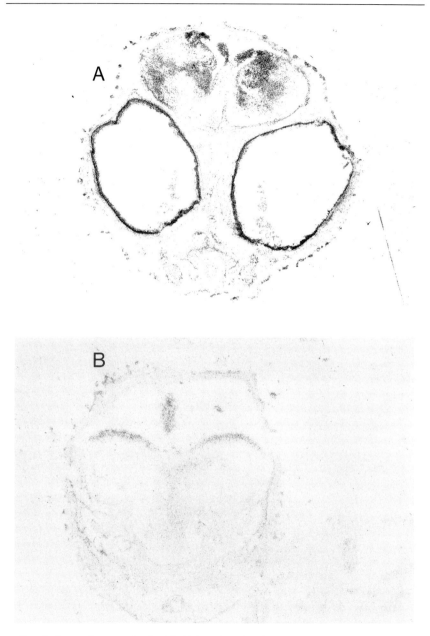

FIG. 3. Autoradiogram of a day 12 chick embryo head. Coronal sections (15 μm) of chick embryo head were incubated with [^{125}I]IGF-I (0.5 ng/ml) alone (A) or with an excess of unlabeled IGF-I (B) under incubation conditions as described.[3]

[^{125}I]IGF-II to localize receptors to specific sites in the rat brain, we have verified the presence of these receptors using radioimmunocytochemical techniques.[1]

Insulin. Characterization and distribution of insulin receptors in rat brain were previously determined by a similar protocol.[3] The protocol differed only in that the incubation time was 90 min.

[4] *In Vivo* Assay of Neuron-Specific Effects of Nerve Growth Factor

By MARK H. TUSZYNSKI and FRED H. GAGE

Introduction

Nerve growth factor (NGF) effects *in vivo* have been demonstrated in a number of model systems.[1,2] The models may be broadly divided into peripheral and central nervous system models. Included among peripheral nervous system models are effects of NGF on sympathetic and sensory neuron survival and outgrowth during development, and effects on neurite regeneration after nerve transection in the adult.[1-4] In the central nervous system, the effects of NGF during development are becoming increasingly apparent. NGF influences the development of neuronal populations in the spinal cord (including motor neurons), cerebellum, brain stem, basal forebrain, and neocortex (for review, see Ref. 5). However, the demonstration that NGF also affects neuronal function in the central nervous system (CNS) of adult mammals[6,7] was a surprise to many investigators.

In the adult mammal, NGF receptors have been demonstrated on cholinergic neurons of the basal forebrain, including the medial septal nucleus, vertical limb of diagonal band, and nucleus basalis of Meynert (NBM).[8-13] Selected cholinergic neurons of the basal ganglia and brain

---

[1] H. Thoenen and Y. A. Barde, *Physiol. Rev.* **60**, 1284 (1980).
[2] R. Levi-Montalcini, *Science* **237**, 1154 (1987).
[3] H. Thoenen and D. Edgar, *Science* **229**, 238 (1985).
[4] R. W. Gundersen and J. N. Barrett, *J. Cell Biol.* **87**, 546 (1980).
[5] F. Hefti, J. Hartikka, and B. Knusel, *Neurobiol. Aging* **10**, 515 (1989).
[6] F. Hefti, *J. Neurosci.* **8**, 2155 (1986).
[7] F. H. Gage, D. M. Armstrong, L. R. Williams, and S. Varon, *J. Comp. Neurol.* **269**, 147 (1988).
[8] P. E. Batchelor, D. M. Armstrong, S. M. Blaker, and F. H. Gage, *J. Comp. Neurol* **284**, 187 (1989).
[9] P. M. Richardson, V. M. K. Verge Isse, and R. J. Riopelle, *J. Neurosci.* **6**, 2312 (1986).
[10] M. Taniuchi and E. M. Johnson, *J. Cell Biol.* **101**, 1100 (1985).

stem also express NGF receptors after lesions.[5] NGF is synthesized in the target regions of innervation of the medial septum, vertical limb of diagonal band, and nucleus basalis, and it is retrogradely transported from the target regions to the cholinergic cell bodies of their projecting axons, where NGF exerts its effects in an as yet undetermined fashion.[5,14,15] NGF appears to be required for either the survival or normal function of these neuronal populations.

In vivo assays for NGF effects on neurons in the CNS have been best developed in three models: (1) lesions of the medial septal–hippocampal projection (fimbria/fornix transections),[7,16–18] (2) lesions of the NBM–cortical projection,[19–21] and (3) age-related cholinergic neuron degeneration in both the medial septum[22,23] and NBM. In addition, NGF effects have been studied after striatal lesions[24] and after grafts of adrenal medullary cells to the brain. Since a detailed review of all of these models is not feasible here, we concentrate on the fimbria/fornix (FF) lesion model; for detailed treatments of other models, see Ref. 5.

Septohippocampal Projection in Rat: Fimbria/Fornix Lesions

The medial septal nucleus of the rat contains a population of cholinergic neurons that project to the hippocampus via the fimbria/fornix fiber bundle (Fig. 1) and supracallosal striae.[25–31] This projection supplies the dorsal

---

[11] F. Hefti, J. Hartikka, A. Salvatierra, W. J. Weiner, and D. C. Mash, *Neurosci. Lett* **69**, 37 (1986).
[12] G. C. Schatteman, L. Gibbs, A. A. Lanahan, P. Claude, and M. Bothwell, *J. Neurosci.* **8**, 860 (1988).
[13] J. H. Kordower, R. T. Bartus, M. Bothwell, G. Schatteman, and D. M. Gash, *J. Comp. Neurol.* **277**, 465 (1988).
[14] M. E. Schwab, U. Otten, Y. Agid, and H. Thoenen, *Brain. Res.* **168**, 473 (1979).
[15] M. Seiler and M. E. Schwab, *Brain Res.* **300**, 33 (1984).
[16] F. Hefti, *J. Neurosci.* **6**, 2155 (1986).
[17] L. R. Williams, S. Varon, G. M. Peterson, K. Wictorin, W. Fischer, and A. Bjorklund, *Proc. Natl. Acad. Sci. U.S.A.* **83**, 9231 (1986).
[18] L. F. Kromer *Science* **235**, 214 (1987).
[19] V. Haroutunian, P. D. Knof, and K. L. Davis, *Brain Res.* **368**, 397 (1986).
[20] L. Garofalo, D. Maysinger, and A. C. Cuello, *Soc. Neurosci. Abstr.* **14**, 826 (1988).
[21] A. C. Cuello, D. Maysinger, L. Garofalo, *et al.*, in "Receptor–Receptor Interactions," (Proceedings of the Wenner-Gren Symposium) (K. Fuxe and L. F. Agnati, eds.), p. 62. Basingstoke, Macmillan, New York, 1989.
[22] W. Fischer, K. Wictorin, A. Bjorklund, L. R. Williams, S. Varon, and F. H. Gage, *Nature (London)* **329**, 65 (1987).
[23] F. H. Gage and A. Bjorklund, *J. Neurosci.* **6**, 2837 (1986).
[24] F. H. Gage, P. Batchelor, K. S. Chen, D. Chin, G. A. Higgins, S. Koh, S. Deputy, M. B. Rosenberg, W. Fischer, and A. Bjorklund, *Neuron* **2**, 1177 (1989).
[25] D. G. Amaral and J. Kurz, *J. Comp. Neurol.* **420**, 37 (1985).
[26] F. H. Gage, A. Bjorklund, U. Stenevi, and S. B. Dunnett, *Brain Res.* **268**, 39 (1983).

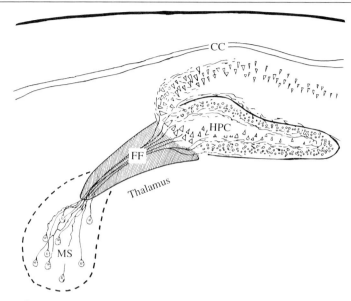

FIG. 1. Septal projection to the hippocampus (HPC) via the fimbria/fornix (FF) fiber bundle in the rat. Aspirative lesions of the FF result in retrograde degeneration of cholinergic cell bodies in the septum, which can be prevented by exogenous infusions of NGF. CC, Corpus callosum; MS, medial septum.

hippocampus with most of its cholinergic innervation. An additional cholinergic projection arising primarily from the diagonal band nuclei projects through a ventral pathway to reach the temporal hippocampus, providing 10% of hippocampal cholinergic input.[27,29,32]

A body of evidence suggests that cholinergic neurons of the medial septal region depend on a continuous supply of NGF derived from its hippocampal target region to function normally.[5,14,15,33–37] Interruption of

---

[27] F. H. Gage, A. Bjorklund, and U. Stenevi, *Brain Res.* **268**, 27 (1983).
[28] P. R. Lewis, C. C. D. Shute, and A. Silver, *J. Physiol.* **191**, 215 (1967).
[29] F. H. Gage and A. Bjorklund, in "The Hippocampus" (R. L. Isaacson and K. H. Pribam, eds.), Vol. 3, p. 33. Plenum, New York, 1986.
[30] J. Storm-Mathisen, *J. Brain Res.* **80**, 119 (1974).
[31] B. H. Wainer, A. I. Levey, D. B. Rye, M. Mesulam, and E. J. Mufson, *Neurosci. Lett.* **54**, 45 (1985).
[32] T. A. Milner and D. G. Amaral, *Exp. Brain Res.* **55**, 579 (1984).
[33] D. L. Sheldon and L. F. Reichardt, *Proc. Natl. Acad. Sci. U.S.A.* **83**, 2714 (1986).
[34] G. R. Auburger, R. Heumann, S. Hellweg, S. Korsching, and H. Thoenen, *Dev. Biol.* **120**, 322 (1987).
[35] S. R. Whittmore, T. Ebendal, L. Larkfors, L. Olson, A. Seiger, I. Stromberg, and H. Persson, *Proc. Natl. Acad. Sci. U.S.A.* **83**, 817 (1986).

the hippocampal source of NGF results in retrograde degeneration of medial septal cholinergic neurons.[38,39] This model system thus provides an *in vivo* assay of the neuron-specific effects of NGF, since NGF is required to maintain choline acetyltransferase (ChAT) and NGF receptor (NGFr) immunoreactivity in medial septal cholinergic neurons.[7,16–18] This *in vivo* assay has three elements: (1) placement of an FF lesion to establish a state of exogenous NGF dependence, (2) infusion of NGF or control solutions via external deliver systems into the brain, and (3) postmortem analysis for effects of NGF infusion (e.g., ChAT and NGFr immunoreactive (IR) labeling, acetylcholinesterase (AChE) histochemistry.

Materials

Male or female rats, weight 200–250 g (young adults)
Small animal stereotaxic apparatus (Kopf Instruments, Tujunga, CA)
Routine surgical instruments plus jeweler's forceps, drill with #2 and #8 drill bits (Roboz Inc.), cyanoacrylate adhesive, and triceps forceps
Aspiration apparatus: A 20-gauge suction catheter may be used; alternatively, suction pipette tips are fashioned from 6-inch glass pipettes (Fisher Scientific, Fairlawn, NJ) that may be curved and heat-stretched to produce a small tip suitable for performing aspirative lesions (1 mm). A 5-mm hole is produced along the dorsal surface of the glass pipette to control suction intensity. The hole may be formed by blowing air through one end of a sealed pipette while heating it over a flame.
Gelfoam (Upjohn, Kalamazoo, MI), to control bleeding
Alzet osmotic minipump Model 2002 (Alza Corp., Palo Alto, CA)
Cannula 28-gauge through which NGF is continuously administered into the lateral ventricle for a 2-week period (Plastic Products Inc., Roanoke, VA)
Vinyl tubing to connect osmotic pump with cannula (Bolab V4 tubing, Lake Havasu City, AZ)
Superglue (sterility not required; Krazy Glue Inc., Itasca, IL)

---

[36] S. R. Whittemore, P. L. Friedman, D. Larhammar, H. Persson, M. Gonzalez-Carvajarl, and V. R. Holets, *J. Neurosci. Res.* **20**, 403 (1988).
[37] C. Ayer-LeLievre, L. Olson, T. Ebendahl, A. Seiger, and H. Persson, *Science* **240**, 1339 (1988).
[38] D. M. Armstrong, R. D. Terry, R. M. Deteresa, G. Bruce, L. B. Hersh, and F. H. Gage, *J. Comp. Neurol.* **264**, 421 (1987).
[39] F. H. Gage, K. Wictorin, W. Ficher, L. R. Williams, S. Varon, and A. Bjorklund, *Neuroscience* **19**, 241 (1986).

Surgery

The animal is anesthetized and placed in the stereotaxic apparatus. A 3-cm incision in the sagittal plane is made along the scalp to expose the skull. Periosteum is scraped away with the knife blade, exposing surface landmarks of the skull. Using bregma as a zero reference point,[40] a hole is drilled with a #2 drill bit at stereotaxic coordinates A/P +0.8 mm, M/L +1.2 mm to receive the cannula for intraventricular infusion. Next, the larger #8 drill bit is used to remove a 2-mm^2 portion of skull immediately caudal to the first hole. This craniotomy may also be performed on the contralateral side of the skull if bilateral FF lesions are to be made, but care must be taken to avoid rupture of the superior saggital sinus over the midline. After the 2 mm^2 square portion of skull is removed, dura mater overlying the cortex is removed with a suction apparatus or jeweler's forceps. The superficial cingulate cortex, deep white matter, and corpus callosum are then aspirated, exposing the ventricular system. The FF is identified, lying between the rostrally located septum and the dorsomedially located hippocampus. The FF is aspirated completely, using extreme caution to avoid neighboring structures yet removing all medially as well as laterally located fibers (Fig. 2). Bleeding may be controlled with one or two pieces of Gelfoam (2 mm^2) that have been presoaked in saline.

The next step is to position and cement in place the NGF infusion apparatus. The osmotic pump, Alzet Model 2002, is first dipped into molten paraffin to cover one-half of its surface area. Paraffin coating reduces the pump flow rate from 0.5 to 0.25 μl/hr. The pump is then filled with 50 μg/ml of NGF (Bioproducts for Science, Indianapolis, IN) in control solution, or control solution alone consisting of artificial cerebrospinal fluid (CSF), 0.1% rat serum (to supply a protein carrier for NGF), and 50 μg/ml gentamicin. The pump is attached to the intracranial portion of the infusion apparatus with a 3-cm length of vinyl tubing (Fig. 3). The joining of tubing to pump and tubing to intraventricular cannula assembly is strengthened with superglue. The pump assembly is fastened to a triceps forceps that has been secured to an arm of the stereotax. The pump assembly is positioned such that is vertically oriented in all planes, and it is lowered through the smaller, rostral craniotomy opening until it contacts the surface of the skull. The proper ventral/dorsal distance has been predetermined by the 5-mm length of the intraventricular cannula. A drop of superglue is placed around the base of the pump to adhere it to the skull, and then a more substantial base of support is created using cyanoacrylate.

---

[40] G. Paxinos and C. Watson, "The Rat Brain in Stereotaxic Coordinates." Academic Press, San Diego, California, 1986.

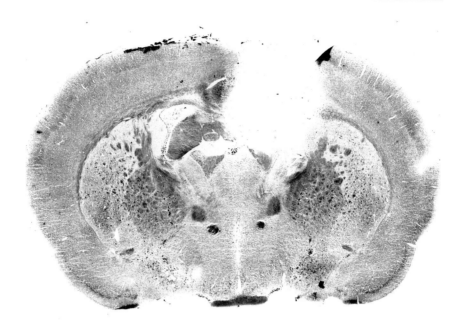

FIG. 2. Acetylcholinesterase stain showing a complete unilateral aspirative lesion of the FF. A normal FF is seen on left-hand side of the photomicrograph. Magnification: ×35.

Finally, finishing drops of superglue are placed around the base of cyanoacrylate to firmly anchor the cannula assembly to the skull. A subcutaneous pocket is created in the neck of the animal and the infrascapular space by blunt dissection to receive the osmotic pump. The incision is closed with wound clips.

Technical Aspects of Fimbria/Fornia Lesion Model

Unilateral versus Bilateral Fimbria/Fornia Lesions

The decision to perform a unilateral or bilateral FF lesion in a given experimental protocol depends on several factors. Advantages of the unilateral FF lesion model include the fact that the lesioned and unlesioned septum can be compared in the same animal, offering an "internal" control through which to assess the effects of the FF lesion. Changes in the cholinergic cell population can be simply quantified by comparing the number of ChAT-labeled or AChE-stained cells on the lesioned side of the

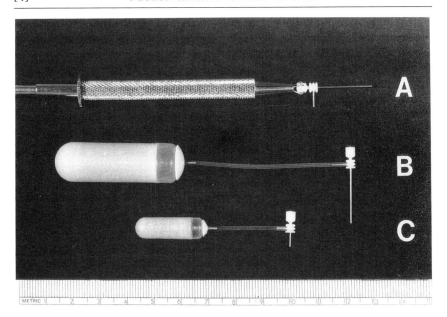

FIG. 3. Cannula–tubing–osmotic pump assembly. (A) Intraventricular cannula held in triceps forceps to stabilize and guide the assembly while being placed on the skull. The triceps forceps is mounted on a stereotax sidearm. (B) Larger intraventricular cannula–tubing–pump assembly used in primate NGF infusion studies. (C) Intraventricular cannula–tubing–pump assembly used in rat NGF infusion studies.

septum to the intact, contralateral side, expressed as a percentage. The chief disadvantage of unilateral FF lesions is their technical difficulty. It is of crucial importance to completely interrupt the FF in the unilateral lesion model, since any remaining fibers will artifactually elevate the number of remaining cholinergic cell bodies in the medial septum. In addition, remaining cholinergic axons may undergo compensatory collateral sprouting in the dorsal hippocampus, thereby restoring some cholinergic innervation to the target. Removing all medial fibers of the FF without damaging the contralateral FF is a technical challenge requiring repeated practice and verification by both Nissl staining and AChE histochemistry to visualize the lesion directly and to verify depletion of cholinergic fibers from the dorsal hippocampus, respectively.

Bilateral FF lesions are technically simpler and more reproducible than unilateral FF lesions, since the medial aspect of the FF is more easily approached from a bilateral craniotomy, and one need not attempt to spare the contralateral FF since it will also be removed. In addition, bilateral FF lesions offer the advantage of resulting in quantifiable behavioral deficits, including loss of mnemonic acquisition and retention.[26,27] Thus, interventions directed toward restoration of functional deficits are possible with the bilateral FF lesion model. Disadvantages of the bilateral FF model include the facts that the contralateral septum can no longer be employed as a simple control and that bilateral lesions are slightly more traumatic to the animal (more bleeding, prolonged recovery).

Distance of Axotomy from Cell Body

It is important to place FF lesions in a consistent and reproducible manner in this model to reduce subsequent variability in quantification of cholinergic neuron numbers. The distance of axotomy from the cell body influences the severity of trauma to the cell; axotomy distal from the cell body is generally a less severe insult to the cell than more proximal axotomy.[41,42] Proximal lesions of the FF result in a greater degree of cholinergic cell degeneration than distal lesions,[43] although this effect in the region of the medial septum is not pronounced. We attempt to place our FF lesions at the junction of the caudal FF and rostral hippocampus, since this region is a readily identifiable landmark *in vivo*. We also use as small a tip on the aspiration apparatus as possible (5 mm) to avoid uncontrolled, inadvertantly large lesions. Reproducibility in lesion size and location reduces subsequent variability.

Parenchymal Damage from Infusion Apparatus

Cerebral parenchymal damage has been reported after intracerebroventricular infusions if large-bore cannulas are used (25 gauge or larger), or if pump flow rates are high (0.5 μl/hr).[44] In addition, the possibility has been raised that some pump contents may be toxic.[45] The problem of parenchymal necrosis is substantially reduced by embedding one-half of the surface area of the osmotic pump in paraffin prior to use as noted above and by using cannulas that are 28 gauge in size. Cannulas smaller than 30 gauge may become obstructed in as many as 30% of cases.[44]

Interpretation of Results

With the above model in the young rat, 2 weeks after the FF lesion 65 to 90% of septal cholinergic cells will undergo retrograde degeneration.[38] Thus, the neuron-specific effects of NGF can be determined after 2 weeks

[41] A. Torvik, *Neuropathol. Appl. Neurobiol.* **2**, 423 (1976).
[42] A. R. Lieberman, *Int. Rev. Neurobiol.* **14**, 49 (1971).
[43] M. V. Sofroniew and O. Isacson, *J. Chem. Neuroanat.* **1**, 327 (1988).
[44] L. R. Williams, H. L. Vahlsing, T. Lindamood, S. Varon, F. H. Gage, and M. Manthorpe, *Exp. Neurol* **95**, 743 (1987).
[45] H. L. Vahlsing, S. Varon, T. Hagg, B. Fass-Holmes, A. Dekker, M. Manley, and M. Manthorpe, *Exp. Neurol.* **105**, 233 (1989).

of NGF infusion. Some experimental paradigms (e.g., attempts to promote axonal regeneration) may require longer NGF infusion periods, in which case pumps may be changed in the animal every 2 or 3 weeks. We have generally found that the animal tolerates three pump changes before suffering significant skin breakdown.

Following the NGF infusion period, the animal is transcardially perfused with paraformaldehyde for 20 min, followed by overnight postfixation in formalin. The brain is placed in phosphate buffer containing 18% (w/v) sucrose for an additional 2 days to prevent sectioning artifacts. Brains are cut at intervals of 30–40 μm on a sliding microtome in series of four or six sections, with sections alternately processed for Nissl stain, AChE histochemistry, and ChAT or NGFr immunohistochemistry.[38]

If the FF lesion is complete, then total depletion of cholinergic fibers in the dorsal hippocampus will be seen on AChE staining, with the possible exception of a small region of residual fibers in the most medial portion of the CA1 region (Fig. 4). These residual fibers are derived from sprouting of midline magnocellular neurons located above the corpus callosum and do not pass through the FF.

In the medial septum, ChAT-immunoreactive labeling will show a reduction in the number of neuronal profiles after FF lesions in animals lacking NGF infusions. Assuming that the brain has been sectioned at 40-μm intervals, with one in four sections processed for ChAT-immunoreactivity, a total of five to seven sections through the region of the medial septum will be available for quantification of ChAT-immunoreactive cell profiles. We choose three sections located approximately 160, 320, and 480 μm rostral to the decussation of the anterior commissure for quantification because they are densely populated with ChAT-positive neurons, provide a major source of cholinergic innervation to the hippocampus, and are clearly separate from cholinergic neurons lying in the ventrally positioned vertical limb of the diagonal band. If necessary, an imaginary line is drawn through the dorsal aspect of the anterior commissure to prevent contamination of cell quantification with neurons in the vertical limb of the diagonal band. The sections are analyzed with a 10× objective using a 0.5 × 0.5 mm counting grid. At this magnification each side of the septum occupies 12 optical fields. Cells labeled positively with peroxidase reaction product and possessing either (1) a cell body with emerging fiber or (2) a cell body with well-defined nucleus are counted.

Employing these quantification methods, control animals that lack an NGF source after the FF lesion will show a loss of ChAT labeling in 65 to 90% of neurons in the septum ipsilateral to the side of the FF lesion compared to the contralateral, unlesioned side of the septum (Fig. 5).[38] This variability in loss of labeling among control animals may be attributable in part to the distance of axotomy from the cell body, quality and intensity

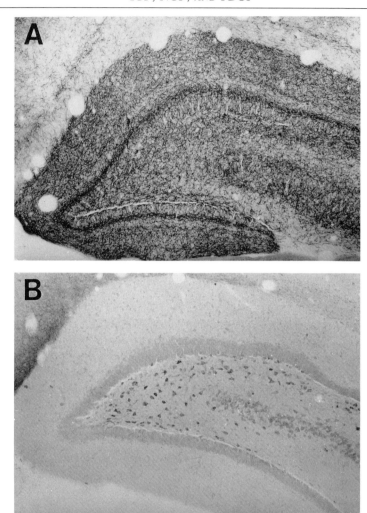

FIG. 4. Acetylcholinesterase (AChE) stain of the dorsal hippocampus in (A) a normal animal and (B) an animal after FF lesion. Depletion of cholinergic fibers is seen in the lesioned animal. Magnification: ×160.

of the immunocytochemical label, time elapsed since the FF lesion, and completeness of the FF lesion. Animals that receive NGF infusions after FF lesions, on the other hand, will show persistent ChAT labeling in 80 to 100% of neurons in the region of the medial septum (Fig. 5).[7,16–18] Some of this variability in the degree of ChAT savings may be accounted for by the above factors, as well as differences in dose of NGF used in various

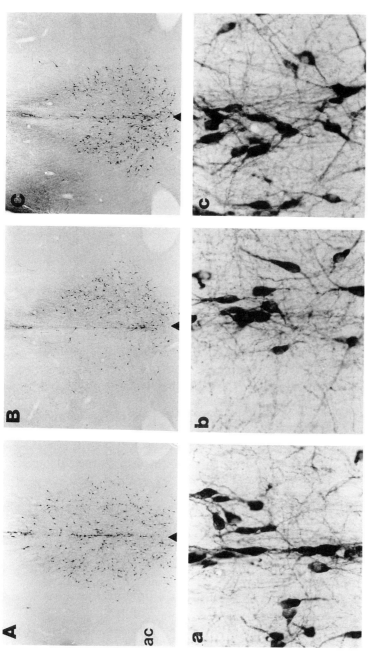

FIG. 5. Choline acetyltransferase immunoreactivity (ChAT-IR) in the medial septal region. (a and A) Normal distribution of cholinergic cell bodies on both sides of medial septum in control, unlesioned animals at low and high magnification. (b and B) Retrograde degeneration of cholinergic cell bodies ipsilateral to the side of the unilateral FF lesion (left-hand side) compared to the contralateral, intact side of the septum (right-hand side). (c and C) Prevention of retrograde cholinergic neuron degeneration in unilaterally FF lesioned animals after receiving 2-week infusions of NGF intracerebroventricularly. The FF lesion was performed on the left-hand side. Magnifications: (A–C) ×106, (a–c) ×264.

studies (from 25 to 120 µg/ml), method of NGF infusion (continuous versus intermittent ventricular infusion), rate of NGF infusion, NGF stability in the osmotic pump (to date, measurements of NGF activity by two-site ELISA after 2 weeks *in vitro* reveal maintenance of 50–100% of activity), and presence or absence of parenchymal necrosis induced by the infusion apparatus (e.g., septal damage from the intraventricular cannula could actually elevate NGF levels reaching the medial septum). Thus, quantification of cholinergic perikarya in the medial septum provides one means of quantifying neuron-specific effects of NGF.

Another means of evaluating NGF effects on cholinergic neurons, namely, the measurement of ChAT enzyme biochemical activity,[39,46] has been less helpful than ChAT-immunoreactive cell labeling in detecting NGF effects on neurons. After an initial drop of activity detected 1 day following an FF lesion, ChAT biochemical activity measured in whole septum returns to normal levels 2 weeks later.[39]

A second means of quantifying the neuron-specific effects of NGF is determination of the number of NGFr-immunoreactive neuron profiles remaining in the medial septum after an FF lesion. Cholinergic and NGFr cell populations overlap to a 90–95% extent in the medial septum,[8,13] so NGFr cell quantification provides an independent parameter for cholinergic neuron quantification that correlates with ChAT quantification. Procedures for NGFr cell quantification are identical to those presented for ChAT. In control animals following FF lesions, approximately 65–90% of NGFr-immunoreactive neurons degenerate, whereas NGF infusion preserves approximately 75–100% of neurons compared to the contralateral, unlesioned side of the septum. In addition to AChE histochemistry, ChAT biochemical activity, and ChAT or NGFr immunohistochemistry, measurements of NGFr mRNA levels can be examined *in vivo* using *in situ* hybridization for NGF mRNA[47] or ChAT mRNA.

Finally, it should be noted that whereas clear degenerative changes occur in septal cholinergic neurons following FF lesions, whether the end result of these changes is permanent cell atrophy or cell death remains controversial. Two studies using fluorescent cell labels suggest that cell death occurs after FF lesions, although the time course for this death may be protracted over several weeks.[48,49] On the other hand, if infusions of NGF are delayed several weeks after the FF lesion, then ChAT immunore-

[46] F. Fonnum, *J. Biochem (Tokyo)* **115**, 465 (1969).
[47] G. A. Higgins, S. Koh, K. S. Chen, and F. H. Gage, *Neuron* **3**, 247 (1989).
[48] M. H. Tuszynski, D. A. Armstrong, and F. H. Gage, *Brain Res.* **508**, 241 (1990).
[49] T. S. O'Brien, C. N. Svendsen, O. Isacson, and M. V. Sofroniew, *Brain Res.* **508**, 249 (1990).

activity is restored to some, but not all, septal cholinergic neurons.[50] Loss of Nissl-labeled neurons occurs in the medial septum after FF lesions, suggesting that cells may indeed die. In contrast, cholinergic neurons in other regions of the CNS may initially lose ChAT staining after axotomy and later recover it after prolonged periods.[51] In these cases, however, Nissl staining is not lost. The cholinergic projection from the nucleus basalis of Meynert to neocortex also relies on retrograde transport of NGF from target to cell body to maintain normal cell function. In this case, cortical lesions that result in NGF deprivation to basal nucleus cell bodies result in atrophy but not loss of ChAT-immunoreactive labeling. Thus, the available evidence suggests that FF lesions result in the death of some cholinergic neurons in the medial septum, the atrophy of others which may eventually result in death, and the persistent survival of some neurons.

Nerve Growth Factor Effects on Noncholinergic Neurons

Other neuron populations present in the medial septum include γ-aminobutyric acid (GABA), substance P, galanin, and possibly other neuropeptides. GABA neurons, like cholinergic neurons, also degenerate after FF lesions.[52] Whether NGF infusions prevent loss of GABAergic neurons in a manner analogous to that of cholinergic neurons is unclear.[5] Changes in other medial septal neuron populations after FF lesions have not yet been clarified.

Nerve Growth Factor Infusions in Nonhuman Primates

A model similar to that used in the rat has been developed in primates to demonstrate dependence of medial septal cholinergic neurons on NGF after unilateral fornix lesions.[53] That similar findings have been shown in primates is not surprising, since primate basal forebrain cholinergic neurons possess NGF receptors.[12,13] In primates, approximately 45% of medial septal ChAT-immunoreactive neurons remain labeled after unilateral fornix lesions, while NGF infusions result in the persistence of 80% of labeled ChAT-immunoreactive neurons.[53] NGFr-immunoreactive neurons show similar changes after unilateral fornix lesions, with a persistence of 42% of neurons in control animals and 79% in NGF-treated animals.

[50] T. Hagg, M. Manthorpe, H. L. Vahlsing, and S. Varon, *Exp. Neurol.* **101**, 303 (1988).
[51] B. E. Lams, O. Isacson, and M. V. Sofroniew, *Soc. Neurosci. Abst.* **14**, 366 (1988).
[52] G. M. Peterson, L. R. William, S. Varon, and F. H. Gage, *Neurosci. Lett* **76**, 140 (1987).
[53] M. H. Tuszynski, H.-S. U, D. G. Amaral, and F. H. Gage, *J. Neurosci.* in press (1990).

Other Means of Nerve Growth Factor Delivery to Brain

Recently, fibroblasts have been genetically modified *in vitro* to produce and secrete NGF. Using a retroviral vector to transmit the NGF gene, fibroblast cell lines and primary fibroblasts were genetically modified *in vitro* to produce and secrete NGF at rates of up to 3 ng/hr/10^5 cells. When these cells were grafted to the brains of animals that had undergone FF lesions, retrograde degeneration of cholinergic neurons in the medial septal region was prevented.[54] Control grafts with fibroblasts lacking the NGF gene failed to prevent retrograde degeneration of cholinergic neurons. Thus, neuron-specific effects of NGF can also be used to test the efficacy of genetic, cellular, or mechanical manipulations.

Conclusion

Neuron-specific effects of NGF *in vivo* can be demonstrated after FF lesions in rats and primates. Other models also demonstrate neuron-specific effects of NGF in the rat, including lesion-induced changes in the striatum and brain stem and age-related degenerations of basal forebrain cholinergic neurons.

[54] E. Recio-Pinto and D. N. Ishii, *J. Neurosci. Res.* **19**, 312 (1988).

[5] Detection and Assay of Nerve Growth Factor mRNA

By MARGARET FAHNESTOCK

Introduction

Nerve growth factor (NGF) is a polypeptide hormone essential for the growth and maintenance of certain peripheral and central nervous system (CNS) neurons.[1-3] NGF is found in high concentrations in the submaxillary glands of the male mouse[4] and the male and female African rat (*Mastomys*)[5,6] and in the snake venom gland.[7] Its physiological role in these

[1] H. Thoenen and Y.-A. Barde, *Physiol. Rev.* **60**, 1284 (1980).
[2] L. A. Greene and E. M. Shooter, *Annu. Rev. Neurosci.* **3**, 353 (1980).
[3] S. R. Whittemore and Å. Seiger, *Brain Res. Rev.* **12**, 439 (1987).
[4] A. C. Server and E. M. Shooter, *Adv. Protein Chem.* **31**, 339 (1977).
[5] T. L. J. Darling and M. Fahnestock, *Biochemistry* **27**, 6686 (1988).
[6] L. Aloe, C. Cozzari, and R. Levi-Montalcini, *Exp. Cell Res.* **133**, 475 (1981).
[7] R. A. Hogue-Angeletti, W. A. Frazier, J. W. Jacobs, H. D. Niall, and R. A. Bradshaw, *Biochemistry* **15**, 26 (1976).

tissues is unknown. However, in most other tissues, particularly those in which NGF is known to function in neuronal development or survival, the levels of NGF are extremely low.

cDNA and genomic clones for NGF have been isolated from a variety of species and exhibit a high degree of sequence conservation.[8-14] These clones have been used as hybridization probes to demonstrate the presence of NGF mRNA in a number of species and tissues.[15,16] Modulation of NGF mRNA levels during development, in response to injury, and in different regions of nervous and other tissues has been shown to be important for the regulation of NGF biological activity.[16-20] Detection of small quantities of NGF mRNA in tissue necessitates use of the most sensitive hybridization methods available. These methods are generally applicable to detection of any mRNA found in small amounts for which a nucleic acid probe is available.

Detection and assay of NGF mRNA is carried out in four steps: (1) purification of poly(A)$^+$ mRNA from tissue or cells; (2) preparation of a Northern and/or dot blot; (3) preparation of the probe; and (4) hybridization, autoradiography, and quantitation.

Purification of mRNA

Two different methods for mRNA purification have proved effective in our laboratory. The first relies on the presence of guanidine thiocyanate

---

[8] J. Scott, M. Selby, M. Urdea, M. Quiroga, G. J. Bell, and W. J. Rutter, *Nature (London)* **302,** 538 (1983).
[9] A. Ullrich, A. Gray, C. Berman, and T. J. Dull, *Nature (London)* **303,** 821 (1983).
[10] M. Fahnestock and R. A. Bell, *Gene* **69,** 257 (1988).
[11] R. Meier, M. Becker-Andre, R. Gotz, R. Heumann, A. Shaw, and H. Thoenen, *EMBO J.* **5,** 1489 (1986).
[12] T. Ebendal, D. Larhammar, and H. Persson, *EMBO J.* **5,** 1483 (1986).
[13] S. R. Whittemore, P. L. Friedman, D. Larhammar, H. Persson, M. Gonzalez-Carvajal, and V. R. Holets, *J. Neurosci. Res.* **20,** 403 (1988).
[14] M. A. Schwarz, D. Fisher, R. A. Bradshaw, and P. J. Isackson, *J. Neurochem.* **52,** 1203 (1989).
[15] D. Shelton and L. F. Reichardt, *Proc. Natl. Acad. Sci. U.S.A.* **81,** 7951 (1984).
[16] R. Heumann, S. Korsching, J. Scott, and H. Thoenen, *EMBO J.* **3,** 3183 (1984).
[17] T. H. Large, S. C. Bodary, D. O. Clegg, G. Weskamp, U. Otten, and L. F. Reichardt, *Science* **234,** 352 (1986).
[18] D. L. Shelton and L. F. Reichardt, *J. Cell Biol.* **102,** 1940 (1986).
[19] R. Heumann, D. Lindholm, C. Bandtlow, M. Meyer, M. J. Radeke, T. P. Misko, E. M. Shooter, and H. Thoenen, *Proc. Natl. Acad. Sci. U.S.A.* **84,** 8735 (1987).
[20] D. L. Shelton and L. A. Reichardt, *Proc. Natl. Acad. Sci. U.S.A.* **83,** 2714 (1986).

and 2-mercaptoethanol during tissue homogenization to inhibit RNases[21,22] and on lithium chloride precipitation to purify total RNA[23]; oligo(dT)-cellulose chromatography is then used to separate poly(A)$^+$ mRNA from the poly(A)$^-$ fraction.[24,25] This method is useful when a poly(A)$^-$ fraction is needed for checking the RNA integrity and for negative controls on blots. The second method, a modification of the procedure of Badley et al.,[26] uses proteinase K and sodium dodecyl sulfate (SDS) to inhibit RNA degradation during tissue homogenization. Batch absorption of poly (A)$^+$ mRNA directly from the homogenate results in a poly(A)$^+$ fraction useful for blots and a fraction containing all other cellular material. This method is much more rapid than the first and often gives higher yields with greater purity. However, the non-poly(A)$^+$ fraction is only suitable for checking RNA integrity after additional phenol extraction and ethanol precipitation, and we have not used it as control material on blots because of the large amount of DNA contamination.

For both procedures, it is essential that all glassware, plasticware, and solutions be RNase-free. Methods for preparation and handling of RNase-free material have been published elsewhere in this series.[27] We have found that, in most cases, autoclaving of glassware, rather than baking, is sufficient if the glassware is kept separate for RNA use only. Autoclaving is preferable to repeated baking at high temperatures, because heating above 200° will weaken Corex centrifuge tubes. Also, we no longer use diethyl pyrocarbonate (DEPC) in our solutions; its omission has not impaired isolation of intact RNA, and in some cases has improved it.

mRNA Purification Using Guanidine Thiocyanate[21-23]

Reagents

> Lysis solution: Mix 2.5 ml of 1 M Tris-HCl, pH 7.5, 2 ml of 250 mM EDTA, pH 8.0, and 29.5 g guanidine thiocyanate (Fluka, Ronkonkoma, NY). Add distilled water to 46 ml. Filter-sterilize (cellulose nitrate). Add 4 ml 2-mercaptoethanol just before use.

[21] J. M. Chirgwin, A. E. Przybyla, R. J. MacDonald, and W. J. Rutter, *Biochemistry* **18,** 5294 (1979).
[22] R. J. MacDonald, G. H. Swift, A. E. Przybyla, and J. M. Chirgwin, this series, Vol. 152, p. 219.
[23] G. Cathala, J. F. Savouret, B. Mendez, B. L. West, M. Karin, J. A. Martial, and J. D. Baxter, *DNA* **2,** 329 (1983).
[24] H. Aviv and P. Leder, *Proc. Natl. Acad. Sci. U.S.A.* **69,** 1408 (1972).
[25] A. Jacobson, this series, Vol. 152, p. 254.
[26] J. E. Badley, G. A. Bishop, T. St. John, and J. A. Frelinger, *BioTechniques* **6,** 114 (1988).
[27] D. D. Blumberg, this series, Vol. 152, p. 20.

4 M LiCl, 6 M LiCl, 2 M Tris-HCl, pH 7.5, 3 M sodium acetate, pH 5.2, and distilled water, autoclaved stock solutions

TES solution: 10 mM Tris-HCl, pH 7.5, 1 mM EDTA, pH 8.0, 0.1% SDS; autoclave

3 M LiCl–3 M urea solution: To 50 ml autoclaved 6 M LiCl add 18 g urea (Schwarz/Mann, Cleveland, OH, Ultra Pure), then add autoclaved distilled water to 100 ml

Phenol–chloroform: To 500 g phenol (solid, melted at 37°) add 100 ml of 2 M Tris-HCl, pH 7.5, autoclaved, 130 ml distilled water, autoclaved, 25 ml m-cresol, 1 ml 2-mercaptoethanol, and 500 mg 8-hydroxyquinoline. Add chloroform in a 1 : 1 ratio with the phenol phase. Mix well and allow phases to separate at room temperature. Store at 4°.

Procedure. The method works equally well for fresh or frozen tissue or cultured cells. Freeze tissue in liquid nitrogen, store it at −80°, and perform all weighing and homogenization steps rapidly in the cold room. Frozen tissue must not be allowed to thaw at all before homogenization. The lysis solution should be prechilled to 4°; however, the guanidine thiocyanate tends to crystallize out at low temperatures, so do not store the solution for prolonged periods of time in the cold. If the guanidine crystallizes out, it will redissolve on warming but must be rechilled again before use. 2-Mercaptoethanol inhibits crystallization at 4°.

Place the frozen tissue in a prechilled RNase-free glass container; we use a Corex tube for small amounts of tissue and a 100-ml pharmaceutical graduate for larger amounts. For tissue containing low concentrations of NGF mRNA, start with 5–10 g tissue. Add 3.5 ml lysis buffer per gram wet weight frozen tissue and homogenize immediately with a Brinkmann Polytron, using from one to four 15-sec bursts at full speed. For some tissues, homogenization will be complete; but for other tissues, chunks of unhomogenized tissue or a lot of fatty tissue may remain. In this case, centrifuge for 10 min at 4° at 10,000 g. Pipette off the lower solution, discarding pelleted debris and floating foam and lipids. Add 7 volumes of chilled 4 M LiCl to the homogenate and mix by inversion.

For cells grown in monolayer culture, remove the medium, rinse the cells with PBS (50 mM sodium phosphate, 0.15 M NaCl, pH 7.2), and add 2 ml lysis solution to each 150-mm dish. After incubation for 5 min at room temperature, remove the viscous solution with a Pasteur pipette. Rinse each plate with 8 ml of 4 M LiCl and add the rinse to the lysis solution, mixing gently. For cells grown in spinner culture, rinse the cells with PBS and add 7 ml lysis solution per milliliter of packed cells. After allowing lysis to occur (5 min at room temperature), add 4 volumes of 4 M LiCl.

Incubate the lysate/LiCl solution overnight at 4° to precipitate RNA.

Centrifuge the suspension at 4° at 10,000 g. For most tissues, a 30-min spin is sufficient. However, in a few cases such as brain tissue, 90 min is required to obtain a sufficiently packed pellet. Discard the supernatant and resuspend the pellets thoroughly in 15 ml cold LiCl–urea solution. Vortexing may be necessary. Pellet the RNA once again by centrifugation at 4° at 10,000 g. Once again, the centrifugation time to obtain a hard pellet may vary from 30 to 120 min. Discard the supernatant.

Resuspend the pellets thoroughly in 5 ml TES solution at room temperature. Extract with an equal volume of phenol–chloroform and mix by inversion to minimize foaming. Separate the phases by centrifugation for 10 min at 7500 g. Remove the aqueous layer, then back-extract the organic layer and interface with one-half volume of Tris-EDTA (no SDS), vortexing to mix this time. Add this material to the aqueous phase. Repeat the phenol–chloroform extraction once more, vortexing to mix and omitting the back extraction. See Ref. 28 for additional information. For some tissues, such as brain tissue, the interface is often quite large. In this case, it is preferable to add phenol first, mix by inversion, add the chloroform, and mix again before centrifugation. If the interface is still significant, remove the phenol–chloroform layer and add another volume of chloroform to the interface; vortex and recentrifuge.

Precipitate the RNA with sodium acetate and ethanol.[29] Resuspend the pellet in 1–2 ml distilled water. Determine the concentration by reading the optical density of the solution at 260 nm in an RNase-free cuvette.

Oligo (dT)-Cellulose Chromatography[24,25]

Reagents

> Low-salt buffer: 10 mM Tris-HCl, pH 7.5, 1 mM EDTA, 0.1% N-lauroylsarcosine (Sarkosyl) (Filter-sterilize a 10% stock solution of Sarkosyl in distilled water and add 1/100th volume to the autoclaved Tris-EDTA)
>
> High-salt buffer: Same as above plus 0.5 M NaCl

Procedure. Weigh out an appropriate amount of oligo(dT)-cellulose Type 3 (Collaborative Research, Bedford, MA). One gram of the resin binds approximately 100 OD$_{260}$ units, and poly(A)$^+$ mRNA usually represents a maximum of 3% of the total RNA. Allowing for a large excess of resin, a convenient amount is 50 mg of resin per RNA preparation. Suspend the resin in 3 ml of low-salt buffer and then pour it into a 1-ml plastic sterile

---

[28] D. M. Wallace, this series, Vol. 152, p. 33.
[29] D. M. Wallace, this series, Vol. 152, p. 41.

syringe plugged with a small amount of siliconized, autoclaved glass wool. Wash the column with 20 column volumes of high-salt buffer. The column is run at room temperature.

Adjust the RNA to 10 mM Tris-HCl, pH 7.5, 1 mM EDTA, 0.1% Sarkosyl, 0.5 M NaCl by the addition of sterile stock solutions. Heat the samples for 2–3 min at 95°, then quick-chill on ice before bringing them to room temperature. Pass the RNA through the column slowly (one drop per 10 sec) so as to facilitate binding of all the poly(A)$^+$ mRNA. Pass the flow-through solution through the column once or twice more. Wash the column with 10–20 column volumes of high-salt buffer. Elute the poly(A)$^+$ mRNA with four 400-μl aliquots of low-salt buffer and collect the eluate in sterile microcentrifuge tubes. Add 50 μl of 3 M sodium acetate and 900 μl of ethanol to precipitate the fractions; resuspend the washed pellets in 0.5 ml distilled water. The yield, measured by absorbance at 260 nm, varies greatly depending on the tissue source (Table I). The purity, estimated by the absorbance ratio at 260/280 nm, varies between 1.6 and 2.0, with 2.0 being ideal.

mRNA Purification Using Proteinase K[26]

Reagents

Lysis buffer: 0.2 M NaCl, 0.2 M Tris-HCl, pH 7.5, 1.5 mM MgCl$_2$. Autoclave. Add SDS (sterile-filtered or autoclaved separately) to 2%. Add proteinase K to 200 μg/ml (from 20 mg/ml frozen stock) just before use.

5 M NaCl, autoclaved

Binding buffer: 10 mM Tris-HCl, pH 7.4, 0.5 M NaCl; autoclave

Elution buffer: 10 mM Tris-HCl, pH 7.4; autoclave

Procedure. Follow directions given above for homogenization of frozen tissue, but use 10 ml lysis buffer per gram wet weight of tissue. Do not centrifuge. Incubate the homogenate in a shaking incubator (350 rpm) at 45° for 2 hr. It is important that the temperature be exactly 45° and that the homogenate be shaken vigorously.

Suspend 0.2 g oligo(dT)-cellulose Type 3 per 10 g tissue in 5 ml elution buffer. Wash the resin by centrifugation in a clinical centrifuge and resuspension of the pellet in 5 ml of binding buffer. Centrifuge again and resuspend the pellet in 1 ml binding buffer.

Adjust the tissue homogenate to 0.5 M NaCl by addition of 5 M stock solution. Add the lysate to the oligo(dT)-cellulose and incubate 20 min at room temperature with rocking. Centrifuge the resin and pour off the supernatant. (The supernatant may be extracted with phenol–chloroform

TABLE I
YIELD OF POLY(A)$^+$ mRNA FROM VARIOUS SPECIES AND TISSUES USING EITHER
GUANIDINE THIOCYANATE OR PROTEINASE K PROCEDURE

Tissue	Yield [μg poly(A)$^+$ mRNA/g wet weight tissue]
Rat submaxillary gland	135.0^a
Rat liver	7.8^a
Rat brain	13.0^a
Rat sciatic nerve	0.5^a
Regenerating rat sciatic nerve	1.4^a
Mastomys (African rat) submaxillary gland	222.0^a
Mastomys kidney	2.9^a
Human frontal cortex (normal)^b	7.5^a
	8.4^a
	4.2^a
	6.7^c
	9.0^c
Human frontal cortex (Alzheimer's)	7.1^a
	11.0^a
	11.7^c
	7.1^c
Human hippocampus (normal)	5.4^c
Human hippocampus (Alzheimer's)	3.7^a
	11.2^c
	4.3^c
Human nucleus basalis (normal)	5.7^c
	5.6^c
Human nucleus basalis (Alzheimer's)	1.5^c
Human hypothalamus (normal)	5.4^c

^{a} Guanidine thiocyanate procedure.
^{b} Where several yields are listed, these are from different samples that may differ in postmortem time, extent of tissue damage, etc.
^{c} Proteinase K procedure.

and ethanol precipitated to allow the ribosomal RNA to be examined by gel electrophoresis.) Resuspend the resin in binding buffer. Repeat the centrifugation and resuspension steps 2 to 3 times.

Pour the resin into a 1-ml plastic, sterile syringe containing siliconized, autoclaved glass wool. Wash the column with binding buffer until the A_{260} of the eluate is less than 0.05, and preferably approaches 0.01. Let the column run dry and elute the poly(A)$^+$ mRNA with four 400-μl fractions of elution buffer. Determine the yield by reading the A_{260} and A_{280} in sterile cuvettes, as described above.

Preparation of Northern Blots

Denaturing Agarose Gel Electrophoresis[30]

Reagents

10× Running buffer: 0.1 M sodium phosphate, pH 7.0, 50 mM sodium acetate, 10 mM EDTA, autoclave

RNA sample buffer: 1× running buffer, 50% formamide (deionized by stirring 30 min with AG501-X8(D) mixed-bed resin from Bio-Rad, Richmond, CA), 2.2 M formaldehyde

Procedure. The gel is 0.8% agarose, 2.2 M formaldehyde in 1× running buffer. The gel may be run in MOPS buffer instead of sodium acetate.[31] Add the formaldehyde (28.1 ml/150-ml gel) after the buffered agarose solution has cooled to 60°, and pour the gel in the hood to minimize fumes. Denature the samples, which should include both poly(A)$^+$ and poly(A)$^-$ fractions, by heating 5 min at 65° before adding bromphenol blue and loading the sample onto the gel. For tissues containing low concentrations of NGF mRNA, load up to 20 μg poly(A)$^+$ mRNA per lane. The gel is run in 1× buffer, either 8–10 hr at 30 V (2V/cm) or 6 hr at 50 V (3.6 V/cm), always with buffer circulation. The bromphenol blue should be near the bottom of the gel at the end of the run. Stain the gel with 2.5 μg/ml ethidium bromide in distilled water for 30 min, rinse with distilled water, then destain 30 min or more in 1 mM MgSO$_4$, which helps to remove background staining. Photograph the gel alongside a ruler with the origin at the wells so that a molecular weight standard curve can be drawn using the 18 S and 28 S ribosomal bands that appear in the poly(A)$^-$ lanes. Ethidium bromide staining has been reported to interfere with RNA transfer to the blot; therefore, a parallel lane of poly(A)$^-$ RNA or RNA or DNA molecular weight standards may be run and stained instead of staining the entire gel.

Northern Blotting[31]

Reagents

20× SSPE: 3.6 M NaCl, 0.2 M sodium phosphate, pH 7.7, 20 mM EDTA; filter-sterilize

We use Schleicher and Schuell (Keene, NH) BA85 nitrocellulose membranes. Nylon membranes do not always provide the sensitivity of nitrocellulose, although we have not tested many different brands. Gene-Screen

---

[30] N. Rave, R. Crkvenjakov, and H. Boedtker, *Nucleic Acids Res.* **6**, 3559 (1979).

[31] P. S. Thomas, *Proc. Natl. Acad. Sci. U.S.A.* **77**, 5201 (1980).

Plus (New England Nuclear, Boston, MA) gives particularly high backgrounds with our technique, whereas Hybond-N (Amersham, Arlington Heights, IL) has very low backgrounds and may be appropriate for some uses. We have recently used Schleicher and Schuell BAS85, a reinforced nitrocellulose membrane, which has the strength of nylon and does not shatter or break apart during repeated hybridizations, while still providing the sensitivity of nitrocellulose.

Soak the gel in 20× SSPE for 20–30 min. Wet the membrane with water and soak it at least 5 min in 20× SSPE. Capillary blotting has been described elsewhere.[31] Line up the top of the membrane with the wells of the gel so that an accurate molecular weight may be determined from the autoradiogram. Blot for 18–24 hr in 20× SSPE. The RNA can be fixed to the membrane either by baking for 1–2 hr at 80° in a vacuum oven or by ultraviolet cross-linking in a Stratalinker oven (Stratagene, La Jolla, CA). We have not compared the efficiency of the two RNA fixing methods.

Probe Preparation

For tissues in which the NGF mRNA concentration is extremely low, the sensitivity of the probe often determines the success of the experiment. In these cases, we have had limited success with nick-translated DNA probes, oligonucleotide probes, or RNA probes, although the successful use of RNA probes has been reported.[14] We have had the greatest success with single-stranded DNA probes prepared by synthesis of ^{32}P-labeled cDNA using Klenow polymerase from an NGF/M13 template.[15] The labeled strand is separated from the unlabeled template by electrophoresis in a denaturing, low-melting-temperature agarose gel, and the excised gel piece containing the probe is melted, counted, and added directly to the hybridization buffer.

Reagents

10× Klenow buffer: 200 mM Tris-Cl, pH 7.4, 100 mM MgCl$_2$, 10 mM dithiothreitol (DTT)

Procedure. The template DNA is mouse NGF cDNA in M13, obtained from Dr. L. F. Reichardt (University of California, San Francisco). It is convenient to prime enough template for 10 reactions at a time. Ethanol precipitate 7 μg of M13–NGF DNA with 120 ng of an M13 primer (catalog #403-2, New England Biolabs, Beverly, MA). Wash the pellet with 70% ethanol, dry, resuspend the pellet in 40 μl of 1× Klenow buffer, boil for 2 min to denature the template, and slowly cool to room temperature over a period of 30 to 60 min to anneal the primer to the template. Use 4 μl of this material in each Klenow reaction. The remainder of the primed template can be stored frozen.

For the Klenow reaction, combine 250 μCi [α-^{32}P]dCTP, dried down (>3000 Ci/mmol, Amersham), 4 μl primed DNA (700 ng annealed primer/template), 2 μl of 10× Klenow buffer, 1.5 μl of a mixture of dATP, dGTP, dTTP, 10 mM each, 10.5 μl distilled water, 2 μl Klenow polymerase from *Escherichia coli*, 5 units/μl (New England Biolabs). Incubate for 5 min at room temperature, then 90 min at 30°. Add 80 μl Tris–EDTA–0.2% SDS and extract with phenol–chloroform. Add 5 μg tRNA as a carrier, 1/10 volume sodium acetate, and precipitate with ethanol. Resuspend in 5 μl distilled water, boil for 2 min, and chill on ice. Add 5 μl of 2× alkali loading buffer (100 mM NaOH, 2 mM EDTA, 5% Ficoll, 0.05% bromphenol blue or bromocresol green).

Pour a 1.4% alkaline agarose gel[32] using Sea-Plaque low-melting-temperature agarose (FMC, Rockland, ME) in 50 mM NaCl, 1 mM EDTA, pH 8.0. Allow the gel to set in the cold, then add cold buffer (30 mM NaOH, 1 mM EDTA) and equilibrate at least 30 min in the cold. Load the sample and run at 100 V (8 V/cm) for 1.5 hr in the cold room, with buffer circulation through cooling coils. The bromphenol blue will gradually disappear but will provide a rough estimate of how the gel is running.

At the end of the run, wrap the gel in plastic wrap. Mark the location of the top of the gel and the wells with tape and radioactive ink, then expose the gel for 2–5 min to Kodak XAR-5 film. Using the developed film as a guide, cut the labeled band from the gel and add distilled water to a volume compatible with addition of the probe to the hybridization buffer. Boil the gel for 5 min, count 1 μl, and add an appropriate number of counts per minute (cpm) directly to the hybridization buffer (see below).

Hybridization, Autoradiography, Quantitation, and Reprobing

Reagents

1× Denhardt's solution: 0.02% bovine serum albumin, 0.02% Ficoll 400, 0.02% poly(vinylpyrrolidone) (360K)
Prehybridization buffer: 50% formamide (deionized), 5× SSPE, 5× Denhardt's, 200 μg/ml denatured salmon sperm DNA, 0.1% SDS
Hybridization buffer: 50% formamide, 5× SSPE, 1× Denhardt's, 200 μg/ml denatured salmon sperm DNA, 0.1% SDS, 2.5–5 × 10^7 cpm ^{32}P-labeled probe

[32] T. Maniatis, E. F. Fritsch, and J. Sambrook, "Molecular Cloning, A Laboratory Manual." Cold Spring Harbor Laboratory, Cold Spring Harbor, New York, 1982.

Procedure. Using a volume of approximately 1 ml of prehybridization buffer per square centimeter of membrane surface, prehybridize the blot for a minimum of 4 hr, and preferably overnight, at 50° in a Seal-A-Meal (Dazey) bag with gentle rocking. Drain the bag of prehybridization buffer, add hybridization buffer, and hybridize the blot for 48 hr at the same temperature. Wash twice in 250 ml of 2× SSPE–0.1% SDS for 15 min at room temperature, then twice in 250 ml of 0.1× SSPE–0.1% SDS for 15 min at 60°. Wrap the blot in plastic wrap, mark the edges of the blot with radioactive ink for orientation purposes, and expose to XAR-5 film at −80° for 3–7 days in the presence of an intensifying screen. NGF mRNA migrates at 1.3 kilobases (kb).

If reprobing the blot is desirable, it is important to remove the signal immediately after obtaining an autoradiogram, since once the blot is dried or stored for some time the probe will be difficult or impossible to remove. To remove the probe, pour boiling water over the blot and incubate for 1–2 hr at 65° with gentle shaking. Expose to XAR-5 film to check that all probe has been removed. Store the stripped blot in plastic wrap or in a sealed bag at 4°.

Dot Blots[33]

Dot blots are prepared by spotting serial dilutions of a sample in 20× SSPE buffer directly onto a nitrocellulose or nylon membrane using a dot-blotting manifold (Bethesda Research Laboratories, Gaithersburg, MD) and then fixing the RNA to the membrane as described for Northern blots. Dot blots are much easier to quantitate than Northern blots, because the dots are not subject to streaking as are samples run on a gel, and no losses are incurred due to the transfer process. However, dot blots are much more prone to nonspecific hybridization because of the lack of molecular weight separation. For this reason it is essential to include negative controls such as yeast tRNA and poly(A)$^-$ RNA on all dot blots. In addition, depending on the purpose of the experiments, it is often desirable to quantitate the amount of NGF mRNA per given amount of poly(A)$^+$ mRNA in each sample. This can be done by including a known standard (such as mouse submaxillary gland mRNA) on the dot blot and hybridizing the blot first to ^{32}P-labeled oligo(dT)$_{18}$. This probe is stripped from the blot, and the blot is hybridized with the NGF probe. The NGF signal can then be compared with the signal from mouse submaxillary gland and normalized for the amount of poly(A)$^+$ mRNA in each sample.

Treat dot blots as described for Northern blots except that for hybrid-

---

[33] B. A. White and F. C. Bancroft, *J. Biol. Chem.* **257**, 8569 (1982).

ization to oligo(dT)$_{18}$ the prehybridization and hybridization are carried out at 4°, the buffers are pH 6.5 instead of 7.7, Denhardt's solution is 0.05% of each component, and the probe is prepared using the kinase reaction.[32] The blot is washed twice at room temperature in 2× SSPE, and twice at 37° in 1× SSPE. The dot blot is exposed to X-ray film, and the autoradiogram can be quantitated using a computer-controlled kinetic microtiter plate reader (Molecular Devices Corporation, Menlo Park, CA) or a scanning densitometer.

Polymerase Chain Reaction

The amount of NGF mRNA in many tissues is extremely low, and, in order to load sufficient mRNA on Northern or dot blots to generate a signal, one must start with many grams of tissue. This requirement has made the study of NGF in smaller structures of the brain, for example, impossible using the methods described above.

A new method of detecting small amounts of NGF mRNA in tissue is to use the polymerase chain reaction (PCR).[34] Starting with small amounts of tissue, NGF mRNA can be detected by cDNA synthesis and PCR amplification of NGF cDNA. This procedure requires only 1 µg of mRNA, as opposed to 15–20 µg of mRNA loaded on a Northern blot. The disadvantage of this method is that amplification of NGF cDNA is not strictly quantitative.

Any set of primers specific for NGF sequences of any species can be used. Information on optimization of primer sequences has been published.[35] The primers can be synthesized using any commercial DNA synthesizer. The primers may be purified, preferably by polyacrylamide gel electrophoresis.[32] The NGF primers used in our laboratory are complementary to both mouse and human NGF sequences. This primer set amplifies a 593-base pair (bp) fragment from both human and mouse cDNA. A unique *Xba*I site (TCTAGA) is present in the mouse 593-bp fragment but not in the corresponding human PCR product, allowing coamplification of mouse (control) and human samples in the same PCR reaction.

NGF primer 1 (upstream): 5' CTT CAG CAT TCC CTT GAC AC 3'
NGF primer 2 (downstream): 5' AGC CTT CCT GCT GAG CAC ACA 3'

Procedure. Starting with 1 µg mRNA, synthesize the first-strand cDNA using any desirable method. We use the cDNA kit from Amersham [oligo(dT) priming] in a 20-µl reaction.

---

[34] K. B. Mullis and F. A. Faloona, this series, Vol. 152, p. 335.
[35] R. Saiki, in "PCR Technology" (H. A. Erlich, ed.), p. 7. Stockton Press, New York, 1989.

Use 10 µl of the first strand cDNA reaction mixture in a 50-µl PCR reaction, or use the entire 20 µl of cDNA in a 100-µl PCR reaction. PCR buffer [10× PCR buffer: 100 mM Tris-HCl, pH 8.3, 500 mM KCl, 15 mM $MgCl_2$, 0.1% (w/v) gelatin] may be substituted for first strand reverse transcriptase buffer. Always include a negative control consisting of the first strand cDNA reaction without reverse transcriptase. This will control for PCR products synthesized from contaminating DNA. Small amounts of contaminating NGF DNA resulting in PCR products have been a recurring problem in RNA samples containing very low amounts of NGF mRNA. Another way to solve this problem is to design PCR primers spanning an intron. Thus, the product size will reveal whether the template was cDNA or DNA.

The PCR reaction mixture contains 10 µl cDNA reaction mixture, 5 µl of 10× PCR buffer, 8 µl dNTPs (mixture of 1.25 mM each), NGF primer 1 (0.5 µM), NGF primer 2 (0.5 µM); adjust to 49.75 µl with distilled water. Heat the mixture for 5 min at 95°, then add 0.25 µl (2.5 U) *Taq* polymerase (*Amplitaq* DNA polymerase from Perkin-Elmer Cetus, Norwalk, CT) and 50–100 µl light mineral oil to minimize evaporation. Perform as few as 25 amplification cycles for tissues containing moderate amounts of NGF mRNA, and up to 45 cycles for tissues containing very low levels of NGF mRNA. The cycles are typically 1 min at 94°, 2 min at 55°, and 3 min at 72°. Extract the reaction mixture with 100 µl chloroform; the aqueous phase will float on top of the chloroform/oil phase, allowing removal of the aqueous phase. Analyze 5–10 µl of the PCR mixture by electrophoresis in an 0.8% agarose gel.

NGF mRNA from human tissue may be analyzed using mouse NGF mRNA as an internal standard. The 593-bp mouse NGF PCR product can be completely digested at a single *Xba*I site into two bands migrating at 404 and 189 bp, whereas human NGF does not contain this *Xba*I site. To destroy the *Taq* polymerase, which will interfere with *Xba*I digestion, the PCR product must be heated to 100° for 6 min, extracted with phenol–chloroform, and ethanol-precipitated before performing the *Xba*I digestion.

PCR products may be quantitated by EtBr staining in an agarose gel or by autoradiography and densitometry if radioactively labeled primers are used.[36] Alternatively, PCR reaction products may be transferred to a filter, probed, and quantitated as described above for Northern or dot blots.

---

[36] K. Hayashi, M. Orita, Y. Suzuki, and T. Sekiya, *Nucl. Acids Res.* **17**, 3605 (1989).

Acknowledgments

I thank Michael S. Cole, Jeffrey K. Glenn, and Lori B. Taylor for technical contributions, Dennis O. Clegg for NGF/M13 DNA and probe labeling protocols, Paul D. Coleman for tissue from the Rochester Alzheimer's Disease Program, and Paul H. Johnson for critical reading of the manuscript. This work was supported by National Institutes of Health Grant AGO3644 and by SRI International Research and Development funds.

[6] Detection of Nerve Growth Factor Receptors after Gene Transfer

By MOSES V. CHAO

Introduction

The nerve growth factor (NGF) receptor is a integral membrane glycoprotein, with a single membrane-spanning domain, a negatively charged extracellular domain rich in cysteine residues, and a cytoplasmic domain of 155 amino acids.[1,2] The receptor is encoded by a single-copy gene that gives rise to a single 3.8-kilobase (kb) mRNA.[3] Although the mode of signal transduction for NGF has not been fully clarified, it is probable that an auxillary protein(s) closely associated with the receptor is necessary to mediate the effects of NGF in responsive cells.[4,5] These effects ultimately result in increases in gene expression, activation of neurotransmitter enzymes, and neurite outgrowth.

Further molecular and mechanistic studies of NGF action will depend on rapid and reliable methods of detecting the receptor in order to assess its role in signal transduction. This chapter reviews accessible ways in which cell surface expression of the NGF receptor is determined after gene transfer of the cloned gene. Many of the techniques cited below are applicable to a wide number of receptor systems provided that the ligand and/or monoclonal antibodies are readily available.

---

[1] D. Johnson, T. Lanahan, C. R. Buck, A. Sehgal, C. Morgan, E. Mercer, M. Bothwell, and M. Chao, *Cell* (*Cambridge, Mass.*) **47**, 445 (1986).
[2] M. J. Radeke, T. P. Misko, C. Hsu, L. Herzenberg, and E. M. Shooter, *Nature* (*London*) **325**, 593 (1987).
[3] M. V. Bothwell, A. H. Ross, H. Koprowski, A. Lanahan, C. R. Buck, and A. Sehgal, *Science* **232**, 418 (1986).
[4] M. Hosang and E. M. Shooter, *J. Biol. Chem.* **260**, 655 (1985).
[5] B. L. Hempstead, L. S. Schleifer, and M. V. Chao, *Science* **243**, 373 (1989).

Gene Transfer

DNA-mediated gene transfer of high molecular weight DNA, followed by rescue and cloning of transformed sequences, led to the isolation of the NGF receptor gene.[2,3] Monoclonal antibodies directed against the receptor were extremely instrumental in the detection and purification of transfected cell lines. The two most commonly used are the IgG-192 antibody, which recognizes the rat receptor,[6] and the ME20.4 antibody, specific for the human receptor.[7] Neither antibody displays any cross-reactivity, but each can be used selectively to immunoprecipitate labeled receptor.[2,3,5]

For initial analysis of NGF receptors in cultured cells, *in situ* rosetting and indirect immunofluorescence provide straightforward and accurate methods. These techniques can be applied to cultured cells or to transfection experiments with the NGF receptor gene. Figure 1 shows the general scheme for detecting stable expression after transfection. A typical eukaryotic expression vector, pMV7, containing murine Moloney sarcoma viral sequences[8] has been employed, owing to its versatility as an double gene vector. An internal promoter from the herpesvirus thymidine kinase gene is used to express the selectable marker, aminoglycoside phosphotransferase, which detoxifies the antibiotic neomycin (G418). The viral 5' long terminal repeat (LTR) is used to transcribe a full-length human receptor cDNA sequence.

For a variety of cell types, the calcium phosphate precipitation procedure is widely used, as described in detail by Wigler *et al.*[9] For some cells such as the rat pheochromocytoma PC12 cell line, stable transformants are difficult to obtain, partly because of the slower rate of cell division and the nonadherent properties of the cells. However, treatment of PC12 cells with 20% (v/v) glycerol after exposure to transforming DNA[10] enhances the efficiency of transfection. Other methods such as electroporation, lipofectin, and DEAE-dextran can also improve the efficiency of transfection.

The availability of stable transformants permits a more detailed analysis of the biochemical and kinetic properties of the ligand and transfected

---

[6] C. E. Chandler, L. M. Parsons, M. Hosang, and E. M. Shooter, *J. Biol. Chem.* **259**, 6882 (1984).

[7] A. H. Ross, P. Grob, M. A. Bothwell, D. E. Elder, C. S. Ernst, N. Marano, B. F. D. Ghrist, C. C. Slemp, M. Herlyn, B. Atkinson, and H. Koprowski, *Proc. Natl. Acad. Sci. U.S.A.* **81**, 6681 (1984).

[8] P. T. Kirschmeier, G. M. Housey, M. D. Johnson, A. S. Perkins, and I. B. Weinstein, *DNA* **7**, 219 (1988).

[9] M. Wigler, A. Pellicer, S. Silverstein, R. Axel, G. Urlaub, and L. Chasin, *Proc. Natl. Acad. Sci. U.S.A.* **76**, 1373 (1979).

[10] E. S. Schweitzer and R. B. Kelly, *J. Cell Biol.* **101**, 667 (1985).

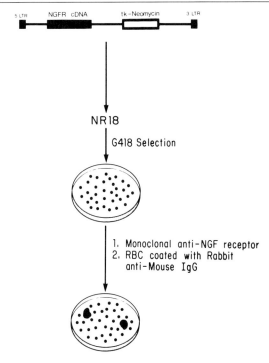

FIG. 1. Detection of NGF receptors after $CaPO_4$-mediated gene transfer. The recipient cell line is a PC12 variant which lacks endogenous receptors [M. A. Bothwell, A. L. Schechter, and K. M. Vaughn, *Cell (Cambridge, Mass.)* **21**, 857 (1980)]. Receptors are detected after sequential incubation with anti-NGF receptor antibodies and red blood cells coupled to rabbit anti-mouse IgG.

receptor. Transient transfection permits high level expression of cloned genes that can be assayed 12 to 80 hr later. In this case, plasmid sequences are maintained extrachromosomally. For most cell lines, 1–10% of the cells can give rise to significant expression.

Transfection

1. Plate cells at a density of 1×10^6 cells/100 mm dish 1 day before transfection.
2. Add dropwise 1.0 ml of $CaPO_4$ precipitate per dish.[9] Each milliliter of precipitate contains 20 μg of DNA, including high molecular weight carrier and plasmid DNA.
3. Leave precipitate on cells from 6 to 12 hr in a CO_2 incubator.
4. Add 3 ml of 20% glycerol in DME (Dulbecco's modified Eagle's medium) for 1 min at room temperature.
5. Wash once with phosphate-buffered saline (PBS).

6. Feed cells with nonselective medium.
7. Place into selective medium 1–2 days after transfection.

Since expression of cell surface receptors is not amenable to metabolic selection, transfer of cloned genes depends on cotransfection with genes carrying selectable markers, most usually dominant acting genes. Neomycin (G418; GIBCO, Grand Island, NY) is the most widely used, and guanosine phosphoribosyltransferase (gpt), dihydrofolate reductase (dhfr), and hygromycin phosphotransferase (hygro) markers are also applicable. Resistance to drug selection will require 10–12 days in culture. For transient transfections, cells can be assayed (see below) 12–80 hr after transfection.

Retroviral Infection

Low rates of gene transfer in cultured cells using conventional techniques can be overcome by retroviral infection with replication-defective vectors. Virus stocks are made by introduction of recombinant proviral DNA as plasmid DNA into packaging cells by using the standard calcium phosphate method outlined above (see Brown and Scott[11]). Packaging lines contain a mutated helper virus that cannot replicate but can encode viral proteins required for production of encapsulated virions. Cultured supernatants produced from these packaging cells can be then used as a source of infectious retrovirus.

The pMV7 vector is capable of generating neomycin-resistant colonies with a high probability of concomitant receptor expression, and it can also give rise to a stock of recombinant virus in packaging cell lines such as the amphotropic PA317 and the ecotropic ψ2 line. These cells, transfected either stably or transiently with pMV7, can produce viral particles that can be used to infect susceptible target cells. The following procedure indicates the way in which infection with viral stocks is carried out.[11]

1. Plate cells at 1×10^6 cells per 100 mm dish 1 day before infection.
2. Aspirate the medium and add viral supernatant [$>10^3$ plaque-forming units (pfu)/ml] and 8 μg/ml Polybrene (Aldrich, Milwaukee, WI) for 2 hr.
3. Add fresh medium for 1–2 days.
4. Place cells in selection. For most cells, G418 selection takes place at an optimum range between 0.6 and 0.8 mg/ml. Cell death will occur approximately 3–4 days after selection commences.

[11] A. M. C. Brown and M. R. D. Scott, *in* "DNA Cloning, A Practical Approach" (D. M. Glover, ed.), Vol. 3, p. 189. IRL, Oxford, 1987.

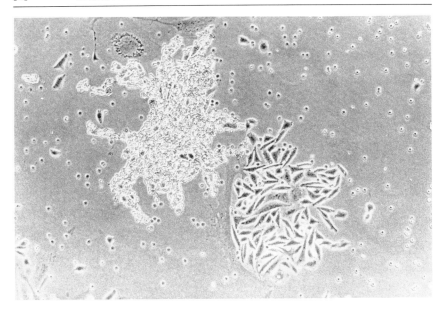

FIG. 2. Positive rosette. Red blood cells coupled to affinity-purified goat anti-mouse IgG bind specifically to mouse fibroblasts expressing the human NGF receptor. A negative colony can be seen on the right side.

Erythrocyte Rosette Assay

Cultured cells expressing specific cell surface antigens can be detected using monoclonal antibodies, and erythrocytes which are coupled to a second antibody. This binding can be visualized easily. Large numbers of colonies can be screened directly on a plate. Figure 1 shows the strategy for gene transfer and rosette assay for identification of cells expressing the human NGF receptor. The recipient cells (NR18) are a derivative of PC12 cells that are deficient in endogenous NGF receptors.[12] The assay made use of a monoclonal antibody directed against the human melanoma NGF receptor.[7]

The rosetting assay is a fast and sensitive technique for detecting the presence of cell surface proteins. Cultured cells expressing less than 1000 receptors per cell can be visualized. Figure 2 shows a strongly rosetting mouse Ltk$^+$ colony expressing the human NGF receptor. Positives can be easily distinguished by the color of the coupled red blood cells (RBC) over an illuminated light source. Many tens of thousands of transformants

[12] M. A. Bothwell, A. L. Schechter, and K. M. Vaughn, *Cell* (*Cambridge, Mass.*) **21,** 857 (1980).

can potentially be screened in only a few hours. When untransfected or receptor-negative cells are assayed in the same manner, no rosetting is observed. Since the monoclonal antibody recognizes only the human NGF receptor molecule, the expression of the NGF receptor in the transformants is not due to an activation of the endogenous NGF receptor.

Rosette Procedure

1. Wash plates 1–2 times with PBS (140 mM NaCl, 2.7 mM KCl, 8 mM Na$_2$HPO$_4$, 1.5 mM KH$_2$PO$_4$).
2. Add antibody in PBS plus 5% fetal calf serum (FCS). (One must determine the appropriate titer on control cells. In general, ascites fluid is diluted at least 1:1000 and hybridoma supernatant 1:10.) Use 5 ml per plate. The ME20.4 hybridoma cell line is available from the ATCC (Rockville, MD).
3. Incubate for 1 hr at room temperature.
4. Wash 3 times with PBS.
5. Add 5–7 ml of coupled RBC. A 2% stock solution is diluted 1:10 in PBS plus 5% FCS (see below).
6. Incubate for 1 hr at room temperature. Positive rosettes can be seen after 1 hr, and positive colonies may be physically picked using cloning cylinders (Belco, Vineland, NJ). The primary antibody can be reused.

Red Blood Cell Coupling

The coupling of rabbit or goat anti-mouse IgG antibodies to human red blood cells is carried out according to the method of Goding[13] with several modifications. Human red cells are separated from their buffy coat using heparin, and chromic chloride is used to couple the antibody to red blood cells. The mixture is washed extensively in sterile saline and can be stored at 4° for up to 2 weeks in Hanks' balanced salt solution.

1. Draw human red blood cells in heparin tube. Spin for 5 min at room temperature (2000 rpm) to remove the buffy coats.
2. Wash the RBC 5–7 times in sterile saline (0.15 M NaCl) in a 50-ml tube. Do not use phosphate-buffered saline (PBS) buffer.
3. To 1 ml of packed RBC, add 2 ml of rabbit anti-mouse IgG antibody (Cappel, Malvern, PA, 0611-0082). Resuspend gently on vortex. The antibody has been previously dialyzed in saline at a concentration of 1 mg/ml. Affinity-purified antibodies are essential for specific rosette formation.

[13] J. W. Goding, *J. Immunol. Methods* **10**, 61 (1976).

4. Add 2 ml working solution of $CrCl_3$ (C325; Fisher, Fairlawn, NJ) per milliliter of packed RBC, *dropwise* while the cells are constantly mixing (slow-speed vortex). The working solution is made from a 1:3 dilution of the stock solution (1 mg/ml in saline). The pH is adjusted to 5.0 with 1 M NaOH.

5. Rotate mixture at room temperature for 7 min.

6. Wash 5 times as in Step 2; however, now use cold (4°) saline.

7. Wash once in Hanks' balanced salt solution (HBSS; GIBCO).

8. Suspend 1 ml of packed RBC into 50 ml of HBSS to a final concentration of 2%. This concentration is equivalent to a 10× stock.

Immunofluorescence

Immunofluorescent analysis of transfected cells is easily carried out on glass coverslips. The cells can be directly exposed to a calcium phosphate precipitate of DNA on glass coverslips, or be replated after transfection. Indirect staining of human NGF receptors on transfected mouse fibroblasts is shown in Fig. 3. The cells were assayed transiently after gene transfer. The following procedure is modified after Rodriquez-Boulan.[14]

1. Plate cells on glass coverslips (Fisher) in 24-well plates.

2. Grow to 50–70% confluence.

3. Wash once with PBS.

4. Fix cells with 2% paraformaldehyde (in PBS) for 30 min (or overnight at 4°).

5. Wash twice with PBS.

6. Quench with 50 mM NH_4Cl in PBS for 20 min.

7. Wash twice with PBS plus 0.1% bovine serum albumin (BSA).

8. Incubate with the ME20.4 monoclonal antibody for 20 min.

9. Wash 3 times with PBS/BSA solution.

10. Incubate with fluorescently tagged rabbit anti-mouse antibody (1:25 dilution) for 20 min. Antibodies are either $F(ab')_2$ fragments or whole antibody coupled to fluorescein isothiocyanate (FITC) (Cappel).

11. Wash 3 times with PBS/BSA solution.

12. Wash once with PBS.

13. Dip in water.

14. Mount on glass slides with a drop of Gelvatol (Monsanto Corp., St. Louis, MO). Gelvatol is polyvinyl alcohol (2.4 g in 6 ml glycerol, 6 ml H_2O, and 2 ml 0.2 M Tris, pH 8.5).

---

[14] E. Rodriguez-Boulan, this series, Vol. 98, p. 486.

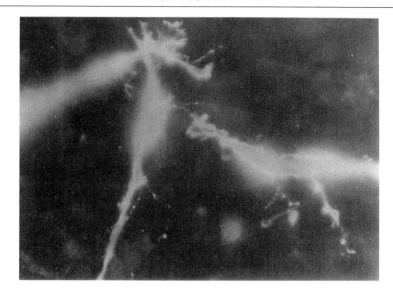

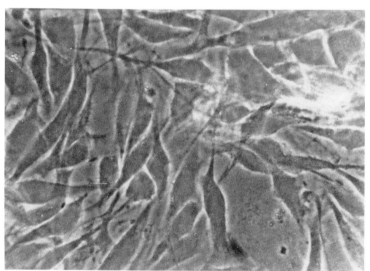

FIG. 3. Indirect immunofluorescence of mouse fibroblasts transiently expressing human NGF receptors. (Top) Fluorescent cells following incubation with ME20.4 antibody[7] and FITC-labeled rabbit anti-mouse IgG. (Bottom) Phase-contrast micrograph of the same cells.

Affinity Labeling

Affinity cross-linking of the receptor to [^{125}I]NGF has become an effective and reproducible means of verifying the appearance of NGF receptor at the cell surface. Because NGF can be labeled to high specific activity by iodination, cell lines can be rapidly screened for NGF receptors. The most versatile cross-linking agent for detection of NGF receptors has been ethyldimethylaminopropylcarbodimide (EDAC), which reacts with carboxylic acid groups and primary amine groups. The major radioiodinated complex from the EDAC reaction of [^{125}I]NGF bound to Schwann cells, neurons, pheochromocytoma, neuroblastoma, and melanoma cells has an apparent molecular weight of 100,000.[15-17] Subtraction of the molecular weight of the β subunit of NGF from the cross-linked complex gives an apparent molecular weight for the receptor of 75,000–85,000. The variation in the apparent molecular weight of the receptor is due to a considerable amount of glycosylation. A 200,000 MW protein found in most cross-linked receptor preparations has been shown to be a homodimer by peptide mapping.[18]

Equilibrium NGF binding experiments of mouse submaxillary [^{125}I]NGF indicates that responsive neurons from the superior cervical or dorsal root ganglia and also cultured cells such as PC12 cells possess two different affinity forms, a high (K_d 10^{-11} M) and a low (K_d 10^{-9} M) affinity site. A second species (M_r 135,000–160,000) can be identified by cross-linking [^{125}I]NGF with the receptor using the heterofunctional and relatively lipophilic cross-linking agent hydroxysuccinimidyl 4-azidobenzoate (HSAB). This species probably represents an accessory protein that cooperates with the receptor to give rise to high affinity NGF binding.[5] For most studies, EDAC cross-linking is far more efficient and less difficult to achieve than cross-linking with HSAB.[4,17] Disuccinimidyl suberate (DSS; Pierce, Rockford, IL), widely used in detecting other receptors, is also applicable to NGF receptor cross-linking.

Since NGF is particularly unstable and binds to numerous surfaces and molecules, care must be taken in radiolabeling with iodine. Lactoperoxidase-catalyzed iodination is the most effective and nonevasive means of labeling NGF. The following protocol is an adaptation of the procedure

---

[15] P. M. Grob, C. H. Berlot, and M. A. Bothwell, *Proc. Natl. Acad. Sci. U.S.A.* **80**, 6819 (1985).

[16] M. Taniuchi, E. M. Johnson, P. J. Roach, and J. C. Lawrence, *J. Biol. Chem.* **261**, 13342 (1986).

[17] S. H. Green and L. A. Greene, *J. Biol. Chem.* **261**, 15316 (1986).

[18] S. Buxser, P. Puma, and G. L. Johnson, *J. Biol. Chem.* **260**, 1917 (1985).

of Green and Greene.[17] An excellent description of current methods of purification of NGF can be found in Longo et al.[19]

Preparation of ^{125}I-Labeled Nerve Growth Factor

1. Purified NGF or commercially obtained NGF (Bioproducts for Science, Indianapolis, IN) is aliquoted and kept at $-70°$ at a concentration of 1 mg/ml.
2. Add 10 μl of 1 mg/ml NGF to iodine^{125} (1 mCi/10 μl) from Amersham (IMS-30).
3. Add 6 μl of 50 μg/ml lactoperoxidase (from ICN, Costa Mesa, CA).
4. Add 25 μl of 0.017% H_2O_2 (Sigma, St. Louis, MO), diluted in 0.2 M sodium phosphate, pH 6.
5. Incubate for 5 min at room temperature and vortex. Incubate for additional 5 min.
6. Quench the reaction with 50 μl of stop buffer [0.1 M NaI, 0.02% NaN_3, 0.1% BSA, 0.1% cytochrome c, 0.05% phenol red, 0.3% acetic acid (added before use)].
7. Assess relative incorporation by trichloroacetic acid (TCA) precipitation of 2-μl aliquots in 1 ml of 10% TCA.
8. Load the reaction on 0.7 × 25 cm P-100 (Bio-Rad, Richmond, CA) column equilibrated with 50 mM sodium phosphate, pH 7.4, 100 mM NaCl, 0.5 mg/ml protamine sulfate (Sigma 4020), and 1 mg/ml bovine albumin (Sigma 7888).

Aliquots of 0.2 ml volume are collected from the column. Labeled NGF appears at roughly twice the migratory distance as the phenol red indicator dye. A shoulder running ahead of the NGF peak represents a small amount of iodinated lactoperoxidase. The labeled NGF can be stored at 4° and must be used within 2 weeks.

Affinity Cross-Linking of Nerve Growth Factor Receptors

1. Wash cell monolayers twice with PBS. Cells are removed with PBS containing 1 mM EDTA, counted, and pelleted.
2. Resuspend cells in 1 ml of PBS, pH 6.5, to a final concentration of 2-4 × 10^6 cells/ml.
3. [^{125}I]NGF is added to the cell suspension, and the mixture is rotated at 37° for 30 min. A typical concentration of NGF is 25 ng/ml; however, the concentration may vary according to experiment. As a control for the

[19] F. M. Longo, J. E. Woo, and W. C. Mobley, in "Nerve Growth Factors" (R. A. Rush, ed.), p. 3. Wiley, New York, 1989.

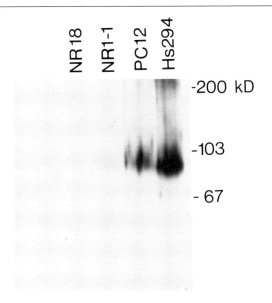

FIG. 4. Affinity cross-linking analysis of the NGF receptor. [^{125}I]NGF was incubated with cultured cells, and the receptor-NGF complex was cross-linked with the carbodiimide cross-linker EDAC. The detergent extract was subjected to polyacrylamide gel electrophoresis and autoradiography. Size markers are in kilodaltons.

specificity of the cross-linking, 2 μg of unlabeled NGF is incubated with an identical sample of cells with [^{125}I]NGF.

4. EDAC (Pierce), made up in PBS, pH 6.5, is added to each sample to a final concentration of 4 mM, and the reaction allowed to proceed for 15 min at room temperature.

5. Cells are pelleted and washed once in PBS, pH 6.5.

6. Final pellets are resuspended in sodium dodecyl sulfate (SDS) loading buffer (plus 5% 2-mercaptoethanol), and the cross-linked products are resolved on 6–8% polyacrylamide gels.

The utility of the cross-linking reaction is demonstrated in Fig. 4. Four cell lines were incubated with 25 ng/ml [^{125}I]NGF and cross-linked with EDAC. The "nonresponsive" NR18 line is a mutant line derived by mutagenesis of PC12 cells.[12] In agreement with previous observations, this cell line does not express any endogenous receptors similar to its parental PC12 line. A retrovirally transformed NR18 line called NR1-1 has a small but detectable number of receptors,[5] whereas human Hs294 melanoma and PC12 cells manifest many more receptors than the NR1-1 cell line. Each cell line displays a cross-linked product with the same apparent molecular size consistent with the size of the fully processed receptor conjugated with [^{125}I]NGF. These cross-linking results indicate that the primary 75,000–85,000 dalton receptor species appears in cells with both high and low affinity forms of the receptor and supports the conclusion that the cloned NGF receptor gene is capable of giving rise to both high and low affinity receptors.[5]

The heightened interest in the role of NGF in the central nervous system and in neurodegenerative diseases such as Alzheimer's dementia requires molecular and biochemical techniques to characterize the interaction of NGF and its receptor. The detection of NGF receptors by *in situ* rosetting, immunofluorescence, and affinity cross-linking provides the opportunity to use *in vitro* mutagenesis of NGF and its receptor to map the binding site and to delineate which structural features of the receptor are responsible for the functional responses of NGF.

[7] Cloning and Expression of Human Platelet-Derived Growth Factor α and β Receptors

By LENA CLAESSON-WELSH, ANDERS ERIKSSON, BENGT WESTERMARK, and CARL-HENRIK HELDIN

Platelet-derived growth factor (PDGF), a major mitogen for connective tissue cells, is composed of disulfide-bonded A and B polypeptide chains. The three isoforms of PDGF have been found to bind to two distinct receptor types with different affinities (reviewed in Refs. 1 and 2). The PDGF β receptor (also denoted PDGF B-type receptor) binds PDGF-BB with high affinity, PDGF-AB with somewhat lower affinity, but appears not to bind PDGF-AA. The PDGF α receptor (also denoted PDGF A-type receptor) binds all PDGF isoforms with high affinity. Both receptors possess an intrinsic kinase activity, which becomes activated after ligand binding.

Cloning and Expression of Human Platelet-Derived Growth
  Factor β Receptor

Partial amino acid sequence information of tryptic fragments from the purified murine PDGF β receptor was used to clone the corresponding

[1] C.-H. Heldin and B. Westermark, *Trends Genet.* **5**, 108 (1989).
[2] R. Ross, E. W. Raines, and D. F. Bowen-Pope, *Cell (Cambridge, Mass.)* **46**, 155 (1986).

cDNA from murine fibroblast and placenta libraries.[3] Using synthetic oligomers based on the murine receptor cDNA sequence as probes, we isolated a number of cDNA clones from a human foreskin fibroblast λgt10 cDNA library.[4] The RNA was isolated from human AG 1523 foreskin fibroblasts using the LiCl method,[5] and mRNA was collected through oligo(dT) affinity chromatography according to standard procedures. cDNA was synthesized and cloned using commercially available kits (Amersham International Ltd., Buckinghamshire, UK). The resulting cDNA, cloned into λgt10, was divided into 11 pools which were amplified independently. About 300,000 plaque-forming units (pfu) from each pool were seeded on 23 × 23 cm agar plates and transferred in duplicate to nitrocellulose filters.[6] The probes used for hybridization were constructed by annealing two partially overlapping synthetic oligomers, each 40 base pairs long. After a fill-in reaction using Klenow polymerase and [α-^{32}P]dCTP, a ^{32}P-labeled 70-mer was obtained.

Two different probes, both located in regions of the cDNA corresponding to conserved areas in the intracellular tyrosine kinase domain, were used for the two replica filters from each agar plate. Using a high stringency protocol for hybridization [50% formamide, 5× SSC (1× SSC is 15 mM sodium citrate, 150 mM NaCl, pH 7.0), 5× Denhardt's solution,[7] 0.1% sodium dodecyl sulfate (SDS), and 0.1 mg/ml salmon sperm DNA, at 37°] and mild washing conditions (2× SSC, 0.1% SDS at 37° for 30 min), we identified four unique positive plaques in the first round of screening. A third 70-mer corresponding to a region in the extracellular domain was used to identify one of the isolated clones as covering most of the open reading frame (λhPDGFR-2A3). This clone was used in a second round of screening of a new seeding of agar plates with aliquots from the 11 cDNA library pools. One new unique clone, denoted λhPDGFR-8, was identified, which based on the size of the insert, was close to full-length. The nucleotide sequence of both strands of λhPDGFR-8 was determined using M13 dideoxy sequencing.[8] After restriction mapping of the cDNA subcloned into pUC19,[9] conveniently sized fragments were cloned into M13. A sche-

---

[3] Y. Yarden, J. A. Escobedo, W.-J. Kuang, T. L. Yang-Feng, T. O. Daniel, P. M. Tremble, E. Y. Chen, M. E. Ando, R. N. Harkins, U. Francke, V. A. Fried, A. Ullrich, and L. T. Williams, *Nature (London)* **323**, 226 (1986).
[4] L. Claesson-Welsh, A. Eriksson, A. Morén, L. Severinsson, B. Ek, A. Östman, C. Betsholtz, and C.-H. Heldin, *Mol. Cell. Biol.* **8**, 3476 (1988).
[5] C. Auffrey and F. Rougeon, *Eur. J. Biochem.* **107**, 303 (1980).
[6] T. Maniatis, E. F. Fritsch, and J. Sambrook, in "Molecular Cloning: A Laboratory Manual." Cold Spring Harbor Laboratory, Cold Spring Harbor, New York, 1982.
[7] D. T. Denhardt, *Biochem. Biophys. Res. Commun.* **23**, 641 (1966).
[8] F. Sanger, S. Nicklen, and A. R. Coulson, *Proc. Natl. Acad. Sci. U.S.A.* **74**, 5463 (1977).
[9] C. Yannisch-Perron, J. Vieira, and J. Messing, *Gene* **33**, 103 (1985).

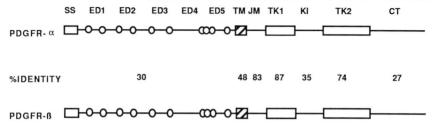

FIG. 1. Schematic representation of the two PDGF receptor types. The receptors are drawn to indicate the presence of 10 evenly distributed cysteine residues in the extracellular parts and the split tyrosine kinase domains in the intracellular parts of the receptors. The level of amino acid identity between the various domains of the molecules are indicated. SS, Signal sequence; ED, extracellular domain; TM, transmembrane domain; JM, juxtamembrane domain; TK, tyrosine kinase domain; CT, carboxy-terminal tail.

matic representation of the deduced structure of the PDGF β receptor is shown in Fig. 1.

In order to express the β receptor stably, the full-length cDNA was subcloned into an SV40-based expression vector.[10] Chinese hamster ovary (CHO) cells were transfected with the cDNA, mixed with a plasmid containing the neomycin resistance gene to allow selection for neomycin-resistant cell colonies coexpressing the PDGF β receptor. Calcium phosphate-mediated transfection[11] was performed in order to introduce the β-receptor cDNA into the CHO cells.

To screen for cells expressing the PDGF β receptor, 20 colonies were picked 3 weeks after transfection, during which time the cells were cultivated in the presence of 0.5 mg/ml of Geneticin 418 (neomycin). The colonies were expanded in duplicate 12-well dishes, and, when confluency was reached, the cells were analyzed for their ability to bind ^{125}I-labeled PDGF-BB. Clones which expressed 10^5 cell surface receptors per cell, that is, a number similar to that of human fibroblasts, were obtained and analyzed with regard to their ability to bind and respond to different isoforms of PDGF.[4]

[10] M. A. Truett, R. Blacher, R. L. Burke, D. Caput, C. Chu, D. Dina, K. Hartog, C. H. Kuo, F. R. Masiarz, J. P. Merryweather, R. Najarian, C. Pachl, S. J. Potter, J. Puma, M. Quiroga, L. B. Rall, A. Randolph, M. S. Urdea, P. Valenzuela, H. H. Dahl, J. Favalaro, J. Hansen, O. Nordfang, and M. Ezban, *DNA* **4**, 333 (1985).

[11] M. Wigler, R. Sweet, G. K. Sim, B. Wold, A. Pellicier, E. Lacy, T. Maniatis, S. Silverstein, and R. Axel, *Cell (Cambridge, Mass.)* **16**, 777 (1979).

Cloning and Expression of Platelet-Derived Growth Factor α Receptor

In order to structurally characterize the PDGF α receptor, a panel of cell lines were analyzed for binding of the PDGF isoforms, with the aim of obtaining one that expressed the α receptor, but not the β receptor. One glioma cell line, U-343 MGa 31L, bound ^{125}I-labeled PDGF-AA as well as ^{125}I-labeled PDGF-BB. Unlike the situation in human fibroblasts, however, all the binding of ^{125}I-labeled PDGF-BB could be competed for by PDGF-AA. This indicated that these cells expressed the PDGF α receptor only. A rabbit antiserum raised against purified porcine PDGF β-receptor preparations[12] was found to react with a 170-kDa component from the U-343 MGa 31L cells. The 170-kDa component was downregulated when the cells were exposed to PDGF-AA, indicating that it was the mature PDGF α receptor. The cross-reactivity of the antiserum, and the fact that the α and β receptors yielded similar peptide maps, indicated that the two receptors were structurally similar.[13] We therefore examined whether a β-receptor cDNA probe would react with any potential α-receptor transcript in the U-343 MGa 31L cell line. Whereas no cross-reactivity could be seen under high stringency conditions, a 6.5-kilobase (kb) glioma cell transcript was detected when the intracellular part, but not the extracellular part, of the β receptor was used as a probe in Northern blotting, under lower stringency conditions [hybridization performed in 40% (v/v) formamide, 5× SSC, 10× Denhardt's solution, 50 mM phosphate buffer, pH 6.5, 0.1% SDS, and 0.1 mg/ml salmon sperm DNA; washing in 2× SSC, 0.1% (w/v) SDS, 2 times, 20 min each, at 45°].

In order to clone the α receptor, a cDNA library was constructed from mRNA purified from the glioma cells, again using cDNA cloning and synthesis kits from Amersham. From 1×10^6 plaques, seeded without prior amplification of the cDNA library and screened using the tyrosine kinase part of the β-receptor cDNA as a probe and the lower stringency conditions described above, we isolated one positive clone (phPDGFRA1). M13 dideoxy sequencing of the 4.5-kilobase pair (kbp) insert revealed that the most 5' segment contained an open reading frame with a sequence which was up to 87% similar to that of the tyrosine kinase domain of the β receptor.

The cDNA library made from the U-343 MGa 31L cells was then screened with the most 5' part of the phPDGFRA1 clone as a probe;

---

[12] L. Rönnstrand, M. P. Beckmann, B. Faulders, A. Östman, B. Ek, and C.-H. Heldin, *J. Biol. Chem.* **262**, 2929 (1987).

[13] L. Claesson-Welsh, A. Hammacher, B. Westermark, C.-H. Heldin, and M. Nistér, *J. Biol. Chem.* **264**, 1742 (1989).

however, since we failed to find extended clones after having screened about 5×10^6 pfu, we constructed a new cDNA library. The previous libraries were made using synthetic linkers with internal EcoRI sites to tail the cDNA. Instead, we now employed synthetic, double-stranded adaptors, with one nonphosphorylated EcoRI overhang and one blunt end. In this way we avoided having to cleave the cDNA at internal EcoRI sites. We purified mRNA from AG 1518 human foreskin fibroblasts; the cDNA kit from Pharmacia-LKB (Uppsala, Sweden) was used for the first and second cDNA strand synthesis. The adaptors were ligated to the cDNA using T4 DNA ligase from New England Biolabs (Beverly, MA); gel chromatography using Sephadex G-50 equilibrated in 1 mM Tris-HCl, pH 7.5, 1 mM EDTA was performed to remove free adaptors and short cDNA molecules. After ethanol precipitation, the cDNA ends were phosphorylated using components in the Pharmacia-LKB kit. After a second ethanol precipitation, the cDNA was ligated to dephosphorylated λgt10 arms purchased from Promega Corp. (Madison, WI), at a ratio of 100 ng cDNA to 2 μg λgt10 arms, in a final volume of 20 μl, using T4 DNA ligase from New England Biolabs. The ligation mixture was packaged using components from Amersham; 3.4×10^6 pfu/100 ng cDNA were thus obtained.

One million plaque-forming units from the adaptor-tailed library were seeded without prior amplification. The most 5' segment from the phPDGFRA1 clone was labeled using [α-^{32}P]dCTP to a specific activity of about 10^8 counts/min (cpm)/μg using the multiprime labeling kit from Amersham; 1×10^6 cpm of the probe was added per milliliter of hybridization solution. The nitrocellulose filters were hybridized and washed under high stringency conditions. Phage DNA was prepared from 16 positive clones. From two of these clones four fragments were released on EcoRI cleavage. The combined size of these fragments was 6.5 kbp for the clone denoted phPDGFRA15. The individual fragments were subcloned and hybridized one by one to U-343 MGa 31L mRNA, to ensure that they originated from the same 6.5-kb transcript. M13 dideoxy sequencing showed that this new clone covered and extended the first clone. A schematic representation of the structure of the molecule encoded by the cDNA, based on the deduced amino acid sequence, is shown in Fig. 1. As described below, expression of the cDNA and PDGF-binding analyses showed that it encoded the PDGF α receptor.[14] The degree of identity between the two PDGF receptors ranges from 27% identity in the C-terminal tails to 87% in the first part of the tyrosine kinase domains, as indicated in Fig. 1.

[14] L. Claesson-Welsh, A. Eriksson, B. Westermark, and C.-H. Heldin, *Proc. Natl. Acad. Sci. U.S.A.* **86**, 4917 (1989).

For expression of the α receptor, a 3.5-kbp subclone covering the entire open reading frame of the α receptor was cloned into the same SV40-based expression vector used for β-receptor expression, to yield a clone denoted pSV7d15.1+5. Calcium phosphate-mediated transfection was then performed to achieve transient expression in COS-1 cells. The transient expression was reproducible enough to allow Scatchard analyses of binding of the PDGF isoforms; all three PDGF isoforms bound to COS-1 cells transfected with pSV7d15.1+5 with high affinities. Immunoprecipitation of metabolically labeled, transfected COS-1 cells also yielded results supporting the conclusion that the 6.5-kbp cDNA clone encoded the PDGF α receptor.

To achieve stable expression of the PDGF α receptor, we subcloned the parts of the clone containing the open reading frame, as well as 1100 bp of the 3' untranslated region. The 3' part was included since it facilitated cloning of the cDNA into a retroviral expression vector, containing the murine leukemia virus long terminal repeats and the neomycin resistance gene. Electroporation[15] rather than calcium phosphate-mediated transfection was used to introduce the DNA into CHO cells. We picked 20 neomycin-resistant colonies and screened for binding of ^{125}I-labeled PDGF-AA. A cell line expressing about 0.2×10^5 α-receptor molecules on the cell surface was obtained. Unlike the case of β-receptor-expressing CHO cells, we found it to be crucial to grow the α-receptor-expressing CHO cells in the absence of serum overnight before performing the binding assay. If this was not done, the amount of binding of ^{125}I-labeled PDGF-AA was low, probably owing to downregulation of the α receptor by constituents of the fetal calf serum.

Conclusions

To facilitate cloning of long cDNA molecules, it is necessary to ensure a high quality of mRNA preparations. We have limited the processing of the mRNA to as few steps as possible and do not enrich for long transcripts, to avoid exposure to degrading enzymes. Instead, a larger number of recombinants have to be screened, preferentially using a nonamplified library. The availability of high quality modifying and restriction enzymes and the use of adaptors rather than linkers for the cDNA tailing reaction make it possible to construct successfully a high quality cDNA library without much previous experience.

[15] D. Rabussay, L. Uher, G. Bates, and W. Plastuch, *Focus* **9,** 1 (1987).

[8] Identification and Quantification of Polyphosphoinositides Produced in Response to Platelet-Derived Growth Factor Stimulation

By LESLIE A. SERUNIAN, KURT R. AUGER, and LEWIS C. CANTLEY

Introduction

Within the past several years, our laboratory has identified and characterized a new phosphoinositide kinase activity that is intimately involved in cell growth and transformation. This kinase serves as a direct biochemical link between a novel phosphatidylinositol (PI) pathway and certain growth factors including platelet-derived growth factor (PDGF),[1] insulin,[2] colony-stimulating factor-1 (CSF-1),[3] and several oncogene products (e.g., the polyomavirus middle T/pp60^{c-src} complex,[4] v-src,[5,6] and v-abl^{6a}) that have intrinsic or associated protein-tyrosine kinase activity.[7,8]

This phosphoinositide kinase, called PI 3-kinase, associates specifically with both antiphosphotyrosine and antigrowth factor receptor immunoprecipitates from a variety of growth factor-stimulated cells[1-3] and with anti-middle T immunoprecipitates from polyomavirus middle T-transformed cells.[9,10] PI 3-kinase phosphorylates the D-3 position of the inositol ring of PI to produce a novel product, phosphatidylinositol 3-phosphate

---

[1] K. R. Auger, L. A. Serunian, S. P. Soltoff, P. Libby, and L. C. Cantley, *Cell (Cambridge, Mass.)* **57**, 167 (1989).

[2] N. B. Ruderman, R. Kapeller, M. F. White, and L. C. Cantley, *Proc. Natl. Acad. Sci. U.S.A.* **87**, 1411 (1990).

[3] L. Varticovski, B. Drucker, D. Morrison, L. Cantley, and T. Roberts, *Nature (London)* **342**, 699 (1989).

[4] D. R. Kaplan, M. Whitman, B. Schaffhausen, L. Raptis, R. L. Garcea, D. Pallas, T. M. Roberts, and L. Cantley, *Proc. Natl. Acad. Sci. U.S.A.* **83**, 3624 (1986).

[5] Y. Sugimoto, M. Whitman, L. C. Cantley, and R. L. Erikson, *Proc. Natl. Acad. Sci. U.S.A.* **81**, 2117 (1984).

[6] Y. Fukui and H. Hanafusa, *Mol. Cell. Biol.* **9**, 1651 (1989).

[6a] L. Varticovski, G. Daley, P. Jackson, D. Baltimore, and L. Cantley, *Mol. Cell. Biol.*, in press (1991).

[7] D. R. Kaplan, M. Whitman, B. Schaffhausen, D. C. Pallas, M. White, L. Cantley, and T. M. Roberts, *Cell (Cambridge, Mass.)* **50**, 1021 (1987).

[8] M. Whitman and L. Cantley, *Biochim. Biophys. Acta* **948**, 327 (1988).

[9] M. Whitman, D. R. Kaplan, B. Schaffhausen, L. Cantley, and T. M. Roberts, *Nature (London)* **315**, 239 (1985).

[10] M. Whitman, C. P. Downes, M. Keeler, T. Keller, and L. Cantley, *Nature (London)* **332**, 644 (1988).

(PI-3-P), that is distinct from phosphatidylinositol 4-phosphate (PI-4-P), the predominant monophosphate form of cellular PI.[10] Moreover, recent work has demonstrated that PI 3-kinase activity phosphorylates not only PI, but also PI-4-P and phosphatidylinositol 4,5-bisphosphate (PI-4,5-P_2) to generate two additional novel phospholipids: phosphatidylinositol 3,4-bisphosphate (PI-3,4-P_2) and phosphatidylinositol 3,4,5-trisphosphate (PIP_3), respectively.[1-3,11]

PI-3-P has been identified in many different cell types,[1,10,12] including early *Xenopus* embryos[13] and *Saccharomyces cerevisiae*.[14] Of interest, the highly phosphorylated PI-3,4-P_2 and PIP_3 products are completely absent in quiescent cells, but they appear within minutes after the addition of growth factors such as PDGF.[1] Both PI-3,4-P_2 and PIP_3 are also detected in human neutrophils activated with formyl peptide and other physiological agonists.[15,16] In addition, these polyphosphoinositides are detected in nontransformed, proliferating cells but are not found in high density, growth-arrested cells.[11] Polyomavirus middle T antigen-transformed cells maintain high levels of these lipids at densities beyond confluence.[11]

Both *in vivo* and *in vitro,* PI 3-kinase activity correlates with the transforming ability of a number of polyomavirus mutants that have been investigated (Refs. 4, 11, and L. Ling and L. Serunian, unpublished results, 1989). Furthermore, mutational analyses of the PDGF receptor indicate that PI 3-kinase activity is an essential component of the cellular proliferative response to PDGF.[17] Thus, in intact cells, PI 3-kinase is activated by PDGF, by other serum factors, and by particular viral oncoproteins to produce novel phospholipids that are detectable only after growth factor stimulation, after transformation, or during cell proliferation.

Neither PI 3-kinase nor its novel polyphosphoinositide products is directly involved in the traditional pathway for generating the second messenger, inositol 1,4,5-trisphosphate. In addition, none of the lipids phosphorylated at the 3-hydroxyl position of the inositol ring are substrates for several different isozymes of phospholipase C that have been purified

---

[11] L. A. Serunian, K. R. Auger, T. Roberts, and L. C. Cantley, *J. Virol.* **64**(10), 4718 (1990).

[12] L. Stephens, P. T. Hawkins, and C. P. Downes, *Biochem. J.* **259**, 267 (1989).

[13] M. Whitman and D. A. Melton, *Science* **244**, 803 (1989).

[14] K. R. Auger, C. L. Carpenter, L. C. Cantley, and L. Varticovski, *J. Biol. Chem.* **264**, 20181 (1989).

[15] A. E. Traynor-Kaplan, A. L. Harris, B. L. Thompson, P. Taylor, and L. A. Sklar, *Nature (London)* **334**, 353 (1988).

[16] A. E. Traynor-Kaplan, B. L. Thompson, A. L. Harris, P. Taylor, G. M. Omann, and L. A. Sklar, *J. Biol. Chem.* **264**, 15668 (1989).

[17] S. R. Coughlin, J. A. Escobedo, and L. T. Williams, *Science* **243**, 1191 (1989).

and characterized to date.[18,19] However, the appearance of the phospholipids produced by PI 3-kinase activity correlates strongly with growth factor- and oncoprotein-dependent protein tyrosine phosphorylation and with cell proliferation and transformation. These results suggest that the novel polyphosphoinositides play a critical role in transmitting a growth signal to the cell.

Until recently, the inability of other laboratories to detect PI-3-P, PI-3,4-P_2, and PIP_3 *in vivo* has been due to the failure of conventional methods to separate these novel isomers from the well-known and more abundant lipids that are phosphorylated at the D-4 position of the inositol ring. For example, PI-3-P and PI-4-P comigrate on nearly all one- and two-dimensional thin-layer chromatography (TLC) systems used; PI-3,4-P_2 and PI-4,5-P_2 also migrate as one enlarged spot on conventional TLC systems. In addition, high-performance liquid chromatography (HPLC) procedures that are commonly used to separate the inositol polyphosphate isomers cannot resolve the various phospholipid isomers.

The identification of the novel polyphosphoinositides has also been hampered by the fact that these lipids are present in relatively low abundance in almost all cells and tissues investigated. Therefore, to detect these minor, novel phospholipids in intact cells, it is necessary to incorporate high levels of radioisotope into the total cellular phosphoinositide pool. Here we describe methods developed in our laboratory in collaboration with others to identify the novel products of PI 3-kinase activity *in vitro* and in PDGF-stimulated cells labeled *in vivo* with either *myo*-[^{3}H]inositol or ortho[^{32}P]phosphate. The procedures developed and reproducibly used to examine and quantify these phospholipid products are presented.

myo-[^{3}H]Inositol and Inorganic ^{32}P Labeling of Human Vascular Smooth Muscle Cells, Platelet-Derived Growth Factor Stimulation, and Extraction of Cellular Lipids

Human vascular smooth muscle cells (SMC) are isolated and cultured as described by Warner *et al.*[20] SMC are growth arrested and are used at quiescence as previously described.[21,22] For labeling, cells are plated at

---

[18] L. A. Serunian, M. T. Haber, T. Fukui, J. W. Kim, S. G. Rhee, J. M. Lowenstein, and L. C. Cantley, *J. Biol. Chem.* **264**, 17809 (1989).
[19] D. L. Lips, P. W. Majerus, F. R. Gorga, A. T. Young, and T. L. Benjamin, *J. Biol. Chem.* **264**, 8759 (1989).
[20] S. J. C. Warner, K. R. Auger, and P. Libby, *J. Exp. Med.* **165**, 1316 (1987).
[21] P. Libby and K. V. O'Brien, *J. Cell. Physiol.* **115**, 217 (1983).
[22] P. Libby, S. J. C. Warner, and G. B. Friedman, *J. Clin. Invest.* **81**, 487 (1988).

2×10^4 cells/cm^2 in 10-cm tissue culture plates and then are placed at 37° in a humidified atmosphere containing 5% CO_2. After the cells have attached to the substratum, 5 to 6 ml of inositol-free Dulbecco's modified Eagle's medium (DMEM) (GIBCO, Grand Island, NY) supplemented to contain 10% (v/v) dialyzed calf serum (MW cutoff 3400) and 10 to 20 μCi/ml of *myo*-[^{3}H]inositol ([^{3}H]Ins, ARC, 15 Ci/mmol) (American Radiolabeled Chemicals, St. Louis, MO) are added to each plate. Cells are incubated for 24 hr in this medium at 37° and then are placed in serum-free conditions: 5 ml of inositol-free DMEM and 10 to 20 μCi of [^{3}H]Ins with 1 μM insulin and 5 μg/ml transferrin. Cells are incubated for 24 to 72 hr under serum-free conditions in the presence of [^{3}H]inositol to induce quiescence.

Cells and cell lines other than those from smooth muscle are routinely radiolabeled to detect the synthesis of PI-3-P, PI-3,4-P_2, and PIP$_3$. In order to ensure that these less abundant, novel polyphosphoinositides are adequately labeled *in vivo* in all cell types, radioisotope is added while cells are still in the exponential growth phase. By allowing the cells to reach quiescence or high density in the presence of the label, a significant amount of radioactivity is incorporated into all of the inositol-containing cellular phospholipids.

Smooth muscle cells are labeled with inorganic ortho[^{32}P]phosphate in a manner similar to that described above for [^{3}H]inositol labeling. However, for ortho[^{32}P]phosphate labeling, quiescent cells are washed from their initial plating medium and placed in 5 ml of phosphate-free DMEM (GIBCO) supplemented to contain 0.5% albumin and 1 to 2 mCi of carrier-free ortho[^{32}P]phosphate (Du Pont NEN, Boston, MA, specific activity 8500–9120 Ci/mmol). The plates are incubated behind a protective plexiglass shield for 1 to 2 hr at 37° in a humidified atmosphere containing 5% CO_2.

For PDGF stimulation, recombinant PDGF (PDGF-BB homodimers, Amgen Biologicals, Thousand Oaks, CA) is added to the cell culture medium at a final concentration of 10 ng/ml for a specific period of time prior to washing the labeled cells and harvesting and extracting the lipids. With the exception of time course studies of growth factor stimulation, PDGF is usually added to plates of cells for 5 to 15 min to detect maximum elevation of the novel polyphosphoinositides. Cells are incubated with PDGF at 37° in a humidified incubator containing 5% CO_2.

To extract cellular lipids, individual plates of cells are washed 3 times with ice-cold phosphate-buffered saline (PBS) to remove unincorporated radioisotope (and growth factor). Cells are harvested with a cell scraper in 750 μl of methanol–1 M HCl (1:1, by volume) and placed in a 1.5-ml microcentrifuge tube. After vortexing vigorously, 380 μl of chloroform is added, and the tube is vortexed and incubated on a rocker platform for 15 min at room temperature. The samples are centrifuged briefly to separate the phases. The upper aqueous phase is carefully removed, placed in a clean microfuge tube, and dried by rotoevaporation *in vacuo*. Approximately 1.0 ml of distilled, deionized water is then added to the tube, and the contents are dried again under vacuum. The dried aqueous sample is stored at $-70°$ until further use. In contrast, the organic phase and the material at the interface are extracted 2 times with an equal volume of methanol–0.1 M EDTA (1:0.9, by volume) to remove traces of divalent cations.[23] The entire organic phase is carefully removed, placed in a clean microfuge tube, and stored at $-70°$ under nitrogen gas until deacylation (described below).

Deacylation of Cellular Lipids

Previous work using reversed-phase HPLC showed that both the novel and the classic phosphatidylinositol phosphates have a similar distribution of fatty acyl side chains[10]; thus, these isomers cannot be separated on the basis of their fatty acid structures. However, the glycerophosphoinositol phosphates (gPIs) derived from the novel polyphosphoinositides can be chromatographically separated from the gPIs derived from the conventional lipids by the use of strong anion-exchange HPLC combined with a very shallow salt gradient. The latter HPLC system resolves the gPIs based on structural differences in their inositol head groups and requires the removal of fatty acyl groups (deacylation) from the phospholipids prior to analysis.

The deacylation procedure is performed as follows: Labeled cellular lipids that have been extracted in chloroform and washed 2 times with an equal volume of methanol–0.1 M EDTA (1:0.9, by volume) (see above) are placed in a glass, screw-capped scintillation vial (20 ml capacity) and then dried under a stream of nitrogen gas. In a fume hood, 1.8 ml of methylamine reagent (42.8% of 25% methylamine in distilled, deionized water, 45.7% methanol, 11.4% *n*-butanol, stored at 4° is added to hydrolyze the dried lipids, and the tightly capped vial is incubated in a 53° water bath for 50 min. After incubation, the contents of the vial are cooled to room temperature, transferred to a 2.0-ml microfuge tube, and dried *in vacuo*. To prevent methylamine vapor damage to the vacuum pump and lyophilizer apparatus, a flask containing concentrated sulfuric acid (~200 ml) is placed in dry ice and attached to the rotoevaporator to serve as an acid "trap." Optimally, the acid is cooled until it becomes a slurry but is not completely frozen. A second flask attached to the acid trap is half-filled with NaOH

[23] N. G. Clarke and R. M. C. Dawson, *Biochem. J.* **195**, 301 (1981).

pellets to sequester and neutralize any acid in the system. Distilled, deionized water (1.8 ml) is added to the dried contents of the tube, the sulfuric acid trap is removed, and the tube contents are dried again under vacuum. The dried samples are resuspended in 2.0 ml of distilled, deionized water, transferred to a glass tube, and extracted 2 times with an equal volume of n-butanol–light petroleum ether–ethyl formate (20:4:1, by volume) to remove fatty acyl groups. After these extractions, the lower, aqueous phase is dried *in vacuo* and stored at $-70°$ until analysis by HPLC.

Immunoprecipitation of Phosphatidylinositol 3-Kinase Activity from Platelet-Derived Growth Factor-Stimulated Smooth Muscle Cells and Production of Novel Polyphosphoinositides

Quiescent SMC are stimulated with PDGF (10 ng/ml) for 5 to 15 min as described above. In all cases, nonstimulated, control cells are analyzed in the same manner as are the growth factor-stimulated cells. Briefly, SMC are washed 3 times in ice-cold PBS, lysed, and immunoprecipitated with antiphosphotyrosine antibody or anti-PDGF receptor antibody as described by Whitman *et al.*[9] and by Auger *et al.*[1] Immune complexes are collected on protein A-Sepharose CL-4B that has been prewashed in 10 mM Tris, pH 7.5, 1% bovine serum albumin and stored in PBS containing 0.02% sodium azide. Immune complexes are washed several times as described by Auger *et al.*[1] For some experiments involving specific antiphosphotyrosine antibodies coupled directly to Sepharose beads,[24] phosphoproteins are eluted from the beads with 10 mM phenylphosphate buffer before performing the phosphoinositide kinase assays.[1] However, for experiments using antireceptor antibodies and all other antiphosphotyrosine antibodies, phosphoinositide kinase assays are routinely performed directly on the beads as described by Whitman *et al.*[9]

In many cases, all three phospholipid substrates, namely, PI (obtained from Avanti Polar Lipids, Birmingham, AL), PI-4-P (obtained from Sigma, St. Louis, MO), and PI-4,5-P$_2$ (obtained from Sigma) are used. The final concentration of the PI, PIP, and PIP$_2$ substrates is 0.03 mg/ml each, in a carrier of phosphatidylserine at a final concentration of 0.1 mg/ml. The phospholipids stored in chloroform are placed in a 1.5-ml microfuge tube, dried under a stream of nitrogen gas, and are resuspended in 20 mM HEPES buffer, pH 7.5, 1 mM EDTA, pH 7.5, prior to sonication. To ensure that the phospholipid substrates are resuspended adequately, a cup horn sonicator (Heat Systems Ultrasonics, Inc., Farmingdale, NY) is used instead of a microprobe. After proper sonication (5 to 10 min at room

---

[24] A. R. Frackelton, A. H. Ross, and H. N. Eisen, *Mol. Cell. Biol.* **3**, 1343 (1983).

temperature on setting 6.5), the lipid suspension, which is initially cloudy, becomes clear. Oversonication results in oxidized, precipitated lipids.

To initiate the PI 3-kinase reaction, [γ-^{32}P]ATP (final concentration 10 to 20 μCi, Du Pont NEN, specific activity 3000 Ci/mmol), Mg^{2+} (10 mM final concentration), and ATP (40 μM final concentration) are added to the washed immune complexes that have been preincubated for 5 min with the sonicated phospholipid substrates. The enzyme reaction is incubated at room temperature for 5 to 10 min. After stopping the reaction with 1 M HCl ($\sim$2 volumes) and extracting the lipids with an equal volume of chloroform–methanol (1 : 1, by volume), the ^{32}P-labeled phospholipid products are resolved on TLC plates (MCB Reagents, Merck, Darmstadt, Germany, silica gel 60, 0.2 mm thickness) that have been precoated with 1% potassium oxalate and baked at 100° for 30 min to 1 hr immediately before use. Unlabeled phospholipid standards are always run in parallel to monitor lipid migration and are visualized by exposure to iodine vapor.

In order to separate the highly phosphorylated PIP_3 from the nonspecific radioactivity remaining at the origin of the TLC plate, an acidic solvent system of n-propanol–2.0 M acetic acid (13 : 7, by volume) is used instead of the more commonly used chloroform–methanol–2.5 M NH_4OH (9 : 7 : 2) solvent system. To achieve maximum resolution of each of the phospholipids from each other and from the material at the origin, the solvent is allowed to migrate nearly to the top of a 20-cm TLC plate, a process that is routinely accomplished in 5 to 6 hrs.

Deacylation of Thin-Layer Chromatography-Purified Lipids

In some instances, phospholipids labeled *in vivo* and separated by TLC are deacylated for HPLC analysis. In addition, ^{32}P-labeled phospholipids generated by immunoprecipitated PI 3-kinase activity and separated by TLC are also deacylated prior to analysis by HPLC (see below). To this end, all TLC-purified lipids are deacylated in the following manner. The region of the TLC plate that contains a particular phospholipid spot is carefully excised, placed in a glass, screw-capped scintillation vial, and treated with methylamine reagent exactly as described above for extracted cellular lipids labeled *in vivo*. (Oxalate treatment of the TLC plate obviates the need to extract with methanol–0.1 M EDTA to remove divalent cations.) After the incubation in methylamine, the solution is removed from the vial and placed in a 2.0-ml microfuge tube; the piece of TLC plate is discarded. The remaining steps in the deacylation procedure are then completed as detailed above. The TLC-purified and deacylated glycerophosphoinositides are then analyzed by HPLC (see below).

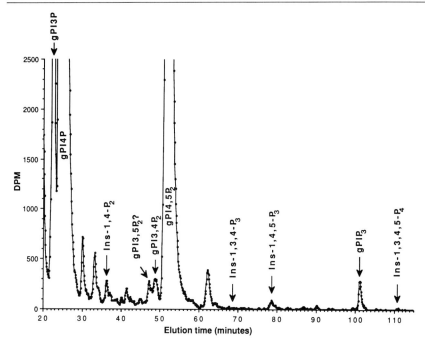

FIG. 1. Anion-exchange HPLC analysis of deacylated myo-[^{3}H]inositol-labeled cellular lipids. A representative HPLC elution profile of [^{3}H]inositol-labeled glycerophosphoinositides from stimulated cells is presented. The migration positions of the novel gPI-3-P, gPI-3,4-P$_2$, and gPIP$_3$ are shown relative to those of the conventional gPI-4-P and gPI-4,5-P$_2$ glycerophosphoinositides. Another novel glycerophosphoinositide, tentatively identified as gPI-3,5-P$_2$ (gPI3,5P$_2$?), elutes just prior to gPI-3,4-P$_2$. The elution positions of ^{3}H-labeled inositol 1,4-bisphosphate (Ins-1,4-P$_2$), inositol 1,3,4-trisphosphate (Ins-1,3,4-P$_3$), inositol 1,4,5-trisphosphate (Ins-1,4,5-P$_3$), and inositol 1,3,4,5-tetrakisphosphate (Ins-1,3,4,5-P$_4$) obtained commercially are indicated by solid arrows. A few of the [^{3}H]inositol-containing peaks are as yet unidentified. Unincorporated myo-[^{3}H]inositol and ^{3}H-labeled phosphatidylinositol elute from the HPLC column between 0 and 20 min.

High-Performance Liquid Chromatography Analysis of Deacylated Phospholipids: Glycerophosphoinositides

Baseline separation of all of the glycerophosphoinositides is achieved using an HPLC high resolution 5-μm Partisphere SAX (strong anion-exchange) column (Whatman, Clifton, NJ) and a shallow, discontinuous salt gradient (Fig. 1). Dried, deacylated samples are resuspended in 0.1 to 0.5 ml of 10 mM (NH$_4$)$_2$HPO$_4$, pH 3.8, and a small aliquot is counted to obtain the total amount of radioactivity in the sample prior to loading onto the HPLC column. To ensure that the minor glycerophosphoinositides are detected, it is necessary to apply at least 1×10^6 dpm (several million disintegrations per minute is ideal) of radiolabeled sample to the column. It should be mentioned that the ^{3}H-labeled inositol phosphates contained in the aqueous phase (see the above section on extraction of cellular lipids) can also be analyzed. To this end, the dried aqueous samples are resuspended as described for the deacylated samples and are applied to the HPLC column.

Routinely, lipid standards labeled with a different radioisotope and deacylated exactly as described above are coinjected with each sample. For example, ^{32}P-labeled polyphosphoinositides produced *in vitro* by immunoprecipitated PI 3-kinase activity are deacylated and applied to the HPLC column simultaneously with all ^{3}H-labeled and deacylated samples. In the same manner, commercially available [^{3}H]PI-4-P, [^{3}H]PI-4,5-P$_2$, and ^{3}H-labeled inositol bis-, tris-, tetrakis-, and hexakisphosphates (Du Pont NEN) are deacylated and coinjected with the ^{32}P-labeled samples. The unlabeled nucleotides, ADP and ATP, are also added to every sample to monitor the reproducibility of elution times among all of the HPLC runs. The samples are loaded onto the HPLC column using a Hamilton syringe fitted with a blunt-end needle.

The HPLC column is washed for 10 min with water and is eluted with $1\ M$ (NH$_4$)$_2$HPO$_4$, pH 3.8, at 1 ml/min. A gradient is established from 0 to $1.0\ M$ (NH$_4$)$_2$HPO$_4$, pH 3.8, over 130 min. To develop the shallow gradient required for these separations, dual pumps are used: pump A contains distilled, deionized water; pump B contains $1.0\ M$ (NH$_4$)$_2$HPO$_4$, pH 3.8. B is run at 0% for 10 min, to 25% B with a duration of 60 min, and then to 100% B over 50 min. For better separation of the different inositol trisphosphate isomers, a second gradient is used: 0% B for 5 min, to 18% B with a duration of 45 min, isocratic elution for 40 min, and then to 100% B over 30 min.

The eluate from the HPLC column is connected to an on-line continuous flow liquid scintillation detector (Flo-One/Beta CT, Radiomatic Instruments, Tampa, FL) that can monitor and quantify two different radioisotopes simultaneously. In addition, a UV monitor is attached directly to the HPLC system to detect the elution of the added ADP and ATP standards, as mentioned above. Data are transmitted to a Macintosh SE or IIx for graphing and further analysis.

Quantification of Data

HPLC samples containing either double- or single-label are quantified by using the microprocessor and internal computer components of the radioactive flow detector (Radiomatic Instruments). The continuous flow scintillation detector and data processor inherent in this system allow the immediate visualization and quantification of each glycerophospholipid peak as it elutes from the HPLC column. Raw counts per minute (cpm) are processed into net cpm by subtraction of a previously determined background value with a sampling time of 6 sec. To convert net cpm to disintegrations per minute (dpm), the counting efficiency of the flow detector is determined by performing a normal HPLC run using ^{3}H-labeled external standards. The counting efficiency values thus obtained are used by the computer to calculate the disintegrations per minute of each sample. Disintegrations contained in each radiographic glycerophosphoinositide peak are integrated, and specific peak areas are determined. In addition, the internal computer calculates the retention time of each phospholipid peak and the total disintegrations in each chromatographed sample.

The integrated disintegrations per minute in each peak reveal the amount of a specific, labeled phosphoinositide that is present in a sample. Based on the exact comigration of the labeled standard lipids injected with every HPLC sample, the novel and the classic glycerophosphoinositides are quickly and reproducibly identified. In addition, the elution positions of the ^{3}H-labeled inositol phosphate standards facilitate the identification of all of the glycerophosphoinositide isomers.

To compare the levels of a particular lipid in different ^{3}H-labeled samples, all data are normalized to total disintegrations per minute incorporated into phosphatidylinositol, the major ^{3}H-labeled phospholipid in cells. It is possible to collect the eluate contained in a specific glycerophosphoinositide peak by attaching a fraction collector to the automatic flow detector. Glycerophosphoinositides obtained in this fashion can be used for further (structural) analyses.

In the event that an automatic radioactive flow detector is unavailable, it is possible to analyze radiolabeled phospholipid samples by attaching a fraction collector to the HPLC column and collecting individual eluate fractions (~100–200 μl) at 5- to 10-sec intervals over the entire gradient or in the region of interest. Scintillation vials are used instead of collection tubes to avoid transferring samples. The appropriate volumes of water and scintillation fluid are added to the eluate fractions in each vial, and the vials are capped, vortexed, and counted in a scintillation counter. The counts per minute obtained are frequently converted to disintegrations per minute and are entered into a computer software program that is capable of graphing the data. Peak areas are then calculated to obtain values for the amount of radioactivity that is contained in each particular glycerophosphoinositide peak. Although this alternative method is quite reliable, it is labor-intensive and slow because of both the vast number of fractions that must be counted and the actual counting time per vial.

Section II

Fibroblast Growth Factor

[9] Biaffinity Chromatography of Fibroblast Growth Factors

By YUEN SHING

Heparin affinity chromatography was first used to purify a tumor-derived angiogenic endothelial mitogen in 1984.[1] It has since been widely used for the purification of fibroblast growth factors (FGF) from a large variety of tissue sources (for review, see Refs. 2–4). We subsequently found that FGF also contains a separate binding site for copper. This led to the development of heparin–copper biaffinity chromatography,[5,6] which was an improvement over previous methods owing to its capacity to resolve multiple forms of FGF. This chapter describes the procedures for the purification of basic and acidic FGF using this novel chromatographic technique.

Principle of Method

The principle of heparin–copper biaffinity chromatography is illustrated diagrammatically in Fig. 1. The success of the method depends on thorough rinsing of the biaffinity column with alternate solutions of 2 M NaCl and 10 mM imidazole in the presence of 0.6 M NaCl before starting the NaCl/imidazole gradient. In addition, the optimal gradient concentrations of NaCl and imidazole for purifying a certain specific form of FGF might have to be determined empirically based on the individual affinities for heparin and copper.

Growth Factor Assays

Fibroblast growth factors stimulate DNA synthesis in both mouse BALB/c 3T3 cells and bovine capillary endothelial cells. Measurement of the stimulation of DNA synthesis in 3T3 cells is a relatively simple but nonspecific assay. Methods for this assay have been reviewed in a previous volume in this series[7] and are not discussed here. The capillary endothelial

---

[1] Y. Shing, J. Folkman, R. Sullivan, C. Butterfield, J. Murray, and M. Klagsbrun, *Science* **223**, 1296 (1984).
[2] J. Folkman and M. Klagsbrun, *Science* **235**, 442 (1987).
[3] D. Gospodarowicz, *Crit. Rev. Oncogenesis* **1**, 1 (1989).
[4] W. H. Burgess and T. Maciag, *Annu. Rev. Biochem.* **58**, 575 (1989).
[5] Y. Shing, *J. Biol. Chem.* **263**, 9059 (1988).
[6] Y. Shing, J. Folkman, and M. Klagsbrun, *Ann. N.Y. Acad. Sci.* **556**, 166 (1989).
[7] Y. Shing, S. Davidson, and M. Klagsbrun, this series, Vol. 146, p. 42.

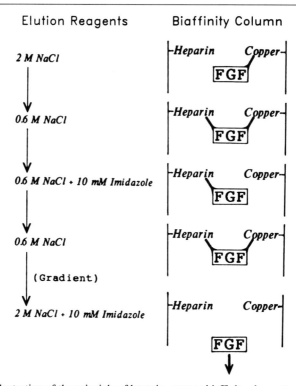

FIG. 1. Illustration of the principle of heparin–copper biaffinity chromatography. When the column as shown at right is initially rinsed with 2 M NaCl after the FGF-containing sample has been loaded, most of the heparin-binding proteins, including FGF, are detached from the heparin moiety but FGF remain bound to the column because of their affinity for copper. This step eliminates most of the heparin-binding proteins except those which are being bound to copper, such as FGF. On the other hand, when the column is subsequently rinsed with 10 mM imidazole in 0.6 M NaCl, most of the copper-binding proteins including FGF are detached from the copper moiety but FGF remain bound to the column because of their affinity for heparin. This step eliminates most of the copper-binding proteins. At this point, virtually all of the non-FGF proteins should have been eluted from the column. The remaining FGF can then be eluted with a linear NaCl/imidazole gradient, and the various forms of FGF are eluted according to their respective affinity for both heparin and copper. (From Shing.[5])

cell DNA synthesis assay is much more specific. This method is particularly useful in the study of endothelial cell-specific growth factors and is presented here.

Measurement of DNA Synthesis in Capillary Endothelial Cells. Capillary endothelial cells are prepared from bovine adrenal glands and grown on gelatin-coated dishes as described by Folkman et al.[8] For assay, endothelial cells are trypsinized and resuspended in Dulbecco's modified Eagle's medium (DMEM, 1 g/liter glucose) supplemented with 10% calf serum, 2 mM glutamine, 100 U/ml penicillin, and 100 µg/ml streptomycin (Irvine Scientific, Santa Ana, CA) and plated sparsely (10,000 cells/0.4 ml/well) into gelatinized 48-well microtiter plates (Costar, Cambridge, MA). On the following day, the medium in each well is replaced by 0.4 ml of DMEM containing 2% calf serum, bovine serum albumin (BSA; 5 mg/ml), and thymidine (0.2 µg/ml). One day later, 5–30 µl of test samples is added. (Samples from the biaffinity column are usually diluted 20- to 40-fold with 0.1% BSA in saline before assaying.) Sixteen hours later, 10 µl (0.6 µCi/well) of [*methyl*-³H]thymidine (85 Ci/mmol, NEN, Boston, MA) is added. Four hours later, cells are fixed by washing the microtiter plates consecutively with phosphate-buffered saline (PBS), methanol (twice, 5 min each), 5% cold trichloroacetic acid (twice, 10 min each), and water. The cells are lysed by addition of 200 µl of 0.3 N NaOH, and the lysates are transferred to scintillation vials. Three milliliters of scintillation fluid (Ecolume, ICN, Costa Mesa, CA) is added, and the vials are counted in a Beckman Model LS1800 scintillation counter. Tritiated thymidine incorporation in this assay varied from batch to batch of cultures for reasons that are not totally clear. In general, the background incorporation with saline is about 2000 counts/min (cpm), and the maximal stimulation obtained by FGF is about 24,000 cpm.

Isolation of Growth Factor

Bovine hypothalami (100 g) obtained from Pel-Freez (Rogers, AR) are homogenized in 300 ml of 0.15 M $(NH_4)_2SO_4$ at pH 6 and extracted by stirring at 4° for 2 hr. The crude extract is centrifuged at 15,000 g for 1 hr, and the supernatant solution is diluted with about 100 ml of 0.15 M $(NH_4)_2SO_4$ and centrifuged once again to remove the remaining residue. The clear supernatant solution is loaded directly onto a heparin-Sepharose column (1.5 × 12 cm) preequilibrated with 0.6 M NaCl in 10 mM Tris, pH 7. The column is rinsed with 300 ml of 0.6 M NaCl in 10 mM Tris, pH 7. FGF are subsequently eluted with 40 ml of 2 M NaCl in the same buffer. This represents the starting material for the purification procedures described in this chapter.

---

[8] J. Folkman, C. Haudenschild, and B. R. Zetter, *Proc.Natl. Acad. Sci. U.S.A.* **76**, 5217 (1979).

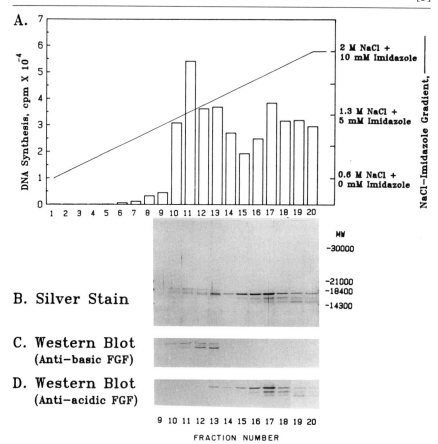

FIG. 2. Heparin–copper biaffinity chromatography of FGF isolated from hypothalamus. Samples of bovine hypothalamic FGF partially purified by heparin-Sepharose were loaded directly onto a heparin–copper biaffinity column. The column was rinsed with alternate solutions and then eluted with a linear NaCl/imidazole gradient. Aliquots of the eluates were assayed for growth factor activity by measurement of DNA synthesis in 3T3 cells (A). Proteins contained in the indicated fractions were analyzed by SDS-gel electrophoresis followed by silver staining (B). Identical samples from all indicated fractions were also analyzed by immunoblotting (Western blot), being probed with antibodies raised against basic FGF (C) and with antibodies raised against acidic FGF (D). (Adapted from Shing.[5])

Heparin–Copper Biaffinity Chromatography

The heparin–copper biaffinity column is prepared by mixing 3.5 ml each of heparin-Sepharose (Pharmacia, Piscataway, NJ) and chelating Sepharose (Pharmacia) that has been saturated with copper(II) chloride. The sample (40 ml) of FGF partially purified by batchwise adsorption to

heparin-Sepharose is applied directly to the blue-colored biaffinity column
(1 × 9 cm) preequilibrated with 2 M NaCl, 10 mM Tris, pH 7. The column
is rinsed consecutively with 40 ml each of the following four reagents in
10 mM Tris, pH 7: (i) 2 M NaCl, (ii) 0.6 M NaCl, (iii) 0.6 M plus 10 mM
imidazole, and (iv) 0.6 M NaCl. Finally, FGF are eluted at a flow rate of
20 ml/hr with a linear NaCl/imidazole gradient from 100 ml of 0.6 M NaCl
without imidazole to 100 ml of 2 M NaCl plus 10 mM imidazole in 10 mM
Tris, pH 7. Fractions eluted from the column are analyzed by sodium
dodecyl sulfate-gel electrophoresis[9] with silver staining[10] and immunoblotting[11] using site-specific antibodies against basic and acidic FGF.[12] A
typical result as shown in Fig. 2 demonstrates that it is possible for this
biaffinity chromatography to resolve from hypothalamus at least two basic
FGF species (with M_r values of 19,000 and 18,000) and three acidic FGF
species (with M_r values of 18,000, 16,400, and 15,600).

Comments

The principle of biaffinity chromatography apparently can also be applied to purify many other proteins which have affinities for more than
one ligand. For example, isolation of three urokinase-related proteins has
been reported by the use of benzamidine–zinc biaffinity chromatography.[13] Furthermore, it has been shown recently that basic FGF can be
purified from a tumor source by the use of tetradecasulfated β-cyclodextrin–copper biaffinity column.[14] The physiological significance of the existence of structurally similar FGF in various tissues is not clear. The
observation that they can be separated based on their differential affinities
for heparin and copper may eventually lead to the development of chromatographic conditions that allow the further identification and purification of varied FGF molecules in various tissue and species sources.

[9] U. K. Laemmli, *Nature (London)* **227**, 680 (1970).
[10] B. R. Oakley, D. R. Kirsch, and N. R. Morris, *Anal. Biochem.* **105**, 361 (1980).
[11] M. Klagsbrun, J. Sasse, R. Sullivan, and J. A. Smith, *Proc. Natl. Acad. Sci. U.S.A.* **83**, 2448 (1986).
[12] M. Wadzinski, J. Folkman, J. Sasse, K. Devey, D. Ingber, and M. Klagsbrun, *Clin. Physiol. Biochem.* **5**, 200 (1987).
[13] D. Zhu, J. Fu, J. Wu, and Y. Shing, "Intracellular Proteolysis, Proceedings of the 7th ICOP Meeting," p. 150. 1989.
[14] Y. Shing, J. Folkman, P. B. Weisz, M. M. Joullie, and W. R. Ewing, *Anal. Biochem.* **185**, 108 (1990).

[10] Cloning, Recombinant Expression, and Characterization of Basic Fibroblast Growth Factor

By STEWART A. THOMPSON, ANDREW A. PROTTER, LOUISE BITTING, JOHN C. FIDDES, and JUDITH A. ABRAHAM

Introduction

Basic fibroblast growth factor (bFGF) is a single-chain protein characterized both by a strong affinity for the sulfated glycosaminoglycan heparin and by potent mitogenic activity on a wide variety of mesoderm- and neuroectoderm-derived cell types.[1-5] Recent results indicate that bFGF is also mitogenic for some epithelial cell types, including keratinocytes.[6,7] Additional activities reported for bFGF include stimulating the migration of vascular endothelial cells and inducing the production of proteases such as collagenase and plasminogen activator in these cells.[8,9] Consistent with these *in vitro* observations, bFGF has been shown *in vivo* to promote new capillary growth (angiogenesis)[2] and wound healing[10-13] in a variety of animal models.

Concurrent with the purification and characterization of bFGF in the late 1970s and early 1980s, the related mitogen acidic fibroblast growth

---

[1] D. Gospodarowicz, N. Ferrara, L. Schweigerer, and G. Neufeld, *Endocrinol. Rev.* **8**, 95 (1987).
[2] J. Folkman and M. Klagsbrun, *Science* **235**, 442 (1987).
[3] R. R. Lobb, *Eur. J. Clin. Invest.* **18**, 321 (1988).
[4] W. H. Burgess and T. Maciag, *Annu. Rev. Biochem.* **58**, 575 (1989).
[5] D. B. Rifkin and D. Moscatelli, *J. Cell Biol.* **109**, 1 (1989).
[6] E. J. O'Keefe, M. L. Chiu, and R. E. Payne, Jr., *J. Invest. Dermatol.* **90**, 767 (1988).
[7] G. D. Shipley, W. W. Keeble, J. E. Hendrickson, R. J. Coffey, Jr., and M. R. Pittelkow, *J. Cell. Physiol.* **138**, 511 (1989).
[8] D. Moscatelli, M. Presta, and D. B. Rifkin, *Proc. Natl. Acad. Sci. U.S.A.* **83**, 2091 (1986).
[9] M. Presta, D. Moscatelli, J. Joseph-Silverstein, and D. B. Rifkin, *Mol. Cell. Biol.* **6**, 4060 (1986).
[10] J. M. Davidson, M. Klagsbrun, K. E. Hill, A. Buckley, R. Sullivan, P. S. Brewer, and S. C. Woodward, *J. Cell Biol.* **100**, 1219 (1985).
[11] G. S. McGee, J. M. Davidson, A. Buckley, A. Sommer, S. C. Woodward, A. M. Aquino, R. Barbour, and A. A. Demetriou, *J. Surg. Res.* **45**, 145 (1988).
[12] P. A. Hebda, C. K. Klingbeil, J. A. Abraham, and J. C. Fiddes, *J. Invest. Dermatol.*, in press (1990).
[13] C. K. Klingbeil, L. B. Cesar, and J. C. Fiddes, *in* "Clinical and Experimental Approaches to Dermal and Epidermal Repair: Normal and Chronic Wounds" (A. Barbul, M. Caldwell, W. Eaglstein, T. Hunt, D. Marshall, E. Pines, and G. Skover, eds.) in press. Alan R. Liss, New York, 1990.

factor (aFGF) was also identified and purified.[1-4] Comparison of the complete amino acid sequences of a 146-residue form of bovine pituitary bFGF[14] and a 140-residue form of bovine brain aFGF[15,16] demonstrated that the two factors share 55% absolute sequence identity. Although some differences in the effects of these two factors have been noted (e.g., on melanocytes[17] and keratinocytes[7]), aFGF generally appears to have the same wide range of biological activities as does bFGF.[1-4]

Like bFGF, aFGF displays strong affinity for heparin.[1-5] Because of this affinity, aFGF and bFGF are sometimes referred to, respectively, as heparin-binding growth factors 1 and 2 (HBGF-1 and HBGF-2).[4,18] Recently, it has been shown that these factors are members of a larger protein family (the FGF family or HBGF family) which includes at least five other members: the protein encoded by a putative murine oncogene, *int2*[19]; the products of the human oncogenes FGF-5[20] and *hst*[21] (also known as K-*fgf*[22]); the product of a gene (*FGF-6*) isolated by homology to *hst*/K-*fgf*[23]; and keratinocyte growth factor (KGF).[24]

In this chapter, we describe some of the methods we have used to isolate bovine and human bFGF gene sequences and to express recombinant human bFGF. The isolation of bFGF clones, by ourselves[25,26] and others,[27-29] employed many standard methods, which have been covered

---

[14] F. Esch, A. Baird, N. Ling, N. Ueno, F. Hill, L. Denoroy, R. Klepper, D. Gospodarowicz, P. Böhlen, and R. Guillemin, *Proc. Natl. Acad. Sci. U.S.A.* **82**, 6507 (1985).

[15] G. Gimenez-Gallego, J. Rodkey, C. Bennett, M. Rios-Candelore, J. DiSalvo, and K. Thomas, *Science* **230**, 1385 (1985).

[16] F. Esch, N. Ueno, A. Baird, F. Hill, L. Denoroy, N. Ling, D. Gospodarowicz, and R. Guillemin, *Biochem. Biophys. Res. Commun.* **133**, 554 (1985).

[17] R. Halaban, S. Ghosh, and A. Baird, *In Vitro Cell. Dev. Biol.* **23**, 47 (1987).

[18] R. Lobb, J. Sasse, R. Sullivan, Y. Shing, P. D'Amore, J. Jacobs, and M. Klagsbrun, *J. Biol. Chem.* **261**, 1924 (1986).

[19] C. Dickson and G. Peters, *Nature (London)* **326**, 833 (1987).

[20] X. Zhan, B. Bates, X. Hu, and M. Goldfarb, *Mol. Cell. Biol.* **8**, 3487 (1988).

[21] T. Yoshida, K. Miyagawa, H. Odagiri, H. Sakamoto, P. F. R. Little, M. Terada, and T. Sugimura, *Proc. Natl. Acad. Sci. U.S.A.* **84**, 7305 (1987).

[22] P. Delli-Bovi, A. M. Curatola, K. M. Newman, Y. Sato, D. Moscatelli, R. M. Hewick, D. B. Rifkin, and C. Basilico, *Mol. Cell. Biol.* **8**, 2933 (1988).

[23] I. Marics, J. Adelaide, F. Raybaud, M.-G. Mattei, F. Coulier, J. Planche, O. de Lapeyriere, and D. Birnbaum, *Oncogene* **4**, 335 (1989).

[24] P. W. Finch, J. S. Rubin, T. Miki, D. Ron, and S. A. Aaronson, *Science* **245**, 752 (1989).

[25] J. A. Abraham, A. Mergia, J. L. Whang, A. Tumolo, J. Friedman, K. A. Hjerrild, D. Gospodarowicz, and J. C. Fiddes, *Science* **233**, 545 (1986).

[26] J. A. Abraham, J. L. Whang, A. Tumolo, A. Mergia, J. Friedman, D. Gospodarowicz, and J. C. Fiddes, *EMBO J.* **5**, 2523 (1986).

[27] T. Kurokawa, R. Sasada, M. Iwane, and K. Igarashi, *FEBS Lett.* **213**, 189 (1987).

[28] A. Sommer, M. T. Brewer, R. C. Thompson, D. Moscatelli, M. Presta, and D. B. Rifkin, *Biochem. Biophys. Res. Commun.* **144**, 543 (1987).

in detail in other volumes of this series (see, e.g., Volume 152). We have, therefore, concentrated on discussing in detail only the methods used to obtain the first bFGF clone (encoding bovine bFGF[25]), since the approach involved the use of a "homology-choice" probe. This novel method of probe design, wherein each codon choice in the probe is dictated as much as possible by the codon used in that position in a homologous gene, may prove useful to others for the isolation of genes encoding proteins sharing regions of high amino acid sequence similarity.

Recombinant expression of bFGF has been reported in mammalian cells,[27,29-35] yeast,[36] and the bacterium *Escherichia coli*.[36-39] We describe here an *E. coli* expression vector, a purification protocol, and analytical techniques that have worked well in our hands for research-scale production and characterization of recombinant human bFGF and analogs thereof.

Isolation of cDNA Clone Encoding Bovine Basic Fibroblast Growth Factor

The amino-terminal amino acid sequences of bovine brain aFGF (140-residue form) and pituitary bFGF (146-residue form) were determined by Böhlen *et al*.[40,41] From the first 34 amino acids of the aFGF protein se-

---

[29] H. Prats, M. Kaghad, A. C. Prats, M. Klagsbrun, J. M. Lelias, P. Liauzun, P. Chalon, J. P. Tauber, F. Amalric, J. A. Smith, and D. Caput, *Proc. Natl. Acad. Sci. U.S.A.* **86**, 1836 (1989).

[30] J. A. Abraham, J. L. Whang, A. Tumolo, A. Mergia, and J. C. Fiddes, *Cold Spring Harbor Symp. Quant. Biol.* **51**, 657 (1986).

[31] S. Rogelj, R. A. Weinberg, P. Fanning, and M. Klagsbrun, *Nature (London)* **331**, 173 (1988).

[32] R. Sasada, T. Kurokawa, M. Iwane, and K. Igarashi, *Mol. Cell. Biol.* **8**, 588 (1988).

[33] S. B. Blam, R. Mitchell, E. Tischer, J. S. Rubin, M. Silva, S. Silver, J. C. Fiddes, J. A. Abraham, and S. A. Aaronson, *Oncogene* **3**, 129 (1988).

[34] G. Neufeld, R. Mitchell, P. Ponte, and D. Gospodarowicz, *J. Cell Biol.* **106**, 1385 (1988).

[35] R. Z. Florkiewicz and A. Sommer, *Proc. Natl. Acad. Sci. U.S.A.* **86**, 3978 (1989).

[36] P. J. Barr, L. S. Cousens, C. T. Lee-Ng, A. Medina-Selby, F. R. Masiarz, R. A. Hallewell, S. H. Chamberlain, J. D. Bradley, D. Lee, K. S. Steimer, L. Poulter, A. L. Burlingame, F. Esch, and A. Baird, *J. Biol. Chem.* **263**, 16471 (1988).

[37] M. Iwane, T. Kurokawa, R. Sasada, M. Seno, S. Nakagawa, and K. Igarashi, *Biochem. Biophys. Res. Commun.* **146**, 470 (1987).

[38] C. H. Squires, J. Childs, S. P. Eisenberg, P. J. Polverini, and A. Sommer, *J. Biol. Chem.* **263**, 16297 (1988).

[39] G. M. Fox, S. G. Schiffer, M. F. Rohde, L. B. Tsai, A. R. Banks, and T. Arakawa, *J. Biol. Chem.* **263**, 18452 (1988).

[40] P. Böhlen, A. Baird, F. Esch, N. Ling, and D. Gospodarowicz, *Proc. Natl. Acad. Sci. U.S.A.* **81**, 5364 (1984).

[41] P. Böhlen, F. Esch, A. Baird, and D. Gospodarowicz, *EMBO J.* **4**, 1951 (1985).

quence, two "codon-choice"[42] oligonucleotide probes were designed: a unique-sequence oligonucleotide 48 bases in length and a 2-fold degenerate oligonucleotide 51 bases in length.[25] Using techniques described in detail elsewhere,[42,43] a bovine genomic library was screened in duplicate with these two probes.[25] DNA sequence analysis[44,45] of one of the phage (λBA2) hybridizing to both probes confirmed that this phage contained an amino-terminal coding exon of the bovine aFGF gene.[25]

Using degenerate and codon-choice oligonucleotide probes designed from the amino-terminal amino acid sequence of bovine bFGF, several unsuccessful attempts were made to obtain clones for this factor through the screening of a number of different cDNA and genomic libraries. An alternative strategy was therefore developed, which took advantage of the amino acid sequence similarity between aFGF and bFGF.[14-16] In comparing the amino-terminal sequences of the two proteins, a region of particularly strong similarity was observed (10 of 14 residues, see Fig. 1A). It was reasoned that if the genes for the two factors were ancestrally related, then the codon choices used in the aFGF gene might be conserved in the bFGF gene. As shown in Fig. 1A, a probe for the bovine bFGF gene sequence encoding the strongly similar region was accordingly designed[25] based on two assumptions: (i) where an amino acid residue in bFGF is identical to the aFGF residue at the corresponding position, the codon for that residue in the bFGF gene is the same as the codon used in the aFGF gene (as determined from λBA2), and (ii) at positions where the amino acid residues differ between bFGF and aFGF, the codon for the bFGF amino acid is the one representing the minimum number of nucleotide changes from the aFGF codon. The resulting homology-choice probe was synthesized as the complement of the coding sequence (Fig. 1A), to allow for possible use in screening for mRNA encoding bFGF.

Probes constructed in this manner would be expected, under appropriate stringency conditions, to hybridize to the homologous DNA sequence used in the probe design (in this case, the aFGF gene sequence). Southern blot analyses[46] were therefore carried out to determine (i) whether the probe shown in Fig. 1A would indeed hybridize to the aFGF gene, and/or to any other sequences in the bovine genome, and (ii) whether

---

[42] W. I. Wood, this series, Vol. 152, p. 443.
[43] W. I. Wood, D. J. Capon, C. C. Simonsen, D. L. Eaton, J. Gitschier, B. Keyt, P. H. Seeburg, D. H. Smith, P. Hollingshead, K. L. Wion, E. Delwart, E. G. D. Tuddenham, G. A. Vehar, and R. M. Lawn, *Nature (London)* **312**, 330 (1984).
[44] F. Sanger, A. R. Coulson, B. G. Barrell, A. J. H. Smith, and B. A. Roe, *J. Mol. Biol.* **143**, 161 (1980).
[45] J. Messing, this series, Vol. 101, p. 20.
[46] E. M. Southern, *J. Mol. Biol.* **98**, 503 (1975).

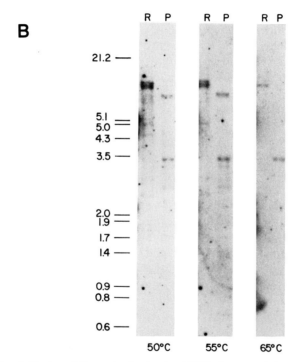

FIG. 1. Design of the probe for bovine bFGF clones, and determination of screening conditions. (A) The bFGF "homology-choice" probe. A portion of the amino-terminal sequence of bovine bFGF (residues 18 to 31 of the 146-residue form) is shown aligned with the corresponding homologous region of the amino-terminal sequence of bovine aFGF (residues 9 to 22 of the 140-residue form). Nonhomologous amino acids are in bold lettering. The nucleotide sequence known to encode this region of the aFGF protein is shown below the amino acid sequences, as is the sequence of the oligonucleotide probe. The probe was designed to match the complement of the aFGF coding sequence, except at nucleotide positions that had to be changed to reflect amino acid differences in the bFGF protein (underlined nucleotides; see text). (B) Southern blot analyses of bovine genomic DNA. The

appropriate stringency conditions could be established in which the probe no longer detected the aFGF gene but still bound to other unique sequences (which would presumably represent bFGF gene fragments). For these Southern analyses (Fig. 1B), bovine genomic DNA was first digested with EcoRI or PstI. Multiple sets of the EcoRI and PstI digests were electrophoresed in parallel on a single 0.8% agarose gel and transferred to nitrocellulose. Strips containing individual sets of the digests were cut from the nitrocellulose and prehybridized at 42° with 20% formamide buffer [20% formamide, 50 mM sodium phosphate (pH 6.8), 6× SSC (1× SSC is 0.15 M NaCl, 15 mM sodium citrate), 5× Denhardt's solution (0.1% Ficoll, 0.1% polyvinylpyrrolidone, 0.1% bovine serum albumin), and 100 μg/ml boiled herring sperm DNA].[42,43] The homology-choice probe shown in Fig. 1A was radiolabeled with ^{32}P using T4 polynucleotide kinase and [γ-^{32}P]ATP,[47] and all the strips were then hybridized overnight at 42° with 2 × 10^6 counts/min (cpm) of the ^{32}P-labeled probe per ml in 20% formamide buffer containing 10% dextran sulfate.

Variations in the stringency of the Southern blot screening were achieved by varying the temperature of the solution used to wash the blot strips. After the overnight hybridization, all strips were washed 3 times at room temperature in 1× SSC, 0.1% sodium dodecyl sulfate (SDS) (15 min per wash). Individual strips were then washed in the same buffer for 10 min at 37°, 45°, 50°, 55°, or 65° and exposed for 5 days at −80° to autoradiographic film backed with an intensifying screen. The results for the highest three wash temperatures are shown in Fig. 1B. Using either 50° or 55° as the wash temperature, a number of hybridizing fragments were detected, including a doublet of approximately 10 kilobases (kb) in the EcoRI-digested DNA. One of the bands of the doublet presumably corresponded to the 10.6-kb EcoRI fragment of the bovine aFGF gene predicted from the restriction map of the genomic clone λBA2 (it was expected that no PstI fragment from the aFGF gene would be detected under these conditions, since there is a PstI site in the aFGF gene that divides the probe region approximately in half). The strip washed at 65° showed only one of the two approximately 10-kb EcoRI fragments, as well

[47] R. B. Wallace and C. G. Miyada, this series, Vol. 152, p. 432.

---

genomic DNA (10 μg per lane) was digested with EcoRI (R) or PstI (P), fractionated on a 0.8% agarose gel, and then transferred to nitrocellulose. Strips of the nitrocellulose were hybridized to the homology-choice probe as described in the text, then washed in 1× SSC, 0.1% SDS at the temperature indicated. Autoradiographs of the washed filter strips are shown. Size markers (bacteriophage λ DNA digested with EcoRI and HindIII) are in kilobases. (Reprinted from Abraham et al.[25]; copyright 1986 by the AAAS.)

as a 3.5-kb *Pst*I fragment. It was assumed that these fragments corresponded to portions of the bFGF gene.

Using the conditions established from the Southern analyses (hybridization at 42° in 20% formamide buffer containing 10% dextran sulfate; final wash at 65° in 1 × SSC, 0.1% SDS), a bovine pituitary cDNA library[25] constructed in the bacteriophage vector λgt10 was screened with the ^{32}P-labeled homology-choice probe. A single hybridizing phage was detected out of 10^6 recombinants screened; this phage (λBB2) was isolated, and fragments of the cDNA insert were subcloned into M13 vectors[48,49] for sequencing by the dideoxynucleotide method.[44,45] The results of the nucleotide sequencing (Fig. 2) demonstrated that the clone contains a potential translation initiation (ATG) codon at nucleotides 104–106, followed by a 154-codon open reading frame terminating with a TGA codon at nucleotides 569–571.

Direct amino acid sequencing of bovine pituitary bFGF by Esch *et al.*[14] had initially indicated that the mature protein consists of 146 residues (corresponding to residues 10–155 in the predicted primary translation product shown in Fig. 2). Subsequent experiments, however, demonstrated that a longer, amino-terminally blocked form of bFGF could be obtained from the pituitaries if protease inhibitors were present during the isolation.[50] Amino acid composition data indicated that the blocked form consists of 154 residues (amino acids 2–155 in Fig. 2). Klagsbrun *et al.*[51,52] also reported the existence of a longer, amino-terminally blocked form of bFGF, when neutral extraction conditions were used on the human hepatoma cell line SK-HEP-1. By analogy with aFGF,[53,54] and from the results of recombinant bFGF studies in the yeast *Saccharomyces cerevisiae*,[36] it is likely that the blocking group is an amino-terminal acetylation. In contrast, an unblocked amino-terminally extended form of bFGF (initiating Ala-Ala-Gly-Ser-Ile-) was obtained when the protein was extracted in the presence of protease inhibitors from human prostate tissue.[55] Taken

---

[48] J. Messing and J. Vieira, *Gene* **19**, 269 (1982).
[49] C. Yanisch-Perron, J. Vieira, and J. Messing, *Gene* **33**, 103 (1985).
[50] N. Ueno, A. Baird, F. Esch, N. Ling, and R. Guillemin, *Biochem. Biophys. Res. Commun.* **138**, 580 (1986).
[51] M. Klagsbrun, J. Sasse, R. Sullivan, and J. A. Smith, *Proc. Natl. Acad. Sci. U.S.A.* **83**, 2448 (1986).
[52] M. Klagsbrun, S. Smith, R. Sullivan, Y. Shing, S. Davidson, J. A. Smith, and J. Sasse, *Proc. Natl. Acad. Sci. U.S.A.* **84**, 1839 (1987).
[53] W. H. Burgess, T. Mehlman, D. R. Marshak, B. A. Fraser, and T. Maciag, *Proc. Natl. Acad. Sci. U.S.A.* **83**, 7216 (1986).
[54] J. W. Crabb, L. G. Armes, S. A. Carr, C. M. Johnson, G. D. Roberts, R. S. Bordoli, and W. L. McKeehan, *Biochemistry* **25**, 4988 (1986).
[55] M. T. Story, F. Esch, S. Shimasaki, J. Sasse, S. C. Jacobs, and R. K. Lawson, *Biochem. Biophys. Res. Commun.* **142**, 702 (1987).

```
                                                      NcoI
CCGGGGCCGC GCCGCGGAGC GCGTCGGAGG CCGGGGCCGG GGCGCGGCGG CTCCCCGCGC GGCT
CCAGGG GCTCGGGGAC CCCGCCAGGG CCTTGGTGGG GCC ATG GCC GCC GGG AGC ATC ACC
                                               Met Ala Ala Gly Ser Ile Thr
                                                1

ACG CTG CCA GCC CTG CCG GAG GAC GGC GGC AGC GGC GCT TTC CCG CCG GGC CAC
Thr Leu Pro Ala Leu Pro Glu Asp Gly Gly Ser Gly Ala Phe Pro Pro Gly His
         10                          20
                                                                HhaI
TTC AAG GAC CCC AAG CGG CTG TAC TGC AAG AAC GGG GGC TTC TTC CTG CGC ATC
Phe Lys Asp Pro Lys Arg Leu Tyr Cys Lys Asn Gly Gly Phe Phe Leu Arg Ile
             30                                          40

CAC CCC GAC GGC CGA GTG GAC GGG GTC CGC GAG AAG AGC GAC CCA CAC ATC AAA
His Pro Asp Gly Arg Val Asp Gly Val Arg Glu Lys Ser Asp Pro His Ile Lys
                     50                                          60

CTA CAA CTT CAA GCA GAA GAG AGA GGG GTT GTG TCT ATC AAA GGA GTG TGT GCA
Leu Gln Leu Gln Ala Glu Glu Arg Gly Val Val Ser Ile Lys Gly Val Cys Ala
                             70

AAC CGT TAC CTT GCT ATG AAA GAA GAT GGA AGA TTA CTA GCT TCT AAA TGT GTT
Asn Arg Tyr Leu Ala Met Lys Glu Asp Gly Arg Leu Leu Ala Ser Lys Cys Val
 80                              90

ACA GAC GAG TGT TTC TTT TTT GAA CGA TTG GAG TCT AAT AAC TAC AAT ACT TAC
Thr Asp Glu Cys Phe Phe Phe Glu Arg Leu Glu Ser Asn Asn Tyr Asn Thr Tyr
            100                              110
                    ACC
CGG TCA AGG AAA TAC TCC AGT TGG TAT GTG GCA CTG AAA CGA ACT GGG CAG TAT
Arg Ser Arg Lys Tyr Ser Ser Trp Tyr Val Ala Leu Lys Arg Thr Gly Gln Tyr
                 120                                     130
            TCC
AAA CTT GGA CCC AAA ACA GGA CCT GGG CAG AAA GCT ATA CTT TTT CTT CCA ATG
Lys Leu Gly Pro Lys Thr Gly Pro Gly Gln Lys Ala Ile Leu Phe Leu Pro Met
                     140                                         150
                                                         CTT
TCT GCT AAG AGC TGA TCTTAATGGC AGCATCTGAT CTCATTTTAC ATGAAGAGGT ATATTTC
Ser Ala Lys Ser
```

FIG. 2. Partial nucleotide sequence of the cDNA insert in the bovine bFGF clone λBB2. Only 618 bases of the 2122-base pair (bp) insert are shown (the bases not shown represent 3'-untranslated sequence). Given below the nucleotide sequence is the amino acid sequence of the primary translation product predicted from the 155-codon open reading frame initiating with the ATG codon at nucleotides 104–106. Boxed sequences indicate changes made in the insert by *in vitro* mutagenesis during construction of the human bFGF expression vector (see text). The underlined region between the indicated *Nco*I and *Hha*I restriction sites was replaced by the synthetic sequence shown in Fig. 4A in a subsequent step in the construction of the vector.

together, these observations indicate that the 146-residue form of bFGF and other shorter forms are generated during the course of purification, while the 154-residue form of the protein (with or without acetylation) represents a true mature bFGF species (see Fig. 3).

Structure of Human Basic Fibroblast Growth Factor Coding Region

Using unique-sequence probes derived from the λBB2 bovine cDNA insert, human bFGF clones were obtained from a number of cDNA and

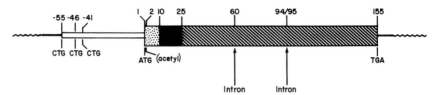

FIG. 3. Schematic diagram of the translated region in human bFGF mRNA. Codons within the translated sequence are numbered relative to the proposed initiating ATG codon at position 1. A TGA translation stop signal occurs immediately following codon 155. Current evidence suggests that translation of the open reading frame extending from codons 1–155 yields a primary translation product that is processed to mature bFGF through the removal of the initiating methionine (and, in at least some cases, acetylation of the new amino terminus). Forms of bFGF with amino termini corresponding to positions 10 or 25[1,14] probably result from proteolytic digestion during purification. Recent results[29,35] have shown that translation can also initiate at three CTG codons, lying at positions −41, −46, and −55 in the continuing open reading frame upstream from the ATG start at codon 1. In the human bFGF gene, the coding region is interrupted by two large introns, one lying within codon 60 and one lying between codons 94 and 95.[26]

genomic libraries.[26] Subsequently, the isolation of human clones was also reported (i) using partial codon-choice oligonucleotide probes, based on the bovine bFGF amino acid sequence, to screen a human foreskin fibroblast cDNA library[27] and (ii) using fully degenerate, short oligonucleotide probes, based on amino acid sequence data from human placental bFGF, to screen an SK-HEP-1 cDNA library.[28] Initial sequence analyses[26] indicated that, as with bovine bFGF, the human bFGF coding region predicts a 155-residue primary translation product (Fig. 3). This product differs by only two amino acids from the proposed bovine primary translation product (at residues 121 and 137; Fig. 2).

In the bovine and human bFGF cDNA sequences, the open reading frame extends for a considerable distance 5′ to the proposed initiating ATG indicated in Figs. 2 and 3. One argument nonetheless supporting the proposal of a 155-residue primary translation product for human bFGF was that no other in-frame ATG codons exist in this 5′ region before an in-frame translation termination codon is encountered.[26] However, subsequent results from a number of different groups demonstrated that forms of bFGF exist that are longer than 155 residues.[3-5] Recently, two groups[29,35] have presented evidence indicating that these longer forms arise through unusual translation initiations occurring at CTG (leucine) codons, lying 41, 46, and 55 codons 5′ to the proposed ATG start. Since these experiments also indicated that translation initiation can occur at the proposed ATG start, there appear to be four possible primary translation products for human bFGF, extending 155, 196, 201, and 210 residues in length (Fig. 3). No differences in bioactivity have yet been demonstrated for the various forms. Surprisingly, despite the existence of cell-surface receptors for bFGF (suggesting extracellular interaction between the receptor and bFGF molecules), none of the primary translation products for this factor has a classic secretion signal sequence.

Recombinant Expression of Human Basic Fibroblast Growth Factor

Vector for Expression of Human Basic Fibroblast Growth Factor in Escherichia coli

We have used plasmid vectors with the general structure shown in Fig. 4B to produce various lengths and analogs of recombinant human bFGF in soluble form in *E. coli,* often at levels of expression exceeding 5% of the total protein in crude cell lysates. The first form of human bFGF produced was the 155-residue primary translation product that initiates at the ATG codon indicated in Fig. 3. The cDNA sequence used to encode this primary translation product was derived through a series of modifications from the bovine bFGF cDNA insert carried in the recombinant phage λBB2 (Fig. 2). In the first modification step, *in vitro* mutagenesis[56] was used to alter the codons for amino acid residues 121 and 137 [from TCC (serine) to ACC (threonine), and CCC (proline) to TCC (serine), respectively], so that the cDNA would encode human bFGF (see Fig. 2). A second round of *in vitro* mutagenesis was then carried out to create a *Hin*dIII restriction site (AAGCTT) 34 bp downstream from the translation stop codon. As a result of these steps, the coding region for the 155-residue form of human bFGF could be isolated as a 503-bp *Nco*I–*Hin*dIII fragment (Fig. 2). Since the amino-terminal portion of this coding region had a high G/C content (70.4% G + C in the first 125 bp), the coding region was digested with *Hha*I, and the amino-terminal portion was replaced with the synthetic sequence shown in Fig. 4A (encoding the same amino acids, but lowering the G/C content to 54.4%). In the synthetic sequence, the initiating ATG of the coding region is contained within an *Nde*I site, and the final bFGF coding sequence was therefore isolated as a 503-bp *Nde*I–*Hin*dIII fragment.

To create the vector pTsF-9dH3 (Fig. 4B), the 5' end of the 503-bp *Nde*I–*Hin*dIII bFGF fragment was first ligated to a synthetic 86-bp fragment (stippled box in Fig. 4B), which consists of bases −56 to +22 of the *trp* promoter/operator region[57] flanked by *Eco*RI and *Nde*I linkers on the

---

[56] M. J. Zoller and M. Smith, this series, Vol. 100, p. 468.
[57] G. N. Bennett, M. E. Schweingruber, K. D. Brown, C. Squires, and C. Yanofsky, *J. Mol. Biol.* **121,** 113 (1978).

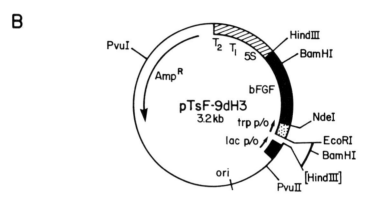

Fig. 4. Expression vector for recombinant bFGF production in *E. coli*. (A) Synthetic nucleotide sequence used to replace the G/C-rich amino-terminal end of the bFGF coding region in the clone λBB2 (underlined sequence in Fig. 2). The amino acids encoded by this sequence are the same as those encoded by the replaced segment. (B) The expression plasmid pTsF-9dH3, in which expression of the 155-residue precursor form of human bFGF (long solid box) is controlled by the *E. coli trp* operon promoter/operator sequence (stippled box). The *rrnB* 5 S ribosomal RNA gene and transcription terminators T_1 and T_2 (hatched box), as well as the 5' end of the β-lactamase gene (Amp^R), are derived from the plasmid pKK233-2. The remainder of the plasmid is derived from pUC9. The expanded segment indicates the polylinker region of pUC9; in pTsF-9dH3, the *Hin*dIII restriction site in the polylinker has been eliminated.

5' and 3' ends, respectively. The 3' end of the bFGF fragment was ligated to a *Hin*dIII–*Pvu*I fragment of pKK233-2[58] containing the two transcription termination signals from the 3' end of the *rrnB* locus (hatched box in Fig. 4B). The resulting composite sequence was then ligated between the *Eco*RI and *Pvu*I sites of a derivative of pUC9[59] (in which the *Hin*dIII site in the polylinker had been destroyed by filling in the cohesive ends of the cleaved site and religating the blunt ends).

Analogs of the 155-residue form of human bFGF have been expressed

---

[58] E. Amann and J. Brosius, *Gene* **40**, 183 (1985).
[59] J. Vieira and J. Messing, *Gene* **19**, 259 (1982).

by replacing the bFGF cDNA segment in pTsF-9dH3 with DNA segments encoding the altered protein sequences of interest. The results presented below to illustrate research-scale production, purification, and analytical techniques for recombinant bFGF were derived from experiments on one such analog, in which the second amino acid of the 155-residue form of the factor had been deleted. The modified version of pTsF-9dH3 encoding this analog is referred to as pTsF-9dH3-154.

Various bacterial host strains have been used to express the human bFGF encoded by pTsF-9dH3 and its derivatives. In the experiments described below, the host strain used was *E. coli* B (American Type Culture Collection, strain 23848, Rockville, MD).

Growth and Lysis of Escherichia coli Cells Expressing Human Basic Fibroblast Growth Factor

To initiate research-scale production of the 154-residue primary translation product described above, a single colony of *E. coli* B containing pTsF-9dH3-154 is used to inoculate 50 ml of LB medium[60] supplemented with 50 μg/ml ampicillin. The culture is grown overnight at 30° to stationary phase, and 10 ml/liter is then used to inoculate four 1-liter batches of supplemented minimal medium (M9 medium[60] containing 0.4% glucose, 2 μg/ml thiamin, 0.5% casamino acids, 0.1 mM CaCl$_2$, 0.8 mM MgSO$_4$, and 50 μg/ml ampicillin). The 1-liter cultures are incubated with shaking in triple-baffled 2.8-liter Fernbach flasks at 30° until the optical density of the cultures (monitored at 550 nm) reaches 0.5–0.7. At this point, 50 mg of 3β-indoleacrylic acid (Sigma, St. Louis, MO) in 10 ml of ethanol is added to each culture to induce the *trp* promoter[61] on the expression plasmid. The cultures are then incubated with shaking for an additional 16 to 24 hr at 30°.

The cells are collected by centrifugation for 15 min at 5000 rpm (7000 g) in a Sorvall RC-3B centrifuge equipped with an H-6000A rotor (Du Pont, Wilmington, DE). The cell pellet from each 1-liter culture is resuspended in a minimum volume of deionized water (~20 ml). The cell suspensions from the four 1-liter cultures are then combined and centrifuged for 15 min at 9000 rpm (10,000 g) in a Sorvall SS-34 rotor. If desired, the resulting cell pellet may be quick-frozen in liquid nitrogen and stored at −80° for purification at a later time.

A crude cell lysate is prepared from the fresh or frozen cell pellet by first resuspending the pellet in 100 ml of 20 mM sodium phosphate (pH

---

[60] J. Sambrook, E. F. Fritsch, and T. Maniatis, "Molecular Cloning: A Laboratory Manual." Cold Spring Harbor Laboratory Press, Cold Spring Harbor, New York, 1989.
[61] W. F. Doolittle and C. Yanofsky, *J. Bacteriol.* **95,** 1283 (1968).

7.0) containing 5 mM EDTA and 1 mM phenylmethylsulfonyl fluoride (Boehringer Mannheim, Indianapolis, IN). Lysozyme (Sigma) is then added to a final concentration of 0.5 mg/ml, and the suspension is incubated on ice for 30 min. In some cases, the mixture has been observed to become so viscous that it appears semisolid; in these cases, more resuspension buffer is added. After lysozyme treatment, the cell suspension is sonicated on ice using an Ultrasonics (Farmingdale, NY) Model W-225R cell disruptor (set to a power level of 4; 50% pulsed cycle; 0.5-inch diameter probe). The sonication is carried out in ten 1-min intervals with a 1-min cooling period between each interval. Bovine pancreatic RNase A (Sigma) and bovine pancreatic DNase I (Boehringer-Mannheim) are each added to a final concentration of 1 μg/ml, and the solution is incubated for 30 min on ice. Cell debris is then removed by centrifugation in an SS-34 rotor at 15,000 rpm (29,000 g) for 30 min at 4°. The pellet is discarded, and the supernatant (crude lysate) is retained at 4° for chromatography.

Purification of Recombinant Basic Fibroblast Growth Factor

Once the crude lysate has been prepared, the purification of bFGF can be accomplished by a two-step procedure. The first step, ion-exchange chromatography carried out on SP-Sephadex C-25 resin (Pharmacia, Piscataway, NJ), results in a dramatic reduction in contaminating proteins. The protein is eluted from the column in a stepwise fashion, as gradient elution has not been observed to improve the degree of purification. Heparin-Sepharose (Pharmacia) is used for the second step in the purification. This chromatographic resin has been previously established as an efficient tool in the isolation of bFGF from natural sources,[1-4,62,63] and its use in this procedure results in a final purification of the protein to greater than 97%, as judged by SDS–polyacrylamide gel electrophoresis (SDS–PAGE).

To prepare the SP-Sephadex C-25 column, the resin is first rehydrated according to the manufacturer's recommendations. In addition, we routinely precycle the resin with acid and base washes before use to ensure that the resin is in the fully ionized form.[64] After precycling, the resin is equilibrated in column buffer [20 mM sodium phosphate (pH 7.0), 5 mM EDTA, 0.1 M NaCl].

The crude lysate supernatant obtained from 4 liters of the expression cell culture (i.e., from ~28 g of cell paste) is fractionated using 14 ml of SP-Sephadex in a 2.5 cm diameter column. This chromatography step and all subsequent purification steps are carried out at 4°. The crude cell lysate

[62] M. Klagsbrun, R. Sullivan, S. Smith, R. Rybka, and Y. Shing, this series, Vol. 147, p. 95.
[63] D. Gospodarowicz, this series, Vol. 147, p. 106.
[64] T. G. Cooper, "The Tools of Biochemistry," p. 144. Wiley, New York, 1977.

is loaded onto the column with a peristaltic pump at a flow rate of 2 ml/min, and the eluate is monitored at 280 nm with an in-line UV detector. After the crude lysate has been completely loaded, the column is washed with the initial column buffer until the absorbance of the eluate returns to baseline level. The eluant is changed to 20 mM sodium phosphate (pH 7.0) containing 5 mM EDTA and 0.6 M NaCl, and the peak of eluting material as judged by absorbance is collected.

For the next step of the purification, 10 to 15 ml of heparin-Sepharose, prepared according to the manufacturer's recommendations, is packed into a 2.5 cm diameter column. The column is equilibrated with 20 mM sodium phosphate (pH 7.0) containing 5 mM EDTA and 0.6 M NaCl. The material that eluted from SP-Sephadex at 0.6 M NaCl is loaded onto the column at 2 ml/min with a peristaltic pump. Protein elution from the column is monitored at 280 nm as described above. After loading, the column is washed with 20 mM sodium phosphate (pH 7.0) containing 5 mM EDTA and 0.6 M NaCl until the absorbance of the eluate returns to baseline. The recombinant bFGF bound to the column is then eluted with 20 mM sodium phosphate (pH 7.0) containing 5 mM EDTA and 2.0 M NaCl.

We have found that the above procedure is effective for the purification of bFGF and most mutant forms of the protein we have tested, including a variety of amino-terminal truncations and extensions as well as many point mutations. It should be noted, however, that single codon changes in the bFGF coding sequence can result in dramatic differences in the level of expression of the recombinant protein. The protocol described here is able to accomodate the purification of at least 150 mg of recombinant bFGF from 4 liters of starting culture.

Analysis of Recombinant Basic Fibroblast Growth Factor

Three methods of analysis are used for determining the purity of the recombinant protein: reversed-phase HPLC, heparin-TSK HPLC, and SDS-PAGE. Although reversed-phase HPLC and SDS-PAGE are commonly used laboratory procedures, they are discussed in this chapter because the sulfhydryl chemistry of recombinant human bFGF as produced in *E. coli* is such that the protein can appear artifactually impure by these two techniques (and by heparin-TSK HPLC), owing to apparent disulfide-mediated heterogeneity.[39,65,66]

In addition to these physical analyses for purity, recombinant bFGF

---

[65] S. A. Thompson, J. W. Rose, J. Hatch, T. M. Palisi, D. I. Blumenthal, K. Y. Sato, J. C. Fiddes, and A. A. Potter, unpublished observations (1989).

[66] M. Seno, R. Sasada, M. Iwane, K. Sudo, T. Kurokawa, K. Ito, and K. Igarashi, *Biochem. Biophys. Res. Commun.* **151**, 701 (1988).

preparations are also routinely tested for bioactivity in a proliferation assay utilizing capillary endothelial cells.[67]

Reversed-Phase HPLC. Reversed-phase HPLC analysis of the protein is generally performed using a Vydac C_4 column purchased from The Separations Group (Hesperia, CA), but recombinant bFGF behaves in a similar fashion on a C_{18} HPLC column. The Vydac C_4 HPLC column (25 cm × 4.6 mm) is equilibrated with 30% acetonitrile in 0.1% trifluoroacetic acid, using a flow rate of 1 ml/min. The eluate is monitored at 220 nm. Recombinant bFGF (20–100 μg) is applied to the column and is then eluted with a 30-min gradient of 30 to 45% acetonitrile in 0.1% trifluoroacetic acid. Under these conditions, the protein will elute at approximately 29 min. Small peaks of absorbance detected at earlier times in the gradient elution appear to be due at least in part to disulfide-mediated microheterogeneity rather than to the presence of contaminants, since treatment of the sample with 20 mM dithiothreitol for 15 min at room temperature eliminates the majority of these species.

Heparin-TSK HPLC. The purity of the protein can also be judged using heparin-TSK HPLC, as has been noted by others.[66] A liquid chromatography system [such as the fast protein liquid chromatography (FPLC) system from Pharmacia] capable of accommodating high sodium chloride concentrations is advisable for this type of chromatography, in order to avoid damage to the pumping system. For analysis of the recombinant human bFGF, a heparin-TSK column (7.5 cm × 7.5 mm, Novex, Encinitas, CA) is equilibrated at a flow rate of 1 ml/min in 20 mM Tris-HCl (pH 7.5) containing 0.72 M NaCl. The eluate is monitored at either 220 or 280 nm. A sample (generally 20 μg) of recombinant bFGF is loaded onto the column and then eluted with a multilinear NaCl gradient in 20 mM Tris-HCl, pH 7.5 (0.72 to 1.2 M NaCl over 1 min, then 1.2 to 3 M NaCl over 23 min). The monomeric recombinant protein elutes at approximately 13 min in this system (Fig. 5A). Before reuse, the column is washed for 5 min with the 3 M NaCl buffer and is then reequilibrated with the 0.72 M NaCl buffer.

With storage at 4° in buffers such as 20 mM Tris-HCl (pH 7.5) containing 1.5 M NaCl, the structure of the recombinant bFGF has been observed to change such that progressively less of the protein elutes from the heparin-TSK HPLC column as the 13-min peak, and peaks of more retained species appear in the analysis (Fig. 5B). The heterogeneity appears to originate through disulfide bond formation, since the more retained species can be reduced or eliminated by treatment with dithiothrei-

---

[67] D. Gospodarowicz, S. Massoglia, J. Cheng, and D. K. Fujii, *J. Cell. Physiol.* **127,** 121 (1986).

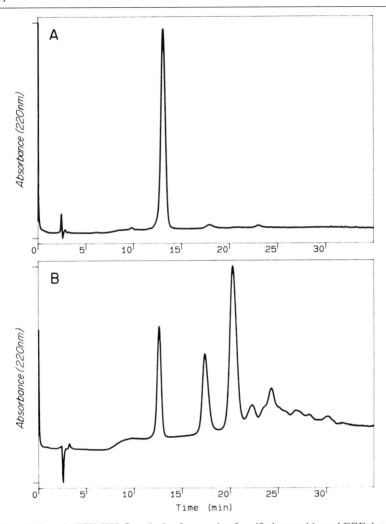

FIG. 5. Heparin-TSK HPLC analysis of a sample of purified recombinant bFGF that has undergone a significant degree of multimerization. Chromatography was carried out on a 7.5 cm × 7.5 mm column either after (A) or before (B) reduction with dithiothreitol. The sample analyzed here was placed for several days at 4° in 20 mM Tris-HCl (pH 7.5), 1.5 M NaCl to allow the multimerization to occur; the product obtained directly from the purification procedure described in this chapter generally chromatographs almost entirely as the 13-min (monomer) peak.

tol (compare Fig. 5B with 5A) or by mutation of the recombinant bFGF to eliminate the cysteines at residues 78 and 96 (see Fig. 2).[39,65,66] Nonreducing SDS–PAGE analyses conducted after iodoacetamide treatment of the samples (see below), along with the results of gel-filtration chromatography studies,[65] have indicated that the later-eluting species in the heparin-TSK HPLC analysis represent multimers of bFGF molecules (dimers elute at ~18 min, and trimers at 21 min; subsequent peaks are heterogeneous mixtures of forms of trimer size or greater). Taken together, these observations indicate that the heterogeneity detected by heparin-TSK HPLC is due to intermolecular disulfide bond formation.

SDS–Polyacrylamide Gel Electrophoresis. SDS–PAGE is carried out with a 15% acrylamide gel containing 0.5% bisacrylamide according to the procedure of Laemmli.[68] The gel dimensions are normally 5 × 8.2 cm (length by width), with a thickness of 0.75 mm. The amount of protein loaded onto the gel will vary according to the staining technique, but 10 μg of protein for a Coomassie-stained gel and 1 μg for a silver-stained gel is sufficient to determine purity.

Two points should be kept in mind when analyzing recombinant bFGF by this technique. First, protocols for the use of SDS–PAGE generally call for heating the protein in SDS–PAGE sample loading buffer for 3 to 5 min at 95°. We have found that recombinant bFGF will partially degrade to lower molecular weight fragments under these conditions, so that after staining of the gel the protein sample appears to contain contaminating species (Fig. 6, lane 3). Partial degradation after heating in SDS-containing buffers has also been reported in other proteins.[69] This problem can be circumvented by eliminating the heating step prior to loading of the sample on the gel (Fig. 6, lane 2).

The second precaution to be taken when using SDS–PAGE concerns interpretation of the results of gel analyses when reductants have not been included in the SDS–PAGE sample loading buffer. As mentioned above, intermolecular disulfide bonds appear to form between recombinant bFGF molecules under some conditions,[39,65,66] generating higher molecular weight, multimeric forms of the protein. With other proteins, comparisons of nonreducing and reducing SDS–PAGE have commonly been used to identify such forms. With recombinant bFGF, however, we have found that in the absence of reducing agents the heterogeneous forms can still appear predominantly as monomeric bFGF on SDS–PAGE analysis, presumably owing to thiol–disulfide rearrangements occurring once the protein has been unfolded by the SDS in the sample loading buffer. These

---

[68] U. K. Laemmli, *Nature (London)* **227,** 680 (1970).
[69] J. Rittenhouse and F. Marcus, *Anal. Biochem.* **138,** 442 (1984).

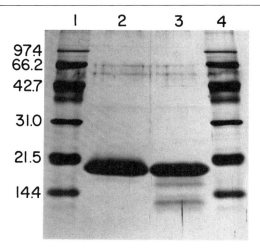

FIG. 6. Heat-induced degradation of bFGF. Samples were fractionated by reducing SDS–PAGE and then visualized by silver staining of the gel. Lanes 1 and 4, molecular weight markers [phosphorylase B (97.4 kD), bovine serum albumin (66.2 kD), ovalbumin (42.7 kD), carbonic anhydrase (31.0 kD), soybean trypsin inhibitor (21.5 kD), and lysozyme (14.4 kD)]; lane 2, 6 μg of recombinant bFGF, loaded in SDS–PAGE sample loading buffer without heat treatment; lane 3, 6 μg of recombinant bFGF, heated to 95° for 10 min in SDS–PAGE sample loading buffer prior to application to the gel.

rearrangements can be blocked by the inclusion of 500 mM iodoacetamide in the sample loading buffer, which will carboxymethylate the free cysteines that otherwise might participate in thiol–disulfide exchange reactions. Lower concentrations of iodoacetamide may be used successfully in this type of analysis, but less than 200 mM is not recommended.

If it is desirable to precipitate a recombinant bFGF sample with trichloroacetic acid (TCA) prior to addition of nonreducing SDS–PAGE loading buffer (e.g., to concentrate the protein or to reduce the salt concentration in the sample), then carboxymethylation of the protein in the presence of a denaturant should be performed prior to the precipitation step. Such carboxymethylations are carried out by mixing the protein sample with an equal volume of 1 M iodoacetamide in 8 M urea and incubating for 5 min at room temperature. TCA can then be added to a final concentration of 10% to precipitate the bFGF. After 3 min on ice, the protein is pelleted by centrifugation in a microcentrifuge for 3 min. The pellet is washed with 0.2 ml of acetone and is then resuspended in nonreducing SDS–PAGE sample buffer.

An example of this technique is shown in Fig. 7. The sample analyzed in this experiment was isolated from a preparation of recombinant bFGF

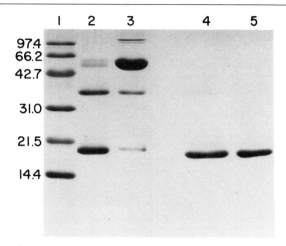

Fig. 7. SDS–PAGE analysis of the form of bFGF eluting at 21 min from the heparin-TSK HPLC column, with or without treatment with iodoacetamide prior to electrophoresis. Lane 1, molecular weight markers, applied to the gel in nonreducing sample loading buffer; lane 2, bFGF sample precipitated with TCA and resuspended in nonreducing SDS–PAGE sample loading buffer before electrophoresis; lane 3, bFGF sample treated with iodoacetamide as described in the text, prior to TCA precipitation and resuspension in nonreducing SDS–PAGE sample loading buffer; lane 4, bFGF sample treated as in lane 3, except that 10% 2-mercaptoethanol was added to the sample loading buffer; lane 5, bFGF sample treated as in lane 2, except that the precipitated sample was resuspended in the sample loading buffer containing 10% 2-mercaptoethanol.

that had undergone a significant degree of multimerization, as judged by heparin-TSK chromatography. The protein eluting from the heparin-TSK column at approximately 21 min (third peak; see Fig. 5B) was collected and shown by heparin-TSK HPLC analysis to reelute at the same position in the gradient. When a sample of this 21-min peak form of bFGF was precipitated with TCA and resuspended in nonreducing loading buffer prior to SDS–PAGE analysis, the protein migrated predominantly at the apparent molecular weights expected for the monomer and dimer forms of bFGF (Fig. 7, lane 2). In contrast, when the sample was first treated with iodoacetamide and urea as described above before precipitation with TCA, the protein migrated predominantly at the molecular weight expected for a trimeric form of bFGF (Fig. 7, lane 3). To demonstrate that the higher molecular weight forms observed in lanes 2 and 3 are due to the presence of disulfide bonds in the loaded protein sample, the bFGF isolated as the 21-min peak from heparin-TSK chromatography was again precipitated with TCA, with or without prior iodoacetamide treatment as in lanes 3 and 2, but was resuspended in sample loading buffer containing 10% 2-

mercaptoethanol before electrophoresis on the SDS gel (lanes 4 and 5, respectively). Under these conditions, the bFGF in both cases migrated almost entirely as the monomeric form.

Bioassay. The activity of the recombinant bFGF is measured in a cell proliferation assay. The target cells used in the assay, bovine adrenal cortex capillary endothelial (ACE) cells, were obtained from D. Gospodarowicz.[67] For passaging, 2.5×10^5 cells (~2.5% of the cells from a confluent T-75 flask) are seeded in a T-75 flask in 15 ml of growth medium [Dulbecco's modified Eagle's medium (DMEM-21; Mediatech, Washington, D.C.) supplemented with 10% calf serum (Hyclone, Logan, UT), 2 mM L-glutamine (GIBCO, Grand Island, NY), 50 IU/ml penicillin (GIBCO), and 50 μg/ml streptomycin (GIBCO)]. The flask is then incubated at 37° in a humidified atmosphere containing 10% CO_2. Basic FGF (1 ng/ml) is added on the day of seeding (day 0) and 2 days later. By day 4, the cells have reached confluence. No further additions of bFGF are made until the cells are split again (on day 7).

For the assay, the day 7 cells are removed from the T-75 flask by first exposing the monolayer for 2–3 min to 1 ml of trypsin/EDTA in Hanks' buffered salt solution (calcium- and magnesium-free; GIBCO) and then adding 9 ml of growth medium to rinse the cells from the flask surface. An aliquot of the resulting cell suspension is diluted with growth medium to a concentration of 5×10^3 cells/well and used to seed 6-well tissue culture plates (2 ml/well). The cells are then incubated for at least 1 hr prior to addition of bFGF samples.

Before addition to the cells, samples to be assayed are serially diluted at 4° in phosphate-buffered saline (PBS) containing 0.2% gelatin (Sigma). Aliquots (10 μl/well) of each dilution are then added to duplicate wells on the day of cell seeding (day 0) and are added again 2 days later. Control wells receive additions of the diluent alone (PBS containing 0.2% gelatin) to determine background growth of the cells in the absence of added bFGF. On day 4, the cells are exposed to 0.5 ml of trypsin/EDTA at 37° for 2–3 min, transferred to a vial containing 10 ml of isotonic saline solution, and counted in a Coulter particle counter. The cell counts from duplicate wells are averaged, and the averaged count from the background wells is substracted. For each sample, the average cell count for each dilution is divided by the highest average cell count obtained in that dilution series, so that the data are expressed as the percentage of the maximal proliferation induced by the sample. The data are then graphed as the percent maximal proliferation versus the log of the concentration of bFGF added to the wells on day 0. Relative activity between any two samples is judged by comparing the ED_{50} values (i.e., the concentration of each sample needed in the assay wells to elicit 50% of maximal proliferation).

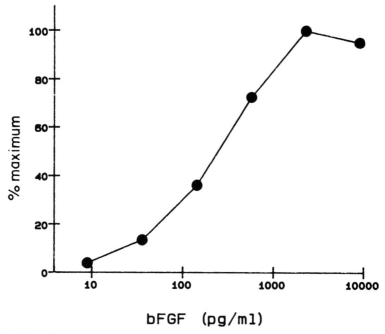

FIG. 8. Endothelial cell proliferation assay of recombinant bFGF. Aliquots of serial dilutions of the bFGF analog described in the text were added to assay wells on the first day of the assay (day 0) and two days later; the amount of proliferation in each well was then determined on day 4 by counting the cells in a Coulter counter. Each point represents the average of the results from duplicate wells. Results are graphed as the percent of the maximal proliferation induced by the sample vs. the concentration of the recombinant bFGF added to the assay wells on day 0.

An example of the results obtained from assaying the recombinant bFGF analog described above (154-residue primary translation product) is given in Fig. 8. In this experiment, the ED_{50} of the analog was calculated to be 242 pg/ml.

Acknowledgments

Much of the work described in this chapter was supported by Small Business Innovation Research Grants GM36762 and HL39348 from the National Institute of General Medical Sciences and the National Heart, Lung, and Blood Institute, respectively.

[11] Construction and Expression of Transforming Gene Resulting from Fusion of Basic Fibroblast Growth Factor Gene with Signal Peptide Sequence

By SNEZNA ROGELJ, DAVID STERN, and MICHAEL KLAGSBRUN

Introduction

Basic fibroblast growth factor (bFGF) is a potent mitogen for cells of mesodermal origin.[1] Basic FGF lacks a signal peptide and is not secreted by cells in culture. Many cell types, for example, endothelial cells, produce bFGF and have bFGF receptors. Yet these cells appear normal and do not seem to undergo FGF-mediated autocrine transformation. NIH 3T3 fibroblasts transfected with bFGF cDNA do not secrete bFGF; they appear normal in culture and are not tumorigenic.[2,3] It has been suggested that lack of bFGF secretion prevents bFGF–receptor interaction with the result that bFGF-producing cells have a normal, nontransforming phenotype. On the other hand, NIH 3T3 cells transfected with a construct of bFGF fused to a signal peptide sequence (spbFGF) are highly transformed morphologically, grow as nonadherent aggregates, and are highly tumorigenic and metastatic.[2–4] Thus, bFGF altered by the addition of a singal sequence is a transforming protein. Although spbFGF transforms NIH 3T3 cells, there is no evidence of bFGF secretion, suggesting that transformation in these cells is the result of an internal autocrine loop.

Overall Strategy

The transforming bFGF expression vector pspbFGF is constructed by fusing a sequence encoding the N-terminal 19 amino acid mouse heavy chain immunoglobulin signal sequence[5] upstream to the 154 amino acid bovine brain bFGF sequence[6] (Fig. 1C). These sequences are placed under the transcriptional control of the Moloney leukemia virus long terminal

---

[1] J. Folkman and M. Klagsbrun, *Science* **235**, 442 (1987).
[2] S. Rogelj, R. A. Weinberg, P. Fanning, and M. Klagsbrun, *Nature (London)* **331**, 173 (1988).
[3] S. Rogelj, R. A. Weinberg, P. Fanning, and M. Klagsbrun, *J. Cell. Biochem.* **39**, 13 (1989).
[4] A. Yayon and M. Klagsbrun, *Proc. Natl. Acad. Sci. U.S.A.* **87**, 5346 (1990).
[5] D. Y. Loh, A. L. M. Bothwell, M. E. White-Scharf, T. Imanishi-Kari, and D. Baltimore, *Cell (Cambridge, Mass.)* **33**, 85 (1983).
[6] J. A. Abraham, A. Mergia, J. L. Whang, A. Tumolo, J. Friedman, K. A. Hjerrild, D. Gospodarowicz, and J. C. Fiddes, *Science* **233**, 545 (1986).

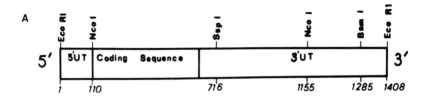

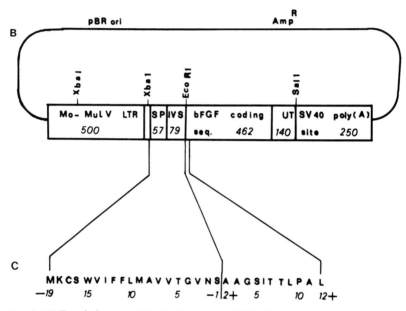

FIG. 1. (A) Restriction map of bovine brain basic FGF cDNA. (B) pspbFGF plasmid: The Moloney murine leukemia virus long terminal repeat (Mo-MuLV LTR) drives constitutive expression of the chimeric signal peptide–bFGF fusion protein. (C) Amino terminus of the predicted primary translation product of the signal peptide–basic FGF protein. The 19 amino acid mouse immunoglobulin heavy chain signal peptide is fused to the second amino acid (alanine) of bovine brain bFGF.

repeat (LTR) and are followed by RNA processing signals derived from the SV40 genome. Cloning of this functional composite gene is performed in a pUC13-derived vector which contains the pBR322 origin of replication and the ampicillin resistance gene, thus allowing for growth of the mammalian vector in bacteria in the presence of ampicillin.

The pspbFGF vector is introduced into NIH 3T3 fibroblasts by the calcium phosphate transfection procedure. Transforming potential is measured by the focus forming assay. Clonal cell lines are obtained by cotransfecting the pspbFGF vector with the dominant selectable marker pSV-neo. G418-resistant colonies are picked and expanded into cell lines which are assayed for the production of growth factor activity as measured by the ability of cell extracts and/or conditioned media to stimulate the incorporation of [^{3}H]thymidine into BALB/c 3T3 DNA. Cell lines positive for the synthesis of growth factor activity are analyzed for specific bFGF expression by heparin affinity chromatography and by immunodetection (immunoprecipitation and Western blot) using specific anti-bFGF antibodies.[2,7]

Materials

A bovine bFGF cDNA clone was obtained from Dr. Judith Abraham (CalBio, Mountain View, CA).[6] pSV2-*neo* was obtained from Dr. Robert Weinberg (The Whitehead Institute for Biomedical Research, Cambridge, MA). Restriction enzymes, mung bean nuclease, and polymerases were obtained from New England BioLabs (Beverly, MA). G418 is obtained from GIBCO Laboratories (Grand Island, NY). Heparin-Sepharose is obtained from Pharmacia (Piscataway, NJ).

Construction of pspbFGF Expression Vectors

A mammalian vector directing synthesis of a chimeric signal peptide–bFGF is constructed by replacing the epidermal growth factor (EGF) sequence for the bFGF sequence in the pUCDS3 vector. This vector was originally designed in collaboration with Dr. David L. Hare (Amgen Development Corp., Boulder, CO), for the expression and secretion of synthetic EGF fused to a signal peptide sequence in mammalian cells.[8] pUCDS3 consists of a Moloney murine leukemia virus long terminal repeat (MuLV LTR) that contains promoter/enhancer sequences, an SV40 polyadenylation site, sequences encoding a mouse immunoglobulin heavy chain signal peptide,[5] and chemically synthesized human EGF-encoding sequences. The signal peptide-encoding sequences are included so that the protein product is translocated to the endoplasmic reticulum (ER)/Golgi compartment for subsequent secretion or expression on the cell surface. pUCDS3 was designed so that the coding sequences to be expressed are joined at the 5' end to a unique *Eco*RI site located just above

---

[7] I. Vlodavsky, J. Folkman, R. Sullivan, R. Friedman, R. Ishai-Michaeli, J. Sasse, and M. Klagsbrun, *Proc. Natl. Acad. Sci. U.S.A.* **84,** 2292 (1987).
[8] D. F. Stern, D. L. Hare, M. A. Cecchini, and R. A. Weinberg, *Science* **235,** 321 (1987).

the signal peptidase cleavage site of the signal peptide and to a unique 3' SalI site located 5' to the polyadenylation site.

The pUCSD3 vector was constructed as described previously.[8] Briefly, the signal peptide containing immunoglobulin heavy chain cDNA clone 17.2.25 is modified by oligonucleotide-directed mutagenesis to create an EcoRI site within the last two codons of the signal peptide sequence. This change in nucleotide sequence maintains the original amino acid sequence of the signal peptide. A 145-base pair (bp) DNA fragment encoding all of the signal peptide sequence, which includes a 79-bp intron, is excised from this modified 17.2.25 clone by cleavage at the newly created EcoRI site at the 3' end and at the AvaII site, 9 nucleotides 5' to the signal peptide initiating ATG codon. Using AvaII–pseudo-SalI and EcoRI–HindIII adapters, this fragment is ligated into a SalI- and HindIII-cleaved pFB1δRI vector to produce pUCDS2. The pFB1δRI vector consists of a partial Sau3A fragment containing the Moloney leukemia virus LTR cloned into the BamHI site of pUC13 that has had the single EcoRI site destroyed by cleaving with EcoRI restriction enzyme, filling-in with the Klenow fragment of DNA polymerase I, and religating with T4 DNA ligase. To provide for the RNA polyadenylation signal, the 155-bp HpaI–BamHI fragment of the SV40 polyadenylation site was linked to synthetic HpaI–SalI and BamHI–HindIII adapters and cloned into the SalI and HindIII restriction enzyme-digested pUC8 3' to the EGF gene. The EcoRI–HindIII fragment containing the EGF gene and the polyadenylation site was then cloned into the EcoRI- and HindIII-digested pUCDS2 to create pUCDS3.

Basic FGF coding sequences are prepared as follows. The EcoRI-flanked bFGF cDNA sequence is digested with NcoI restriction enzyme to remove all of the 5' end and some 350 bp of the 3'-end noncoding sequences (Fig. 1A). Cleavage with this enzyme yields a 1045-bp fragment with a 4-nucleotide overhang. The second, third, and fourth nucleotide of the overhang at the 5' end of the coding sequence specify for the proposed translation initiating ATG codon. The nucleotide overhang is removed by blunting with mung bean nuclease, and the fragment is ligated to 8-mer EcoRI linkers and digested with the EcoRI enzyme to create EcoRI sticky ends. At the 5' end of the gene this removal of the NcoI overhang results in elimination of the sequences encoding for the initiating methionine, whereas the EcoRI site containing linker added to the blunted 5' end encodes for a serine which reconstructs the terminal amino acid of the signal sequence. This fragment is then further digested with a unique SspI restriction enzyme to remove an additional 439 bp from the 3'-end noncoding sequence and to create a 3' blunt end. The resulting 608-bp fragment contains 140 bp of the 3'-end noncoding sequence, a 4-nucleotide overhang and 2 bp from the EcoRI linker at the 5' end, and 462 bp of the coding sequence for the entire bFGF protein except the initiating methionine which has been removed.

To construct an expression vector directing synthesis of a chimeric, immunoglobulin signal peptide containing bFGF protein (spbFGF), the 608-bp bFGF-encoding DNA fragment described above is cloned into the PUCDS3 vector from which the synthetic EGF sequences are removed by digestion with *Eco*RI and *Sal*I restriction enzymes (Fig. 1B). The fusion joins the immunoglobulin signal peptide to alanine, which is the second amino acid residue of the putative open reading frame of the 18-kDa bFGF molecule. The amino terminus of the primary translation product of the chimeric spbFGF protein is illustrated in Fig. 1C. After translation, the signal peptide should be cleaved off in the endoplasmic reticulum, yielding an 18-kDa bFGF protein which differs from native bFGF only by lacking the first amino acid (Met), thus initiating with an alanine.

Control vector pUCDS5[8] that expresses only the signal peptide sequence was constructed by cleavage of the pUCDS3 vector at the unique *Eco*RI site, filling-in with the large fragment of DNA polymerase I (Klenow fragment), and religating, resulting in a 4-nucleotide insertion.

Transfection

The transforming potential of the expression vectors is ascertained by introducing the plasmids into NIH 3T3 cells using the calcium phosphate transfection procedure of Graham and van der Eb.[9] The NIH 3T3 cell line is chosen because it responds mitotically to bFGF treatment and is known to express bFGF receptors. A day prior to transfection, 100-mm dishes are seeded with 1×10^6 cells in Dulbecco's modified Eagle's medium (DMEM) supplemented with 10% bovine calf serum, penicillin, and streptomycin. Milligram quantities of plasmid DNAs suitable for transfection are prepared by the alkaline lysis method, followed by a further purification on a CsCl gradient as detailed elsewhere.[10]

Transfections are carried out using 0.1 μg of the dominant selectable-marked pSV2-*neo* and 0.8, 8.0, or 40 μg of pspbFGF plasmid DNA. Carrier NIH 3T3 DNA is added to a final concentration of 75 μg of DNA per 2×10^6 cells (2 dishes). Positive control for cellular transformation is obtained using 0.8 μg of pEJ6.6 plasmid DNA.[11] After a 4-hr exposure to calcium phosphate DNA precipitate, the cells are glycerol-shocked for 3 min with

[9] F. L. Graham and A. J. van der Eb, *J. Virol.* **52**, 456 (1973).
[10] T. Maniatis, E. F. Fritsch, and J. Sambrook, "Molecular Cloning: A Laboratory Manual," p. 86. Cold Spring Harbor Laboratory, Cold Spring Harbor, New York, 1982.
[11] C. Shih and R. A. Weinberg, *Cell (Cambridge, Mass.)* **29**, 161 (1982).

15% glycerol in phosphate-buffered saline, washed, fed with DMEM containing 10% calf serum, and incubated in 5% CO_2 for further 24 hr. The transfected cells are then trypsinized and split in a ratio of 1 : 10 into 100-mm dishes. One-half of the dishes are used for the focus forming assay[12] and the other half for the selection of G418-resistant monoclonal cell lines.[13]

Focus Forming Assay

For the focus forming assay, the cells are fed with DMEM containing 10% calf serum and antibiotics, refed every 5 days, and the foci counted 14, 21, and 28 days after seeding. The control plasmid pUCDS5 is unable to induce focus formation. In contrast, the pspbFGF plasmid induces foci visible to the naked eye within 10 days of transfection. These early foci have a characteristic spindly morphology but are present only at a low frequency of approximately 40 foci per 8 μg of pspbFGF DNA per 10^6 cells. However, after an additional 2 weeks in culture, about 20 times more foci with the same characteristic morphology appear. The delayed foci do not arise through seeding of cells from the original foci, as the second wave of focus formation occurs even when the monolayer is overlayed with soft agar (0.6%, w/v) immediately after splitting of the transfected cells. When transfection is carried out using higher quantities of pspbFGF plasmid, fewer early and second-wave colonies are observed, suggesting that pspbFGF has a toxic effect on the transfected cells. For example, when 2×10^6 cells are transfected with 40 μg of pspbFGF DNA, only 4–7 early colonies are obtained. The number of second-wave foci is still about 20 times the number of the original foci. Control Ha-*ras* oncogene, which induces focus formation at about 150 colonies per microgram of pEJ6.6 plasmid DNA per 10^6 cells, does not display such toxicity when larger quantities of plasmid DNA are used for transfection. The nature of this toxicity and delayed focus formation is not understood.

Selection of Cell Lines Expressing Chimeric Fibroblast Growth Factor

To obtain clonal cell lines synthesizing the chimeric signal peptide–bFGF proteins, transfected cells are placed into DMEM containing 10% calf serum supplemented with 0.5 mg/ml of the drug G418. The selective medium is changed every 3 days. Colonies of resistant cells are observed macroscopically after 10 to 14 days. Clonal cell lines are derived

[12] H. Land, L. F. Parada, and R. A. Weinberg, *Nature* (*London*) **304**, 596 (1982).
[13] P. J. Southern and P. Berg, *J. Mol. Appl. Genet.* **1**, 327 (1982).

from these colonies by placing glass cylinders around single colonies, trypsinizing the cells within the cylinder, and transferring the cells to 30-mm dishes. Upon confluency the clonal lines are expanded further.

Again, the toxicity of pspbFGF clone is measurable. Addition of 8 μg instead of 0.8 μg of pspbFGF DNA to 0.1 μg of transfected pSV2-*neo* DNA causes a 10-fold decrease in the surviving, G418-resistant colonies. A similar inhibitory effect by an oncogenic plasmid has been observed for the Abelson leukemia virus transforming gene.[14]

The morphology of some of the G418-resistant pspbFGF-transfected clones is identical to the morphology of the foci seen in the focus forming assay and differs dramatically from the control pSV2-*neo*-transfected colonies. However, in about 90% of the G418-resistant, pspbFGF-transfected colonies, the spbFGF-induced morphological transformation is delayed. When these colonies are first isolated, they display the normal morphology and express barely detectable bFGF activity. However, during the first few rounds of passage, foci of stable transformed morphology arise in cultures grown from the original clonal cell line. When the cells from these foci of transformed cells are expanded into cell lines, they express higher levels of bFGF than the parental clonal line. This delayed focus formation in the pspbFGF-transfected, G418-resistant clonal cell lines is reproducible and is not caused by cross-contamination with the transformed cells since it occurs in clonal cell lines picked from plates that do not contain any transformed colonies.

Analysis of Basic Fibroblast Growth Factor Expression

Basic FGF is measured by a combination of heparin affinity chromatography, bioassay on BALB/c 3T3 cells and/or endothelial cells, and immunoreactivity. These methods have been described in detail.[2,7,15] Briefly, samples are applied to columns of heparin-Sepharose or fast protein liquid chromatography (FPLC) TSK-heparin which are subsequently eluted with a gradient of NaCl. Basic FGF elutes at about 1.5–1.7 M NaCl. Fractions are analyzed for the ability to stimulate DNA synthesis (incorporation of tritiated thymidine) in confluent monolayers of BALB/c 3T3 cells and/or for the ability to stimulate the proliferation of endothelial cells. Active fractions are further analyzed in a Western blot for the ability to interact with anti-bFGF antibodies. Alternatively, cells are labeled with [^{35}S]methionine and the bFGF detected by immunoprecipitation. Signal peptide–

---

[14] S. F. Ziegler, C. A. Whitlock, S. P. Goff, A. Gifford, and O. N. Witte, *Cell (Cambridge, Mass.)* **27**, 477 (1981).

[15] R. Sullivan and M. Klagsbrun, *J. Tissue Cult. Methods* **10**, 125 (1986).

bFGF-transfected NIH 3T3 cells usually express about 0.8 units of bFGF per 10^4 cells. It is thought that higher levels of expression are toxic to cells.

Tumorigenicity

Signal peptide-bFGF-transformed cells are highly tumorigenic in syngeneic mice.[3] Large tumors (~1 cm in diameter) appear within 2 weeks in 100% of the animals injected subcutaneously with 2×10^6 cells. The size of the tumors and their rate of growth are comparable to Ha-*ras*-transfected NIH 3T3 cells. When the tumor tissue is cultured, G418-resistant cell lines with the characteristic transformed morphology are obtained. These tumor-derived cell lines still express immunoprecipitable, heparin-binding, biologically active bFGF.[2,3]

Acknowledgments

The authors are grateful to Robert A. Weinberg in whose laboratory most of this work was done. We thank David L. Hare who collaborated in design and construction of pUCDS3. Financial support was provided in part by grants from the National Institutes of Health (EY05321 to R.A.W., CA37392 and CA45548 to M.K., and CA07813 to S.R.).

[12] Identification and Characterization of Fibroblast Growth Factor-Related Transforming Gene *hst-1*

By TERUHIKO YOSHIDA, KIYOSHI MIYAGAWA, HIROMI SAKAMOTO, TAKASHI SUGIMURA, and MASAAKI TERADA

Introduction

In contrast to many other classic peptide growth factors, the *hst-1* transforming gene rather than its protein was identified first.[1] The fact that the gene encodes a growth factor was then presumed from a deduced amino acid sequence[2] and from its remarkable sequence similarity to fibroblast growth factors (FGF).[3] Finally, growth factor activity was con-

[1] H. Sakamoto, M. Mori, M. Taira, T. Yoshida, S. Matsukawa, K. Shimizu, M. Sekiguchi, M. Terada, and T. Sugimura, *Proc. Natl. Acad. Sci. U.S.A.* **83**, 3997 (1986).
[2] M. Taira, T. Yoshida, K. Miyagawa, H. Sakamoto, M. Terada, and T. Sugimura, *Proc. Natl. Acad. Sci. U.S.A.* **84**, 2980 (1987).
[3] T. Yoshida, K. Miyagawa, H. Odagiri, H. Sakamoto, P. F. R. Little, M. Terada, and T. Sugimura, *Proc. Natl. Acad. Sci. U.S.A.* **84**, 7305 (1987).

```
Met-Ser-Gly-Pro-Gly-Thr-Ala-Ala-Val-Ala-Leu-Leu-Pro-Ala-Val-Leu-Leu-Ala-Leu-Leu-
Ala-Pro-Trp-Ala-Gly-Arg-Gly-Gly-Ala-Ala-Ala-Pro-Thr-Ala-Pro-Asn-Gly-Thr-Leu-Glu-
Ala-Glu-Leu-Glu-Arg-Arg-Trp-Glu-Ser-Leu-Val-Ala-Leu-Ser-Leu-Ala-Arg-Leu-Pro-Val-
Ala-Ala-Gln-Pro-Lys-Glu-Ala-Ala-Val-Gln-Ser-Gly-Ala-Gly-Asp-Tyr-Leu-Leu-Gly-Ile-
Lys-Arg-Leu-Arg-Arg-Leu-Tyr-Cys-Asn-Val-Gly-Ile-Gly-Phe-His-Leu-Gln-Ala-Leu-Pro-
Asp-Gly-Arg-Ile-Gly-Gly-Ala-His-Ala-Asp-Thr-Arg-Asp-Ser-Leu-Leu-Glu-Leu-Ser-Pro-
Val-Glu-Arg-Gly-Val-Val-Ser-Ile-Phe-Gly-Val-Ala-Ser-Arg-Phe-Phe-Val-Ala-Met-Ser-
Ser-Lys-Gly-Lys-Leu-Tyr-Gly-Ser-Pro-Phe-Phe-Thr-Asp-Glu-Cys-Thr-Phe-Lys-Glu-Ile-
Leu-Leu-Pro-Asn-Asn-Tyr-Asn-Ala-Tyr-Glu-Ser-Tyr-Lys-Tyr-Pro-Gly-Met-Phe-Ile-Ala-
Leu-Ser-Lys-Asn-Gly-Lys-Thr-Lys-Lys-Gly-Asn-Arg-Val-Ser-Pro-Thr-Met-Lys-Val-Thr-
His-Phe-Leu-Pro-Arg-Leu
```

FIG. 1. Deduced amino acid sequence of the *hst-1* protein (206 amino acids, 22 kDa). A hydrophobic core sequence of a putative signal peptide is shaded.

firmed by synthesizing a recombinant *hst-1* protein.[4] In this chapter, we summarize these processes of identification and characterization of *hst-1*.

High molecular weight DNA was extracted from 37 surgical specimens of gastric cancer and from 21 noncancerous portions of gastric mucosa. Three of the 58 samples, including one from noncancerous gastric tissues, showed transforming activity when transfected into NIH 3T3 cells. A genomic library was constructed from the DNA of a secondary transformant, T361-2nd-1, and *Alu*-containing portions were physically mapped. Three contiguous repetitive-free fragments were then identified to hybridize to the mRNA of the T361-2nd-1 transformant but not to the mRNA of the parental NIH 3T3 cells. One of the fragments was employed to screen a cDNA library of the T361-2nd-1 cells, and a novel transforming gene, which we designated *hst* (human stomach), was identified.[1,2] Afterward the gene was renamed *hst-1* when a close homolog of *hst*, termed *hst-2*, was cloned by cross-hybridization.[5] The transforming activity of the *hst-1* cDNA clone and various mutated derivatives was evaluated with an SV40-based eukaryotic expression vector. An open reading frame, termed ORF1, from nucleotides 239 to 856 of the cDNA was thus identified as encoding a transforming protein of 206 amino acids (Fig. 1).[2]

All three independent NIH 3T3 foci induced by human gastric DNAs were found to be transformed by *hst-1*. Moreover, a number of subsequent reports have shown the presence of the "transforming" *hst-1* gene in nongastric cancers, such as colon cancer,[6] hepatoma,[7,8] Kaposi sarcoma[9]

---

[4] K. Miyagawa, H. Sakamoto, T. Yoshida, Y. Yamashita, Y. Mitsui, M. Furusawa, S. Maeda, F. Takaku, T. Sugimura, and M. Terada, *Oncogene* **3**, 383 (1988).

[5] H. Sakamoto, T. Yoshida, M. Nakakuki, H. Odagiri, K. Miyagawa, T. Sugimura, and M. Terada, *Biochem. Biophys. Res. Commun.* **151**, 965 (1988).

[6] T. Koda, A. Sasaki, S. Matsushima, and M. Kakinuma, *Jpn. J. Cancer Res.* **78**, 325 (1987).

[7] H. Nakagama, S. Ohnishi, M. Imawari, H. Hirai, F. Takaku, H. Sakamoto, M. Terada, M. Nagao, and T. Sugimura, *Jpn. J. Cancer Res.* **78**, 651 (1987).

[8] Y. Yuasa and K. Sudo, *Jpn. J. Cancer Res.* **78**, 1036 (1987).

[9] P. Delli Bovi and C. Basilico, *Proc. Natl. Acad. Sci. U.S.A.* **84**, 5660 (1987).

(where *hst* was referred to as *KS* or *K-fgf*), melanoma,[10] and osteosarcoma.[11] These reports indicate *hst-1* is the most prevalent non-*ras* transforming gene.

Identification of *hst-1* Gene from Normal Genomic Libraries

Cloning of *hst-1* Gene from Cosmid Libraries

Since the *hst-1* cDNA clone was derived from an NIH 3T3 transformant, it was possible that the *hst-1* amino acid sequence decoded from the ORF1 of the cDNA was a product of an artificial gene rearrangement during the transfection and not a genuine human protein. To address this question, an *hst-1* genomic fragment was cloned from a normal human library.[5] Peripheral blood leukocytes were taken from a healthy male that had no previous or family history of serious illness. High molecular weight DNA was extracted, partially digested with *Sau*3AI, and fractionated by sucrose density gradient centrifugation. DNA fragments of 30–50 kilobase pairs (kbp) were ligated to the *Bam*HI site of the cosmid vector LoristB[12] and packaged *in vitro*. About 1×10^5 clones were screened with a probe corresponding to the ORF1 of the *hst-1* cDNA under stringent conditions at 42° in a hybridization buffer containing 50% formamide.

Two groups of cosmids were obtained, with two different patterns of hybridization to the ORF1 probe as shown in Fig. 2A; six of nine cosmids have three *Eco*RI fragments of 5.8, 2.8, and 0.8 kbp, whereas each of the remaining three clones has one 8.0-kbp *Eco*RI fragment hybridized to ORF1. These two groups of clones represented *hst-1* and its close homolog *hst-2*, respectively. Partial physical maps are presented for these clones and for the clone L361-Hu3, which is an *hst-1* clone derived from the original NIH 3T3 transformant T361-2nd-1 (Fig. 2B).

Transforming Potential of Various hst-1 Genes

All six *hst-1* cosmids from the normal human library had transforming activity equivalent to that of L361-Hu3. A 4.0-kbp *Taq*I fragment, YT4.0, which contains a 1.3-kbp region upstream of TATA box (Fig. 2B) was fully transforming by itself. A mouse *hst-1* fragment derived from the genome of NIH 3T3 cells showed a similar transforming activity.

Although the cosmids containing *hst-2* did not transform NIH 3T3 cells

---

[10] J. Adelaide, M.-G. Mattei, I. Marics, F. Raybaud, J. Planche, O. De Lapeyriere, and D. Birnbaum, *Oncogene* **2**, 413 (1988).
[11] X. Zhan, A. Culpepper, M. Reddy, J. Loveless, and M. Goldfarb, *Oncogene* **1**, 369 (1987).
[12] S. H. Cross and P. F. R. Little, *Gene* **49**, 9 (1986).

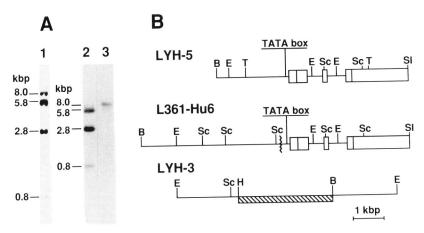

FIG. 2. (A) Southern blot analysis of *Eco*RI-digested DNAs hybridized with the ORF1 probe. Lane 1, human placental genomic DNA; lane 2, LYH-5 containing the *hst-1* gene; lane 3, LYH-3 corresponding to the *hst-2* gene. (B) Partial physical maps of LYH-5, L361-Hu6, and LYH-3. Boxes indicate exons of *hst-1* on LYH-5 and L361-Hu6, with the coding sequences (ORF1) stippled. The wavy line on the L361-Hu6 map indicates the rearrangement site in the genome of the NIH 3T3 transformant, T361-2nd-1. On LYH-3, a region which cross-hybridizes to ORF1 is hatched. B, *Bam*HI; E, *Eco*RI; T, *Taq*I; Sc, *Sac*I; SI, *Sal*I; H, *Hin*dIII. The *Taq*I sites are indicated only on LYH-5.

morphologically in an ordinary focus assay, we found recently[13] that they induce distinctive foci of transfected NIH 3T3 cells in a defined medium lacking platelet-derived growth factor (PDGF) and FGF.[14] Another group of investigators cloned a gene, designated FGF6, which has transforming and tumorigenic activities.[15] The gene was identified by cross-hybridization with the *hst-1* gene, and a partial nucleotide sequence revealed that FGF6 is identical to *hst-2*.[16]

Sequence analysis of the *hst-1* genomic fragment showed that the coding sequence was completely identical to the ORF1 of the cDNA clone from the T361-2nd-1 transformant.[3] Thus, it was suggested that the mechanism of activation of the transforming potential of *hst-1* is its transcriptional deregulation rather than a structural aberration. As described later, expression of *hst-1* seems to be tightly suppressed in normal adult cells. One may speculate that such transcriptional silence is main-

[13] H. Sakamoto, unpublished results (1990).
[14] X. Zhan and M. Goldfarb, *Mol. Cell. Biol.* **6,** 3541 (1986).
[15] I. Marics, J. Adelaide, F. Raybaud, M.-G. Mattei, F. Coulier, J. Planche, O. de Lapeyriere, and D. Birnbaum, *Oncogene* **4,** 353 (1989).
[16] K. Naito and M. Terada, unpublished data (1989).

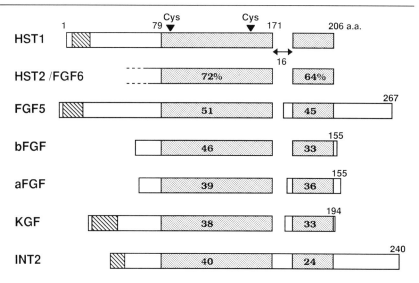

FIG. 3. Heparin-binding growth factor family. Amino acid sequences of currently known members of the family were aligned to that of the *hst-1* protein. Gaps were introduced to achieve the best alignment. Hatched boxes indicate putative signal peptides, and two cysteine residues conserved among the family members are indicated by arrowheads. Stippled areas have significant sequence similarity with the *hst-1* protein as shown by percentage identity. Amino-terminal sequencing is not yet completed for the *hst-2*/FGF6 protein.

tained by gene methylation and/or by some flanking silencer sequence, both of which can be invalidated by transfection or gene cloning. This probably explains why *hst-1* is so frequently found as a transforming gene in a variety of sources of human DNA.

Homology with Fibroblast Growth Factors and Related Molecules

Upon realizing that the amino acid sequence decoded from ORF1 represents an authentic human *hst-1* protein, we proceeded to search for homology among DNA and protein databases. As shown in Fig. 3, significant homology was noted with acidic and basic fibroblast growth factors (aFGF and bFGF, respectively), the *hst-2*/FGF6 protein, the *int-2* protein, FGF5, and keratinocyte growth factor (KGF). These constitute a potentially large and growing family of growth factors and oncogene products, the heparin-binding growth factor (HBGF) family.

Fibroblast growth factors are potent and ubiquitous mitogens with diverse functions in a variety of target cells (reviewed in Ref. 17). They are known as angiogenic factors *in vivo* and *in vitro*. The factors also regulate the differentiation pathway of several cells, and bFGF acts as a mesoderm-inducing factor in *Xenopus* embryos.[18] *int-2* was initially identified as one of the cellular genes activated transcriptionally by proviral insertion in mouse mammary tumor virus (MMTV)-induced breast cancers in mice.[19] Aside from these murine mammary tumors, expression of *int-2* has been detected only in embryos and in some teratocarcinomas.[20] FGF5 was identified by its ability to support growth of the transfected NIH 3T3 cells in a serum-free defined medium[11] and was found to be expressed in neonatal brain.[21] KGF is a growth factor expressed in fetal and adult fibroblasts and is considered to be a mitogen specific for epithelial cells.[22]

Synthesis of Recombinant *hst-1* Protein

Baculovirus and Silkworm Cells

One of the potential advantages of the use of cells from multicellular organisms such as *Bombyx mori* (silkworm) over prokaryotes to synthesize human proteins is that we can expect posttranslational modifications to occur in a way more akin to human cells, such as glycosylation and recognition and cleavage of a signal peptide. Previous successes with the synthesis of α-interferon and interleukin 3 showed that biologically active materials were secreted into the culture medium of silkworm cells.[23,24] This also facilitates recovery and purification of the protein. Protein produced in *Escherichia coli,* on the other hand, is in many cases resistant to solubilization. The disadvantage of the baculovirus system is that the yield is lower and the procedure is more demanding than in the *E. coli* systems.

[17] D. Gospodarowicz, this series, Vol. 147, p. 106.
[18] J. M. W. Slack, B. G. Darlington, J. K. Heath, and S. F. Godsave, *Nature (London)* **326**, 197 (1987).
[19] G. Peters, S. Brookes, R. Smith, and C. Dickson, *Cell (Cambridge, Mass.)* **33**, 369 (1983).
[20] A. Jakobovits, G. M. Shackleford, H. E. Varmus, and G. R. Martin, *Proc. Natl. Acad. Sci. U.S.A.* **83**, 7806 (1986).
[21] X. Zhan, B. Bates, X. Hu, and M. Goldfarb, *Mol. Cell. Biol.* **8**, 3487 (1988).
[22] P. W. Finch, J. S. Rubin, T. Miki, D. Ron, and S. A. Aaronson, *Science* **245**, 752 (1989).
[23] S. Maeda, T. Kawai, M. Obinata, H. Fujiwara, T. Horiuchi, Y. Saeki, Y. Sato, and M. Furusawa, *Nature (London)* **315**, 592 (1985).
[24] A. Miajima, J. Schreurs, K. Otsu, A. Kondo, K. Arai, and S. Maeda, *Gene* **58**, 273 (1987).

Generation of Recombinant Baculovirus

ORF1 with minimal flanking cDNA sequences (nucleotides 1 to 916 of the cDNA[2]) is cloned into an EcoRI site of a baculovirus transfer vector, pBM030,[25] a gift from Dr. M. Furusawa (Daiichi Seiyaku Research Institute, Tokyo). The construct is placed the coding sequence of hst-1 under the strong promoter of a nuclear inclusion body polyhedrin gene of the B. mori nuclear polyhedrosis virus (BmNPV).[23] The vector is designed to express the insert as a nonfusion protein. Twenty-five micrograms of the plasmid is cotransfected with 5 μg of BmNPV DNA into 5×10^5 of silkworm-derived BmN cells in a 25-cm^2 flask (Corning Glass Works, Corning, NY) by a standard calcium phosphate-mediated DNA transfection procedure. The viral DNA for cotransfection is prepared as described.[26] Recombinant virus harboring hst-1 is generated by homologous recombination between the transfer vector and the wild-type BmNPV genome in the transfected BmN cells.

The recipient BmN cells were provided by Dr. S. Maeda (Tottori University, Japan) and are grown in TC-10 medium[27] (Table I) supplemented with heat-inactivated 10% fetal calf serum (FCS, Böhringer, Mannheim, Germany) in a 27° air incubator with a tightly closed Corning flask cap. The medium is replaced every 4–5 days, and confluent cultures are transferred by gentle scraping with a silicon rubber policeman at a dilution ratio of 1:3. Freeze–thaw cycles of the BmN cells for storage are not recommended.

The medium is replaced with 5 ml of fresh TC-10 on the next day of the transfection. A week later, the medium is harvested, and the recombinant virus in it is cloned by limiting dilution as follows: 200 μl of the 10^{-5}, 10^{-6}, and 10^{-7} dilutions of the virus-containing medium are mixed with 100 to 200 fresh BmN cells and plated on a 96-well dish. Infected cells show a bizarre irregular shape after 1 week and are easily recognized among the round, smooth-surfaced uninfected cells under a phase-contrast microscope. In sharp contrast to cells infected with the wild-type virus, cells infected with the recombinant virus are identified by their lack of the characteristic massive nuclear inclusion body, polyhedrin, which has a different refractile index from normal cellular components. One more round of the limiting dilution is performed to obtain purified recombinant virus stocks, each derived from single clones. Ten to 20 μl of the purified

---

[25] Y. Marumoto, Y. Sato, H. Fujiwara, K. Sakano, Y. Saeki, M. Agata, M. Furusawa, and S. Maeda, J. Gen. Virol. **68,** 2599 (1987).
[26] S. Maeda, in "Invertebrate Cell System and Applications" (J. Mitsuhashi, ed.), p. 167. CRC Press, Boca Raton, Florida, 1989.
[27] G. R. Gardiner and H. Stockdale, J. Invertebr. Pathol. **25,** 363 (1975).

TABLE I
PREPARATION OF TC-10 MEDIUM[a]

Component	Amount	Component	Amount
NaCl	0.5 g	10× Soluble amino acids[b]	100 ml
KCl	2.87 g	10× Insoluble amino acids[c]	100 ml
$CaCl_2 \cdot 2H_2O$	1.32 g	1000× Vitamins[d]	1 ml
$MgCl_2 \cdot 6H_2O$	2.28 g	$NaH_2PO_4 \cdot 2H_2O$	0.89 g
$MgSO_4 \cdot 7H_2O$	2.78 g	$NaHCO_3$	0.35 g
Tryptose	2.0 g	Kanamycin	60 mg
Glucose	1.1 g	Distilled water	to 900 ml
L-Glutamine	0.3 g		

[a] This preparation yields 1000 ml of TC-10 medium (pH 6.30–6.35, 315 mOsm). Filter-sterilize over Millipak 20 (0.22 μm, Millipore, Bedford, MA) and add 100 ml of heat-inactivated FCS before use.

[b] 10× Soluble amino acids (1000 ml):

L-Arginine-HCl	7.00 g	L-Isoleucine	0.50 g
L-Aspartic acid	3.50 g	L-Leucine	0.75 g
L-Asparagine	3.50 g	L-Lysine-HCl	6.25 g
L-Alanine	2.25 g	L-Methionine	0.50 g
β-Alanine	2.00 g	L-Proline	3.50 g
L-Glutamic acid	6.00 g	L-Phenylalanine	1.50 g
L-Glutamine	3.00 g	DL-Serine	11.00 g
Glycine	6.50 g	L-Threonine	1.75 g
L-Histidine	25.00 g	L-Valine	1.00 g

[c] 10× Insoluble amino acids (1000 ml): L-cysteine, 0.25 g; L-tryptophan, 1.00 g; L-tyrosine, 0.50 g. Dissolve in 500 ml of 40 mM HCl by heating and then increase to 1000 ml with water.

[d] 1000× Vitamins (100 ml):

Thiamin-HCl	2.0 mg	Folic acid	2.0 mg
Riboflavin	2.0 mg	Nicotinic acid	2.0 mg
Calcium D-pantothenate	2.0 mg	myo-Inositol	2.0 mg
Pyridoxine-HCl	2.0 mg	Biotin	1.0 mg
p-Aminobenzoic acid	2.0 mg	Choline chloride	20.0 mg

virus stock is then used to infect approximately 10^5 *Bm*N cells in a Corning 75-cm^2 flask for the amplification of the virus stock.

An aliquot of the amplified virus stock is used to infect a total of 2×10^8 *Bm*N cells in 20 Corning 150-cm^2 flasks with 30 ml per flask of TC-10 medium with 10% FCS. The amount of the virus stock used for infection is titrated, so that the yield of the protein attains a maximum on the fourth day of infection. Then the medium is harvested, centrifuged at

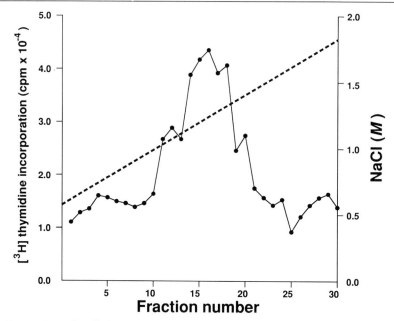

FIG. 4. Heparin affinity chromatography of the concentrated medium of BmN cells infected with the recombinant baculovirus. The dashed line represents the NaCl gradient, and circles indicate the mitogenic activity of each fraction as determined by a [^{3}H]thymidine incorporation assay with NIH 3T3 cells.

1000 rpm for 5 min to remove cells, and concentrated to around 60 ml by an Amicon (Danvers, MA) YM5 membrane.

Purification of Recombinant hst-1 Protein

The concentrated medium is loaded onto an Affi-Gel heparin (Bio-Rad, Richmond, CA) column (1.6 × 5 cm), which is preequilibrated at 4° by a buffer consisting of 10 mM Tris-Cl (pH 7.0)/0.6 M NaCl and washed in the same buffer. The *hst-1* protein is eluted with 60 ml of a 0.6–2.0 M linear gradient of NaCl in 10 mM Tris-Cl (pH 7.0) at a flow rate of 30 ml/hr at 4°. Two-milliliter fractions are collected and stored at −70°. One microliter each of the fractions diluted 50-fold in phosphate-buffered saline (PBS)/0.1% bovine serum albumin (BSA) is assayed by [^{3}H]thymidine incorporation using NIH 3T3 cells (see below). Typically, the biologically active *hst-1* fractions are eluted at 1.0–1.2 M NaCl (Fig. 4) and are concentrated to about 100 μl by a Centricon-10 apparatus (Amicon).

The second step of the purification is reversed-phase high-performance liquid chromatography (HPLC). A 0.46 × 25 cm Vydac C_4 column with

5 μl particle size and 300 Å pore size (The Separation Group, Hesperia, CA) is equilibrated with 0.1% trifluoroacetic acid (TFA) and 30% acetonitrile on a Tosoh CCPM 8000 gradient liquid chromatography system. The sample is eluted by a 1.0 ml/min gradient of 30–60% acetonitrile in 0.1% TFA for 90 min at room temperature. Fractions of 1.0 ml are collected in siliconized Eppendorf tubes and stored at $-70°$. One-microliter aliquots of each fraction diluted 1:10 in PBS/0.1% BSA are analyzed in the [^{3}H]thymidine incorporation assay. The HPLC-purified, biologically active *hst-1* protein is about 18 kilodaltons (kDa) in size on SDS–PAGE. The final yield of the purified *hst-1* protein is approximately 5 μg. Amino acid sequence analysis reveals cleavage of the N-terminal 58 amino acids containing the signal peptide.[28]

Growth Factor Activity of Recombinant *hst-1* Protein

Stimulation of DNA Synthesis in NIH 3T3 Cells

The [^{3}H]thymidine incorporation assay is performed essentially as described,[29] and the conditions for *hst-1* are summarized here briefly. NIH 3T3 cells are cloned by limiting dilution to select those with good contact inhibition, well-aligned flat morphology, and high efficiency of transformation by transfected *hst-1* expression vectors. The selected clones are expanded, stored in liquid nitrogen, and, once thawed, discarded after 1 month of culture. Approximately 1×10^4 cells in 0.5 ml of Dulbecco's modified Eagle's medium (DMEM) with 10% calf serum are incubated for 24 hr, then the medium is changed to DMEM with 0.5% calf serum and incubated for another 24–72 hr. As the pH of the HPLC eluent in acetonitrile/TFA is very low, test samples should be diluted at least 4-fold by PBS before addition to the medium. Cells are then incubated for 16 hr, and [^{3}H]thymidine is added to final concentration of 18.5 kBq (0.5 μCi)/ml. Eight hours later, the trichloroacetic acid-precipitable radioactivity is measured.

As shown in Fig. 5, the purified *hst-1* protein shows a potent mitogenic effect on NIH 3T3 cells, with half-maximal stimulation observed at 220 pg/ml in the absence of heparin and at 80 pg/ml in the presence of heparin. The potency in the presence of heparin is comparable to that of bovine pituitary bFGF (Takara-Shuzo, Kyoto, Japan). The morphological changes of NIH 3T3 cells in response to the *hst-1* protein are characterized by a highly refractile, criss-cross appearance with long process formation.

[28] K. Miyagawa, unpublished observations, 1989.
[29] K. A. Thomas, this series, Vol. 147, p. 120.

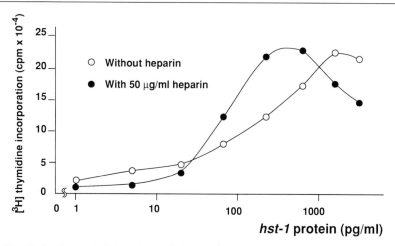

FIG. 5. hst-1 protein-induced stimulation of [^{3}H]thymidine incorporation by NIH 3T3 cells. The mitogenic activity was measured in the absence and presence of 50 µg/ml of heparin.

The changes appear 12 hr after the addition of the protein, reached maximum after 24–48 hr and reverted after 72 hr.

Stimulation of HUVE Cell Growth

Human umbilical vein endothelial (HUVE) cells are isolated as described[30] and cultured in MCDB107 medium (Kyokuto, Tokyo, Japan) with 15% heat-inactivated fetal calf serum (FCS), 2.5 ng/ml of aFGF and 5 µ/ml of heparin (Sigma, St. Louis, MO) on fibronectin-coated flasks. HUVE cells are seeded onto 12-well plates at a density of 2 × 10^3 cells/cm^2 in 2 ml of medium per well. Twenty-four hours later, the medium is replaced by MCDB107 with 15% FCS and various concentrations of test samples. After 48 hr, the treatment is repeated once more, and cells are incubated for another 48 hr. The cell number in each well is determined with a Coulter particle counter after trypsinization. Half-maximal stimulation of HUVE cell growth is observed at about 30 pg/ml of *hst-1* protein.

Anchorage-Independent Growth of NIH 3T3 Cells

A soft agar assay is done as described,[31] using 5 × 10^3 NIH 3T3 cells in the overlayer with 0.4% Bacto-agar (Difco, Detroit, MI) poured over the base layer of 0.9% agar, both in DMEM with 5% calf serum. The

[30] T. Imamura and Y. Mitsui, *Exp. Cell Res.* **172**, 92 (1987).
[31] A. Rizzino, this series, Vol. 146, p. 341.

number of colonies larger than 2000 μm^2 is counted after 2 weeks of incubation at 37° in 5% CO_2. The *hst-1* protein induces anchorage-independent growth in a dose-dependent manner. An approximately 10-fold higher concentration of the protein is required to support soft agar colony formation of NIH 3T3 cells as compared to stimulation of DNA synthesis of the same cells.

These data on the functions of the recombinant protein prove that *hst-1* actually encodes a heparin-binding growth factor, and, as for FGFs, angiogenic activity is also expected for the *hst-1* protein. Recently, we observed a potent *in vivo* angiogenic activity of the recombinant *hst-1* protein using the chorioallantoic membrane (CAM) assay and the rat cornea assay.[32]

Expression of *hst-1*

Expression of hst-1 in Embryos and in Germ Cell Tumors

Northern blot analysis did not detect the *hst-1* message in about 80 cancerous and noncancerous cells and tissues.[33] These samples included gastric cancers and Kaposi's sarcomas, in which types of cancers the transforming *hst-1* gene was identified previously by transfection assays. To the best of our knowledge, the expression of *hst-1* is confined to embryos and to some germ cell tumors (Fig. 6). The sizes of the *hst-1* transcripts are 3.0 and 1.7 kilobases (kb) for human, and 3.0 kb for mouse.

Differential Expression of hst-1 and int-2

The embryo-specific pattern of expression is also noted for *int-2*,[20] but *hst-1* and *int-2* are regulated in different ways. *In vitro* induction of differentiation of a mouse teratocarcinoma cell line, F9, is thought to simulate early stages of embryonic differentiation.[20] F9 stem cells are grown on ungelatinized tissue culture dishes in DMEM supplemented with 15% heat-inactivated FCS. The cells should be transferred before they get to confluence and discarded 1 month after thawing out from liquid nitrogen storage. Retinoic acid (Sigma, type XX) and dibutyryl cyclic AMP (Böhringer) are added to final concentrations of 10^{-7} and 10^{-3} M, respectively, to 5×10^5 F9 cells carefully dispersed in a 100-mm dish (Falcon 3003) with 12 ml of the medium. The medium is renewed every 48 hr, and, after 5

---

[32] T. Yoshida, unpublished observations, 1990.
[33] T. Yoshida, M. Tsutsumi, H. Sakamoto, K. Miyagawa, S. Teshima, T. Sugimura, and M. Terada, *Biochem. Biophys. Res. Commun.* **155**, 1324 (1988).

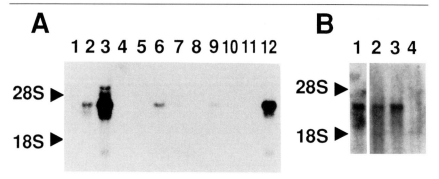

FIG. 6. (A) Northern blot analysis showing expression of *hst-1* in human germ cell tumors. Lanes 1 to 9 contained RNA from surgical specimens of testicular germ cell tumors, lanes 10 and 11 RNA from noncancerous portions of testes, and lane 12 RNA from an immature teratoma cell line, NCC-IT. Three micrograms of poly(A)$^+$ RNA was hybridized with the 3' one-third of the ORF1 probe. The positions of the 28 and 18 S rRNAs are indicated by arrowheads. (B) Expression of *hst-1* in mouse embryos. Lane 1, day 11 embryos; lane 2, heads of day 14 embryos; lane 3, bodies of day 14 embryos with livers removed; lane 4, livers of day 14 embryos. Five micrograms of poly(A)$^+$ RNAs was hybridized with probe M1.8, a mouse genomic fragment of an *hst-1*-containing exon.

days, the cells are entirely differentiated to parietal endodermal cells, which are harvested for RNA extraction.[34]

hst-1 is preferentially expressed in F9 stem cells and dramatically downregulated upon induction of parietal endodermal differentiation; in contrast, transcription of *int-2* is very low in undifferentiated F9 cells, whereas it markedly increases after differentiation. We surmise that these related oncogenes are functionally coupled and convey distinct signals during embryogenesis.

Chromosomal Localization of *hst-1* Gene

Human *hst-1* and *int-2* were mapped on the same chromosome band, 11q13.3,[35] and cosmid mapping showed[36] that they are only 35 kbp apart, with the same transcriptional orientation (Fig. 7). This is also the case with the mouse genome, in which the two related oncogenes are within 17

---

[34] T. Yoshida, H. Muramatsu, T. Muramatsu, H. Sakamoto, O. Katoh, T. Sugimura, and M. Terada, *Biochem. Biophys. Res. Commun.* **157**, 618 (1988).

[35] M. C. Yoshida, M. Wada, H. Satoh, T. Yoshida, H. Sakamoto, K. Miyagawa, J. Yokota, T. Koda, M. Kakinuma, T. Sugimura, and M. Terada, *Proc. Natl. Acad. Sci. U.S.A.* **85**, 4861 (1988).

[36] A. Wada, H. Sakamoto, O. Katoh, T. Yoshida, J. Yokota, P. F. R. Little, T. Sugimura, and M. Terada, *Biochem. Biophys. Res. Commun.* **157**, 828 (1988).

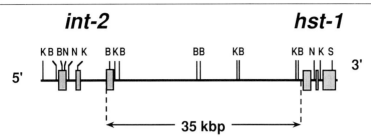

FIG. 7. Physical map of the human *int-2–hst-1* region at 11q13.3. Exons are indicated by boxes. K, *Kpn*I; B, *Bam*HI; N, *Not*I; S, *Sal*I.

kbp of one another on chromosome 7.[37] Their closeness on the genomes prompted the following two observations: first, *hst-1* was also found to be activated transcriptionally in some MMTV-induced murine mammary tumors with or without *int-2* activation.[37] Second, the genes are coamplified in some human cancers. The incidence of coamplification is relatively high in breast cancers (12–22%)[38,39] and very high in esophageal cancers (47% in primary tumors and 100% in metastatic foci).[40] When the chromosomal localization of the *hst-1* gene was determined, the committee on human gene nomenclature registered this gene officially as *HSTF1* (heparin-binding secretory transforming factor 1).[35]

Perspectives

The HBGF family is a large family, probably larger than currently known, encompassing numerous members with diverse functions and patterns of expression. Among the members of this family, *hst-1* and *int-2* are unique and especially interesting in that their expressions are tightly regulated, and specific roles in embryogenesis are expected. The receptors for HBGFs may also constitute a family, as genes homologous but not identical to the recently isolated bFGF receptor are being cloned in several

---

[37] G. Peters, S. Brookes, R. Smith, M. Placzek, and C. Dickson, *Proc. Natl. Acad. Sci. U.S.A.* **86,** 5678 (1989).

[38] H. Tsuda, S. Hirohashi, Y. Shimosato, T. Hirota, S. Tsugane, H. Yamamoto, N. Miyajima, K. Toyoshima, T. Yamamoto, J. Yokota, T. Yoshida, H. Sakamoto, M. Terada, and T. Sugimura, *Cancer Res.* **49,** 3104 (1989).

[39] J. Adnane, P. Gaudray, M.-P. Simon, J. Simony-Lafontaine, P. Jeanteur, and C. Theillet, *Oncogene* **4,** 1389 (1989).

[40] T. Tsuda, E. Tahara, G. Kajiyama, H. Sakamoto, M. Terada, and T. Sugimura, *Cancer Res.* **49,** 5505 (1989).

laboratories.[41,42] Evidently, one of the most exciting targets of the current biological research is analysis of the HBGF ligand–receptor systems, which may be involved in a number of important biological processes.

[41] P. L. Lee, D. E. Johnson, L. S. Cousens, V. A. Fried, and L. T. Williams, *Science* **245**, 57 (1989).
[42] Y. Hattori and M. Terada, unpublished results, 1990.

[13] Phosphorylation and Identification of Phosphorylated Forms of Basic Fibroblast Growth Factor

By JEAN-JACQUES FEIGE and ANDREW BAIRD

Introduction

Basic fibroblast growth factor (FGF) is a potent mitogen for endothelial cells and for a wide variety of mesoderm- and neuroectoderm-derived cells.[1-4] It has recently been suggested that the tight association ($K_d \sim 10$ nM) of basic FGF with the basement membrane regulates its bioavailability and controls its interaction with its high-affinity (K_d 10–20 pM) receptors.[5-7] It has thus been important to determine whether posttranslational changes in basic FGF participate in the regulation of its activity or modulate its extracellular and intracellular localization. Although many mechanisms might exist, we have studied the potential role of protein phosphorylation. Careful analysis of the primary structure of basic FGF revealed the presence of consensus sequences for the phosphorylation of basic FGF by both protein kinase C and cyclic AMP-dependent protein kinases (Table I).[8,9] Accordingly, we established that basic FGF is a substrate for these kinases and is synthesized as a phosphoprotein by bovine capillary endo-

[1] W. H. Burgess and T. Maciag, *Annu. Rev. Biochem.* **58**, 575 (1989).
[2] D. Gospodarowicz, N. Ferrara, L. Schweigerer, and G. Neufeld, *Endocr. Rev.* **8**, 95 (1987).
[3] A. Baird and P. Böhlen, in "Handbook of Experimental Pharmacology" (M. B. Sporn and A. B. Roberts, eds.), p. 163. Academic Press, New York, 1990.
[4] A. Baird, F. Esch, P. Mormède, N. Ueno, N. Ling, P. Böhlen, S. Y. Ying, W. Wehrenberg, and R. Guillemin, *Recent Prog. Horm. Res.* **42**, 143 (1986).
[5] A. Baird and N. Ling, *Biochem. Biophys. Res. Commun.* **142**, 428 (1987).
[6] J. Folkman and M. Klagsbrun, *Science* **235**, 442 (1987).
[7] A. Baird and P. A. Walicke, *Br. Med. J.* **45**, 438 (1989).
[8] J.-J. Feige and A. Baird, *Proc. Natl. Acad. Sci. U.S.A.* **86**, 3174 (1989).
[9] J. J. Feige, J. D. Bradley, K. Fryburg, J. Farris, L. C. Cousens, P. Barr, and A. Baird, *J. Cell Biol.* **109**, 3105 (1989).

TABLE I
POTENTIAL SITES OF PHOSPHORYLATION OF HUMAN BASIC FIBROBLAST GROWTH FACTOR BY PROTEIN KINASE C AND cAMP-DEPENDENT PROTEIN KINASE[a]

	Protein kinase C	cAMP-dependent protein kinase
Consensus	-**Ser/Thr**-X-Lys/Arg-	-Arg/Lys-Arg/Lys-X-**Ser/Thr**- or -Arg/Lys-Arg/Lys-X-Y-**Ser/Thr**- 112
Basic FGF	-Arg-Gly-Val-Val-**Ser**-Ile -Lys-Gly-Val- 64 108 -Asn-Thr-Tyr-Arg-**Ser**-Arg-Lys-Tyr-Thr- 143 -Phe-Leu-Pro-Met-**Ser**-Ala-Lys-Ser	-Thr-Arg-Lys-Tyr-**Thr**-Ser-Trp-Tyr-Val-

[a] Consensus sequences for protein kinase C [J. R. Woodgett, K. L. Gould, and T. Hunter, *Eur. J. Biochem.* **161**, 177 (1986)] and cAMP-dependent protein kinase [E. G. Krebs and J. A. Beavo, *Annu. Rev. Biochem.* **48**, 923 (1979); G. M. Carlson, P. J. Bechtel, and D. J. Graves, in "Advances in Enzymology and Related Areas of Molecular Biology" (A. Meister, ed.), p. 41. Wiley, New York, 1979] and potential sites of phosphorylation present in the human basic FGF molecule are reported. Target amino acids are in bold characters whereas basic amino acids that contribute to the definition of the specific site are underlined. The numbering system corresponds to the 146 amino acid sequence reported by Esch *et al.* [F. Esch, A. Baird, N. Ling, N. Veno, F. Hill, L. Denoroy, R. Klepper, D. Gospodarowicz, P. Böhlen, and R. Guillemin, *Proc. Natl. Acad. Sci. U.S.A.* **82**, 6507 (1985)]. All these sequences are conserved in the bovine basic FGF molecule, except for residue 112 which is a serine instead of a threonine in the human sequence. Serine-64 and threonine-112 are the targets for PK-C and PK-A, respectively, when the entire basic FGF molecule was phosphorylated by the purified kinases.[8,9]

thelial cells and human hepatoma cells in culture.[8] Although the posttranslational modification of basic FGF, which is modulated by components of the extracellular matrix,[9] has no known physiological function, it has been postulated to play a role in the control of basic FGF bioavailability.[7-9] The methods which were used to demonstrate that basic FGF is phosphorylated are the subject of this chapter.

Source of Basic Fibroblast Growth Factor and Protein Kinases

Basic FGF has been purified from a variety of sources, and recombinant forms have now been made available. We use routinely recombinant human basic FGF expressed in yeast or *Escherichia coli*. Stock solutions (10 mg/ml in 100 mM Tris-HCl, pH 7.4, buffer) are stored as aliquots at −80° in Eppendorf microtubes and thawed only once to maintain stability and to avoid adsorption of the growth factor to the plastic tubes. Even then, irreversible polymerization of basic FGF to a 35-kDa species can be observed and must be monitored carefully by SDS–PAGE under reducing conditions. Protein kinase C (PK-C) is purified from bovine brain according to the procedure outlined by Walton *et al.*[10] The catalytic subunit of cyclic AMP-dependent protein kinase (PK-A) is purified from porcine skeletal muscle according to Zoller *et al.*[11] The specific activity of both enzyme preparations is 1000 units/mg protein, where 1 unit is defined as 1 nmol phosphate incorporated per minute into histones H_I (PK-C) or histones H_{IIA} (PK-A).

Phosphorylation of Basic Fibroblast Growth Factor *in Vitro* by Purified Kinases

Principle. The phosphorylation of recombinant human basic FGF by purified PK-C and PK-A is based on the ability of these enzymes to catalyze the transfer of the γ-phosphate of ATP into specific amino acids (serine, threonine) of the substrate (see Table I). The growth factor can thus be radiolabeled with [γ-^{32}P]ATP and then separated from unreacted [γ-^{32}P]ATP by SDS–PAGE or by trichloroacetic acid (TCA) precipitation.

Reagents

10 mM ATP in 10 mM Tris-HCl, pH 7.4
[γ-^{32}P]ATP (3000 Ci/mmol), 10 mCi/ml
5 mM CaCl$_2$

---

[10] G. M. Walton, P. J. Bertics, L. G. Hudson, T. S. Vedvick, and G. N. Gill, *Anal. Biochem.* **161**, 425 (1987).
[11] M. J. Zoller, A. R. Kerlavage, and S. S. Taylor, *J. Biol. Chem.* **254**, 2408 (1979).

3 M MgCl$_2$

160 μg/ml phosphatidylserine and 3.2 μg/ml dioctanoylglycerol in 100 mM Tris-HCl, pH 7.4 [phospholipids are first dissolved in chloroform/methanol under nitrogen and solubilized into 100 mM Tris buffer (pH 7.4) using sonication at 4° for 10 min]

5× sample buffer [10% sodium dodecyl sulfate (SDS), 50% glycerol (v/v), 5% mercaptoethanol (v/v), 0.25% bromphenol blue (w/v), 250 mM Tris, pH 6.8]

50% TCA

1 N NaOH

Using these stock solutions, a radioactive ATP mix is prepared containing the following: (1) for PK-C, 100 μCi/ml [γ-^{32}P]ATP, 2 × 10^{-5} M ATP, 1.2 mM CaCl$_2$, 20 mM MgCl$_2$; (2) for PK-A, 100 μCi/ml [γ-^{32}P]ATP, 2 × 10^{-5} M ATP, 20 mM MgCl$_2$.

Procedure. When performing the phosphorylation assays, it is important to include the appropriate controls. They include omitting the kinase, its activator (phospholipids for PK-C), or the substrate. The final assay volume is 20 μl and contains 2 μl basic FGF (0.25 mg/ml), 5 μl of the phospholipid mixture, and 3 μl of PK-C (0.02 units) or PK-A (0.06 units). All the reagents are kept on ice while the assay is being set up, and the reaction is started by adding the ATP mix (10 μl), vortexing, and placing the Eppendorf microtube in a 30° water bath. The reaction is stopped by adding of 5 μl of a 5× Laemmli's sample buffer, and the phospho-FGF is analyzed by polyacrylamide gel electrophoresis. After heating at 95° for 3 min, the sample is loaded onto a 0.8-mm thick 0.1% SDS–15% polyacrylamide gel (8 × 10 cm, Idea Scientific, Corvallis, OR), and phosphoproteins are electrophoresed according to Laemmli[12] and visualized by autoradiography of the dried gel. An example of this result is shown in Fig. 1A, where basic FGF phosphorylated by purified PK-C is shown.

Identification of Phosphoamino Acids

Phosphoamino acids are identified in phosphorylated basic FGF using the procedure of Cooper *et al.*[13] Briefly, ^{32}P-labeled basic FGF is extracted from the polyacrylamide gel in 50 mM ammonium bicarbonate (pH 7.3–7.6) supplemented with 0.1% SDS and 1% mercaptoethanol. After precipitation with 50% TCA, the pellet is dissolved in 6 N HCl, and the protein is hydrolyzed for 60 min at 110°. ^{32}P-Labeled amino acids are then

[12] U. K. Laemmli, *Nature (London)* **227,** 680 (1970).
[13] J. A. Cooper, B. M. Sefton, and T. Hunter, this series, Vol. 99, p. 387.

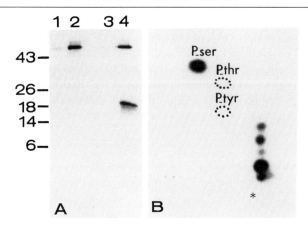

FIG. 1. Detection of phosphorylated human basic FGF. (A) In this example, recombinant human basic FGF was phosphorylated for 10 min by purified PK-C and [γ-^{32}P]ATP as described in the text. The ^{32}P-labeled proteins were separated by SDS-PAGE and visualized by autoradiography. Lane 1, PK-C, no phospholipid, no basic FGF; lane 2, PK-C, phospholipids, no basic FGF; lane 3, PK-C, no phospholipid, basic FGF; lane 4, PK-C, phospholipids, basic FGF. (B) Phosphoamino acid analysis was performed on the 18-kDa band from lane 4 in (A) (corresponding to ^{32}P-labeled basic FGF), which was excised from the gel and hydrolyzed in 6 N HCl as described in the text. The radiolabeled amino acids were mixed with standard phosphoamino acids, analyzed by two-dimensional electrophoresis on cellulose plates, and visualized by autoradiography. The asterisk indicates the position of sample loading. The position of phosphoamino acids is indicated as follows: P-ser, phosphoserine; P-thr, phosphothreonine; P-tyr, phosphotyrosine.

mixed with standard phosphoamino acids and are subsequently separated by two-dimensional electrophoresis on cellulose thin-layer plates. The first dimension is run for 20 min at 1.5 kV in pH 1.9 buffer, and the second dimension is run for 16 min at 1.3 kV in pH 3.5 buffer.[13] Standard phosphoamino acids are revealed by ninhydrin staining and used to identify the radiolabeled amino acids, which are visualized by autoradiography (Fig. 1B).

Kinetic Analyses of Phosphorylation

When the stoichiometry of phosphorylation is to be determined, the reaction volume is scaled up to 200 μl with the final reagent concentrations kept unchanged, except for the concentration of ATP which is raised to 5×10^{-4} M. At appropriate time intervals, 20-μl aliquots are removed from the reaction tube and the radiolabeled basic FGF precipitated by addition of 0.5 ml of 50% TCA. An aliquot of 10 μl of 10 mg/ml bovine

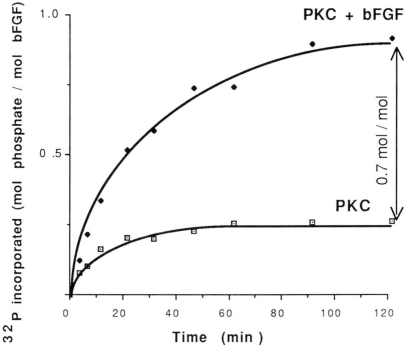

FIG. 2. Stoichiometry of phosphorylation of basic FGF by protein kinase C. Recombinant human basic FGF was phosphorylated by purified PK-C in the presence of phospholipids and [γ-^{32}P]ATP. The total amount of radioactivity incorporated into TCA-precipitable proteins was measured as a function of time as described in the text. A control incubation was done where basic FGF was omitted. The difference between the incorporations measured at the plateau of both these conditions (represented by the arrow) represents the amount of phosphate specifically incorporated into the basic FGF molecule.

serum albumin (BSA) is added as a carrier protein, and the tubes are allowed to stand in an ice bucket until the end of the assay. All precipitates are then centrifuged for 10 min in an Eppendorf microfuge. The pellets are dissolved in 0.1 ml sodium hydroxide and immediately reprecipitated with TCA. Since phosphoserine and phosphothreonine are alkali-labile, it is important to proceed quickly after the first solubilization of the pellet. After the second precipitation, the pellet of radiolabeled proteins is dissolved in sodium hydroxide and counted in a β-counter after addition of scintillation fluid. The amount of phosphate incorporated per mole of basic FGF can be estimated from the amount of radioactivity incorporated into basic FGF at the plateau of the kinetic reaction (90–120 min, as shown on Fig. 2). Because the kinases will also autophosphorylate during the reaction, it is important to run a parallel experiment where basic FGF is omitted and to subtract the radioactivity incorporated into the kinase alone from the total amount of radioactivity incorporated (Fig. 2).

Maximal incorporation of 0.7 mol phosphate/mol basic FGF was routinely obtained with PK-C (Fig. 2) and of 0.8 mol phosphate/mol basic FGF with PK-A. Taking recovery into account, this result is in agreement with the observation that a single residue in the basic FGF molecule is modified by each of these kinases.[9] In our experience, previous treatment of recombinant basic FGF with alkaline phosphatase has little effect on the amount of phosphate that can be incorporated into the molecule. By inference then, over 90% of recombinant basic FGF expressed by yeast is the dephospho form.

Phosphorylation of Basic Fibroblast Growth Factor by Cells in Culture

Phosphorylated forms of basic FGF have been detected in two cell lines: the human hepatoma cells SK-HEP and the bovine adrenocortical capillary endothelial cells (ACE).[8] The protocol that we describe in this chapter can be adapted to other cell types as well.

Reagents

Ortho[^{32}P]phosphate

Phosphate-free Dulbecco's modified Eagle's medium (DMEM) supplemented with 2% dialyzed calf serum and 5% normal DMEM (labeling buffer)

1% Nonidet P-40, 1% sodium deoxycholate, 0.1% SDS, 0.1 M NaCl, 50 mM NaF, 10 μg/ml aprotinin (RIPA buffer)

Rabbit polyclonal antiserum 773 (an antiserum raised against the peptide basic FGF(1–24) coupled to bovine serum albumin (Ab 773)

Recombinant human basic FGF

Phosphate-buffered saline (PBS)

Protein A-Sepharose, 50% (v/v) in RIPA buffer above

2% SDS, 10% glycerol, 1% β-mercaptoethanol, 0.05% bromphenol blue, 50 mM Tris, pH 6.8 (sample buffer)

Procedure. Cells are plated at a density of 6 × 10^6 cells/10-cm diameter plate and grown for 24–48 hr in standard serum-supplemented medium. We routinely use three 10-cm plates per labeling, and, before metabolic labeling, the plates are rinsed twice with phosphate-free labeling medium. After addition of 4 ml labeling medium containing 0.4 mCi/ml ortho[^{32}P]phosphate, the cells are incubated overnight at 37° in a CO_2 incubator. At the end of the labeling period, the cells are washed with PBS. It is important to collect and dispose of this highly radioactive solution using appropriate caution. The cells are then scraped with a rubber policeman into 0.5 ml RIPA buffer, collected in a screw-capped Eppendorf microtube (1.5 ml), kept on ice for 30 min, and subsequently centrifuged for 10 min in an Eppendorf microcentrifuge. ^{32}P-Labeled basic FGF is immunoprecipitated from this supernatant. Accordingly, all operations must be carried out behind protective plexiglass shielding (2 cm thick).

For the immunoprecipitation, antibasic FGF is added to the extract at an optimal concentration, in this instance at a final dilution of 1/200. Bovine serum albumin (0.5 mg/ml final concentration) is also added to the extracts in order to block nonspecific antibodies present in the rabbit antiserum. After a 2-hr incubation at 4°, 50 μl of a protein-A Sepharose suspension is added to the extracts and the incubation continued for an additional 1 hr at 4° under vigorous rotary shaking. The solution is then centrifuged for 5 min in an Eppendorf microfuge, resuspended in 0.5 ml of RIPA buffer, and washed 3 times in RIPA buffer and twice in 0.1% SDS. The final pellet is resuspended in 50 μl Laemmli[12] sample buffer, heated for 3 min at 95°, and centrifuged, after which the supernatant is analyzed by SDS-PAGE on 0.1% SDS-15% polyacrylamide gels. After fixation, the gel is dried, and the radiolabeled, immunoprecipitated proteins are visualized by autoradiography.

Remarks. In our experience, no phosphorylated forms of basic FGF are recovered from the conditioned culture medium, and phosphorylated basic FGF is always found to be cell-associated. The antibody also precipitates several phosphoproteins, but the one corresponding to phosphobasic FGF is easily identified by its absence when the precipitation is performed in the presence of an excess of cold basic FGF.[8]

Biological Assays of Phosphorylated Forms of Basic Fibroblast Growth Factor

The biological activity of phosphorylated basic FGF can be determined using the same conditions described for the purified mitogen.[14-18] It is important, however, to have control preparations of basic FGF that have undergone the phosphorylation reaction without being phosphorylated

---

[14] M. Klagsbrun, R. Sullivan, S. Smith, R. Rybka, and Y. Shing, this series, Vol. 147, p. 95.
[15] D. Gospodarowicz, this series, Vol. 147, p. 106.
[16] F. Esch, A. Baird, N. Ling, N. Ueno, F. Hill, L. Denoroy, R. Klepper, D. Gospodarowicz, P. Böhlen, and R. Guillemin, *Proc. Natl. Acad. Sci. U.S.A.* **82**, 6507 (1985).
[17] J. J. Feige and A. Baird, *J. Biol. Chem.* **263**, 14023 (1988).
[18] P. A. Walicke, J.-J. Feige, and A. Baird, *J. Biol. Chem.* **264**, 4120 (1989).

(i.e., with no enzyme or ATP). We describe here two assays among the many that are possible.

Heparin-Binding Assay

The following assay is based on the strong affinity of basic FGF for heparin.[14] The growth factor elutes from heparin-Sepharose columns with 1.4 and 1.8 M NaCl. The affinity of phosphorylated forms of basic FGF can be verified by loading the radiolabeled sample onto a 0.5-ml heparin-Sepharose column and then eluting the column with increasing concentrations of NaCl (2 ml of 20 mM Tris-HCl, pH 7.5, containing 0, 0.15, 0.6, 1.2, 1.4, 1.6, 1.8, 2.0, and 3.0 of NaCl, respectively). Aliquots of the radiolabeled reaction mixture and of each elution step are then analyzed by SDS–PAGE on 0.1% SDS–15% polyacrylamide gels. Autoradiography of the gel identifies the salt concentration at which ^{32}P-labeled basic FGF elutes from the heparin-Sepharose column.

Radioreceptor Assay on BHK Cells

There exist at least two binding sites for basic FGF on the surface of target cells. One site (K_d 10^{-10} M) corresponds to cell surface-associated heparan sulfate proteoglycans, and a second, higher affinity site (K_d 10^{-11} M) corresponds to the transmembrane receptor.[19] Binding of basic FGF to heparan sulfate–proteoglycans is sensitive to high salt concentrations, whereas binding to the high affinity receptor is not. The radioreceptor assay for basic FGF is based on these observations. It is a competition assay that is routinely performed on baby hamster kidney (BHK) fibroblasts, which have a relatively high number of high-affinity receptors (20,000–40,000 receptor/cell) compared to many other cell types.[18,19]

Reagents

BHK cells plated in 24-well plates and grown to subconfluence in Ham's F12/DMEM (1 : 1) supplemented with 5% calf serum

Binding medium: DMEM with 25 mM HEPES, 0.15% gelatin, pH 7.5

Phosphate-buffered saline (PBS)

20 mM HEPES, pH 7.5, 2 M NaCl

Lysis solution: 0.1 M sodium phosphate buffer, pH 8.1, 0.5% Triton X-100

^{125}I-Labeled basic FGF [100,000 counts/min (cpm)/ng] radioiodinated by the lactoperoxidase method,[20] purified by heparin-Sepharose chromatography, and stored at 4°

---

[19] D. Moscatelli, *J. Cell. Physiol.* **131**, 123 (1987).
[20] D. Schubert, N. Ling, and A. Baird, *J. Cell. Biol.* **104**, 635 (1987).

Phospho-basic FGF, prepared as described above except that no [γ-^{32}P]ATP is added to the reaction mixture; it is phosphorylated either by PK-C or by PK-A (for controls, it is important to prepare "unphospho-FGF" under identical conditions, but in the absence of enzyme or ATP)

Procedure. BHK cells are rinsed once with 0.5 ml binding buffer and incubated for 2 hr at 4° with 0.3 ml binding medium, containing $1-2 \times 10^5$ cpm ^{125}I-labeled basic FGF and various concentrations of phosphorylated or unphospho-basic FGF. At the end of the incubation, cells are washed twice with PBS and twice with 2 M NaCl in 20 mM HEPES, pH 7.5. The salt washes are collected, counted, and used to measure ^{125}I-labeled basic FGF bound to glycosaminoglycans. The ^{125}I-labeled basic FGF that is bound to the high-affinity receptor is then collected by solubilization of the cells in the lysis buffer and counted for radioactivity on a γ-counter. We routinely observe that PK-A-phosphorylated basic FGF is more potent than unphosphorylated basic FGF at displacing the binding of ^{125}I-labeled basic FGF to its high-affinity receptor. In contrast, PK-C-phosphorylated basic FGF is equipotent.[8,9] No difference has been observed between these different forms in the binding to cell surface heparan sulfate proteoglycans (salt washes).[8]

Summary and Perspectives

Phosphorylation of secreted factors by intact cells is a recent observation which has been reported for atrial natriuretic factor[21] and acidic[22] and basic FGF.[8,9] Several growth factors have been reported to be substrates for purified protein kinases.[8,23] Thus, the techniques described in this chapter may be successfully adapted to other growth factors and should lead to a better understanding of the possible role of phosphorylation in the regulation of their biological activity.

Acknowledgments

We are grateful for the skillful secretarial work of Sonia Lidy and Denise Higgins. This work was supported by National Institutes of Health Grants HD-09690 and DK-18811, The Whittier Institute Angiogenesis Research Program, the Institut National de la Santé et de la Recherche Médicale (INSERM), and the Commissariat à l'Energie Atomique (DRF-G).

---

[21] J. Rittenhouse, L. Moberly, H. Ahmed, and F. Marcus, *J. Biol. Chem.* **263**, 3778 (1988).
[22] F. Mascarelli, D. Raulais, and Y. Courtois, *EMBO J.* **8**, 420 (1989).
[23] H. F. Kung, I. Calvert, E. Bekesi, F. R. Khan, K. P. Huang, S. Oroszlan, L. E. Henderson, T. D. Copeland, R. C. Sowder, and S. J. Wei, *Mol. Cell. Biochem.* **89**, 29 (1989).

[14] Derivation of Monoclonal Antibody to Basic Fibroblast Growth Factor and Its Application

By YOSHINO YOSHITAKE, KOUICHI MATSUZAKI, and KATSUZO NISHIKAWA

Introduction

Basic fibroblast growth factor (bFGF), an endothelial cell mitogen and angiogenesis factor, has been isolated from a wide variety of tissues with the aid of heparin-Sepharose (HS) chromatography and has been used in studies of its biological role in the proliferation and differentiation of various types of cells.[1,2] bFGF was first reported to be a single-chain polypeptide composed of 146 amino acids,[3] but later forms that were truncated or extended at the amino terminus were also isolated. The primary structure is highly conserved in human, bovine, ovine, and rat bFGF.

Monoclonal antibodies (MAb) are useful in studies of growth factors because they are specific, homogeneous, and easy to produce in large amounts. Neutralizing MAb are powerful tools in studies of the physiological roles of these factors. This chapter describes techniques for preparation of MAb to bFGF, for studies on the properties of the MAb, and for their application in *in vivo* studies concerning the physiological role of bFGF.

Preparation of Bovine Basic Fibroblast Growth Factor

Basic FGF is purified from bovine brain by the method of Gospodarowicz[4] involving ammonium sulfate precipitation, CM-Sephadex chromatography, and a modification of HS chromatography.[5] The protein concentration in the eluate with 1.4 M NaCl/10 mM Tris-HCl (pH 7.2) from the HS column is determined with bovine serum albumin (BSA) (Sigma, St. Louis, MO) as a standard with the Pierce (Rockford, IL) BCA protein

---

[1] D. Gospodarowicz, G. Neufeld, and L. Schweigerer, *J. Cell. Physiol.* **5** (Suppl.), 15 (1987).
[2] J. Folkman and M. Klagsbrun, *Science* **235**, 442 (1987).
[3] F. Esch, A. Baird, N. Ling, N. Ueno, F. Hill, L. Denoroy, R. Klepper, D. Gospodarowicz, P. Böhlen, and R. Guillemin, *Proc. Natl. Acad. Sci. U.S.A.* **82**, 6507 (1985).
[4] D. Gospodarowicz, this series, Vol. 147, p. 106.
[5] K. Nishikawa, Y. Yoshitake, and S. Ikuta, this series, Vol. 146, p. 11.

assay reagent, which gives an accurate determination in the presence of high concentrations of NaCl in accordance with the Pierce Chemical Company research report. This fraction is diluted with 0.1% (w/v) bovine serum albumin (BSA) (Sigma A7511, fatty acid-free) in phosphate-buffered saline (PBS) (BSA/PBS), frozen once for sterilization at $-40°$ for at least 16 hr, and used for assays to stimulate both DNA synthesis in BALB/c 3T3 cells and growth of bovine brain capillary endothelial (BCE) cells and for radioimmunoassay (see below). This fraction is also dialyzed against water, lyophilized, dissolved in PBS, and then used for immunization, ELISA, and radioiodination (see below).

Radioiodination of Basic Fibroblast Growth Factor

Purified bovine bFGF or recombinant bovine bFGF (The Radiochemical Centre, Amersham, UK) is labeled with ^{125}I by the chloramine-T method of Kan et al.[6] with some modifications. Bovine bFGF (2 µg) in 40 µl of PBS, 10 µl of a 50 mCi/ml solution of Na^{125}I, 2 µl of 1% 3-[(3-cholamidopropyl)dimethylammonio]-1-propane sulfonate (CHAPS) solution in water, 37 µl of 0.4 M sodium phosphate buffer (pH 7.4), and 10 µl of 1 mg/ml of chloramine-T solution in water are mixed and incubated for 1 min at room temperature. A volume of 10 µl of the same chloramine-T solution is again added, and the reaction mixture is incubated for another 1 min. The reaction is stopped by adding 100 µl of 20 mM dithiothreitol in 50 mM sodium phosphate buffer (pH 7.4), followed by successive addition of 100 µl of 10 mM NaI in water.

The ^{125}I-labeled bFGF is then separated from free ^{125}I by HS affinity chromatography by the method of Neufeld and Gospodarowicz.[7] The reaction mixture is applied to a column (0.25 ml) equilibrated with 0.3 M NaCl/10 mM Tris-HCl (pH 7.2)/0.1% CHAPS, and the column is washed with the same buffer until the radioactivity of the eluate becomes negligible (~7 ml). ^{125}I-Labeled bFGF is eluted with 2 M NaCl/10 mM Tris-HCl (pH 7.2)/0.1% CHAPS, and the eluate can be stored at $-20°$.

The recovery of ^{125}I in labeled bFGF is about 5–60%. The recovery with recombinant bovine bFGF is much better than that with purified bovine brain bFGF; however, the reason for this discrepancy is unclear. ^{125}I-Labeled bFGF (2 × 10^5 cpm/ng) is used for assay after appropriate dilution with BSA/PBS within 4 weeks. After longer storage, ^{125}I-labeled bFGF must be separated on a HS column from free ^{125}I which is liberated

---

[6] M. Kan, D. DiSorbo, J. Hou, H. Hoshi, P. E. Mansson, and W. L. McKeehan, *J. Biol. Chem.* **263,** 11306 (1988).

[7] G. Neufeld and D. Gospodarowicz, *J. Biol. Chem.* **260,** 13860 (1985).

during storage. ^{125}I-Labeled bFGF and unlabeled bFGF have identical biological activities as judged by stimulation of DNA synthesis in BALB/c 3T3 cells. CHAPS (0.02%) does not interfere with the iodination reaction. The detergent improves recovery during chromatography and storage of ^{125}I-labeled bFGF.[8]

Production of Monoclonal Antibodies[9]

Media and Cell Lines[5]

RPMI 1640 + 10% fetal calf serum (FCS) for maintenance of myeloma and hybridoma cell lines: 90% RPMI 1640 (Flow Lab, McLean, VA) supplemented with 2 g/liter of sodium bicarbonate, 100 U/ml of penicillin, 100 μg/ml of streptomycin, 15 mM N-hydroxyethylpiperazine-N'-2-ethanesulfonic acid buffer (HEPES) (pH 7.3), 2 mM glutamine, 1 mM sodium pyruvate, 2.5 mg/ml of glucose, and 50 μM 2-mercaptoethanol; 10% heat-inactivated fetal calf serum (FCS) (Flow Lab). Before use, sterilized and concentrated stock solutions of glutamine (100×), sodium pyruvate (100×), glucose (200×), and 2-mercaptoethanol (2000×) in water are added to RPMI 1640.

RPMI 1640 + 10% FCS + HAT (hypoxanthine, aminopterin, and thymidine) for selection of hybridoma cells[5]

RPMI 1640 + 10% FCS + HT (hypoxanthine and thymidine) for growth of hybridoma cells[5]

Mouse myeloma P3X63Ag8U.1 (P3U1) cells, kept in RPMI 1640 + 10% FCS and subcultured every 3–4 days

Immunization of Mice

Female BALB/c mice, 6 weeks old, are immunized 5 times at 2- to 4-week intervals by subcutaneous injections with 10 μg of bovine bFGF in Freund's complete or incomplete adjuvant. After the third injection, a drop of blood is taken from the retroorbital venous plexus (see below, Application of Monoclonal Antibodies), and the plasma is separated and tested by ELISA for its ability to bind to bFGF as described below. If a further increase of the antibody titer is not observed after further immunization, the mouse is intraperitoneally injected with 10 μg of bFGF in PBS. Four days later, it is sacrificed, and the spleen is excised.

[8] Y. Matuo, N. Nishi, K. Matsumoto, K. Miyazaki, K. Matsumoto, F. Suzuki, and K. Nishikawa, this volume [48].

[9] K. Matsuzaki, Y. Yoshitake, Y. Matuo, H. Sasaki, and K. Nishikawa, *Proc. Natl. Acad. Sci. U.S.A.* **86,** 9911 (1989).

Generation of Hybridomas and Cloning

Because detailed procedures for cell fusion to generate MAb have been described previously in this series,[10] and brief procedures have been also described for anti-human epidermal growth factor (hEGF) MAb,[5,11] only points of modification are mentioned here. Mixtures of 10^8 spleen cells from an immunized mouse and 5×10^7 P3U1 cells are fused in 0.5 ml of 50% (w/v) polyethylene glycol 4000 (Merck, Darmstadt, Germany) in Dulbecco's modified Eagle's medium (DME) for 90 sec at 37°. The fused cells are washed with DME and then RPMI 1640 + 10% FCS by centrifugation and resuspended in RPMI 1640 + 10% FCS + HAT medium. A suspension of the cells equivalent to 5×10^5 original spleen cells in 0.1 ml of medium is distributed into the wells of two 96-well multiwell plates (Falcon, 0.32 cm^2) in which thymocytes from normal mice (about 3×10^5 cells/well) have been cultured in 0.1 ml of the same medium for 1–2 days as the feeder layer. One-half of the culture medium of each well is replaced by fresh medium every other day. Hybridoma cell growth should be observed about 10 days after cell fusion. A sample of 50 μl of the conditioned medium is taken for assay of antibody activity by ELISA.

ELISA for Basic Fibroblast Growth Factor Antibody[12,13]

1. Wells of a Micro Test III flexible assay plate (Falcon 3912, Becton Dickinson & Co., Lincoln Park, NJ) are coated overnight at 4° with 50 μl of a 2 μg/ml solution of purified bovine bFGF in PBS.

2. The liquid is withdrawn, and the wells are filled with 1% BSA (Sigma fraction V) solution in PBS and left to stand for 1 hr at room temperature.

3. The liquid is again withdrawn, and the wells are washed 3 times with 0.3 ml of 0.1% (w/v) Tween 20/PBS.

4. Samples of 50 μl of conditioned media from the hybridoma cultures or diluted plasma are added to the wells, and the plates are incubated overnight at 4° or for 4 hr at room temperature.

5. The liquid is withdrawn, and the wells are washed 4 times with 0.3 ml of 0.1% Tween 20/PBS.

6. Then 100 μl of horseradish peroxidase-labeled rabbit anti-mouse IgG (Dako, Glostrup, Denmark) diluted 1:2000 in PBS is added, and the plates are incubated for 2 hr at room temperature.

7. The liquid is withdrawn, and the wells are washed 4 times with 0.3 ml of 0.1% Tween 20/PBS.

---

[10] G. Galfrè and C. Milstein, this series, Vol. 73, p. 3.
[11] Y. Yoshitake and K. Nishikawa, *Arch. Biochem. Biophys.* **263**, 437 (1988).
[12] S. L. Massoglia, J. S. Kenney, and D. Gospodarowicz, *J. Cell. Physiol.* **132**, 531 (1987).
[13] J. Y. Douillard and T. Hoffman, this series, Vol. 92, p. 168.

8. A volume of 50 μl of peroxidase–substrate solution [4.9 ml of 0.1 M citric acid, 5.1 ml of 0.2 M Na$_2$HPO$_4$, 10 ml water, 8 mg o-phenylenediamine dihydrochloride, and 4 μl of H$_2$O$_2$ (30%), pH 5.0] is added to the wells, and the plates are incubated for 15 min at room temperature in the dark. The reaction is stopped by adding 50 μl of 2 M H$_2$SO$_4$.

9. The peroxidase reaction is measured by the increase in absorbance at 492 nm in an ELISA reader.

Determination of Immunoglobulin Classes

Immunoglobulin classes and subclasses can be determined by the double-immunodiffusion technique (micro-Ouchterlony method) as described in this series,[5] but it is more convenient to use a mouse monoclonal antibody isotyping kit (Amersham). A typing stick carries goat antibodies specific for the different types of mouse immunoglobulin classes and subclasses. Merely 0.5 ml of conditioned medium of hybridoma culture is sufficient to detect antibodies by peroxidase-labeled anti-mouse antibody.

Purification of Monoclonal Antibodies

Since detailed procedures for purification of mouse MAb have been described elsewhere in this series,[5,14] here the description is restricted to that for MAb of the IgG$_1$ class from ascites fluid by protein G-Sepharose. Ascites fluid (5 ml) is diluted with a equal volume of 20 mM sodium phosphate buffer (pH 7.0) and applied to a column of protein G-Sepharose (gel bed, 3 ml) (Pharmacia LKB Biotechnology, Uppsala, Sweden) equilibrated with 20 mM sodium phosphate buffer (pH 7.0). The column is washed with 70 ml of the same buffer, and IgG$_1$ can be eluted with 10 ml of 10 mM glycine-HCl buffer (pH 2.7). The fractions of 1 ml are collected in tubes which contain 70 μl of 1 M Tris-HCl buffer (pH 9.0) for neutralization. Recovery of IgG$_1$ in fractions 4 and 5 is about 80%. The antibody concentration is conventionally determined by measuring the absorbance at 280 nm, assuming an $A_{1\%}$ value of 14. The column can be regenerated by washing with 5 ml of 10 mM glycine-HCl buffer (pH 2.7) and 30 ml of 20 mM sodium phosphate buffer (pH 7.0).

Properties of Monoclonal Antibodies

Determination of Specificity and Dissociation Constant (Radioimmunoassay)

1. The reaction mixture, assembled in a 12 × 75 mm disposable polystyrene tube, contains 0.35 ml of 0.1 M sodium phosphate buffer (pH 7.4),

[14] P. Parham, this series, Vol. 92, p. 110.

0.05 ml of 1–2 ng/ml of ^{125}I-labeled bFGF [8000–15,000 counts/min (cpm)] in BSA/PBS, 0.05 ml of unlabeled bFGF or acidic FGF in BSA/PBS to give a final concentration of 1–200 ng/ml, and 0.05 ml of 0.12–1.5 μg/ml of purified MAb in BSA/PBS. The mixture is incubated for 16 hr at 4°.

2. Then 0.1 ml of 1% normal mouse serum in PBS and 0.1 ml of 0.77 mg/ml goat anti-mouse immunoglobulins (Dako) in PBS are added to the reaction mixture.

3. The mixture is incubated for 4 hr at 4°.

4. Then 1 ml of 2% (w/v) polyethylene glycol 6000 in water is added to the reaction mixture.

5. The mixture is centrifuged at 1200 g for 30 min at 4°, and the supernatant is discarded.

6. The radioactivity in the precipitate is counted in a γ-counter.

The value for nonspecific binding of radioactivity (~9% of the total tracer added) in the absence of MAb is subtracted. Values of B (specific binding of ^{125}I-labeled bFGF in the presence of various concentrations of unlabeled bFGF) divided by B_0 (specific binding in the absence of unlabeled bFGF), that is, B/B_0 values, are plotted against the logarithm of the concentration of unlabeled bFGF. The concentration of MAb in the reaction mixture is adjusted to give a B_0 value of 30–50% of the total tracer added. The apparent K_d value can be calculated by Scatchard plot analysis of the displacement curves of bFGF.

Effects of Monoclonal Antibodies on Biological Activities of Basic Fibroblast Growth Factor

Addition of bFGF to cultured quiescent BALB/c 3T3 cells or cultured endothelial cells stimulates DNA synthesis or proliferation, respectively. Therefore, it is possible to determine whether the MAb inhibit these biological activities of bFGF. Detailed procedures for assay of DNA synthesis in BALB/c 3T3 cells[5] and for growth experiments with endothelial cells[4,15] have been described in this series. Here, a growth experiment using capillary endothelial cells isolated from bovine brain cortex (BCE cells) is described.

BCE cells are isolated and maintained as described by Goetz *et al.*[16] with minor modifications. The procedure involves dispersion of brain tissue by collagenase–dispase, pelleting clusters of BCE cells by centrifugation in medium containing 25% BSA, and retaining the clusters on a column of glass beads. The modifications are as follows: The cells are

[15] M. Klagsbrun, R. Sullivan, S. Smith, R. Rybka, and Y. Shing, this series, Vol. 147, p. 95.
[16] I. E. Goetz, J. Warren, C. Estrada, E. Roberts, and D. Krause, *In Vitro Cell. Dev. Biol.* **21**, 172 (1985).

cultured in RPMI 1640 medium containing heat-inactivated fetal calf serum (FCS) and 1 ng/ml of bovine bFGF in plates that have been coated with type IV collagen (see below). Because a pure cell population of BCE cells can be obtained by isolating a primary colony formed from a single cluster containing 2–4 cells, this detailed method is also described.

Media and Solutions

RPMI 1640 + 20% FCS + bFGF for primary culture of BCE cells (primary culture medium): 80% RPMI 1640 supplemented with 2 g/liter of sodium bicarbonate, 100 U/ml of penicillin, 100 μg/ml of streptomycin, 2.5 μg/ml of amphotericin B, and 15 mM HEPES (pH 7.3); 20% heat-inactivated FCS; 1 ng/ml of bovine bFGF. An appropriate volume of bovine bFGF solution in BSA/PBS should be added to the medium before use.

RPMI 1640 + 10% FCS + bFGF for maintenance of BCE cells (maintenance medium): 90% of RPMI 1640, but without amphotericin B; 10% heat-inactivated FCS; 1 ng/ml of bovine bFGF

Type IV collagen solution (10 μg/ml) for coating plates: 5 mg of type IV collagen (Sigma) is dissolved in 500 ml of PBS containing 100 U/ml of penicillin and 100 μg/ml of streptomycin by stirring overnight. The solution is sterilized by filtration through a 0.22-μm filter.

Trypsin–EDTA solution: 0.05% trypsin in PBS containing 0.02% EDTA

Isolating Progeny of Single Cluster of BCE Cells

1. Falcon plates (100 mm) are coated with 4 ml of a 10 μg/ml type IV collagen solution for 2 hr at 37°, and the solution is withdrawn.

2. The clusters of BCE cells, derived from about 10 g of brain cortex which are separated from glass beads, are suspended in 10 ml of primary culture medium. A volume of 0.2–0.5 ml of the suspension is added to coated plates containing 10 ml of the same medium. The cells are cultured at 37° in a humidified atmosphere of 5% CO_2 in air.

3. On days 5 to 7, the sites of isolated colonies formed from single clusters are marked on the bottom of the plate; 5 to 10 colonies should be selected.

4. The medium is withdrawn, and the plate is washed with 5 ml of PBS.

5. A stainless steel cloning ring of 5–7 mm inside diameter and 10 mm in height, which has been dipped in silicone grease in a glass dish and autoclaved, is placed over the marked colony with forceps.

6. A drop of trypsin–EDTA solution is added to the ring, and the plate is incubated for 2 min at room temperature.

7. A small volume of primary culture medium is added to the ring, and

the cells are gently suspended and transferred to a well of a 24-well multiwell tray (Falcon, 2 cm^2) coated with type IV collagen and containing 1 ml of the same medium. After 4–5 days, the cells grow to subconfluency.

8. The cells are washed with PBS, trypsinized at room temperature for 2–3 min, suspended in maintenance medium, and subcultured in the same medium.

9. Within 5 passages the cells are harvested, centrifuged, and suspended in a maintenance medium containing 8% dimethyl sulfoxide at a concentration of 1–2 × 10^6 cells/ml. The cells are frozen and stored at −80°.

Frozen cells can be thawed, subcultured every 3–4 days, and used for growth experiments within 10 passages. The cells should be checked for the presence of von Willebrand factor (Factor VIII) by immunohistochemical techniques.

Growth Experiment

1. BCE cells in subconfluent culture in maintenance medium are trypsinized.

2. The cells are suspended in maintenance medium which contains 1 ng/ml of bFGF, and inocula of 2 × 10^4 cells in 5 ml of the medium are introduced into 60-mm Falcon plates that have been coated with type IV collagen as described above.

3. Purified MAb (1–5 mg/ml) at a final concentration of 0.1–25 μg/ml is added.

4. After 5 days, the medium is withdrawn, and the cells are washed with PBS and trypsinized in 0.5 ml of trypsin–EDTA solution at 37° for 15 min.

5. The cells are suspended in 9.5 ml of PBS and counted in a Coulter counter.

All measurements should be done at least in duplicate. The cell number after 5 days in the presence of 1 ng/ml of bFGF is about 4–6 × 10^5/plate. Addition of appropriate concentrations of neutralizing MAb to bFGF should inhibit the growth of these cells. When the cells are inoculated at higher density (10^5 cells), the cells can grow even in the absence of bFGF, although the growth rate is somewhat lower. If the neutralizing MAb to bFGF inhibits this growth, an autocrine action of bFGF which is produced by these cells may be indicated.

Application of Monoclonal Antibodies

When a specific MAb to bFGF has been obtained it can be used for various general techniques: identification (Western blot, immunoprecipitation), quantitative determination (radioimmunoassay, ELISA), purification (immunoaffinity chromatography), and determination of localization in cells or tissues (immunohistochemical staining). Furthermore, when a hybridoma cell line which produces a MAb to bFGF that can block its biological activity (neutralizing MAb) has been obtained, it can be used for study of the physiological role of bFGF not only *in vitro* but also *in vivo*. After a large quantity of pure MAb is available, it can be injected into animals; however, high levels of antibody activity in the blood of the animal can also be continuously maintained by injection of the hybridoma cells. Here *in vivo* methods using a hybridoma cell line are described.

Generation of Athymic Mice Having High Level of Anti-Basic Fibroblast Growth Factor Activity in Blood[9]

1. Inocula of 5×10^5 hybridoma cells are introduced into 150-mm Falcon plates with 20 ml of RPMI 1640 + 10% FCS (see above). After 3 days, hybridoma cells grow to subconfluency ($\sim 1.6 \times 10^7$ cells/plate).

2. The cell pellet collected by centrifugation (500 g, 10 min) is suspended in 10 ml of serum-free RPMI 1640, and the cell suspension is again centrifuged.

3. The cell pellet is suspended in an appropriate volume of serum-free RPMI 1640 (5×10^7 cells/ml).

4. A volume of 0.2 ml of the cell suspension (10^7 cells) is injected subcutaneously into the back of an athymic mouse (BALB/c AJcL-nu; 6–10 weeks old) by a syringe with 25-gauge needle.

5. The hybridoma cells grow on the back of the mouse, and a solid tumor appears within 1 week. The tumor-bearing mouse can survive for 3–4 weeks, and the tumor grows to about 15% of the body weight of the host animal at the final stage.

6. At 4- to 6-day intervals, a drop of blood is taken from the retroorbital venous plexus of anesthetized mouse by a heparinized glass capillary ($1.45-1.65 \times 75$ mm). The capillary is sealed with clay at one end and centrifuged at 1200 g for 10 min. The plasma is taken from the capillary by a microsyringe and diluted with BSA/PBS containing 0.02% NaN_3.

Although the solid tumor of the hybridoma cells can be also produced in BALB/c mice, the athymic mouse is more favorable for studies of the *in vivo* growth of tumors from various species including humans.

Assay of Anti-Basic Fibroblast Growth Factor Antibody Activity in Plasma

The reaction mixture contains 0.35 ml of 0.1 M sodium phosphate buffer (pH 7.4), 0.05 ml of 1–2 ng/ml ^{125}I-labeled bFGF (8000–15,000 cpm) in BSA/PBS, 0.1 ml of purified MAb in BSA/PBS to give a final concentration of 0.2–1000 ng/ml or plasma at an appropriate dilution in BSA/PBS containing 0.02% NaN_3. The procedure for the binding experiment is the same as that for radioimmunoassay (see above). Values of radioactivity in the precipitate divided by the total tracer added are plotted against the logarithm of the concentration of purified MAb or plasma. The concentration of MAb in plasma can be calculated from the titration curves with purified MAb as a standard.

Example.[9] We produced two neutralizing MAb to bovine bFGF. The immunoglobulin class and subclass of the MAb are IgG_1/κ. The MAb are named bFM-1 and bFM-2. In a competitive binding assay of the MAb with labeled bovine bFGF, half-maximal displacement is observed with unlabeled bovine bFGF at 6.4 and 64 ng/ml for bFM-1 and bFM-2, respectively. The apparent dissociation constants (K_d) for bovine bFGF are 8.7×10^{-11} (bFM-1) and $1.6 \times 10^{-9}\ M$ (bFM-2). The antibodies are highly specific for bFGF from bovine, human, rat, and mouse sources and do not cross-react with bovine acidic FGF. bFM-2 cross-reacts with heat-inactivated bFGF, whereas bFM-1 does not, suggesting that bFM-1 recognizes the conformation of the bFGF molecule necessary for its biological activity and that bFM-2 recognizes a linear epitope of the bFGF molecule. Therefore, bFM-2 can be used for detection of inactivated bFGF by immunoblotting.

The MAb inhibit growth of cultured BCE cells in both the presence and absence of exogenously added bFGF, indicating not only the neutralizing property of the MAb but also the autocrine action of bFGF in *in vitro* growth of these cells. bFM-1 also inhibits angiogenesis in chorioallantoic membranes of chick embryos induced by bFGF.

Injections of the hybridoma cell lines producing bFM-1 or bFM-2 into the backs of athymic mice result in development of solid tumors and a sustained high level of anti-bFGF activity in the blood of the tumor-bearing mice. The activity in blood increases with an increase in tumor mass. The concentrations of bFM-1 and bFM-2 in the plasma of these mice at the final stage are 2.9–4.3 and 35–50 mg/ml, respectively. These MAb should be useful for studies on bFGF, including the physiological role and the conformation–function relationships of this factor.

Acknowledgments

This work was supported in part by a grant-in-aid for cancer research by the Ministry of Education, Science and Culture of Japan.

[15] Identification and Assay of Fibroblast Growth Factor Receptors

By MIKIO KAN, ER-GANG SHI, and WALLACE L. MCKEEHAN

Introduction

Heparin-binding (fibroblast) growth factors (HBGF) constitute a structurally defined family of polypeptides that are widely distributed in tissues. Although originally described as fibroblast growth factors FGF, the polypeptides exhibit a wide spectrum of biological activities on both mesenchymal and epithelial cells that includes stimulation and inhibition of cell growth and effects on gene expression. Currently the HBGF family consists of seven cloned genes with overall amino acid sequence similarity of 19% (see Ref. 1). At least five members of the family exhibit an affinity for the glycosaminoglycan heparin or similar molecules which affect the structure, activity, and protease sensitivity of the polypeptides.

An important advance in understanding the mechanism of action of the HBGF family has been the identification of specific cell membrane receptors for the polypeptides and dissociation of the receptors from the abundant cell-associated extracellular matrix heparin-like binding sites. This chapter describes methods for iodination and recovery of high specific activity ^{125}I-labeled HBGF and for assay of HBGF receptors. Methods underlying advances in characterization of HBGF receptors are also reviewed.

Iodination of HBGF-1

Preparation of HBGF-1 for Iodination

HBGF-1 is a very sticky polypeptide, especially at neutral pH. Although bovine serum albumin (BSA), heparin, and reducing agents such as dithiothreitol (DTT) prevent HBGF-1 from being sticky (Table I), these additives and low pH interfere with the iodination reaction. Therefore, freshly isolated and reduced HBGF-1 must be prepared for the iodination to carry out the reaction under nonsticky conditions.

Partially purified HBGF-1 is prepared from bovine brain as described previously.[2] One milligram of the bovine brain heparin-Sepharose (0.65 to

[1] W. H. Burgess and T. Maciag, *Annu. Rev. Biochem.* **58,** 575 (1989).
[2] J. W. Crabb and W. L. McKeehan, *Anal. Biochem.* **164,** 563 (1987).

TABLE I
ADSORPTION OF HBGF-1 TO PLASTIC TUBES[a]

Conditions	Adsorption (%)
PBS (pH 7.2)	63
Acid (pH 2.0)	2.8
PBS + BSA (1 mg/ml)	6.8
PBS + heparin (25 μg/ml)	13.0
PBS + DTT (10 mM)	18.6

[a] Approximately 500 pg of ^{125}I-labeled HBGF-1 (specific activity 2.1 × 10^5 cpm/ng) was added in polypropylene tubes containing 1 ml of each indicated solution. After incubation for 15 hr at room temperature, soluble and adsorbed (tube) radioactivity was determined in a γ-scintillation counter.

2 M NaCl) fraction is reduced by adding 10 mM (final) DTT at neutral pH for 1 hr at 37°. This reduction is very important to get good recovery by preventing adsorption of HBGF to reaction tubes (Table II). The solution is immediately acidified by adding trifluoroacetic acid and acetonitrile at final concentrations of 0.1% and 10% (v/v), respectively. After clarification by centrifugation, the solution is injected into C$_4$ reversed-phase HPLC column (25 cm × 4.6 mm Phenomenex, Palos Verdes, CA). HBGF-1 is eluted by linear gradient from 36 to 40% of acetonitrile as described previously.[2] The concentration of purified HBGF-1 is around 20–40 μg/ml. The purified HBGF-1 is immediately concentrated to 100 μg/ml

TABLE II
EFFECT OF REDUCTION ON RECOVERY OF HBGF-1 UNDER DIFFERENT IODINATION CONDITIONS[a]

Reduction of starting material	Reduction after chloramine-T treatment	Recovery (%)
−	−	2.6
−	+	4.9
+	−	4.5
+	+	23.2

[a] Partially purified HBGF-1 was purified by C$_4$ reversed-phase HPLC after treatment with or without DTT (10 mM). After chloramine-T treatment, the reaction was stopped with 10 mM DTT or excess KI and tyrosine.

and used for iodination. The remaining HBGF-1 is stored in the presence of 10 mM (final) DTT at $-20°$ for use as a competitor.

Iodination Reaction

The reaction is carried out at room temperature in an iodination hood in a 2-ml conical siliconized tube. Reagents are added in the following order:

1. 50 μl of HBGF-1 (~5 μg)
2. 100 μl of 0.25 M phosphate buffer (pH 7.0)
3. 20 μl of Na^{125}I (~2 mCi, Amersham, Arlington Heights, IL)
4. 50 μl of rinse of the Na^{125}I vessel with phosphate buffer
5. 30 μl of chloramine-T (400 μg/ml in phosphate buffer, freshly made)

The total reaction volume is 250 μl. After each addition, the reactants are shaken gently every 30 sec. The reaction time is 90 sec. At the end of the reaction time, 250 μl of 10 mM DTT in phosphate buffer (pH 7.0, freshly made) is added to stop the reaction and reduce the iodinated HBGF-1 (total volume is 500 μl). This reduction is also very important to get good recovery (Table II). After a further 10-min incubation, a 5-μl aliquot is removed from the iodination mixture and added to 1 ml of phosphate buffer containing 1 mg/ml BSA for a determination of trichloroacetic acid (TCA) precipitation. Incorporation of ^{125}I into HBGF-1 is determined by mixing equal volumes of cold 20% (w/v) TCA with the solution and precipitating ^{125}I-labeled HBGF-1. TCA-precipitable radioactivity is usually over 50%, and specific activity is around $2.5-5 \times 10^5$ counts/min (cpm)/ng.

Separation of ^{125}I-Labeled HBGF-1 from Free Na^{125}I

To remove unincorporated ^{125}I, the reaction mixture is applied to a heparin-Sepharose column. Heparin-Sepharose beads (400 μl; Pharmacia, Piscataway, NJ) equilibrated by phosphate buffer containing 0.5 M NaCl is poured into a disposable plastic column (2.0 ml, Bio-Rad, Richmond, CA). The column is washed first by 1.5 M NaCl and 1 mg/ml BSA (10-20 ml) and then equilibrated by 0.5 M NaCl in phosphate buffer (10-15 ml). The iodinated mixture (~500 μl) is loaded on the column at a flow rate of about 0.1 ml/1.5 min. After washing the column with 15 ml of 0.5 M NaCl to remove free ^{125}I, the ^{125}I-labeled HBGF-1 is eluted by 1.5 M NaCl at a flow rate of about 0.05 ml/min into conical tubes containing 10 μl of 5 mg/ml BSA. Each collection volume is approximately 0.5 ml, and the total collection volume is 2 ml (four tubes). Five microliters of each of the aliquots is removed to measure recovery. The ^{125}I-labeled HBGF-1 is stored at $-70°$ until use. The recovery is around 20-50%. The ^{125}I-labeled

HBGF-1 is diluted 10-fold with NaCl-free phosphate buffer containing 1 mg/ml BSA prior to use.

Binding Assay of HBGF-1 Cell Surface Receptors

Preparation of Cells

HBGF-1 shows a variety of activities for many types of cells including epithelial types of cells.[1] HepG2 is a well-differentiated human hepatoblastoma cell line[3] in which cell growth is stimulated by HBGF-1 at low concentrations (<1 ng/ml), but inhibited at high concentrations (>1 ng/ml) of HBGF-1.[4] To test whether the effects correlated with specific cell surface receptors, we tried to develop an assay system for HBGF-1 cell surface binding.

HepG2 cells (2×10^4/well) are placed in medium WAJC101[5] containing 1% fetal bovine serum (FBS) into 24-well (2 cm^2) collagen-coated tissue culture plates. The medium is replaced with MCDB107 serum-free medium[6] after overnight incubation. Incubation is continued for 2 to 3 days until the cell number reaches $5-8 \times 10^4$ cells/well. The cells are washed once with 500 μl of binding buffer [phosphate-buffered saline (PBS) or MCDB107, pH 7.2, containing 1 mg/ml BSA and 0.1 mM DTT], and then 250 μl of binding buffer is added to each of the wells.

Binding Assay of HBGF-1

^{125}I-Labeled HBGF-1 and various concentrations of unlabeled HBGF-1 are incubated at 4° (on ice) with HepG2 cells. Unusually high amounts of unlabeled HBGF-1 are required to compete with detergent-extractable (1% Triton X-100) ^{125}I-labeled HBGF-1 (Fig. 1). Although unlabeled HBGF-1 decreases cell-associated ^{125}I-labeled HBGF-1, a 100-fold excess of unlabeled HBGF-1 only reduces the binding to 50%. Furthermore, at high concentrations of ^{125}I-labeled HBGF-1, nonspecific binding (in the presence of 100-fold excess of unlabeled HBGF-1) dramatically increases and finally overcomes the total binding (Fig. 2). We reasoned that the affinity of HBGF-1 for pericellular heparin-like sites might distort the competition binding kinetics of HBGF-1 to specific cell membrane receptor sites. Labeled and unlabeled HBGF-1 may also aggregate and bind to the

---

[3] B. B. Knowles, C. C. Howe, and D. P. Aden, *Science* **209**, 497 (1980).
[4] M. Kan, D. DiSorbo, J. Hou, H. Hoshi, P.-E. Mansson, and W. L. McKeehan, *J. Biol. Chem.* **263**, 11306 (1988).
[5] H. Hoshi and W. L. McKeehan, *In Vitro Cell. Dev. Biol.* **21**, 125 (1985).
[6] H. Hoshi and M. Kan, *J. Tissue Cult. Methods* **10**, 83 (1987).

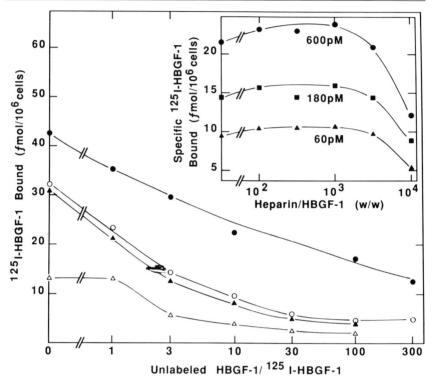

FIG. 1. Competition of unlabeled HBGF-1 with bound ^{125}I-labeled HBGF-1. The indicated amounts of unlabeled and ^{125}I-labeled HBGF (260 pM) were incubated with HepG2 cells at 4° for 4 hr. Assay wells were washed 3 times with binding buffer and then treated with binding buffer alone (●) or binding buffer containing 250 μg/ml heparin (○), 0.5 M NaCl (▲), or 1.5 M NaCl (△) for 1 min at 4°. Cells were then extracted with 1% Trition X-100 and the radioactivity of the extract determined. (Inset) Various amounts (60, 180, and 600 pM) of ^{125}I-labeled HBGF-1 were incubated with HepG2 cells in the presence of the indicated ratios of heparin to both labeled and unlabeled HBGF-1 (w/w) at 4° for 4 hr. Specific HBGF-1 binding was determined as described in text.

heparin-like sites because of the poor solubility of HBGF-1 at high concentrations and pH values above the isoelectric point.[2] Because HBGF-1 heparin binding is dissociated by 1.0–1.5 M NaCl, we treat the cells with 0.5–1.5 M NaCl after binding to remove heparin-like site binding. Extraction of cells with high salt improves the competition curves (Fig. 1). However, the salt extraction affected specific receptor-bound ^{125}I-labeled HBGF-1 (see Fig. 7).

Extraction of cells with excess heparin (250 μg/ml) just after binding assay reveals a fraction of detergent-extractable ^{125}I-labeled HBGF-1 that

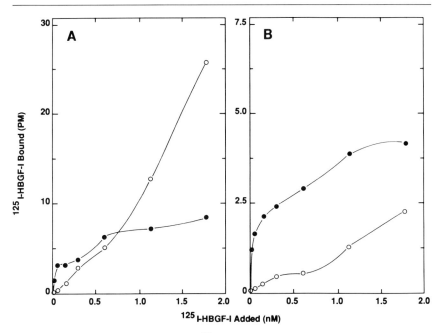

FIG. 2. Concentration dependence of ^{125}I-labeled HBGF-1 binding to HepG2 cells in the absence of heparin. The indicated concentrations of ^{125}I-labeled HBGF-1 were incubated with cells at room temperature for 3 hr in the presence (●) or absence (○) of 100-fold excess unlabeled HBGF-1. At the end of the incubation, the cells were washed with binding buffer (A) or heparin (250 μg/ml) (B), and the ^{125}I-labeled HBGF-1 bound to the cells was extracted with detergent (1% Triton X-100).

is reduced by 30–45% in the presence of an equal amount of unlabeled HBGF-1 (Fig. 1). A 100-fold excess of unlabeled HBGF-1 reduces detergent-extractable ^{125}I-labeled HBGF-1 by about 85%. However, at high concentrations of ligand, nonspecific binding (in the presence of 100-fold excess unlabeled ligand) still shows a relatively high ratio (Fig. 2). To further reduce the interference of heparin-like binding sites during equilibrium binding of HBGF-1 to specific receptor sites and to more closely correlate HBGF binding assays to cell growth assays containing heparin, a constant ratio (1000 : 1, w/w) of heparin to both ^{125}I-labeled and unlabeled HBGF-1 is included during the competition binding assays (Fig. 1, inset). When added in equimolar amounts, unlabeled HBGF-1 reduces detergent-extractable ^{125}I-labeled HBGF-1 binding to the expected 50% levels (Fig. 3), and a 100-fold excess of unlabeled reduces ^{125}I-labeled HBGF-1 binding by 90–98% (Fig. 3, Table III).

Table III compares the heparin- and detergent-extractable fractions of

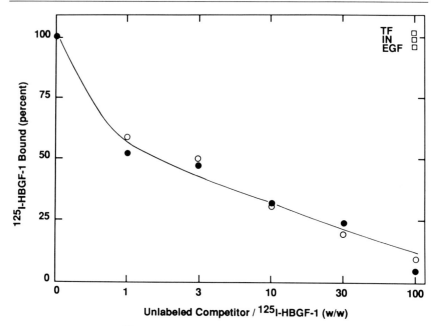

FIG. 3. Specificity of ^{125}I-labeled HBGF-1 binding to detergent-extractable sites. HepG2 cells were incubated for 3 hr at room temperature with 150 pM of ^{125}I-labeled HBGF-1 and the indicated concentrations of unlabeled HBGF-1 and HBGF-2. Heparin was added at 1000-fold (w/w) excess for HBGF-1 (○) and 100-fold (w/w) excess for HBGF-2 (●). Transferrin (TF), insulin (IN), and epidermal growth factor (EGF) were added at 250 μg/ml (□). Cells were washed with heparin (250 μg/ml), and then detergent-extractable (1% Triton X-100) ^{125}I-labeled HBGF-1 was determined.

TABLE III
HEPARIN- AND DETERGENT-EXTRACTABLE ^{125}I-LABELED HBGF-1 ON HepG2 CELLS[a]

^{125}I-Labeled HBGF-1 added (pM)	Heparin in assay	^{125}I-Labeled HBGF-1 extracted (pM)					
		Heparin		Triton		Heparin + Triton	
		Total ^{125}I-labeled	+ Unlabeled HBGF-1	Total ^{125}I-labeled	+ Unlabeled HBGF-1	Total ^{125}-I labeled	+ Unlabeled HBGF-1
50	−	1.0 ± 0.05	1.3 ± 0.02	5.5 ± 0.10	0.34 ± 0.13	6.5	1.6
50	+	0.8 ± 0.13	0.05 ± 0.01	4.7 ± 0.08	0.11 ± 0.06	5.5	0.16
560	−	7.8 ± 1.39	22.2 ± 0.44	11.3 ± 0.18	1.80 ± 0.01	19.1	24.0
560	+	4.4 ± 1.00	0.16 ± 0.03	10.4 ± 0.07	0.51 ± 0.04	14.8	0.67

[a] ^{125}I-Labeled HBGF-1 (50 or 560 pM) was incubated with HepG2 cells at 22°. Cells were sequentially extracted with heparin and Triton X-100. Where indicated, binding assays contained a 100-fold excess of unlabeled to ^{125}I-labeled HBGF-1 and heparin at a ratio of 1000:1 (w/w) to both ^{125}I-labeled and unlabeled HBGF-1.

cell-associated ^{125}I-labeled HBGF-1 at two levels of ^{125}I-labeled HBGF-1 in the absence and presence of soluble heparin and 100-fold excess unlabeled HBGF-1. At 50 pM ^{125}I-labeled HBGF-1, heparin-extractable ^{125}I-labeled HBGF-1 constitutes 15 and 81% of cell-associated ^{125}I-HBGF-1 in the absence and presence of unlabeled HBGF-1, respectively. Inclusion of heparin in the binding reaction has a small effect on total heparin-extractable ^{125}I-labeled HBGF-1 but reduces heparin-extractable ^{125}I-labeled HBGF-1 binding in the presence of unlabeled HBGF-1 by 96%. At 560 pM ^{125}I-labeled HBGF-1, the heparin-extractable binding in the absence of heparin during binding assays completely masks a reduction of total cell-associated ^{125}I-labeled HBGF-1 due to addition of unlabeled HBGF-1. Heparin-extractable ^{125}I-labeled HBGF-1 constitutes 41 and 93% of cell-associated ^{125}I-labeled HBGF-1 in the absence and presence of unlabeled HBGF-1, respectively. Heparin in the binding reaction reduces the total heparin-extractable ^{125}I-labeled HBGF-1 by 44% and the residual heparin-extractable ^{125}I-labeled HBGF-1 in the presence of unlabeled HBGF-1 by 99%. Heparin in the binding assays reduces total and specific detergent-extractable ^{125}I-labeled HBGF-1 binding by less than 15% at both concentrations of ligand, but it reduces nonspecific detergent-extractable binding by 60–70%. In sum, inclusion of heparin at a constant 1000 : 1 (w/w) ratio to HBGF-1 in the binding reaction and the postreaction extraction of cells with excess heparin reveals the fraction of ^{125}I-labeled HBGF-1 specifically bound to detergent-extractable membrane receptor sites with minimal interference by nonreceptor, heparin-like sites.

^{125}I-Labeled HBGF-1 binding to detergent-extractable binding sites on HepG2 cells is unaffected by transferrin, insulin, and epidermal growth factor (EGF). HBGF-2, the homolog of HBGF-1, competes with ^{125}I-labeled HBGF-1 binding equally for HBGF-1 when 100-fold (w/w) heparin is added with HBGF-2 (Fig. 3). Heparin in the binding reaction mixture reduces the effectiveness of HBGF-2 at higher concentrations.

Kinetics of HBGF-1 Binding to Specific HepG2 Cell Surface Receptors

Association of detergent-extractable ^{125}I-labeled HBGF-1 with HepG2 cells is complete after 3 hr of incubation at 4° (Fig. 4). In contrast, ^{125}I-labeled HBGF-1 binding to heparin-extractable sites is complete after 1 hr at 4° (Fig. 4). Analysis of specific detergent-extractable ^{125}I-labeled HBGF-1 binding at 4° as a function of ^{125}I-labeled HBGF-1 concentration and following treatment of the data according to Scatchard[7] indicates primarily a class of 146,000 ± 30,000 specific receptor sites per cell with

---

[7] G. Scatchard, *Ann. N.Y. Acad. Sci.* **51**, 660 (1949).

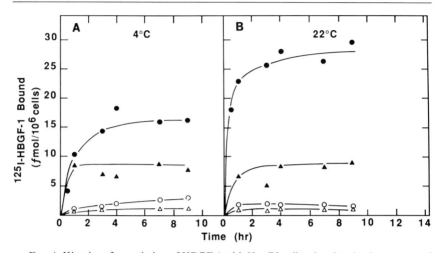

FIG. 4. Kinetics of association of HBGF-1 with HepG2 cells, showing the time course of ^{125}I-labeled HBGF-1 association with HepG2 cells at 4° (A) and 22° (B). HepG2 cells were incubated for the indicated times with 300 pM ^{125}I-labeled HBGF-1. Heparin-extractable (▲, △) and detergent-extractable (●, ○) ^{125}I-labeled HBGF-1 was sequentially determined. Total (●, ▲) and nonspecific (○, △) ^{125}I-labeled HBGF-1 binding was determined.

an apparent K_d of 2.5 ± 1.8 nM (Fig. 5). Data points at low concentrations of ^{125}I-labeled HBGF-1 indicate the possible presence of high affinity binding sites at 4° on HepG2 cells (Fig. 5). Performing the assays at room temperature (22°) reveals a significant increase in specific detergent-extractable ^{125}I-labeled HBGF-1 binding (Fig. 4) without a significant increase in the extent of heparin-extractable or nonspecific detergent-extractable ^{125}I-labeled HBGF-1 binding (Fig. 4).

^{125}I-Labeled HBGF-1 binding to detergent-extractable sites as a function of HBGF-1 concentrations at 22° is clearly biphasic (Fig. 5). Scatchard plots of the binding data at 22° can be resolved into high and low affinity phases of specific ^{125}I-labeled HBGF binding. From the two linear extremes of the Scatchard plot at 22°, an apparent K_d of 9.2 ± 0.9 pM is estimated for about 15,000 ± 900 high affinity receptor sites per cell, and an apparent K_d of 2 ± 0.4 nM can be estimated for about 180,000 ± 18,000 low affinity receptor sites per cell. Association of ^{125}I-labeled HBGF-1 with heparin-extractable sites is also saturable, reversible, but temperature-independent. From a linear Scatchard plot of the heparin-extractable fraction of ^{125}I-labeled HBGF-1 associated with HepG2 cells, an apparent K_d of about 650 ± 60 pM for this type of nonreceptor binding site can be estimated (Fig. 6, inset).

The temperature dependence of the curvilinear binding of ^{125}I-labeled

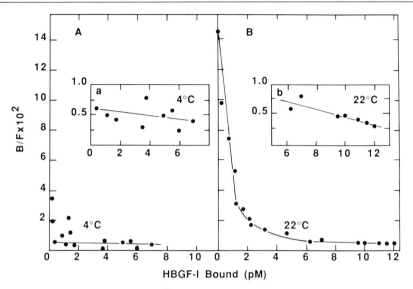

FIG. 5. Scatchard analysis of ^{125}I-labeled HBGF-1 binding to HepG2 cells. Specific detergent-extractable HBGF-1 binding was determined at 4° (A) and at 22° (B). B, Bound ligand; F, free ligand. (Inset a) Expanded plot of binding data from 1 to 7 pM from (A). (Inset b) Expanded plot of the data from 6 to 12 pM from (B).

HBGF-1 to detergent-extractable binding sites on HepG2 cells is in contrast to other cell types which exhibit simple linear Scatchard curves (Fig. 6). For example, Scatchard analysis of HBGF-1 binding data obtained at 22° for the human hepatocellular carcinoma cell line, HUH-7,[8] indicates a single class of about 5600 ± 340 detergent-extractable receptors per cell with an apparent K_d of about 112 ± 13 pM (Fig. 6). Kinetic parameters for the association of ^{125}I-labeled HBGF-1 to heparin-extractable binding sites on HUH-7 cells were similar to those for HepG2 cells (Fig. 6, inset). We interpret the data to indicate two discrete classes of specific HepG2 membrane receptor sites for HBGF-1 defined by both affinity and function. The association of HBGF-1 with high and low affinity classes of receptors correlates with opposite effects on cell growth, and only the low affinity receptor class correlates with stimulation of secretion of IαTI-related antigens.[4] Moreover, cells that exhibit a simple, positive mitogenic response to HBGF-1 exhibit linear Scatchard plots for HBGF-1 with no evidence of a ligand-dependent decrease in receptor affinity.

[8] H. Nakabayashi, K. Taketa, K. Miyano, T. Yamane, and J. D. Sato, *Cancer Res.* **42**, 3858 (1982).

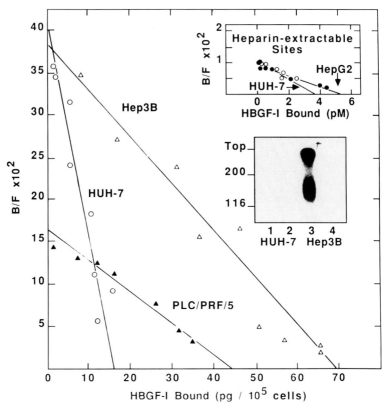

FIG. 6. Scatchard analysis of HBGF-1 binding to Hep3B, HUH-7, and PLC/PRF/5 cells. Cell densities were 77,600 Hep3B, 110,000 HUH-7, and 171,000 PLC/PRF/5 cells/well (2 cm^2). (Inset, top) Scatchard plot of specific heparin-extractable ^{125}I-labeled HBGF-1 binding to HepG3 and HUH-7 cells. (Inset, bottom) Affinity cross-linking of ^{125}I-labeled HBGF-1 to HUH-7 and Hep3B cells was performed as described in text. Lanes 2 and 4 contained a 100-fold excess unlabeled HBGF-1.

Cross-Linking of ^{125}I-Labeled HBGF-1–Receptor Complex by Disuccinimidyl Suberate

Covalent Cross-Linking of ^{125}I-Labeled HBGF-1 to HepG2 Cells

After ligand binding, the cells are washed 2 times with chilled PBS and 1 time with PBS containing 250 μg/ml heparin (Sigma, St. Louis, MO, bovine intestine or lung) followed by washing twice with PBS. The purpose of the heparin wash is to remove HBGF-1 bound to heparin-like sites on the cell surface or on the extracellular matrix. The duration of the heparin wash should not be too long (<5 min is recommended) since overwashing can cause the bound ^{125}I-labeled HBGF-1 to dissociate from the receptor. After washing, the cells are incubated at 4° or at room temperature for an additional 10 min in the presence of a small volume of PBS (e.g., 2.0 ml for the 60-mm culture dish) containing 0.3–1.0 mM disuccinimidyl suberate (DSS). The DSS solution is prepared as a 30–100 mM stock solution in dimethyl sulfoxide. After cross-linking, the reaction is terminated by adding 0.25 M Tris-HCl (pH 7.0) buffer containing 5 mM EDTA. A 60-mm culture dish containing 2.0 ml total volume of 0.3 mM DSS, 200 μl of 0.25 M Tris-HCl–5 mM EDTA buffer is recommended.

After quenching, the cells are washed once with PBS and scraped from the dishes with a rubber policeman in cold PBS containing 10 μg ml leupeptin, 10 μg/ml pepstatin A, and 1 mM phenylmethylsulfonyl fluoride (PMSF). The scraped cells are collected by centrifugation (600 g at 4°). The cross-linked ^{125}I-labeled HBGF-1–receptor complex can be extracted directly with sodium dodecyl sulfate (SDS) sample buffer, followed by boiling for 3 min, or extracted with 0.2–1% Triton X-100 buffer containing protease inhibitors for immunochemical or other studies. The cross-linked ligand–receptor complex can be analyzed by 7.5% SDS–PAGE followed by autoradiography.

Characterization of Specific HBGF-1 Receptor Sites on HepG2 Cells

Extraction of cells after treatment with the homobifunctional cross-linking agent (DSS) and analysis of the cross-linked product by SDS–PAGE followed by autoradiography reveal a broad ^{125}I-labeled band with mean molecular weight of 146,500 ± 7300 (n = 7) (Fig. 7). Omission of reducing agents from samples prior to analysis has no effect on the major cross-linked complex. Neither the presence of heparin in the binding assay nor the exposure of cells to heparin after the binding assay has an effect on the amount of the 147K band that was labeled with ^{125}I-labeled HBGF-1 (Fig. 5). In contrast, exposure of cells to 0.5 M or greater NaCl, an alternative means to reduce ^{125}I-labeled HBGF-1 association with nonreceptor heparin-like sites (Fig. 1), significantly reduces the amount of the HBGF-1-labeled polypeptide. Assuming that the autoradiographic band of 147K included one molecule of molecular weight about 15,200 representing Asn^{21}-HBGF-1, the native receptor molecular prior to covalent cross-linking would have an estimated molecular weight of 131,300 ± 6500 (n = 7).

Despite the evidence for functional high and low affinity classes of HBGF-1 receptor sites on HepG2 cells, a single molecular weight species of 147K appears on SDS–PAGE gels after covalent cross-linking of

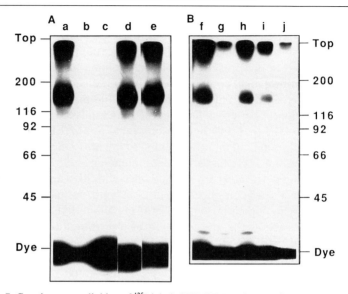

FIG. 7. Covalent cross-linking of ^{125}I-labeled HBGF-1 to the HepG2 cell surface. Binding reactions contained 300 pM of ^{125}I-labeled HBGF-1 and were performed at 22° (A) or at 4° (B). Cells in lanes a to c were extracted with PBS, and lanes d and e were extracted with PBS containing 250 μg/ml heparin for 1 min at 4° before the cross-linking action. The binding reaction for lane b contained 30 nM unlabeled HBGF-1 and 505 μg/ml heparin during the binding reaction. DSS was omitted from the cross-linking reaction for lane c. The binding reaction for lane e contained 5 μg/ml heparin. Cells in lanes f and g were washed with PBS prior to the cross-linking reaction. The binding reaction for lane g contained 30 nM unlabeled HBGF-1. Cells for lanes h, i, and j were washed with PBS containing 0.5, 1.0, and 1.5 M NaCl, respectively, prior to the cross-linking reaction. Gel lanes were excised and counted at 1-cm intervals. The autoradiographic bands at 147K in lanes a, d, and e represented 7160, 6930, and 7330 cpm, respectively. The 147K band in lanes f, h, i, and j contained 1704, 1284, 894, and 294 cpm, respectively. The molecular weight markers were myosin (M_r 200,000), β-galactosidase (M_r 116,250), phosphorylase b (M_r 92,500), BSA (M_r 66,200), and ovalbumin (M_r 45,000).

^{125}I-labeled HBGF to HepG2 cells after binding at 4° and 22° and both low and high concentrations of HBGF-1. Therefore, the determinants of the high and low affinity properties of HBGF-1 receptors on HepG2 cells must lie in differences in receptor structure and conformation in associated proteins or other molecules or in dimerization of receptor molecules that are not detected by the affinity cross-linking and SDS–PAGE procedure. A similar conclusion has been reached for the EGF[9-11] and NGF[12,13] receptors, each of which are the product of a single gene.

[9] J. Schlessinger, *J. Cell Biol.* **103**, 2067 (1986).
[10] J. Schlessinger and Y. Yarden, *Biochemistry* **26**, 1443 (1987).

Conclusions

We developed procedures for iodination of HBGF-1 and for binding assays of cell surface receptors. The preparation of radiolabeled HBGF-1 allowed the detection of HBGF receptors on a number of cell types, and the binding assays have provided a good deal of information on these receptors. For example, the receptor kinetics, at least in some types of cells, are heterogeneous although covalent cross-linking of ^{125}I-labeled HBGF-1 to its receptor, showed only a single broad 147K band. Recently, a putative chicken HBGF-2 (basic FGF) receptor gene, a *fms*-like gene (*flg*),[14] has been identified.[15] The transfection of this gene will provide more detailed knowledge of the mechanisms of heterogeneity of the binding kinetics and the intracellular signals.

---

[11] C. Cochet, O. Kashles, E. M. Chambaz, I. Borrello, C. R. King, and J. Schlessinger, *J. Biol. Chem.* **263**, 3290 (1988).

[12] D. Johnson, A. Lanahan, C. R. Buck, A. Sehgel, C. Morgan, E. Mercer, M. Bothwell, and M. Chao, *Cell (Cambridge, Mass.)* **47**, 545 (1986).

[13] M. J. Radeke, T. P. Misko, C. Hsu, L. A. Herzenberg, and E. M. Shooter, *Nature (London)* **325**, 593 (1987).

[14] M. Ruta, R. Howk, G. Ricca, W. Drohan, M. Zabelshansky, G. Laureys, D. E. Barton, U. Francke, J. Schlessinger, and D. Givol, *Oncogene* **3**, 9 (1988).

[15] P. L. Lee, D. E. Johnson, L. S. Cousens, V. A. Fried, and L. T. Williams, *Science* **245**, 57 (1989).

Section III

Epidermal Growth Factor, Transforming Growth Factor α, and Related Factors

[16] Expression of Epidermal Growth Factor Precursor cDNA in Animal Cells

By BARBARA MROCZKOWSKI

Introduction

Analysis of cDNA clones derived from mRNA transcripts encoding murine or human epidermal growth factor (EGF) has revealed that these peptides are synthesized as large precursor molecules comprised of 1217 and 1207 amino acid residues, respectively.[1,2] These unexpectedly large molecules contain a hydrophobic domain that may serve to anchor the precursor in the plasma membrane, thus making the EGF precursor structurally analogous to a receptor. In the kidney, where the EGF precursor does not undergo proteolytic processing, it is plausible that the precursor molecule functions as a receptor for a yet unidentified ligand.[3] Why EGF is encoded within a much larger membrane-bound precursor remains a matter of speculation.

An interesting characteristic of the precursor molecule is that it contains eight regions of partial sequence similarity to mature EGF. Each repeat unit is comprised of approximately 40 amino acids and includes 6 cysteine residues spaced as in EGF. Whether these EGF-related sequences are ever processed to serve a biological function is not known.

To gain a better understanding of the function of the EGF precursor and the factors resulting in its tissue-specific processing to EGF, we have used a bovine papillomavirus (BPV), mouse metallothionein I promoter-based expression vector, together with a human kidney cDNA encoding the EGF precursor, to produce high levels of this protein in mouse NIH 3T3 and C127 cells. The ability of BPV vectors to propagate extrachromosomally as multicopy plasmids, in conjunction with their ability to express foreign proteins efficiently (i.e., human insulin receptor,[4] human tissue-type plasminogen activator[5]), has enabled us to establish stably transfected

---

[1] A. Gray, T. J. Dull, and A. Ullrich, *Nature (London)* **303,** 722 (1983).
[2] J. Scott, J. Urdea, M. Quiroga, R. Sanchez-Pescador, N. Fong, M. Selby, W. J. Rutter, and G. I. Bell, *Science* **221,** 236 (1983).
[3] L. B. Rall, J. Scott, G. I. Bell, R. J. Crawford, J. D. Penshow, H. D. Niall, and J. P. Coghlan, *Nature (London)* **313,** 228 (1985).
[4] J. Whittaker, A. K. Okamoto, R. Thys, G. I. Bell, D. J. Steiner, and C. A. Hofmann, *Proc. Natl. Acad. Sci. U.S.A.* **84,** 5237 (1987).
[5] V. B. Reddy, A. J. Garramone, H. Sasak, C.-M. Wei, P. Watkins, J. Galli, and N. Hsiung, *DNA* **6,** 461 (1987).

cell lines that express high levels of the hEGF precursor. This chapter presents specific protocols that our laboratory has used for the selection and establishment of stably transfected cell lines using BPV-derived shuttle vectors.

Construction and Preparation of Bovine Papillomavirus Recombinant Expression Vectors

Synthetic *Xho*I oligonucleotide linkers (New England Biolabs, Beverly, MA) are ligated to a 6.4-kilobase (kb) *Sma*I fragment containing the entire coding region and 3' and 5' flanking sequences of the human EGF precursor isolated from a λgt10 adult human kidney cDNA library.[6] After *Xho*I digestion and removal of excess linkers on a low gelling temperature agarose gel (FMC BioProducts, Rockland, ME), the 6.4-kb fragment is ligated into the *Xho*I site of pBPV-MTH-Xho (kindly provided by D. Hamer, National Institutes of Health, Bethesda, MD). This eukaryotic shuttle vector is similar to that described by Pavlakis and Hamer[7] and contains the following structural elements: the 5.5-kb subgenomic fragment of BPV, the mouse metallothionein gene modified by replacement of the *Bgl*II site upstream of the initiator ATG by an *Xho*I site, and the origin of replication and ampicillin resistance gene of pML2.[8] The physical map of this construct, pBPV-MTH-hEGF, and the restriction map of the human prepro-EGF cDNA sequence are shown in Fig. 1. Vector DNAs used for transfection are prepared by a slight modification of the standard lysozyme–Triton method.

Plasmid Purification

The quantities given are per 1 liter culture.

1. Harvest cells, spin 5000 rpm for 10–15 min at 4°.
2. Resuspend in 10 ml sucrose solution [25% sucrose (w/v), 50 mM Tris-Cl (pH 8.0), 1 mM EDTA (pH 8.0)].
3. Add 2 ml of 5 mg/ml lysozyme freshly made up in 0.25 M Tris-Cl (pH 8.0) and 4 ml of 0.25 M EDTA (pH 8.0). Mix gently and incubate on ice for 10 min.
4. Add 16 ml of cold TLM [Triton lysis mix: 50 mM Tris-Cl (pH 8.0), 60 mM EDTA (pH 8.0), 0.1% Triton X-100]. Mix gently and incubate on ice for 15 min.

---

[6] G. I. Bell, N. M. Fong, M. A. Wormsted, D. F. Caput, L. Ku, M. S. Urdea, L. B. Rall, and R. Sanchez-Pescador, *Nucleic Acids Res.* **14**, 8427 (1986).
[7] G. N. Pavlakis and D. H. Hamer, *Proc. Natl. Acad. Sci. U.S.A.* **80**, 397 (1983).
[8] M. Lusky and M. Botchan, *Nature (London)* **293**, 79 (1981).

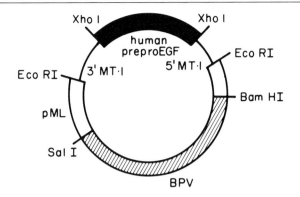

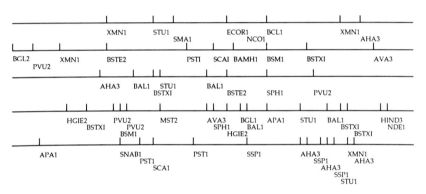

FIG. 1. Expression vector pBPVMTMhEGF and restriction map of the human EGF precursor cDNA sequence.

5. Centrifuge the viscous lysate at 17,000 rpm for 30 min in a Beckman SW27 or SW28 rotor at 4°.

6. To the cleared lysate from Step 5 add one-fourth initial volume of 30% PEG (polyethylene glycol, MW 6000–8000) made up in 1.5 M NaCl. Mix and incubate on ice for several hours or overnight.

7. Centrifuge at 10,000 rpm for 15 min at 4°. Remove the supernatant and wash the pellet with 70% ethanol.

8. Dissolve the DNA in 2.1 ml of TE (10 mM Tris-Cl, pH 7.5, 1 mM EDTA). Add 2.3 g CsCl$_2$ and 0.15 ml of a 10 mg/ml ethidium bromide solution.

9. Centrifuge at 10,000 rpm for 1 hr at 20° to remove cell debris and film. Transfer the solution to a new tube.

10. Supercoiled plasmid DNA is purified by density gradient centrifugation in a VTi 65 rotor (55,000 rpm, 12–15 hr at 20°).

11. Collect the lower, supercoiled band and extract with 2-propanol equilibrated with $CsCl_2$-saturated TE.

12. Dialyze plasmid solution for 24 hr at 4° against four 4-liter volumes of TE.

13. Precipitate the plasmid DNA with 2 volumes of ice-cold 100% ethanol and 1/10 volume of 3 M sodium acetate (pH 6.0). Centrifuge at 10,000 rpm for 20 min at 4° and wash the pellet with 70% ice-cold ethanol prior to resuspension in TE.

Cell Transfection

Mouse NIH 3T3 cells or mouse C127 cells (mammary tumor cell line of an RIII mouse, ATCC CRL 1616) (2–4 × 10^5) are cotransfected with a calcium phosphate precipitate containing 5 μg of pBPV-MTH-hEGF and 0.5 μg of pSV2Neo,[9] followed by osmotic shock with glycerol.[10] Our protocol for transfection is based on the procedure published by Fordis and Howard.[11]

Transfection

On day 1, plate out cells at a density of 2 × 10^5 cells per T-25 Falcon flask. Prepare the DNA for precipitation and transfection as follows. Each T-25 flask is transfected with 5 μg of the BPV expression vector and 0.5 μg of pSV2Neo; we found this to be the optimum DNA concentration and ratio for NIH 3T3 and C127 recipient cells. Ethanol-precipitate the DNA from stock TE solutions by the addition of 1/10 volume of 3 M sodium acetate (pH 6.0) and 2 volumes of 100% ethanol, mix well, and place in dry ice–ethanol bath for 15 min, then collect the precipitate by centrifugation (13,000 g for 15 min) in a microcentrifuge at 4°. Aspirate dry and rinse with 70% ethanol. Add 20 μl of 70% ethanol to the precipitated DNA, mix gently, and evaporate in a Speed-Vac overnight to ensure that the DNA precipitate is dry.

On day 2, prepare the following stock solutions. 10× HEPES-buffered saline (10× HBS) contains 8.18% NaCl (w/v), 5.94% HEPES (w/v), and 0.2% $Na_2HPO_4 \cdot 7H_2O$ (w/v). Prepare a 2× HBS solution from the stock 10× HBS and adjust the pH to 7.12 with 1 N NaOH (pH must be exact). The $CaCl_2$ solution is 2 M $CaCl_2 \cdot 2H_2O$. Prepare 15% glycerol/HBS by

[9] P. J. Southern and P. Berg, *J. Mol. Appl. Genet.* **1**, 327 (1981).
[10] E. Frost and J. Williams, *Virology* **91**, 39 (1978).
[11] C. M. Fordis and B. H. Howard, this series, Vol. 151, p. 382.

mixing 30 ml of 50% glycerol in water (w/v), 50 ml of 2× HBS, and 20 ml doubly distilled water. Filter-sterilize the 2× HBS, 2 M CaCl$_2$, and the 15% glycerol/HBS solutions in a laminar flow hood.

Calcium Phosphate–DNA Coprecipitation

Note that Steps 1–3 are performed in a laminar flow hood. The reaction volumes are given per single T-25 Falcon flask.

1. Resuspend the DNA in a 1:10 dilution of filter-sterilized TE (pH 7.5). For 5 μg DNA resuspend in 50 μl of a 1:10 dilution of TE, 15.5 μl of 2 M CaCl$_2$, and 59.5 μl doubly distilled water.

2. Add the DNA solution dropwise to a tube containing 125 μl of 2× HBS. During addition use a sterile cotton-plugged Pasteur pipette to direct a gentle stream of air in the tube to mix the components. The final concentration of DNA for the calcium phosphate coprecipitation step should be 20 μg DNA/ml.

3. Let the DNA incubate 30 min at room temperature.

4. Wash the flask 2 times with serum-free Dulbecco's modified Eagle's medium (DMEM).

5. Add 5 μg of the calcium phosphate–DNA precipitate to each T-25 flask containing 5.0 ml of DMEM. Return the flask immediately to the incubator to avoid pH changes in the medium.

6. Four hours later, osmotically shock the cells.* Wash the cells 2 times with DMEM. Overlay the cells with 3.0 ml of 15% glycerol/HBS. Incubate for 2 min at 37°. Wash 2 times with DMEM and feed with fresh complete medium.

On day 4, split the transfected cells. A single T-25 flask is split into five large 100-mm tissue culture dishes containing selective medium.

Selection by Neomycin Resistance

Transfected cells are fed twice a week with fresh medium containing the appropriate concentration of Geneticin. The selective medium for NIH 3T3 cells contains 600 μg/ml of Geneticin (G418 sulfate, GIBCO, Grand Island, NY); the selective medium for C127 cells contains 1 mg/ml Geneticin. Geneticin-resistant colonies are readily observed after a 2-week period. Individual clones are isolated by removing the medium and placing a sterile cloning ring around the colony to be isolated. The colonies are

---

* All solutions must be prewarmed to 37°.

then trypsinized[12,13] and subcultured directly into flat-bottomed 24-well plates (Linbro Flow Laboratories, McLean, VA).

Screening of Transformants

Geneticin-resistant cells are subcultured into 12-well plates (Costar, Cambridge, MA), then tested for the expression of EGF precursor mRNA and protein by RNA–RNA hybridization and immunoblot analysis, respectively.

RNA Isolation and Blot Hybridization Analysis

Cytoplasmic RNA is isolated from individual clones by detergent lysis and phenol/chloroform/isoamyl alcohol extractions. Briefly, each well is washed and scraped in 1.0 ml of calcium-free, magnesium-free phosphate-buffered saline. Cells are transferred to an Eppendorf tube and pelleted at 1500 rpm for 10 min. Cell pellets are resuspended in 500 μl of RNA lysis buffer [20 mM Tris-Cl pH (7.5), 0.15 M KCl, 5 mM MgCl$_2$, and 0.5% Nonidet P-40 (NP-40)] and incubated on ice for 15 min. Lysates are spun in an Eppendorf microfuge (13,000 g) for 10 min, and the RNA is isolated from the supernatants by phenol/chloroform/isoamyl alcohol (25 : 24 : 1) extractions followed by ethanol precipitation.

Slot blots are performed using a minifold II slot-blot system (Schleicher and Schuell, Keene, NH) and Nytran membranes (Schleicher and Schuell). The total RNA from a single well is resuspended in 7.5× SSC (1× SSC is 0.15 M NaCl, 15 mM sodium citrate) containing 4.3 M formaldehyde and incubated at 80° for 10 min prior to spotting onto the nylon membrane. Blots are baked at 80° for 30 min under reduced pressure, then prehybridized for 3 hr at 55° in hybridization buffer {50% (v/v) formamide, 0.75 M NaCl, 0.15 M Tris-Cl (pH 8.0), 10 mM EDTA, 0.2 M sodium phosphate (pH 6.8), 1× Denhardt's solution [0.02% bovine serum albumin (BSA), 0.02% (w/v) Ficoll, 0.02% (w/v) poly(vinylpyrrolidone)], 10% (w/v) dextran sulfate (MW 500,000), 0.1% sodium dodecyl sulfate (SDS)}.

Hybridizations are carried out for 16 hr at 55° in fresh hybridization buffer containing [^{32}P]CTP-labeled prepro-EGF RNA. The hybridization probe is a 1015-base pair (bp) *Eco*RI/*Bam*HI fragment (Fig. 1) of prepro-EGF subcloned into the vector pGEM4 (Promega Biotec). High specific activity ^{32}P-labeled probes of both sense and antisense strands are prepared

---

[12] L. C. M. Reid, this series, Vol. 58, p. 152.
[13] R. I. Freshney, "Culture of Animal Cells: A Manual of Basic Technique," Chap. 13, p. 129. Alan R. Liss, New York, 1983.

by using SP6 and T7 RNA polymerases according to the manufacturer's protocol (Promega Biotec).

Following hybridization, the blots are sequentially washed in a shaking water bath in a total volume of 40 ml for 1 hr at 68° in the following solutions[14]:

> Wash buffer I: 0.15 M Tris-Cl (pH 8.0), 0.75 M NaCl, 10 mM EDTA, 25 mM sodium phosphate (pH 6.5), 0.1% sodium pyrophosphate, 0.1% SDS
> Wash buffer II: 30 mM Tris-Cl (pH 8.0), 0.15 M NaCl, 2.0 mM EDTA, 25 mM sodium phosphate (pH 6.5), 1× Denhardt's, 0.1% sodium pyrophosphate
> Wash buffer III: 5 mM Tris-Cl (pH 8.0), 50 mM NaCl, 0.4 mM EDTA, 0.1% sodium pyrophosphate, 0.1% SDS

The membranes are then covered with Saran plastic wrap and autoradiographed at $-70°$, using Kodak XAR-5 film and intensifying screens.

Figure 2A shows a representative blot of RNA isolated from individual Geneticin-resistant clones immobilized onto Nytran membranes using the procedures outlined above. The slot blots are performed in duplicate and represent constitutive levels of prepro-EGF mRNA in the transfected cell lines. Figure 2B represents the relative steady-state levels of prepro-EGF mRNA in parental NIH 3T3 cells and in the transfected Geneticin-resistant cell lines hEGF12, hEGF11, and hEGF19 in the presence or absence of 5 mM sodium butyrate.[15] We have observed that hybridization of the hEGF precursor RNA probe to mRNA isolated from various transfected NIH 3T3 or C127 cell lines results in at least a 10-fold increase in steady-state levels of mRNA after induction. Growth of transfected cells in the presence of heavy metals (i.e., zinc or cadmium) results in only a modest 2- to 3-fold increase in prepro-EGF mRNA. A major advantage of the use of sodium butyrate is that, unlike the heavy metals generally used to induce the metallothionein promoter, sodium butyrate is not toxic to the cells for incubation periods of up to 3 days.

Detection of Recombinant Prepro-EGF

Immunoblot Analysis of Prepro-EGF in Transfected Cell Lines

Polyclonal antibodies directed toward epitopes specific for the mature hEGF peptide are used to identify and quantitate relative levels of the hEGF precursor in transfected cell lines. Various aliquots of NP-40 lysates

---
[14] D. DeLeon, K. H. Cox, L. M. Angerer, and R. C. Angerer, *Dev. Biol.* **100**, 197 (1983).
[15] B. W. Birren and H. R. Herschman, *Nucleic Acids Res.* **14**, 853 (1986).

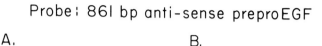
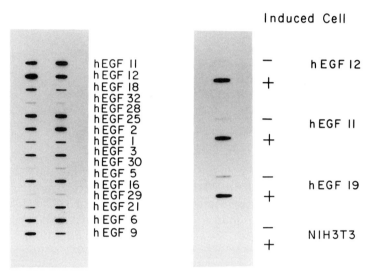

FIG. 2. Slot-blot analysis of prepro-EGF mRNA levels in transfected cells. (A) Analysis of total RNA isolated from duplicate cultures immobilized onto Nytran membranes and hybridized to a ^{32}P-labeled antisense RNA probe. (B) Analysis of 5 μg of RNA isolated from control NIH 3T3 cells and three different transfected cell lines grown in the presence (+) or absence (−) of 5 mM butyric acid.

from transfected NIH 3T3 cells induced for 12 hr with either butyric acid or sodium butyrate reveal the presence of a 170,000-dalton protein recognized by anti-hEGF antiserum (Fig. 3). As observed with mRNA levels, the levels of prepro-EGF protein are found to be consistently much higher in cells exposed to butyric acid or sodium butyrate as opposed to $ZnCl_2$ or $CdCl_2$. The optimum sodium butyrate concentration for the induction of recombinant prepro-EGF in transfected NIH 3T3 or C127 cells is found to be 5 mM.

Immunoprecipitation of Prepro-EGF from Transfected Cells

The synthesis of prepro-EGF in transfected cell lines is examined by immunoprecipitating metabolically radiolabeled EGF precursor from solubilized cells. Briefly, transfected cells are grown to 70–80% confluency in 60-mm dishes prior to induction with 5 mM sodium butyrate for 8–12

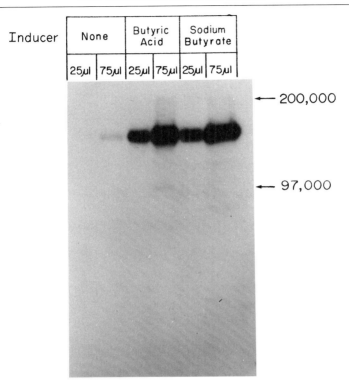

FIG. 3. Immunoblot analysis of the prepro-EGF content in transfected NIH 3T3 cells. Cells were incubated for 12 hr in the presence or absence of either 5 mM butyric acid or sodium butyrate. Aliquots (25 and 75 μl) of the NP-40 cellular lysates prepared from single 100-mm dishes were analyzed by Western blotting after SDS–polyacrylamide gel electrophoresis.

hr. The cells are washed 3 times with phosphate-buffered saline (PBS) and incubated at 37° for 30 min in DMEM lacking cysteine and supplemented with 4 mM glutamine. The medium is then removed and replaced with fresh serum-free and cysteine-free DMEM to which [^{35}S]cysteine (200 μCi/ ml, 940 Ci/mmol; NEN, Boston, MA) is added. After labeling, the cells are cooled to 4°, washed 3 times with PBS, harvested by scraping, and collected by centrifugation. Radioactively labeled cell pellets are solubilized by the direct addition of lysis buffer [20 mM Tris-Cl (pH 7.5), 0.15 M KCl, 5 mM MgCl$_2$, 0.5% Nonidet P-40].

Cells from each 60-mm dish are solubilized in 500 μl of lysis buffer. After a 10-min incubation at room temperature, the lysate is clarified by centrifugation at 13,000 g for 8 min at 4°. Aliquots (100 μl) of the supernatant are immunoprecipitated with 5 μl of rabbit antiserum. Immunoprecipi-

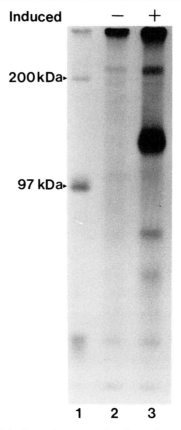

FIG. 4. Immunoprecipitation of prepro-EGF from C127 transfected mouse cells. Transfected C127 cells grown in 60-mm dishes were incubated in the absence (lane 2) or presence (lane 3) of 5 mM sodium butyrate for 12 hr prior to metabolic labeling with [^{35}S]cysteine for 2 hr. Immunoprecipitates were separated electrophoretically on a SDS–polyacrylamide gel and examined by fluorography. Lane 1 represents ^{14}C-labeled markers.

tates of mouse C127 transfected cell lysates incubated in the presence or absence of sodium butyrate indicate that the transfected cells synthesize a 150,000–160,000 dalton immunoreactive protein on induction with sodium butyrate (Fig. 4). As observed with transfected NIH 3T3 cells, no protein of 6.3 kDa, corresponding to mature EGF, is detected. Neither of the mouse cell lines expresses the appropriate proteolytic processing enzymes for production of mature EGF. The levels of expression of prepro-EGF in the stably transfected cell lines established in our laboratory remain constant for periods of up to 1 year in culture; on further passage a substantial decrease in the synthesis of recombinant protein is observed.

Summary

This chapter outlines in detail the optimal conditions for the expression of human recombinant proteins in mouse cells using a bovine papillomavirus-based mammalian expression vector. The procedures we have described were used to successfully express high levels of the human EGF precursor in our laboratory and the human insulin receptor in the laboratory of Whittaker.[4] Using this experimental approach we were able to demonstrate that expression of a cDNA for prepro-EGF produces a glycosylated membrane protein with biological activity.

Acknowledgments

I am grateful to Martha Reich for skilled technical assistance, Jonathan Whittaker for helpful discussions and valuable advice, Susan Heaver for secretarial assistance, and Ann M. Soderquist for critical reading of the manuscript.

[17] Generation of Antibodies and Assays for Transforming Growth Factor α

By CATHERINE LUCAS, TIMOTHY S. BRINGMAN, and RIK DERYNCK

Introduction

Two types of growth factors have been termed transforming growth factors (TGF) because they were discovered in an assay that evaluated their ability to elicit cellular transformation using an immortalized nonneoplastic fibroblast line. One of these, TGF-α, is a single-chain polypeptide structurally related to epidermal growth factor (EGF). Several TGF-α species are proteolytically derived from a transmembrane precursor form. The fully processed form is 50 amino acids long and lacks glycosylation, whereas the larger forms are glycosylated. All these forms and the transmembrane precursor are biologically active and interact with the same receptor as EGF. The best studied activity of TGF-α is its ability to stimulate DNA synthesis and mitosis of the many cell types that contain the corresponding cell surface receptors.[1-3] TGF-β is structurally unrelated

---

[1] R. Derynck, *Cell* (*Cambridge, Mass.*) **54,** 593 (1988).
[2] R. Brachmann, P. B. Lindquist, M. Nagashima, W. Kohr, T. Lipari, M. Napier, and R. Derynck, *Cell* (*Cambridge, Mass.*) **56,** 691 (1989).
[3] S. T. Wong, L. F. Winchell, B. K. McCune, H. S. Earp, J. Teixido, J. Massagué, B. Herman, and D. C. Lee, *Cell* (*Cambridge, Mass.*) **56,** 495 (1989).

to TGF-α and binds to distinct receptors.[4,5] It is described in more detail in [29]–[32] in this volume.

Much interest in both TGF-α and TGF-β has arisen because both types of factors have been shown to be functional in normal physiology and also have been implicated in malignant transformation and tumorigenesis. The availability of antibodies to transforming growth factor α is of utmost importance in studying its biology, not only by neutralizing its activities, but also in developing fast, reliable, and specific assays. This chapter describes the methodology we have used to develop antibodies and assays specific for TGF-α.

Because both TGF-α and EGF bind to the same receptor, a receptor-binding assay would not be able to discriminate between the factors; however, it would have the advantage of measuring only active growth factors in a configuration able to bind the receptor. The lack of specificity for TGF-α in a receptor binding assay thus dictates the need to develop a TGF-α-specific antibody and immunoassay. Ideally such antibodies and assays should not cross-react with EGF. EGF and TGF-α share about a 30% sequence identity, and it is thus imperative that the lack of cross-reactivity be documented. It would also be an advantage if the antibody only reacts with the biologically active molecules and interacts with an epitope formed by the secondary structure of the properly folded factor. Finally, the antibodies should be able to recognize the multiple forms of secreted TGF-α and the larger transmembrane form. A major problem in developing antibodies for TGF-α is the lack of naturally secreted TGF-α. The TGF-α secreted by human cell lines is present in such low concentrations in the medium that it becomes virtually impossible to purify sufficiently large quantities for immunization experiments. We have overcome this difficulty by expressing the 50 amino acid form of TGF-α as a fusion protein in *Escherichia coli*. The fusion protein is then purified, refolded into the proper disulfide configuration, and cleaved to release the 50 amino acid TGF-α.

Preparation of Recombinant Transforming Growth Factor α

The 50 amino acid form of human TGF-α can be obtained as follows. An expression vector is constructed in such a way that it directs the expression of a 69 amino acid TGF-α fusion protein in *E. coli*. The fusion

---

[4] A. B. Roberts and M. B. Sporn, *in* "Peptide Growth Factors and Their Receptors" (M. B. Sporn and A. B. Roberts, eds.), Handbook of Experimental Pharmacology. Springer-Verlag, Heidelberg, in press.

[5] M. B. Sporn, A. B. Roberts, L. M. Wakefield, and B. de Crombrugghe, *J. Cell Biol.* **105**, 1039 (1987).

protein consists of an 18 amino acid leader sequence followed by a unique methionine that precedes the 50 amino acid TGF-α. This fusion protein is isolated and purified from the recombinant *E. coli* and cleaved with cyanogen bromide. This treatment results in a cleavage following the methionine residue, thus releasing TGF-α. The TGF-α is then refolded in an oxidation–reduction buffer, and the biologically active TGF-α with the correct disulfide bridge configuration is purified to apparent homogeneity by HPLC. The bacterial expression of TGF-α and its purification have been detailed.[6,7]

Generation of Antibodies

Polyclonal Antibodies

Antisera against human TGF-α are prepared in rabbits using the purified and refolded recombinant human TGF-α as immunogen. A first injection on day 0 consists of 400 μg TGF-α emulsified in Freund's complete adjuvant and is given intradermally at multiple sites. Subsequent booster injections in Freund's incomplete adjuvant are given subcutaneously at multiple sites on day 60 with 175 μg TGF-α and on days 90, 150, 180, and 210 with 100 μg TGF-α. The antibody titers are measured by indirect ELISA (for a general description of this procedure, see the production of monoclonal antibodies to TGF-β in [29], this volume) at 10 to 14 days after each injection. The antiserum is collected starting at day 190 and is designated 34D. This antiserum blocks the binding of TGF-α to the receptor and immunoprecipitates the 50 amino acid TGF-α. It also reacts with denatured and reduced 50 amino acid TGF-α in Western blots. The antibody is also able to immunoprecipitate the secreted glycosylated TGF-α species and to a lesser extent the nonglycosylated transmembrane precursor. We were not able, however, to immunoprecipitate the solubilized glycosylated transmembrane form using the antibody.[8]

Antipeptide Antibodies

As an alternative approach to produce an anti-TGF-α antiserum we injected rabbits with synthetic peptides that correspond to segments of the 50 amino acid TGF-α sequence. The peptides, synthesized using the

---

[6] R. Derynck, A. B. Roberts, M. E. Winkler, E. Y. Chen, and D. V. Goeddel, *Cell* (Cambridge, Mass.) **38**, 287 (1984).
[7] M. E. Winkler, T. S. Bringman, and B. J. Marks, *J. Biol. Chem.* **261**, 13838 (1986).
[8] T. S. Bringman, P. B. Lindquist, and R. Derynck, *Cell* (Cambridge, Mass.) **48**, 429 (1987).

solid-phase method,[9] are coupled via thioester linkages from the terminal cysteine to soybean trypsin inhibitor at a ratio of 5–10 mol of peptide per mole carrier protein. Rabbits are injected intradermally on day 0 with 1 mg of the conjugated peptide in Freund's complete adjuvant. Subsequent subcutaneous booster injections of 200 μg of the immunogen in Freund's incomplete adjuvant are given on day 30, and injections of 100 μg immunogen are given on days 70, 100, 130, and 160. The antibody titer to each peptide is determined by indirect ELISA at 10 to 14 days after each injection.[8]

The first antiserum, 34AP1, was against a peptide corresponding to the first 10 amino acids of the 50 amino acid TGF-α (CRFLVQEDKP), whereas antiserum 34AP2 was raised against the C-terminal octapeptide (CEHADLLA). Another TGF-α peptide, which was used to raise the antiserum 34AP3, corresponds to a C-terminal loop of TGF-α (KVCHSGYVGARCEHADLLA). This 19 amino acid long peptide was disulfide bonded in order to mimic the TGF-α loop and linked via a N-terminal Lys residue to the carrier protein.[10] Although we were able to raise peptide-specific antibodies, they are only moderately useful to immunoprecipitate TGF-α and are inferior to the above-described 34D antiserum.

Finally an antiserum was raised against the 15-mer RHEKPSALLK-GRTAC, which corresponds to a sequence in the cytoplasmic segment of the TGF-α precursor. This peptide was also coupled via its Cys-residue to the soybean trypsin inhibitor carrier protein. We have used this antiserum 34E successfully to immunoprecipitate the cytoplasmic segment following the cleavage of the precursor as well as both the glycosylated and nonglycosylated transmembrane TGF-α precursors.[8]

Monoclonal Antibodies

One hundred micrograms of purified and refolded human TGF-α in Freund's complete adjuvant is injected intraperitoneally into BALB/c mice. After 5 weeks each animal receives an additional subcutaneous injection with 20 μg TGF-α in Freund's incomplete adjuvant. Sixteen days later, the presence of TGF-α antibodies is demonstrated by indirect ELISA in one of the mice. The mouse then receives an intravenous injection of 25 μg TGF-α, followed by three consecutive daily injections of 100 μg TGF-α. The next day the spleen is removed, and the splenocytes are fused

---

[9] G. Barany and R. B. Merrifield, *in* "The Peptides" (E. Gross and J. Meienhofer, eds.), Vol. 2, p. 1. Academic Press, New York, 1980.
[10] J. W. Littlefield, *Science* **145**, 709 (1964).

with P3X63Ag8.653 plasmacytoma cells[11] as described.[12] The fused cells are plated into the wells of 96-well polystyrene microtiter plates at a density of 2×10^5 cells/well followed by HAT (hypoxanthine, aminopterin, thymidine) selection[10] 1 day after the fusion. The resulting hybridomas are screened for the production of anti-TGF-α antibodies by indirect ELISA (see [29], this volume).

This procedure gave rise to the hybridoma secreting the monoclonal antibody (MAb) TGF-α_1. The hybridoma cells are grown as ascites in pristane-primed BALB/c mice.[13] Characterization of the antibody showed that TGF-α_1 belongs to the IgG$_1$ type and has a K_d of 10^{-9} mol/liter. The antibody has very little cross-reactivity with EGF, since about 30,000-fold more EGF than TGF-α is needed to displace ^{125}I-labeled TGF-α binding in a radioimmunoassay (RIA). MAb TGF-α_1 is able to block effectively the binding of the 50 amino acid TGF-α to the receptor (whether this antibody neutralizes glycosylated TGF-α precursor forms is still unknown). We were not able to accurately determine the antibody epitope of TGF-α; however, the antibody does not bind to the C-terminal loop of TGF-α and does not recognize denatured and reduced TGF-α.

Purification of Antibodies

The immunoglobulin G fraction of the rabbit anti-TGF-α antisera or of the ascites fluid from anti-TGF-α hybridomas is purified by adsorption to protein A-Sepharose. Whereas the rabbit antisera are loaded undiluted, we dilute the ascites fluids 5-fold with 3 M KCl, 0.1 M acetic acid, 0.1 M Tris-HCl (pH 8.5), 25 mM EDTA, 0.1% Tween 20 before loading onto the column. The protein A column is subsequently washed with phosphate-buffered saline (PBS) or with the dilution buffer in the case of the ascites fluid. The bound IgG is eluted with 0.15 M NaCl, 0.1 M acetic acid, neutralized to pH 8.0 with 1 M Tris, and dialyzed against PBS. The antibody preparation is concentrated to 10 mg/ml by ultrafiltration using an Amicon (Danvers, MA) PM10 membrane.

We use peptide affinity column chromatography in order to purify the 34E antiserum. The affinity gel is made by coupling 20 mg of the oligopeptide used as immunogen to 10 ml ω-aminobutyl-agarose (Sigma, St. Louis, MO). The matrix is activated with N-maleimidobenzoyl sulfosuccinimide ester (Pierce, Rockford, IL) and is then allowed to react with the synthetic peptide via its C-terminal cysteine using the manufacturer's suggested

---
[11] J. F. Kearney, A. Radbruch, B. Liesegang, and K. Rajewski, *J. Immunol.* **123**, 1548 (1979).
[12] S. F. De St. Groth and D. Scheidegger, *J. Immunol. Methods* **35**, 1 (1980).
[13] M. Potter, J. G. Humphrey, and J. L. Walter, *J. Natl. Cancer Inst.* **49**, 305 (1972).

protocol. Five milliliters of serum is loaded on a 10-ml column, which is then washed with 50 ml PBS before elution of the bound antibodies with 0.15 M NaCl, 0.1 M acetic acid. The eluted antibody fractions are adjusted to pH 8.0 with 1 M Tris and dialyzed against PBS. Bovine serum albumin (BSA) is then added to 5 mg/ml. A typical purification of 5 ml antiserum yields about 0.6 mg affinity purified antibody.

Assay of Transforming Growth Factor α

Enzyme-Linked Immunosorbent Assay

Quantitation of TGF-α is performed using a double-sandwich ELISA. We find that this assay, outlined below, is superior in both sensitivity and specificity to an RIA using our antibodies. Micro-ELISA plates (Nunc, Roskilde, Denmark) are coated with 200 ng per well of purified TGF-α_1 antibody in PBS for 16 hr at 4°. After the plates are briefly rinsed with PBS–0.05% Tween 20, 100-μl aliquots of the samples in TBS–BSA–Tween (0.15 M NaCl, 50 mM Tris-HCl, pH 7.4, 2 mM EDTA, 5 mg/ml BSA, 0.05% Tween 20) are incubated in the wells for 2 hr at 22°. A dilution series of pure refolded human TGF-α derived from *E. coli* (5 to 0.078 ng/ml) is included in duplicate on every plate. The plates are washed 5 times with PBS–Tween and then incubated for 2 hr at 22° with 100 μl/well of a 1:1000 dilution of the 34D IgG fraction in TBS–BSA–Tween. The plates are again washed 5 times in PBS–Tween and are incubated for 2 hr at 22° with 100 μl/well of a 10,000-fold dilution of horseradish peroxidase-conjugated goat anti-rabbit immunoglobulin in TBS–BSA–Tween. The plates are again washed 5 times and then allowed to react for 30 min with 100 μl/well of 0.2 mg/ml *o*-phenylenediamine, 0.012% (v/v) H_2O_2 in 0.1 M phosphate–citrate (pH 5.0). The reaction is stopped by the addition of 50 μl/well of 2.5 N H_2SO_4. The absorbance is measured using a Titertek Multiscan autoreader interfaced with a Hewlett–Packard integral personal computer with Titercalc software. This allows the conversion of OD_{492} values to TGF-α concentrations. The ELISA does not detect TGF-α with incorrectly formed disulfide bridges, equivalent amounts of rat TGF-α, or 10 μg/ml human EGF. The results obtained from this assay are in agreement with the values determined from an EGF receptor-binding assay. The ELISA allows accurate measurements of TGF-α between 0.1 and 5 ng/ml.

Radioreceptor Assay

The radioreceptor assay measures the binding of properly folded EGF or TGF-α to the common cell surface receptor and thus does not discriminate between the two ligands. Mink lung epithelial-like cells (Mv1Lu, ATCC CCL64) are grown in Dulbecco's modified Eagle's medium (DMEM) supplemented with 10% fetal bovine serum and 1% nonessential amino acids. Cells are seeded at 2×10^5/well in 96-well tissue culture plates with removable wells (Dynatech, Alexandria, VA), for 18 to 24 hr prior to assay. Immediately before the assay the cells are washed twice with DMEM containing 10 g/liter BES [N,N-bis(2-hydroxyethyl)-aminoethanesulfonic acid] buffer, pH 7.0 (Sigma), and 1 g/liter BSA (dilution buffer). The assay samples are serially diluted in the same buffer, and 100-μl aliquots are added to each well. EGF is labeled with ^{125}I by the chloramine-T method to a specific activity of 80–100 μCi/μg. One hundred microliters containing 10^6 counts/min (cpm) of ^{125}I-labeled EGF diluted in the same dilution buffer is added to each well, and the plates are incubated at 22° for 70 min. The wells are washed 4 times with Hanks' balanced salt solution and are then counted with a γ-counter. Under these conditions, 4–9% of the ^{125}I-labeled EGF binds to the cells, and 4×10^{-10} M EGF is needed to displace 50% of the specific binding, as analyzed by Scatchard analysis.

Conclusion

The development of antibodies and assays for transforming growth factor α has been difficult. In particular, the production of antibodies of sufficient titer and specificity, and especially of neutralizing antibodies, was only possible with the availability of larger quantities (recombinant) of the protein in the proper configuration. In spite of these difficulties, reagents of high quality were produced, and reliable procedures were developed for the measurement of TGF-α.

[18] Molecular and Biochemical Approaches to Structure–Function Analysis of Transforming Growth Factor α

By DEBORAH DEFEO-JONES, JOSEPH Y. TAI, and ALLEN OLIFF

Introduction

Transforming growth factor α (TGF-α) is a 50 amino acid peptide hormone. Medical interest in TGF-α stems from its biological activity as a stimulator of tumor cell growth. Transfection of nontransformed murine fibroblasts with a molecular clone of TGF-α cDNA converts these cells to tumorogenic cell lines.[1] In addition, many human tumor cell lines secrete TGF-α, and many cancer patients exhibit elevated levels of TGF-α in the blood and urine.[2-5] These observations suggest a role for TGF-α in the pathophysiology of human malignancies and imply that antagonists of the biological activity of TGF-α may have utility as anticancer agents.[6,7]

The biological effects of TGF-α are mediated via attachment of TGF-α to epidermal growth factor (EGF) receptors on the surface of susceptible cells.[8] This biochemical interaction offers an obvious target for the identification of TGF-α antagonists. As a first step in the design and isolation of TGF-α–EGF receptor binding antagonists, a structure–function analysis of the receptor binding properties of TGF-α is desirable. We have used four independent methods for analyzing the contribution of specific regions of TGF-α to the receptor binding activity of the entire hormone. First, synthetic peptides representing discrete segments of TGF-α were prepared. These peptides were tested for their ability to inhibit hormone attachment to the EGF receptor and to block TGF-α-stimulated mitogenesis in mammalian cells. Second, the same peptides were coupled to thyroglobulin, and the peptide–thyroglobulin conjugates were used to raise antibodies that recognized specific regions of TGF-α. The antibodies were tested for the ability to block the attachment of whole TGF-α to EGF receptors. Third, a synthetic human TGF-α gene was constructed and introduced into bacteria for the production and isolation of recombinant human TGF-α. The synthetic gene was also modified using techniques of site-directed mutagenesis to create a series of TGF-α derivatives with defined changes in the primary amino acid sequences. These derivatives were analyzed for their EGF receptor binding and mitogenic activities. Fourth, a series of recombinant TGF-α derivatives was

[1] A. Rosenthal, P. B. Lindquist, T. S. Bringman, D. V. Goeddel, and R. Derynck, *Cell (Cambridge, Mass.)* **46,** 301 (1986).
[2] B. Gusterson, G. Cowley, J. McIlhinney, B. Ozanne, C. Fisher, and B. Reeves, *Int. J. Cancer* **36,** 689 (1972).
[3] R. Derynck, D. V. Goeddel, A. Ullrich, J. U. Gutterman, R. D. Williams, T. S. Bringman, and W. H. Berger, *Cancer Res.* **47,** 707 (1987).
[4] T. A. Libermann, H. R. Nusbaum, N. Razon, R. Kris, I. Lax, H. Soreq, N. Whittle, M. D. Waterfield, A. Ullrich, and J. Schlessinger, *Nature (London)* **313,** 144 (1985).
[5] S. A. Sherwin, D. R. Twardzik, W. H. Bohn, K. D. Cockley, and G. J. Todaro, *Cancer* **43,** 403 (1983).
[6] J. J. Nestor, S. R. Newman, B. M. DeLustro, and A. B. Schreiber, *in* "Peptides Structure and Function" (C. M. Deber, V. J. Hruby, and K. D. Koppel, eds.), p. 39. Pierce Chemical Co., Rockford, Illinois, 1985.
[7] J. J. Nestor, Jr., S. R. Newman, B. M. DeLustro, G. J. Todaro, and A. B. Schreiber, *Biochem. Biophys. Res. Commun.* **129,** 226 (1985).
[8] J. Massague, *J. Biol. Chem.* **258,** 13614 (1983).

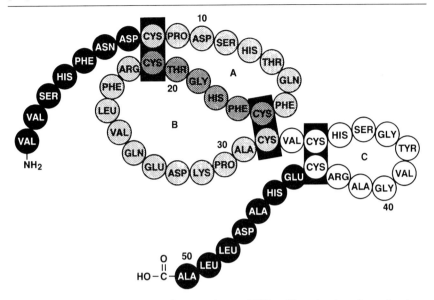

FIG. 1. Amino acid sequence of mature human TGF-α. Note the three loop structures (A, B, and C) formed by three intrachain disulfide bonds. Six synthetic peptides were prepared corresponding to the following residues: 1–7, 8–21, 8–32, 16–32, 34–43, and 44–50. Disulfide bonds were formed in each peptide containing cysteine residues. Elimination of free sulfhydryl groups following disulfide bond formation was assessed with Ellman's reagent.

created containing a single lysine residue at various positions in the primary structure of TGF-α. The mutated TGF-α proteins were isolated and modified by covalent attachment of a p-hydroxybenzimidyl group to the ε-amino group of the lysine residues. This postsynthetic modification of TGF-α generated a series of TGF-α derivatives whose biological activities were evaluated both before and after conjugation with the methyl p-hydroxybenzimidate "blocking group."

Synthetic Peptide Analysis

The primary amino acid sequence and location of disulfide bonds of mature human TGF-α are presented in Fig. 1. One method of conceptualizing how TGF-α functions is to hypothesize that a discrete segment of the molecule is responsible for binding to the EGF receptor while a separate segment is responsible for triggering the biological activity of TGF-α. This view of peptide hormone structure–function relationships underlies the dissection of TGF-α by synthesis of defined peptide segments. Individual peptides representing the N terminus, C terminus, and each of the disulfide-constrained loop structures or combinations of loop structures were synthesized. The peptides were tested in radiolabeled EGF receptor-binding inhibition studies[9] and in [^{3}H]thymidine incorporation assays of the mitogenic activity of TGF-α.[10] In order to identify weakly active species, it is generally necessary to test the peptides at concentrations up to 1000 times the dose of TGF-α required to illicit a maximal biological response.

Synthetic peptides were prepared by standard methods of solid-phase synthesis,[11,12] using an Applied Biosystems Model 430A automated peptide synthesizer. Purification of the newly synthesized peptides was performed by reversed-phase HPLC.[13]

Antipeptide Antibodies

Antibodies raised against whole TGF-α block the attachment of TGF-α to the EGF receptor. The antibodies recognize multiple regions of the TGF-α molecule. It is unclear which, if any, single epitope on TGF-α is responsible for elliciting the receptor-blocking antibodies. One method of defining the relative contribution of different regions of TGF-α to the generation of receptor-blocking antibodies is to raise antisera against specific TGF-α peptides that represent discrete segments of the TGF-α molecules. Antipeptide antibodies can be purified from the antisera and tested for the ability to block the binding of whole TGF-α to EGF receptors. The synthetic peptides described above are excellent candidates for this type of analysis. Unfortunately, small synthetic peptides are frequently only weakly immunogenic. To ensure an adequate immune response, we first couple the peptides to thyroglobulin and then inject the thyroglobulin-peptide conjugates into rabbits. High titer peptide-specific antisera are consistently produced in this manner.

Preparation of Peptide–Thyroglobulin Conjugates

1. Equal amounts (5–10 mg) of the TGF-α peptides and bovine thyroglobulin are dissolved in 0.1 M phosphate buffer, pH 7.0 (20 mg/ml total protein, final concentration).

[9] S. Cohen, H. Ushiro, C. Stoscheck, and M. Chinkers, *J. Biol. Chem.* **257**, 1523 (1982).
[10] M. W. Riemen, R. J. Wegrzyn, A. E. Baker, W. M. Hurni, C. D. Bennett, A. Oliff, and R. B. Stein, *Peptides* **8**, 877 (1987).
[11] R. B. Merrifield, *J. Am. Chem. Soc.* **85**, 2149 (1963).
[12] J. M. Steward and J. D. Young, *in* "Solid Phase Peptide Synthesis," (J. M. Stewart and J. D. Young, eds.), 2nd Ed. Pierce Chemical Co., Rockford, Illinois, 1984.
[13] J. Rivier, R. McClintock, R. Galyean, and H. Anderson, *J. Chromatogr.* **288**, 303 (1984).

2. The protein solution is cooled to 4° on ice.

3. Twenty-five percent (w/v) glutaraldehyde is diluted 1 : 20 with water to give a 0.125 M glutaraldehyde solution. Two hundred fifty microliters of the diluted glutaraldehyde is added slowly to the protein solution over a period of 15–20 min with stirring.

4. The mixture is incubated with stirring at 4° for 18 hr and then dialyzed extensively against phosphate-buffered saline.

Immunization with Peptide–Thyroglobulin Conjugates

1. New Zealand white rabbits are injected intramuscularly with 250 μg of the peptide–thyroglobulin conjugate emulsified in Freund's complete adjuvant.

2. Thirty-five days later the rabbits are bled and then injected subcutaneously with 250 μg of conjugate emulsified in Freund's incomplete adjuvant.

3. On day 49 the rabbits are bled and then injected subcutaneously with 250 μg of conjugate emulsified in Freund's incomplete adjuvant.

4. On day 63 the rabbits are bled out and sacrificed; alternatively, a sublethal bleeding is performed so that further immunizations and bleedings can be continued.

Isolation of Antipeptide Immunoglobulins

Anti-TGF-α peptide IgG is isolated from whole antisera by protein A-agarose affinity column chromatography.

1. One milliliter of rabbit serum is diluted with 1 ml of agarose affinity binding buffer (Pierce Chemical Co., Rockford, IL) and applied to a 1-ml, 0.9 by 2.5 cm diameter protein A-agarose column.

2. The column is washed with 15 ml of binding buffer, and IgG is eluted with 5 ml of agarose affinity column elution buffer. Both the binding and elution buffers are prepared as specified by the manufacturer of the agarose affinity column.

3. The IgG fraction is dialyzed extensively against phosphate-buffered saline.

4. The final IgG preparation is analyzed by SDS–PAGE to assess its purity.

Site-Directed Mutagenesis of Recombinant Transforming
  Growth Factor α

Several aspects of the structure–function relationship of TGF-α can only be examined in the context of the entire TGF-α molecule. For example, proper cysteine pairings and formation of correct disulfide bonds are

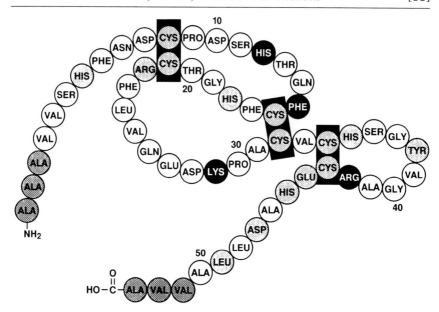

FIG. 2. Amino acid sequence of recombinant human TGF-α made in *Escherichia coli*. The first three and last three residues (cross-hatched) were derived from the precursor of human TGF-α. The codons for these residues were added to the synthetic gene used to produce TGF-α in bacteria. Nonconservative substitutions using alanine were introduced at the diagonally shaded residues 8, 16, 21, 22, 32, 34, 38, 43, 44, 47, and 49. Semiconservative substitutions using lysines were introduced at the diagonally shaded residues 4, 18, 35, and 45. A conservative substitution using lysine was introduced at the diagonally shaded residue 22. Both conservative and nonconservative or semiconservative and nonconservative substitutions were introduced at the blackened residues 12, 15, 29, and 42.

critical to producing fully active TGF-α. Elimination of disulfide bonds by substituting alanines for the cysteine residues in TGF-α dramatically reduces both receptor binding and mitogenic activity.[14] Experiments yielding information on the contribution of individual amino acid residues to the biological activity of TGF-α were performed by site-directed mutagenesis of the gene encoding TGF-α and subsequent expression and isolation of the mutated TGF-α species in bacteria. Conservative, semiconservative, or nonconservative amino acid changes were introduced either singly or as multiple changes in all regions of TGF-α. Figure 2 illustrates the site-specific mutations that we have introduced into TGF-α and analyzed. This

[14] D. Defeo-Jones, J. Y. Tai, R. J. Wegrzyn, G. A. Vuocolo, A. E. Baker, L. S. Payne, V. M. Garsky, A. Oliff, and M. W. Riemen, *Mol. Cell. Biol.* **8**, 2999 (1988).

analysis has identified critical amino acids at several positions in TGF-α, including residues 15, 38, 42, and 48. Nonconservative and in some cases conservative amino acid substitutions at these positions dramatically reduced the receptor binding and mitogenic activities of TGF-α.[15]

Site-Directed Oligonucleotide Mutagenesis

1. The human TGF-α gene is subcloned in M13MP19, and single-stranded DNA is prepared as a template for mutagenesis.[16]

2. Specific, short [~18 base pairs (bp)] oligonucleotides encoding the changes are synthesized using an Applied Biosystems Model 380A DNA synthesizer (Applied Biosystems Inc., Foster City, CA).

3. The specific oligonucleotide is phosphorylated in a reaction mixture containing: 1× kinase buffer, 0.8 μg of each oligonucleotide, 1 mM ATP, and 2.5 units of T4 kinase (New England Biolabs, Beverly, MA) in a total reaction volume of 25 μl. The reaction is carried out at 37° for 40 min and stopped by heating to 65° for 15 min. The kinased oligonucleotide is next annealed to the single-stranded template DNA using 1 to 2 μg of single-strand DNA, 0.16 to 0.2 μg of kinased oligonucleotide, and 1× hybridization buffer in a total reaction volume of 30 μl. All enzyme buffers are prepared as specified by the enzyme manufacturer. Hybridization buffer is 20 mM Tris-HCl, pH 7.5, 10 mM MgCl$_2$, 0.5 M NaCl.

4. The reaction mixture is incubated at 55° for 20 min, then shifted to room temperature for 1 hr.

5. Using the annealed oligonucleotide as a primer, a second-strand synthesis is performed using DNA polymerase (Klenow fragment) under the following conditions: to the 30-μl annealing reaction are added all four deoxynucleotide triphosphates to a final concentration of 0.5 mM, 0.1 M 2-mercaptoethanol, 1 mM ATP, 5 units of T4 ligase, and 5 units DNA polymerase (Klenow fragment). The reaction is adjusted to 1× hybridization buffer conditions and is incubated for 3 hr at room temperature.

6. The reaction is stopped by raising the temperature to 65° for 10 min and then extracting with an equal volume of phenol/chloroform (1:1). After extraction the reaction is precipitated with 2 volumes of 95% ethanol.

7. The mutagenized DNA is resuspended in water and digested with an appropriate restriction enzyme to isolate the mutant gene fragment, which is then ligated into the specific vector DNA required. The resulting plasmid DNA is then transfected into bacteria.

---

[15] D. Defeo-Jones, J. Y. Tai, G. A. Vuocolo, R. J. Wegrzyn, T. L. Schofield, M. W. Riemen, and A. Oliff, *Mol. Cell. Biol.* **9**, 4083 (1989).

[16] P. H. Schien and R. Corese, *J. Mol. Biol.* **129**, 169 (1979).

8. Mutant clones are selected by hybridizing to a radiolabeled oligonucleotide probe containing the mutation. Posthybridization washings [$2 \times$ SSC and 0.1% sodium dodecyl sulfate (SDS)] are performed at a temperature 2°–3° below the determined melt-out temperature for the oligonucleotide used in the mutagenesis procedure. These conditions ensure that only clones carrying the desired change will hybridize to the oligonucleotide probe.

Isolation and refolding of recombinant TGF-α and mutant derivatives of TGF-α proteins are performed as described by Defeo-Jones et al.[14] with the following modification of the final purification step. Partially purified TGF-α proteins are loaded on a 2.2 × 25 cm Vydac preparative C_4 column developed with a linear gradient of 30 to 50% acetonitrile–water in 0.1% trifluoroacetic acid for 60 min. Fractions from the reversed-phase HPLC column containing the purified TGF-α proteins are identified by spectrophotometric absorption at 210 nm.

It should be noted that the synthetic gene employed for producing TGF-α in bacteria encodes three additional N-terminal and three additional C-terminal amino acids derived from the precursor of mature TGF-α. These additional residues appear to stabilize the TGF-α species, enabling consistently high yields of intact proteins to be isolated from bacteria.

Postsynthetic Modification of Recombinant Transforming
  Growth Factor α

In addition to analyzing the biological activity of peptide fragments of TGF-α and TGF-α substitution mutants, we examined the effect of adding chemical "blocking groups" to mature TGF-α at various sites in the molecule. Methyl p-hydroxybenzimidate can be covalently coupled to the ε-amino group of lysine residues in TGF-α, creating a bulky side group on the lysine. Human TGF-α normally has only one lysine residue at position 29. However, by using the techniques of site-directed mutagenesis described above, the lysine at position 29 may be converted to an arginine. The TGF-α Lys-29 to Arg mutant exhibited full receptor binding and mitogenic activities. The TGF-α Lys-29 to Arg mutant was further modified to create a series of TGF-α derivatives that possessed unique lysine residues at various positions around the TGF-α protein. The location of lysine in five of these substitution mutants is indicated in Fig. 2. Each of the mutant TGF-α species was isolated and assayed for receptor-binding activity both before and after coupling with methyl p-hydroxybenzimidate. The addition of the p-hydroxybenzimidyl group to the lysine residues at some positions (e.g., 12, 18, and 35) significantly reduced receptor-binding activity. At other positions conjugation of the lysine residue to methyl *p*-hydroxybenzimidate had little or no effect on the biological activity of TGF-α.

Preparation of Amidinated Mutant Proteins

1. One-half milligram of purified mutant TGF-α is dissolved in 0.5 ml of 0.2 *M* sodium borate buffer, pH 9.5.
2. Fifteen milligrams of methyl *p*-hydroxybenzimidate-HCl (Pierce) is suspended in 100 µl of dimethyl sulfoxide and solubilized in 400 µl of water.
3. The methyl *p*-hydroxybenzimidate solution is titrated to pH 9.5 with 10 *N* NaOH and added to the TGF-α protein solution. The mixture is incubated at room temperature for 18 hr.
4. At the end of this time, the reaction mixture is applied directly to a reversed-phase HPLC semipreparative C_4 column, 1.0 × 25 cm (Vydac, The Nest Group, Southboro, MA) equilibrated in 0.1% trifluoroacetic acid and 30% acetonitrile and water. The amidinated mutant TGF-α proteins are purified with a linear gradient of 30 to 50% acetonitrile–water in 0.1% trifluoroacetic acid at a flow rate of 4 ml/min for 50 min.
5. The amidinated protein peaks are identified by the spectrophotometric absorbance at 210 nm, collected, lyophilized, and stored at −20° until needed for bioassays.

The amidinated protein was characterized by analytical reversed-phase HPLC (C_4, 0.46 × 15 cm). The native and amidinated proteins have distinctly different profiles using this HPLC system. The amidinated protein was further characterized by the disappearance of the single lysine residue in each TGF-α species as determined by amino acid composition analysis using a Beckman System 6300 amino acid analyzer (Beckman Instruments, Fullerton, CA).[17]

Conclusions

We have employed several approaches in analyzing the structure–function relationships of human TGF-α. Each method has provided complementary data that support the following conclusions: (1) Biological activity resides in the entire TGF-α molecule. Peptide segments of TGF-α possess little or no receptor-binding or mitogenic activity. (2) Amino acid residues critical to the biological activity of TGF-α exist in each of the disulfide-constrained loops and in the C terminus of the molecule. An important

[17] F. T. Wood, M. M. Wu, and J. C. Gerhart, *Anal. Biochem.* **69**, 339 (1975).

question that remains unanswered is whether changes at these critical residues affect the conformation of TGF-α or are directly involved as contact points between TGF-α and the EGF receptor. Clarification of this issue will require a detailed structure analysis of native TGF-α as well as several mutant versions of TGF-α by NMR or X-ray crystallography techniques.

[19] Assessment of Biological Activity of Synthetic Fragments of Transforming Growth Factor α

By GREGORY SCHULTZ and DANIEL TWARDZIK

Introduction

Transforming growth factor α (TGF-α) is a mitogenic hormone which appears to play important roles in normal fetal development, tissue regeneration, and tumor growth.[1] Mature TGF-α is a 50 amino acid, single-chain polypeptide (Fig. 1) belonging to the family of structurally related peptide growth factors that includes epidermal growth factor (EGF),[2] vaccinia virus growth factor (VGF),[3] and amphiregulin.[4] TGF-α has substantial (~40%) sequence similarity with the other growth factors in the EGF-like family and, in particular, shares similar placement of three conserved intrachain disulfide bonds. All four structurally related growth factors bind and activate the tyrosine kinase activity of a common 170-kDa membrane receptor.[5] The sequence of the gene for TGF-α suggests that it is synthesized as part of a larger, single-chain, transmembrane glycoprotein of 160 amino acids. The mature 50 amino acid polypeptide hormone is apparently cleaved from the precursor between Ala and Val residues at both the N and C terminals.[6]

A major biochemical objective in studying growth factor–receptor systems is the identification of amino acid sequences which form the specific

---

[1] R. Derynck, *Cell* (*Cambridge, Mass.*) **54**, 593 (1988).
[2] C. R. Savage, Jr., T. Inagami, and S. Cohen, *J. Biol. Chem.* **247**, 7612 (1972).
[3] J. P. Brown, D. R. Twardzik, H. Marquardt, and G. J. Todaro, *Nature* (*London*) **313**, 491 (1985).
[4] M. Shoyab, V. L. McDonald, J. G. Bradley, and G. J. Todaro, *Proc. Natl. Acad. Sci. U.S.A.* **85**, 6528 (1988).
[5] C. S. King, J. A. Cooper, B. Moss, and D. R. Twardzik, *Mol. Cell. Biol.* **6**, 332 (1986).
[6] G. J. Todaro, D. C. Lee, N. R. Webb, T. M. Rose, and J. P. Brown, *Cancer Cells* **3**, 51 (1985).

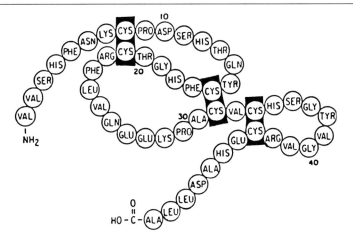

FIG. 1. Primary amino acid sequence of rat TGF-α. TGF-α has substantial sequence similarity with EGF, VGF, and amphiregulin, including the alignment of the three disulfide bonds.

receptor-binding domains of the growth factor. Once the sequences have been identified, a more rational approach can be used to select peptide sequences of the growth factor which might have agonist or antagonist activities, or sequences which might be useful in generating specific antibodies that would neutralize the action of the growth factor. As part of our studies on the role of TGF-α in various biological systems, we utilized standard solid-phase techniques to synthesize a series of peptides that encompassed the entire primary sequence of the mature TGF-α 50-mer.[7] These peptides and several peptides purchased from commercial sources were evaluated in five different assays for TGF-α activity. Previous volumes of this series (Volumes 146 and 147) have dealt extensively with the purification of TGF-α from transformed cells, biochemical assays for TGF-α, identification of receptor proteins for TGF-α, and techniques for solid-phase synthesis of mature TGF-α. We have drawn on many of the techniques discussed in those volumes, and in this chapter we discuss how we used and modified these techniques. Several studies on fragments of TGF-α and EGF also helped to guide our work, and recent publications on the partial solution structures of TGF-α and EGF provide additional insight into putative receptor-binding domains of TGF-α.

---

[7] K. Darlak, G. Franklin, P. Woost, E. Sonnerfeld, D. Twardzik, A. Spatala, and G. Schultz, *J. Cell. Biochem.* **36**, 341 (1988).

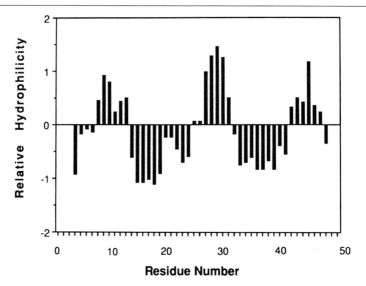

FIG. 2. Hydrophilicity plot of rat TGF-α. TGF-α contains three separate regions of hydrophilicity.

Strategy for Selection of Transforming Growth Factor α Fragment Sequences

Since all the members of the EGF family of peptide growth factors bind to a common membrane receptor protein (170 kDa), one might expect that the receptor-binding domain of TGF-α would lie primarily in the region which has the highest degree of conservation of amino acid sequence between the members of the EGF family of growth factors. In addition, one would expect the receptor-binding domain to have a strong hydrophilic character which would favor its orientation to the surface of the peptide where it could interact with the receptor protein. The third disulfide loop region (34–43) of TGF-α has the highest degree of sequence similarity with the other members of EGF family.[3] In addition, hydrophilicity plots of TGF-α (Fig. 2) indicate three discrete sequences which are theoretically exposed on the surface of the peptide and, thus, are potentially able to interact with the TGF-α receptor.

Initial studies of the activities of synthetic fragments of TGF-α and EGF indicated that different regions of the related growth factors were active in binding to their common receptor. Nestor et al.[8] reported that

[8] J. J. Nestor, Jr., S. R. Newman, B. DeLustro, G. J. Todaro, and A. B. Schreiber, *Biochem. Biophys. Res. Commun.* **129,** 226 (1985).

a fragment of rat TGF-α comprising the C-terminal third disulfide loop (residues 34–43) had low affinity (500-fold less potent than EGF) for EGF receptors of human cells. The fragment also blocked the mitogenic effect of TGF-α and EGF on human fibroblasts and inhibited vaccinia virus infection purportedly by occupying the EGF receptor of target cells.[9] In contrast, Komoriya et al.[10] reported that the C-terminal third disulfide loop of mouse EGF (residues 32–48) had no detectable activity in competing for EGF receptor binding. They concluded that the second disulfide loop region (residues 14–31) constituted a major receptor-binding region based on the results that both the linear and cyclic second disulfide loop region of mouse EGF (residues 14–31) induced several biological responses associated with intact EGF but were 10,000-fold less potent than intact EGF. A third study by Heath and Merrifield[11] reported that a fragment comprising both the second and third disulfide loop region of EGF (residues 15–53) was 10,000 times less potent than the mature EGF (residues 1–50) for binding to the EGF receptor and stimulating DNA synthesis. Furthermore, they found that the C-terminal third disulfide loop of mouse EGF (residues 32–53) was 100,000-fold less potent than EGF in receptor binding and did not stimulate DNA synthesis. These initial studies presented a somewhat confusing and contradictory concept of the major receptor-binding domain for TGF-α and EGF.

Crystal structures for EGF or TGF-α are currently not available, but partial solution structures for both TGF-α and EGF have recently been reported. The key structures detected for human[12,13] and for murine[14,15] EGF are similar and consist of a triple-stranded, antiparallel, β sheet at the N terminal involving bonding between residues 3 and 23, 5 and 21, and residues 18–23 paired with residues 34–28. A small antiparallel β-sheet structure also links residues 37–38 with 45–44 at the C terminal, and an interaction of residue 52 with the edge of the β sheet of the second loop region tends to restrict the position of the C-terminal segment. In contrast, a cyclic heptadecapeptide of the second loop of human TGF-α (residues 16–32) did not form an antiparallel β-sheet structure, as observed for this

[9] D. A. Eppstein, Y. V. Marsh, A. B. Schreiber, S. R. Newman, G. J. Todaro, and J. J. Nestor, Jr., *Nature (London)* **318**, 663 (1985).
[10] A. Komoriya, M. Hortsch, C. Meyers, M. Smith, H. Kanety, and J. Schlessinger, *Proc. Natl. Acad. Sci. U.S.A.* **81**, 1351 (1984).
[11] W. F. Heath and R. B. Merrifield, *Proc. Natl. Acad. Sci. U.S.A.* **83**, 6367 (1986).
[12] K. Makino, M. Morimoto, M. Nishi, S. Sakamoto, A. Tamura, H. Inooka, and K. Akasaka, *Proc. Natl. Acad. Sci. U.S.A.* **84**, 7841 (1987).
[13] R. M. Cooke, A. J. Wilkinson, M. Baron, A. Pastore, M. J. Tappin, I. D. Campbell, H. Gregory, and B. Sheard, *Nature (London)* **327**, 339 (1987).
[14] K. H. Mayo, *Biochemistry* **24**, 3783 (1985).
[15] G. T. Montelione, K. Wuthrich, E. C. Nice, A. W. Burgess, and H. A. Scheraga, *Proc. Natl. Acad. Sci. U.S.A.* **83**, 8594 (1986).

TABLE I
TRANSFORMING GROWTH FACTOR α FRAGMENTS[a]

Peptide number	Sequence
1	TGF-α (34–50)
2	Ac-TGF-α (34–43)-NH$_2$
3	[S-AcmCys^{21,32}]TGF-α (21–33)
4	[Ala^{32}]TGF-α (22–43)-NH$_2$
5	[Ala^{16}]TGF-α (1–21)-NH$_2$
6	[S-AcmCys^{16,32}Ala^{21}]TGF-α (15–33)
7	[Ala^{21}]TGF-α (16–33)
8	TGF-α (1–15)
9	[Tyr^{32}]TGF-α (22–32)
10	TGF-α (34–50)
11	TGF-α (34–43)
12	TGF-α (41–50)
13	TGF-α (26–36)

[a] Peptides 1 through 7 were prepared by the Merrifield method of solid-phase peptide synthesis, and peptides 8 through 13 were prepared by Peninsula Laboratories. (From Ref. 7.)

sequence of intact EGF, but instead assumed an ellipsoidal conformation.[16]

Materials and Methods

Synthesis of Peptide Fragments

The primary amino acid sequence of rat TGF-α (residues 1–50) is shown in Fig. 1. Fragments of TGF-α evaluated for activity are listed in Table I. Peptides 1–7 are prepared by solid-phase peptide synthesis (SPPS) using double coupling with dicyclohexylcarbodiimide (DCC) active ester formation with a Peptides International Synthor 2000 automated synthesizer (Louisville, KY), and peptides 8–13 are prepared by Peninsula Laboratories (San Carlos, CA). An extensive description of SPPS techniques for preparing mature TGF-α (1–50) have been presented in a previous volume of this series by Tam.[17] The synthetic strategies we use for the TGF-α fragments are similar to that used by Tam with the α-amino groups protected with *tert*-butyloxycarbonyl (Boc) group and the following side-

[16] K. H. Han, C. H. Niu, and P. P. Roller, *Biopolymers* **27,** 923 (1988).
[17] J. P. Tam, this series, Vol. 146, p. 127.

chain protecting groups: Arg(Tos), Asp(Cxl), Cys(Acm) or Cys(Meb), Glu(Cxl), His(Bom), Lys(Cl-Z), Ser(Bzl), and Tyr(Dcb). Each synthetic cycle consists of (1) 5 and 25 min deprotection with 40% trifluoroacetic acid (TFA) (v/v)/10% anisole (v/v)/50% dichloromethane (v/v), (2) 5 min neutralization with 10% triethylamine in dichloromethane, and (3) double couplings (3 and 16 hr) with Boc-amino acid in the presence of N-hydroxybenzotriazole (HOBt) and DCC both at 2.5-fold mole excess over amino acid on resin. Asn and Gln are coupled as their preformed p-nitrophenyl esters in the presence of HOBt, and Arg(Tos) and Gly are coupled with DCC omitting HOBt. All couplings are monitored with the Kaiser test. Triethylamine is distilled from ninhydrin; dichloromethane is distilled from potassium carbonate; dimethylformamide is distilled under reduced pressure. Other solvents and reagents are of analytical grade.

Peptides are deprotected and cleaved from the resin with anhydrous HF (1 hr at 0°) in the presence of 5% p-cresol (v/v) and 5% dimethyl sulfide (v/v). After evaporation of HF and scavengers, solid residues are washed with diethyl ether, extracted with 10% acetic acid (v/v), and lyophilized. Residues are dissolved in 30% acetic acid (v/v) and desalted on a Sephadex G-25 column eluted with 30% acetic acid. The major peak of each peptide is pooled, lyophilized, and purified by reversed-phase high-performance liquid chromatography (HPLC) using a C_{18} column (10 × 250 mm, 300 Å pore size, Vydac, Hesperia, CA) with a gradient of 15 to 30% acetonitrile/water containing 0.05% TFA. We have found that the separation of peaks is substantially improved with silica matrices that have large pore sizes, that is, 300 Å, compared to resins that have smaller pore sizes such as 75 Å. In addition, it is extremely important to use new reversed-phase materials to purify fragments of TGF-α. We have found that reversed-phase HPLC columns continue to release small amounts of active TGF-α and EGF even after extensive washing under extremely effective elution conditions. Since the activities of small fragments of TGF-α are very low compared to those of the intact molecule, artifacts can be generated by very low levels of EGF or TGF-α eluting from columns during purification of peptide fragments.

Peptides containing cystine disulfide bonds are produced by washing the residues obtained after HF cleavage with diethyl ether containing 1% 2-mercaptoethanol, then extracting with 10% acetic acid under nitrogen. After lyophilization, residues are dissolved in 0.2% acetic acid saturated with nitrogen, and 2 N aqueous ammonia is added gradually to give a final pH of 7.0 to 7.5. Solutions of the peptides are treated with 20 μM $K_3Fe(CN)_6$ until a permanent yellow color is generated and then stirred for an additional 20 min. Solutions are passed through an anion-exchange column (AG 3-X4, acetate form) to remove excess ferri- and ferrocyanide ions and then lyophilized. Residues are desalted by chromatography on a Sephadex G-15 column eluted with 30% acetic acid and then purified by HPLC as described above.

Competition of Epidermal Growth Factor Binding

Peptides are tested for their ability to compete for ^{125}I-labeled EGF binding using two receptor sources: placental cell microvillus membranes[18] and A431[19] cells. Placental microvillus membrane is our preferred source of EGF/TGF-α receptor because very large amounts of relatively pure receptor can be obtained rapidly at very low cost. Once a large preparation of placental microvillus membranes has been prepared, it can be stored in small aliquots at $-80°$ until thawed with essentially no loss of activity. Placental microvillus membranes are prepared from fresh, normal term human placenta that is immediately cooled to $4°$ and maintained at this temperature throughout the procedure. Placental membranes are removed and the placental lobes cut into small pieces (1 cm^3 cubes) using crossed scalpel blades. The pieces of placental tissue are washed 3 to 5 times in 0.15 M NaCl containing 50 mM CaCl$_2$ until the solution is free of blood. Failure to remove red blood cells results in substantial contamination of microvilli with material that may contribute to nonspecific binding. Washed pieces of tissue are vigorously stirred on a magnetic stirrer for 1 hr at $4°$ in 1.5 volumes of 0.15 M NaCl. This step sheers off microvilli from the plasma membrane of placental cells while minimally disrupting the cells. Large tissue pieces are removed by filtration through a coarse pore nylon membrane (500 μm pores), and the filtrate is centrifuged at 500 g for 10 min. The supernatant is centrifuged at 25,000 g for 30 min, after which the supernatant is discarded and the pellet resuspended in buffer (150 mM NaCl, 2 mM CaCl$_2$, and 50 mM NaH$_2$PO$_4$, pH 7.4) at a high concentration. The microvillus membrane suspension is frozen in 1-ml aliquots at $-80°$ and thawed one time for use in the receptor-binding assay. Aliquots of placental membrane (100 μl containing ~50 μg protein) are incubated in 12 × 75 mm polypropylene tubes with a 100 μl of ^{125}I-labeled EGF [100,000 counts/min (cpm) at a concentration of ~100 pM] and 100 μl of buffer containing increasing concentrations of either unlabeled EGF or peptides for 2 hr at $37°$. Four milliliters of buffer at $4°$ is then added, and the placental microvillus membranes are pelleted by centrifugation at 7000 g for 20 min at $4°$, the supernatant removed by aspiration, and pellets counted with a γ scintillation counter.

[18] Ch. V. Rao, N. Ramani, N. Chegini, B. K. Stadig, F. R. Carman, Jr., P. G. Woost, G. S. Schultz, and C. L. Cook, *J. Biol. Chem.* **260,** 1705 (1985).
[19] J. E. DeLarco and G. J. Todaro, *Proc. Natl. Acad. Sci. U.S.A.* **75,** 4001 (1978).

Competition for binding of ^{125}I-labeled EGF to receptors on A431 cells is performed using fixed monolayers of cells. A431 cells are seeded onto 24-well plates and grown in Dulbecco's modified Eagle's medium (DMEM) containing 10% calf serum (v/v) to a density of approximately 10,000 cells/ well, then washed and fixed by a brief 10-min exposure to 10% formalin in phosphate-buffered saline (PBS). Brief fixation by formalin does not destroy EGF binding, and replicate values are more consistant since fixed cells do not slough off plates as easily as do unfixed cells. Using a procedure similar to that described above, 100 µl of buffer containing ^{125}I-labeled EGF (final concentration ~300 pM) is added to each well along with 100 µl of buffer containing increasing concentrations of either unlabeled EGF or TGF-α peptide fragments. After incubation for 1 hr at 37°, the wells are washed with PBS and the cells solubilized by the addition of 1 ml of 0.5 N NaOH; the radioactivity is measured with a γ scintillation counter.

Induction of DNA Synthesis

Peptides are tested for the ability to stimulate DNA synthesis using incorporation of radioactive analogs of thymidine, 5-[^{125}I]iodo-2'-deoxyuridine (IdU), or tritiated thymidine into trichloroacetic acid-insoluble material of diploid human foreskin fibroblasts (HFF) or mouse 3T3 fibroblast cells. Cultures of HFF are established from explants of newborn foreskin, and HFF are seeded in 96-well plates and grown to confluency in DMEM containing 10% calf serum and antibiotics. Quiescent cultures of HFF which have been held in 0.2% calf serum for 2 days receive 10 ng/ml of TGF-α or 100 ng/ml of peptides. After 8 hr, cultures are labeled with IdU (10 µCi/ml) for 2 hr, then washed with PBS followed by 10% trichloroacetic acid (TCA). Cells are dissolved in 1 ml of 1 N NaOH, the amount of radioactivity incorporated into TCA-insoluble material is measured by γ scintillation counting, and the mean level for triplicate wells is calculated. The effect of the peptides on DNA synthesis in 3T3 fibroblast cells is determined using the procedure described below.

Inhibition of EGF-Induced DNA Synthesis

Confluent cultures of the J-2 clone of mouse 3T3 fibroblasts (H. Green, Harvard University) are washed with PBS and held in chemically defined medium (CDM) (equal parts of DMEM, Medium 199, and Ham's F10 and buffered with 25 mM HEPES to pH 7.4) containing 0.5% calf serum for 24 hr, then harvested with trypsin. Twenty-four-well plates are seeded with 30,000 cells/well in 500 µl of CDM containing 0.5% calf serum and tritiated thymidine (1 µCi/ml, Amersham, Arlington Heights, IL, [*methyl*-1',2'-^{3}H]thymidine, final specific activity 100 µCi/mmol). Five hundred microliters of CDM containing 0.5% calf serum and the indicated levels of serum, EGF, or TGF-α peptides is added and the cells incubated for 72 hr. Culture wells are washed twice with PBS, the DNA is precipitated with 5% trichloroacetic acid (TCA) and washed with methanol, the cells are then dissolved in 1 ml of 1 N NaOH, and the radioactivity is measured with a β scintillation counter.

Stimulation of Anchorage-Independent Cell Growth

A soft agar colony growth assay is performed using normal rat kidney fibroblasts (NRK), clone 49F.[19] Agar plates are prepared in 60-mm petri dishes by first applying a 2-ml base layer of 0.5% agar (w/v) (Difco, Detroit, MI, Agar Noble) in DMEM containing 10% calf serum. Over this basal layer, an additional 2 ml of 0.3% agar in the same medium/calf serum mixture is added which contained 30,000 NRK cells/ml and TGF-α (5 ng/ml) or TGF-α fragment peptides (5 or 50 ng/ml). The cells are incubated at 37° in a humidified atmosphere of 5% CO_2 in air. Colonies are measured unfixed and unstained by using a microscope with a calibrated grid. The number of soft agar colonies represents the number of colonies containing a minimum of 20 NRK cells per six random low power fields 10 days after seeding. Plates of NRK cells treated with TGF-β alone or without TGF-β did not form colonies.

Growth Factor-Induced Phosphorylation

Placental microvillus membranes are isolated as described above, and aliquots (100 μl containing 50 μg) are incubated for 10 min at 22° in phosphorylation buffer (20 mM $MgCl_2$, 1.5 mM $MnCl_2$, 25 mM HEPES, pH 7.4) which contains EGF, TGF-α, or TGF-α peptides at the designated amounts. The membranes are then cooled to 4°, and phosphorylation is initiated by addition of 5 μM [γ-^{32}P]ATP, 100 μM adenyl-5'-yl imidodiphosphate (AMP-PNP), and 10 mM Na_2MoO_4 (which reduces phosphatase activity). The final reaction mixture is 200 μl. After incubation for 10 min at 4°, phosphorylation is stopped by addition of Laemmli[20] sodium dodecyl sulfate (SDS) sample buffer and heating for 15 min at 60°. Samples are chromatographed by SDS-10% (w/v) polyacrylamide gel electrophoresis, the gels are stained with Coomassie blue and dried under vacuum, and autoradiography is performed at −70° using Kodak X-Omat AR film and Lightning Plus intensifying screen (Du Pont, Wilmington, DE). Proteins

[20] U. K. Laemmli, *Nature (London)* **227**, 680 (1970).

of known molecular weights are used as standards to calculate the apparent molecular weight of phosphorylated bands.

Iodination of Epidermal Growth Factor

Recombinant human EGF was provided by Dr. Carlos Nascimento of Chiron Corporation (Emeryville, CA). We have used several different methods for radioactively iodinating EGF including lactoperoxidase/glucose oxidase, Iodogen, Bolton–Hunter, and chloramine-T. All these methods produce ^{125}I-labeled EGF which is suitable for the receptor binding assays described here. The main objectives of the iodination are to obtain specific activities in the range of 100 to 400 μCi/μg and to retain biological activity of the iodinated EGF sample. We routinely use chloramine-T as the oxidizing agent in the iodination reaction because the reaction is the simplest, least expensive, and most consistent in our hands. However, it is important to recognize that the chloramine-T method is also the most vigorous and potentially damaging method, and it may not be compatible with some applications. For example, the rabbit polyclonal antiserum we prepared to TGF-α peptide 1 did not recognize mature TGF-α iodinated by chloramine-T, but the antiserum bound very well with mature TGF-α iodinated by the Bolton–Hunter reagent.

The procedure used to iodinate EGF is modified from Carpenter and Cohen[21] and is carried out using an approximately equal molar ratio of EGF to Na^{125}I. The reaction is performed in a 15-ml conical centrifuge tube mounted on a magnetic stirring plate with a short piece of paper clip as a stirring bar. Twenty microliters of 50 mM phosphate buffer, pH 7.5, containing 2.5 μg EGF is added to 20 μl of 500 mM phosphate buffer, pH 7.5, containing 1 mCi of carrier-free Na^{125}I (13.4 mCi/μg of iodine). Ten microliters of 50 mM phosphate buffer containing 100 μg of chloramine-T is added, and the reaction is stopped after 25 sec by the addition of 100 μl of 50 mM phosphate buffer containing 100 μg of sodium metabisulfite. One hundred microliters of 1% KI (w/v) solution is added followed by 100 μl of 1% γ-globulin (w/v), and the reaction mixture is added to a 1 × 50 cm column of Sephadex G-50 and eluted in 1-ml fractions with column buffer (150 mM NaCl, 1 mM CaCl$_2$, and 50 mM phosphate, pH 7.5). The labeled EGF is pooled and stored at 4°. Specific activities generally range from 100 to 300 μCi/μg, and the labeled EGF is usable for 4 to 6 weeks. Rechromatography on Sephadex G-25 column is not usually necessary prior to receptor competition assays.

---

[21] G. Carpenter and S. Cohen, *J. Cell Biol.* **71**, 159 (1976).

TABLE II
PHYSICOCHEMICAL PROPERTIES OF PEPTIDE FRAGMENTS[a]

Peptide number	R_f		K' (HPLC)
	BAW	BAWP	
1	0.08	0.77	2.90
2	0.05	0.76	3.96
3	0.09	0.78	4.40
4	0.02	0.70	3.39
5	0.15	0.74	4.55
6	0.05	0.78	3.18
7	0.16	0.77	3.92

[a] Purified peptides gave the indicated relative migration in TLC systems: BAW, 1-butanol–acetic acid–water (4:1:5, v/v, upper phase); BAWP, 1-butanol–acetic acid–water–pyridine (15:12:10:3, v/v). Analytical C_{18} reversed-phase HPLC was performed using 20-min linear gradient (flow rate 1 ml/min) from 15 to 40% acetonitrile/water (v/v) containing 0.05% TFA.

Results

Chemical Characterization of Peptide Fragments

The primary amino acid sequence of rat TGF-α is shown in Fig. 1, and the hydrophilicity plot of TGF-α is shown in Fig. 2. Three regions contain high indexes of hydrophilicity and encompass amino acids 8–13, 25–31, and 42–47. These sequences are located essentially within the three disulfide loops. Table I lists the synthetic fragments of TGF-α analyzed. Amino acid compositions of the TGF-α fragments we synthesized (peptides 1–6) were within experimental error of predicted values for all the peptides. Table II lists the physicochemical properties of the TGF-α fragments (peptides 1–6), and all the peptides were greater than 95% pure by HPLC analysis and gave a single spot in two thin-layer chromatography (TLC) systems.

Competition of Epidermal Growth Factor Binding

As shown in Fig. 3, EGF effectively competed for ^{125}I-labeled EGF binding to human placental membranes, with 50% displacement at approximately 2 nM and 90% of the total binding displaced by 10 μM unlabeled

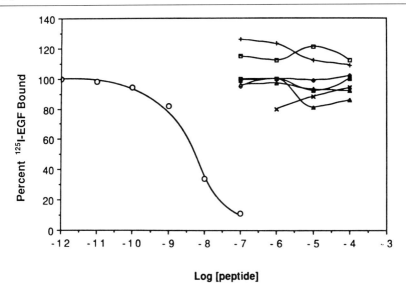

FIG. 3. Competition of EGF binding to human placental membranes. Aliquots of placental membranes (50 μg) were incubated with a constant amount of ^{125}I-labeled EGF (100 pM) and unlabeled EGF or peptides at the indicated levels. After 2 hr at 37°, tubes were centrifuged and pellets counted. Values are the mean of triplicate samples. Peptides are EGF (○), peptide 1 (◇), peptide 2 (△), peptide 3 (▲), peptide 4 (x), peptide 5 (□), peptide 6 (■), and peptide 7 (+).

EGF. In contrast, none of the TGF-α fragments (peptides 1–7) effectively competed for ^{125}I-labeled EGF binding even at 100 μM concentrations. Utilizing A431 cells as the EGF/TGF-α receptor source, the same results were obtained: EGF (10 nM) displaced 92% of ^{125}I-labeled EGF binding whereas none of the TGF-α fragments (peptides 1–5, 7) competed for ^{125}I-labeled EGF binding even at 100 μM.

Induction of DNA Synthesis

TGF-α (10 ng/ml) stimulated incorporation of [^{125}I]IdU 7-fold over control cultures of human foreskin fibroblasts. Fragments of TGF-α (peptides 1–5, 7) tested at 100 ng/ml all failed to stimulate DNA synthesis above control levels. When 3T3 fibroblasts were used for the mitogenesis assay, EGF (1 nM) stimulated thymidine incorporation 2-fold over control cultures and peptides 1 to 4 (100 μM) again failed to stimulate DNA synthesis.

Inhibition of Epidermal Growth Factor-Induced DNA Synthesis

3T3 cells incubated in CDM containing 0.5% calf serum incorporated small amounts of tritiated thymidine. Addition of 10% calf serum increased thymidine incorporation approximately 5-fold, and addition of 1 nM EGF increased thymidine incorporation approximately 2-fold over incorporation in the presence of 0.5% calf serum. Simultaneous addition of 1 nM EGF with each of TGF-α fragments 1 through 4 at 100 μM failed to significantly reduce thymidine incorporation below the level of stimulation measured with EGF alone.

Stimulation of Anchorage-Independent Cell Growth

TGF-α stimulated a large number of colonies of NRK cells in soft agar. In contrast, peptides 1–5 and 7 failed to stimulate colony formation even at 10-fold higher concentration (50 ng/ml).

Growth Factor-Induced Phosphorylation

Autoradiography of SDS-polyacrylamide gels of placental microvillus membranes incubated with EGF or TGF-α (2 μM) and [^{32}P]ATP showed enhanced phosphorylation of the EGF/TGF-α receptor (170,000 daltons) and p35 (35,000 daltons) relative to membranes incubated without the growth factors. Placental microvillus membranes incubated with TGF-α fragments (peptides 1–5, 7) at 200 μM showed no increased phosphorylation of the EGF/TGF-α receptor or p35 above control.

Discussion

Analysis of the peptide fragments of TGF-α for activities characteristic of mature TGF-α gave uniformly negative results in five different assays. The peptides did not compete for ^{125}I-labeled EGF binding to receptors of intact A431 cells or human placental membranes even at 1000 times the concentration of EGF that completely displaced the labeled EGF. Concentrations of the peptides up to 10 μM did not stimulate DNA synthesis of human foreskin fibroblasts or 3T3 cells, nor did they stimulate anchorage-independent growth of NRK cells. Also, the peptides did not inhibit EGF-induced stimulation of DNA synthesis or stimulate phosphorylation of the EGF/TGF-α receptor or p35 even when added at 10 μM.

In addition to the seven peptides we synthesized, six peptides with sequences overlapping our peptides were synthesized and purified by Peninsula Laboratories. These fragments (peptides 8–13) also uniformly failed to produce any response characteristic of TGF-α in the five assays.

Thus, our results of five different assays testing the 13 peptides consistently indicated that none of the individual sequences tested had the ability to bind significantly to the EGF/TGF-α receptor. Based on these results, we conclude either that the major receptor-binding domain of TGF-α is not contained entirely within any of the single peptide sequences reported here or that the conformation of the peptides in solution is sufficiently different from their conformation in intact TGF-α to prevent effective binding to the receptor. Since the solution structures predicted by NMR suggest that the disulfide loops of TGF-α and EGF are folded over each other, we feel that it is likely that the receptor-binding domain is composed of separate regions of the TGF-α sequence which fold into the correct alignment when the hormone assumes its native conformation rather than being formed from a single segment of the sequences. In addition, linear TGF-α, which has the disulfide bonds reduced, and polymeric EGF molecules have 100- to 1000-fold lower potency, respectively, than the correctly folded hormones, even though both structures encompass the entire primary amino acid sequences.[11,22] There are limited data, however, to support the possibility that the conformation of the TGF-α fragments differs significantly from structure of the sequences in the mature proteins. The cyclic peptide comprising the second disulfide loop of TGF-α appears to assume an ellipse structure rather than the β-sheet structure observed in intact TGF-α.[14] As the three-dimensional structure of TGF-α and EGF become more defined, it may be possible to synthesize fragments with constrained conformations that will closely model the conformations of the regions in the intact molecules.

[22] A. R. Hanauske, J. B. Buchok, L. R. Pardue, V. A. Muggia, and D. D. Von Huff, *Cancer Res.* **46,** 5567 (1986).

[20] Purification of Amphiregulin from Serum-Free Conditioned Medium of 12-O-Tetradecanoylphorbol 13-Acetate-Treated Cell Lines

By MOHAMMED SHOYAB and GREGORY D. PLOWMAN

Introduction

We have recently reported the isolation of a novel growth regulatory glycoprotein, termed amphiregulin (AR), from the serum-free conditioned medium of MCF-7 human breast carcinoma cells that had been treated with 12-O-tetradecanoylphorbol 13-acetate (TPA).[1] Amphiregulin inhibits the growth of A431 human epidermoid carcinoma and other human cells but stimulates the proliferation of normal human fibroblasts, murine keratinocytes, and other normal and tumor cells.[1,2] The mature protein exists in two forms: the truncated form contains 78 amino acids, whereas a larger form of AR contains 6 additional amino acids at the amino-terminal end.[2] Amphiregulin is a member of the epidermal growth factor (EGF) family. The carboxyl-terminal half of the molecule exhibits significant sequence identity to both EGF and transforming growth factor α (TGF-α), possessing, in particular, the six essential cysteine residues which determine the positions of the three disulfide bridges.[2-10] The amino-terminal half of AR is extremely hydrophilic, and a tetrapeptide Arg-Lys-Lys-Lys is repeated twice in this region. Such sequences have been reported to serve as nuclear targeting signals.[11] The functional roles played by this sequence and by the entire hydrophilic structure of AR are currently under investigation. Amphiregulin interacts with the EGF receptor but not as efficiently as EGF does. AR fully supplants the requirement for EGF or TGF-α in murine keratinocyte growth.[2] Unlike EGF or TGF-α, AR does not induce anchorage-independent growth of normal rat kidney (NRK) cells in the presence of TGF-β.[2]

Amphiregulin was first isolated and characterized from the serum-free conditioned medium of TPA-treated MCF-7 cells. Untreated cells produce negligible amounts of AR. The original purification method, sometimes with minor modifications, has been successfully used in our laboratory to purify AR from other TPA-induced cells. This method is the subject of this chapter.

---

[1] M. Shoyab, V. L. McDonald, J. G. Bradley, and G. J. Todaro, *Proc. Natl. Acad. Sci. U.S.A.* **85**, 6528 (1988).
[2] M. Shoyab, G. D. Plowman, V. L. McDonald, J. G. Bradley, and G. J. Todaro, *Science* **243**, 1074 (1989).
[3] G. Carpenter and S. Cohen, *Annu. Rev. Biochem.* **48**, 193 (1979).
[4] C. R. Savage, Jr., T. Inagami, and S. Cohen, *J. Biol. Chem.* **247**, 7612 (1972).
[5] C. R. Savage, Jr., J. H. Hash, and S. Cohen, *J. Biol. Chem.* **248**, 7669 (1973).
[6] R. E. Doolittle, D. F. Feng, and M. S. Johnson, *Nature (London)* **307**, 558 (1984).
[7] H. Gregory, *Nature (London)* **257**, 325 (1975).
[8] R. J. Simpson, J. A. Smith, R. L. Moritz, M. J. O'Hare, P. S. Rudland, J. R. Morrison, C. J. Lloyd, B. Crego, A. W. Burgess, and E. C. Nice, *Eur. J. Biochem.* **153**, 629 (1985).
[9] G. Todaro, C. Fryling, and J. DeLarco, *Proc. Natl. Acad. Sci. U.S.A.* **77**, 5258 (1980).
[10] H. Marquardt, M. W. Hunkapillar, L. E. Hood, and G. J. Todaro, *Science* **223**, 1079 (1985).
[11] B. Roberts, *Biochim. Biophys. Acta* **1008**, 263 (1989).

Cell Growth Modulatory Assay

AR inhibits the growth of A431 cells and stimulates the proliferation of human fibroblasts. These growth modulatory properties of AR have been used to quantitate AR. Particularly, the inhibition by AR of ^{125}I-labeled deoxyuridine incorporation into the DNA of A431 cells is routinely utilized in our laboratory for assaying AR during purification.

Reagents

>Cell lines: A clone of A431, termed A431-A3, developed in our laboratory is maintained in Dulbecco's modified Eagle's medium (DMEM) supplemented with 10% (v/v) heat-inactivated fetal bovine serum (FBS), L-glutamine, penicillin, and streptomycin. A human forearm fibroblast line (SS) is similarly maintained.
>^{125}I-Labeled deoxyuridine (Amersham, Arlington Heights, IL)
>Phosphate-buffered saline (PBS)
>Methanol
>Sodium hydroxide, 1 M

Procedure

The assays are performed in flat-bottomed 96-well microtiter plates (Falcon 3072). Human epidermoid carcinoma of the vulva cells (A431) are used as test cells for growth inhibitory activity (GIA), and human foreskin fibroblasts (SS) are used as indicator cells for growth stimulatory activity (GSA). Approximately 3.5 × 10^4 cells in 50 μl of test medium [DMEM supplemented with 5% heat-inactivated FBS, penicillin/streptomycin (PS) and glutamine] are placed in all except peripheral wells. The peripheral wells receive 50 μl PBS. Three hours later, 50 μl of test sample in test medium is added to all wells, except control wells which receive only 50 μl of test medium. Three wells are used for each concentration of test sample. Plates are incubated at 37° for 2–3 days. After this, 100 μl of a solution of [^{125}I]iodo-2′-deoxyuridine {[^{125}I]IUdR; 4 Ci/mg, 0.5 mCi/ml (2 μl/ml in test medium)} is added to each well, and plates are incubated at 37°. After 4–6 hr, the medium is aspirated from the wells, which are then washed once with 200 μl PBS. Methanol (200 μl) is added to each well, the plates are incubated for 10 min at room temperature, and the methanol is removed by aspiration. Two hundred microliters of 1 M sodium hydroxide is added to each well, and the plates are incubated for 30 min at 37°. Sodium hydroxide is removed with Titertek plugs (Flow Labs, McLean, VA). The plugs are transferred to 12 × 75 mm plastic tubes and counted in a γ counter to quantify [^{125}I]IUdR incorporation.

Results can be expressed as GIA units or GSA units, one GIA unit being the amount of factor needed to inhibit 50% of ^{125}I-labeled deoxyuridine incorporation into cells. One GSA unit is defined as the amount of factor required to increase ^{125}I-labeled deoxyuridine incorporation into cells by 100%. One GIA or GSA unit corresponds to 100–500 pg of AR depending on assay conditions.

Cell Culture and Collection of Conditioned Medium

MCF-7 cells are cultured in T150 Corning tissue culture flasks in a total volume of 25 ml of 50% IMDM (Iscove's modified Dulbecco's medium) plus 50% DMEM containing 0.6 μg/ml insulin and 15% heat-inactivated FBS (MCF-7 complete medium). Approximately 1×10^6 cells are seeded per flask and incubated at 37° with 5% CO_2. On day 6, all medium is removed, and 20 ml of fresh MCF-7 complete medium containing 100 ng/ml of TPA is added to each flask. Forty-eight hours later, the medium is removed, each flask is rinsed with 15 ml of 50% IMDM plus 50% DMEM (serum-free medium), and 25 ml of fresh serum-free medium is added to each flask and incubated at 37° with 5% CO_2. Four days later, the conditioned serum-free medium is collected, centrifuged to remove debris, and stored at $-20°$. Cells are again fed with 25 ml/flask of serum-free medium, and conditioned serum-free medium is collected as above every third or fourth day. An aliquot of the conditioned medium from each collection is assayed for growth inhibitory activity (GIA) on A431 human epidermoid carcinoma cells. Usually, three to four rounds of conditioned medium are collected from each batch of TPA-treated MCF-7 cells.

Concentration and Preparation of Acid-Soluble Extract

About 4500 ml of conditioned medium is thawed and centrifuged at 4° for 15 min at 3500 rpm. The supernatant is concentrated at 4° in an Amicon 2-liter concentrator using a YM10 membrane (Amicon, Danvers, MA). When the volume of retenate becomes about 200 ml, 1000 ml of cold Milli-Q water is added, and the mixture is reconcentrated to 200 ml. The concentrate is removed and transferred to a precooled 250-ml Corning centrifuge bottle. Concentrated acetic acid is slowly added with stirring to a final concentration of 1 M acetic acid. The mixture is allowed to stand for 1 hr at 4° and centrifuged for 20 min at 40,000 g at 4° in a Sorval RC-5B centrifuge. The supernatant is removed and stored at 4°. The pellet is suspended in 30 ml of 1 M acetic acid and recentrifuged as described above. The supernatant is again carefully removed and pooled with the first supernatant and then dialyzed against 17 liters of 0.1 M acetic acid in

No. 3 Spectrapor dialysis tubing (molecular weight cutoff approximately 3000). The dialysis buffer is changed 3 times over a 2-day period. The retenate is lyophilized. The dry material is removed, pooled, weighed, and stored at $-20°$ until further use. This material is referred to as crude powder.

Reversed-Phase Liquid Chromatography

Crude powder (950 mg, from ~9 liters of serum-free conditioned medium) is suspended in 300 ml of 0.1% TFA (trifluoroacetic acid) and centrifuged for 20 min at 7000 g. The supernatant is carefully removed and applied to a column of preparative C_{18} (2.54 × 27 cm, 55–105 μm, Waters, Milford, MA) equilibrated with 0.1% TFA in water. The chromatographic support is suspended in acetonitrile (CH_3CN) with 0.1% TFA, the slurry is poured into a column 2.54 cm in diameter, and the column is washed with 400 ml of 0.1% TFA in water. The flow rate is 4 ml/min, and the chromatography is carried out at room temperature. The column is washed with 650 ml of 0.1% TFA in water. The flow-through and wash are collected together. Then stepwise elution is performed as follows: (1) 650 ml of 20% CH_3CN/water with 0.1% TFA, (2) 650 ml of 40% CH_3CN/water with 0.1% TFA, (3) 650 ml of 60% CH_3CN/water with 0.1% TFA, and (4) 650 ml of 100% CH_3CN with 0.1% TFA. An aliquot is taken from each fraction and tested for GIA. About 77% of total growth inhibitory activity appeared in the flow-through and wash fractions.

Flow-through and wash fractions are injected isocratically onto a preparative Partisil 10 ODS-3 column (10 μm, 2.2 × 50 cm, Whatman, Clifton, NJ) attached to a high-performance liquid chromatography (HPLC) system (Waters). The flow rate is set at 4 ml/min. Once the sample has passed onto the column, the column is washed with 250 ml of 0.1% TFA in water. A linear gradient is generated between the primary solvent, 0.1% TFA in water, and the secondary solvent, acetonitrile containing 0.1% TFA. The gradient conditions are as follows: 0 to 15% in 10 min, 15 to 15% in 30 min, 15 to 25% in 150 min, 25 to 65% in 100 min, and 65 to 100% in 10 min. All solvents are HPLC grade. Fractions of 14 ml are collected, and aliquots of each fraction are assayed for GIA. Two broad peaks of activity are seen (Fig. 1A). The early eluting peak (eluted between 20 and 23% acetonitrile) is further purified and characterized.

Fractions 47 to 62 are pooled. Two hundred twenty-four milliliters of 0.1% TFA in water is added to the pooled fractions. The mixture is isocratically injected onto a semipreparative μBondapak C_{18} column (7.8 × 300 mm, Waters) at a flow rate of 2 ml/min at room temperature. The linear gradient conditions are 0 to 17% in 10 min, 17 to 17% in 30 min,

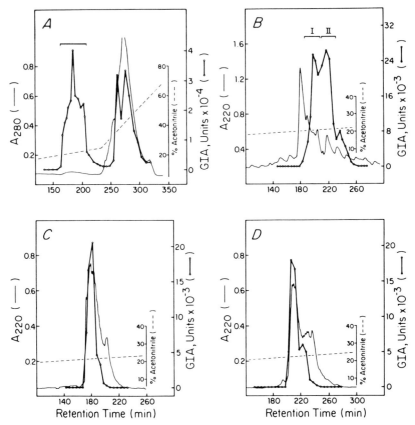

FIG. 1. Purification of amphiregulin by reversed-phase (rp) HPLC. (A) Preparative rpHPLC of break-through and wash fractions. (B) Semipreparative rpHPLC of pooled fractions 47–62 from (A). (C) Analytical rpHPLC of pool I from (B). (D) Analytical rpHPLC of pool II from (B). (From Ref. 1.)

17 to 25% in 320 min, and 25 to 100% in 40 min. The flow rate is 1 ml/min during gradient, and 4-ml fractions are collected. Aliquots are taken and assayed for GIA. Two major peaks of activity are observed eluting at acetonitrile concentrations of approximately 20 and 21%, respectively (Fig. 1B).

Fractions 49–53 are pooled. Twenty milliliters of 0.1% TFA is added to the pooled fractions. The mixture is isocratically applied onto a μBondapak C_{18} column (3.9 × 300 mm, Waters) at a flow rate of 1 ml/min at room temperature. The gradient conditions are 0–18% in 10 min, 18–18% in 30 min, 18–25% in 280 min, and 25–100% in 20 min. The flow rate is

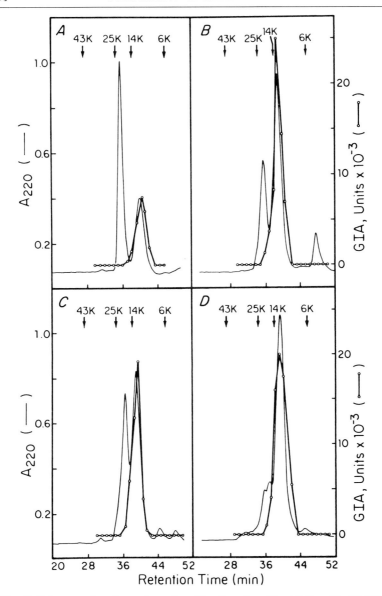

FIG. 2. Gel permeation chromatography of fractions from Fig. 1C,D on Bio-Sil TSK-250 columns. (A) HPLC of concentrated fraction 35 (Fig. 1C). (B) HPLC of concentrated fraction 36 (Fig. 1C). (C) HPLC of concentrated fraction 37 (Fig. 1C). (D) HPLC of concentrated fractions 41 and 42 (Fig. 1D). (From Ref. 1.)

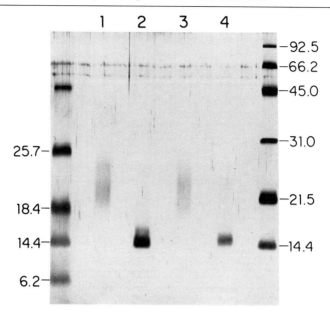

FIG. 3. SDS–polyacrylamide gel electrophoretic analysis of amphiregulin. A 15% SDS–polyacrylamide gel (0.75 mm × 18 cm × 15 cm, Bio-Rad) with a discontinuous buffer system was electrophoresed at a constant current of 30 mA by the method of Laemmli.[12] Proteins were detected by silver staining.[13] The markers were phosphorylase b, bovine serum albumin, ovalbumin, carbonic anhydrase, chymotrypsinogen A, soybean trypsin inhibitor, β-lactoglobulin, lysozyme, aprotonin, and insulin subunit. Lanes: 1, AR; 2, N-glycanase-treated AR; 3, SPE-AR; 4, N-glycanase-treated SPE-AR. Molecular weights ($\times 10^{-3}$) are shown. (From Ref. 1.)

0.4 ml/min, and 2-ml fractions are collected. Most of the activity emerges from the column at about 21.5% acetonitrile (Fig. 1C). Fractions 54–59 (Fig. 1B) are pooled and chromatographed exactly as described above (Fig. 1C). Most of the activity elutes from the column at an acetonitrile concentration of approximately 22.2% (Fig. 1D).

Gel-Permeation Chromatography

Fractions 35 to 38 (Fig. 1C) are individually concentrated to about 70 μl, using a Speed-Vac concentrator (Savant, Farmingdale, NY), to which is added an equal volume of acetonitrile containing 0.1% TFA. This 140-μl sample is injected onto two Bio-Sil TSK-250 columns (7.5 × 300 mm each, Bio-Rad, Richmond, CA) arranged in tandem. The elution is performed isocratically with a mobile phase of 50% acetonitrile/water with 0.1% TFA at room temperature. The flow rate is 0.4 ml/min and the chart speed 0.25 cm/min; 0.4-ml fractions are collected, and aliquots are assayed for GIA. Chromatographic profiles of fractions 35, 36, and 37 (Fig. 1C) are shown in Fig. 2A–C, respectively.

Fractions 41 and 42 (Fig. 1D) are pooled together, concentrated to 70 μl, and then subjected to gel-permeation chromatography as described above. The chromatographic profile is given in Fig. 2D.

Purified AR or S-pyridylethylated AR (SPE-AR) migrate in sodium dodecyl sulfate (SDS)–polyacrylamide gels as a single broad, diffuse band with a median relative molecular weight of 22,500 and a range of 20,000–25,000 (Fig. 3).[12,13] The treatment of AR and SPE-AR with N-glycanase resulted in the faster migration of these proteins. N-Glycanase-treated AR and N-glycanase-treated SPE-AR migrate as single bands with median molecular weights of 14,000 and 14,500, respectively (lanes 2 and 4). Similar results are observed when proteins are electrophoresed in a 15% gel under nonreducing conditions (data not shown). The calculated molecular weights of the larger and truncated forms of AR without any modification are 9759 and 9060, respectively.[2]

Storage

Homogeneous preparations of human AR are stable for at least several months in 50% acetonitrile in water with 0.1% TFA.

---

[12] U. K. Laemmli, *Nature (London)* **227,** 680 (1970).
[13] C. R. Merril, D. Goldman, S. A. Sedman, and M. H. Ebert, *Science* **211,** 1437 (1981).

Section IV

Epidermal Growth Factor Receptor and Related Receptors

[21] Construction and Expression of Chimeric Cell Surface Receptors

By REINER LAMMERS and AXEL ULLRICH

Introduction

Cell surface receptors of the tyrosine kinase family are multidomain polypeptides that are designed to carry out a variety of functions. These include binding of a specific ligand, dimerization, transmembrane activation of the cytoplasmic kinase, autophosphorylation, phosphorylation of cellular substrates, and, as a consequence, the generation of cellular signals. Because of the very similar structural organization of the receptors (see Ref. 1 for review) and their distinct characteristics with regard to some functions, the construction of chimeric receptors promised to be a viable approach in the analysis of structure–function relationships. Initial studies showed that chimeras between receptors for epidermal growth factors (EGF) and insulin are functional and mediate responses that are defined by the cytoplasmic domain.[2] Chimeras between insulin receptor (IR) and insulin-like growth factor I receptor (IGF-I-R) proved to be useful for the demonstration of receptor-specific signaling differences. Both receptors are coexpressed in a wide range of cell types and appear to play significantly different roles *in vivo*. In cultured cell lines, however, the receptors show similar properties in a variety of assays. Interpretation of the data is complicated by the fact that IGF-I and insulin cross-react with each others' receptor, although with different affinities. For analysis of receptor function, the latter effects can be overcome by chimeric molecules that contain an extracellular domain of IR and an intracellular domain of IGF-I-R or vice versa.

Construction of Chimeric Receptors

There are several possible strategies for the construction of chimeric receptors using cloned cDNA sequences. The most straightforward approach utilizes conserved restriction endonuclease cleavage sites within homologous sequences of closely related molecules, such as insulin and IGF-I receptors (Fig. 1A). If such sites are not available, appropriate

[1] Y. Yarden and A. Ullrich, *Annu. Rev. Biochem.* **57**, 443 (1988).
[2] H. Riedel, T. J. Dull, A. M. Honegger, J. Schlessinger, and A. Ullrich, *EMBO J.* **8**, 2943 (1989).

restriction sites localized within about 50 base pairs (bp) of the fusion site on either receptor may be connected using chemically synthesized double-stranded DNA (Fig. 1B). Alternatively, such restriction sites may be artificially created at the chosen fusion site of both receptors by specific oligonucleotide-directed mutagenesis without changing the amino acid sequence. The subsequent exchange of restriction fragments would be as straightforward as that shown in Fig. 1A (Fig. 1C).[3] A fourth possible approach involves the oligonucleotide-directed, specific deletion of sequences within a single-stranded plasmid containing DNA fragments of both receptors. Restriction fragments of both cDNAs are cloned in tandem into a cloning vector containing an f1 origin of replication. An oligonucleotide overlapping the sequences at both sides of the exchange site is then used to delete intervening sequences in the single-stranded DNA (Fig. 1D). Specific descriptions of experimental procedures for examples B and D in Fig. 1 are given below.

Example B

To create a DNA sequence coding for a chimeric receptor with extracellular and transmembrane domains of IGF-I-R and intracellular sequences of IR, an *Rsa*I restriction site at nucleotide position 2910 in the transmembrane region of the IGF-I-R cDNA sequence[4] is connected to a *Bgl*Isite (position 2961) present in the juxtamembrane region of the IR cDNA.[5] A chemically synthesized double-stranded DNA fragment with *Rsa*I and *Bgl*I ends is used to fill the gap and connect IGF-I-R and IR sequences flanking the fusion site (Figs. 1B and 2). Experiments are carried out with both receptor cDNAs cloned into an SV40 promoter-based expression vector also containing the genes for mouse dihydrofolate reductase and neomycin resistance [7.3 kilobases (kb) plus cDNA insert].

1. Two micrograms IGF-I-R plasmid DNA is cut with *Rsa*I, the enzyme is heat-inactivated (5 min, 65°), and the 28/25-mer complementary oligonucleotides are ligated to the fragment mixture. This is done using 5'-nonphosphorylated oligonucleotides at a 10-fold molar excess over cDNA fragments to prevent fragment religation. To remove the oligonucleotides

---

[3] I. Lax, F. Bellot, R. Howk, A. Ullrich, D. Givol, and J. Schlessinger, *EMBO J.* **8**, 421 (1989).

[4] A. Ullrich, A. Gray, A. W. Tam, T. Yang-Feng, M. Tsubokawa, C. Collins, W. Henzel, T. LeBon, S. Kathuria, E. Chen, S. Jacobs, U. Francke, J. Ramachandran, and Y. Fujita-Yamaguchi, *EMBO J.* **5**, 2503 (1986).

[5] A. Ullrich, J. R. Bell, E. Y. Chen, R. Herrera, L. M. Petruzzelli, T. J. Dull, A. Gray, L. Coussens, Y.-C. Liao, M. Tsubokawa, A. Mason, P. H. Seeburg, C. Grunfeld, O. M. Rosen, and J. Ramachandran, *Nature (London)* **313**, 756 (1985).

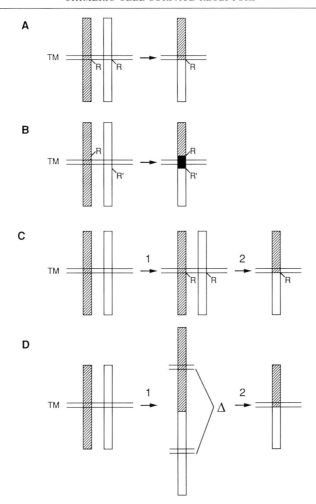

FIG. 1. Schematics of alternative approaches to receptor chimera construction. (A) Exchange of restriction fragments via homologous restriction site. (B) Connecting restriction fragments of two different receptors with an oligonucleotide covering the missing sequence. (C) Exchange of restriction fragments after creation of new restriction sites by point mutation. (D) Cloning restriction fragments of two receptors into an f1 origin-containing vector and subsequent oligonucleotide-directed deletion mutagenesis. TM, Transmembrane domain; R, R', restriction sites.

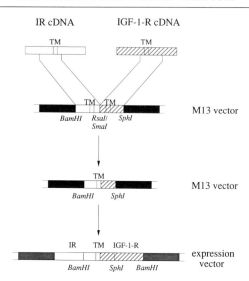

FIG. 2. Construction plan of a receptor chimera by an approach analogous to that of Fig. 1D. TM, Transmembrane domain.

from one side of the *Rsa*I fragment of interest [471 base pairs (bp); bp 2439–2910],[4] the mixture is phenol-extracted, ethanol-precipitated, and digested with *Cla*I to yield a 492-bp fragment (bp 2446–2910; *Rsa*I fragment plus synthetic DNA on one side), which is isolated from a 5% (w/v) polyacrylamide gel.[6] The yield is quantitated by gel electrophoresis and staining of an aliquot.

2. Similarly, 2 μg of the same plasmid is cut with *Cla*I within the IGF-I-R sequence (bp 2446) and at the 5' end of the cDNA within the vector, and the fragment containing the cDNA for the extracellular domain is isolated from a 1% low melting point agarose gel.[7]

3. A fragment containing cytoplasmic sequences of IR is isolated by digesting human insulin receptor (HIR) cDNA plasmid with *Bgl*I and *Bst*EII, and the 1360-bp fragment (bp 2961–4321) containing coding sequences for most of the intracellular domain is isolated.

4. Finally, the expression vector containing IR sequences is digested with *Bst*EII and *Cla*I, and the 7360-bp fragment containing the vector sequences is isolated (*Cla*I and *Bst*EII sites are identical to those in Steps 2 and 3).

[6] T. Maniatis, E. F. Fritsch, and J. Sambrook, "Molecular Cloning: A Laboratory Manual." Cold Spring Harbor Laboratory, Cold Spring Harbor, New York, 1982.

[7] F. Ausubel, R. Brent, R. E. Kingston, D. D. Moore, J. G. Seidman, J. A. Smith, and K. Struhl (eds.), "Current Protocols in Molecular Biology." Wiley, New York, 1987.

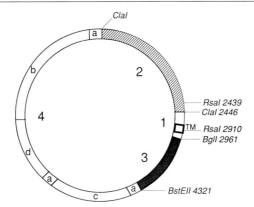

FIG. 3. Schematic of a hybrid receptor cDNA in an expression vector constructed from four different restriction fragments. TM, Transmembrane domain; 1, ClaI–RsaI fragment plus oligonucleotide; 2, ClaI fragment of IGF-I-R cDNA coding for extracellular domain; 3, BglI–BstEII fragment of HIR cDNA coding for intracellular domain; 4, Expression vector and location of (a) SV40 promoter (b) pBR322 sequences, including origin of replication and ampicillin resistance gene, (c) neomycin resistance gene, and (d) dihydrofolate reductase gene.

5. Equimolar amounts of the four isolated fragments are ligated in a 10-µl reaction volume. Subsequently, competent *Escherichia coli* cells are transformed with the ligation mixture and positive clones identified by restriction analysis.

Example D

A chimeric receptor consisting of extracellular ligand-binding sequences and the transmembrane domain of IR and cytoplasmic sequences of IGF-I-R is constructed following the strategy shown in Figs. 1D and 3.

1. Two micrograms each of the plasmids containing the cDNAs of I-R and IGF-I-R are digested with *Bam*HI/*Rsa*I and *Sma*I/*Sph*I, respectively. The resulting restriction fragments [1926–3042 (IR cDNA); 2736–3164 (IGF-I-R cDNA)] are isolated after electrophoresis on a 1.2% low melting point agarose gel as described above, and the yield is estimated by electrophoresing an aliquot on an agarose gel. In parallel, M13mp19RF is digested with *Bam*HI and *Sph*I restriction endonucleases, dephosphorylated with calf intestine phosphatase, phenol-extracted, and precipitated with 1/5 volume of 10 *M* ammonium acetate and 2.5 volumes of ethanol (to remove the *Bam*HI/*Sph*I linker). Equimolar amounts of the two receptor cDNA fragments and the linearized cloning vector are ligated in a volume

of 10 μl and transformed into competent cells of an F⁺ bacterial host. Positive clones containing IR and IGF-I-R cDNA fragments in tandem are isolated and confirmed by restriction analysis, and single-stranded DNA of the M13 clone is isolated following previously published procedures.[8]

2. For the fusion of specific IR and IGF-I-R sequences, an oligonucleotide is designed to enable the deletion of intervening sequences. In order to achieve this, a 30-mer complementary to 15 nucleotides on each side of the fusion point is synthesized and phosphorylated at its 5' end using T4 polynucleotide kinase. The reaction mixture containing 10 μl oligonucleotide (10 pmol/μl), 2 μl of 10× kinase buffer (500 mM Tris-HCl, pH 8, 100 mM MgCl$_2$), 1 μl of 100 mM dithiothreitol (DTT), 2 μl of 10 mM ATP, 4 μl water, and 1 μl of T4 polynucleotide kinase (10 U) is incubated for 30 min at 37°. The enzyme is then activated for 10 min at 70° and the solution stored at $-20°$ until use.

3. For deletion mutagenesis the phosphorylated oligonucleotide is annealed to the single-stranded M13 DNA containing IR and IGF-I-R coding sequences in tandem in a reaction solution containing the following components: 2 μl oligonucleotide (5 pmol/μl), 5 μl of M13 DNA (0.2 μg/μl), 2 μl water, and 1 μl of 10× TM buffer (100 mM Tris-HCl, pH 8, 100 mM MgCl$_2$). The tube (0.5 ml Eppendorf) containing the annealing mixture is placed in boiling water for 5 min and subsequently slowly (~30 min) cooled to room temperature. Then, to synthesize a complete complementary strand containing the mutated sequence, 1 μl of 10× TM buffer, 1 μl of 5 mM d(A,C,G,T)TP, 1 μl of 5 mM ATP, 1 μl of 100 mM DTT, 4 μl water, 1 μl of the Klenow fragment of *Pol*I (1 U), and 1 μl of T4 DNA ligase (1 U) are added and incubated at 14° for 4 hr. After the addition of 180 μl TE, the reaction mixture is frozen at $-20°$. An aliquot of the reaction mixture is used for the transformation of suitable competent host cells.

4. Plasmids containing the desired mutation may be identified by different means. (a) Replicative forms of M13 plaques are isolated and characterized by restriction analysis or size determination.[8] (b) Phage plaques containing correctly mutated sequences may be identified using the mutagenesis oligonucleotide as a hybridization probe under stringent conditions. For this procedure, replicas of the phage lawn are generated on nitrocellulose filters essentially as described previously.[6] Baked filters are prehybridized in 5× SSC, 0.1 g/l boiled salmon sperm DNA, 50 mM sodium phosphate, pH 6.8, 5× Denhardt's, and 100 μM ATP for 1 to 2 hr at 37°. The radioactively labeled oligonucleotide (1 ng/ml; phosphorylated with T4 polynucleotide kinase with a 3-fold molar excess of [γ-³²P]ATP, 5000 Ci/mmol) is then added without removing nonincorporated

---

[8] J. Messing, this series, Vol. 101, p. 20.

[γ-^{32}P] ATP. Hybridization is carried out at 37° overnight; filters are then washed 3 times with 6× SSC at room temperature for 15 min and subsequently twice with 3 M TMA-Cl solution [3 M tetramethylammonium chloride, 50 mM Tris-HCl, pH 8, 1 g/liter sodium dodecyl sulfate (SDS), 2 mM sodium EDTA, pH 8] at a temperature determined according to the procedure of Ullrich et al.[9] The dried filters are wrapped in Saran wrap and exposed for 3 hr at −80° using an intensifying screen (Cronex Lightning Plus, Kodak, Rochester, NY).

5. Positive plaques are localized using the hybridization signals and purified by replating and rescreening as described above. The mutation is verified by DNA sequencing, and the BamHI/SphI fragment containing the IR/IGF-I-R fusion is isolated and transferred into the expression plasmid. In this step, a full-length chimeric receptor cDNA is reconstituted.

Expression and Characterization of Chimeric Receptors

To further examine whether the mutagenesis experiment is successful and results in the construction of a chimeric receptor sequence, we employ a transient expression system initially developed by Eaton et al.[10] This system utilizes a modified 293 human embryonic kidney fibroblast cell line and a cytomegalovirus (CMV) promoter-based expression vector. Chimeric receptor cDNA sequences are cloned into a CMV vector, and CsCl gradient-purified plasmid DNA is prepared. Transfection of 293 cells is carried out as described by Chen and Okayama.[11] The cells are grown to confluence in Dulbecco's modified Eagle's medium/Ham's F12 (DMEM/F12, 50:50; GIBCO, Grand Island, NY, #86-5005), 10% fetal calf serum, 1 mM L-glutamine, and antibiotics, then split 1:5 into 6-well dishes 24 hr before transfection (afternoon of day 1). Each well is transfected with 4 μg of expression plasmid (afternoon of day 2); cells are exposed to calcium phosphate/DNA precipitate for 15 hr, followed by washing with fresh medium, and then 1 ml of medium containing 0.5% FCS is added (morning of day 3).

To examine the expression of chimeric receptors, cells are metabolically labeled with [^{35}S]methionine. The medium is carefully aspirated off the subconfluent cell monolayer, and 1 ml of methionine-free medium containing 50 μCi/ml L-[^{35}S]methionine and 10% (v/v) dialyzed fetal calf serum is added (evening of day 3). The supernatant is removed the next

---

[9] A. Ullrich, A. Gray, W. I. Wood, J. Hayflick, and P. H. Seeburg, *DNA* **3**, 387 (1984).

[10] D. L. Eaton, W. I. Wood, D. Eaton, P. E. Hass, P. Hollingshead, K. Wion, J. Mather, R. M. Lawn, G. A. Vehar, and C. Gorman, *Biochemistry* **25**, 8343 (1986).

[11] C. Chen and H. Okayama, *Mol. Cell. Biol.* **7**, 2745 (1987).

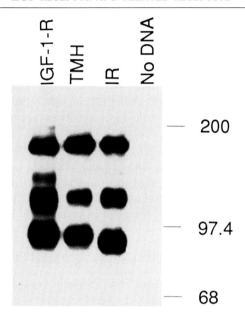

FIG. 4. Human fibroblasts (293) were transfected with expression plasmids containing cDNAs of IGF-I-R, a chimeric receptor TMH (extracellular domain of IR, intracellular domain of IGF-I-R), or IR. Nontransfected cells (No DNA) served as a control. After metabolic labeling with [^{35}S]methionine, cells were lysed, and receptors were immunoprecipitated and analyzed by 7% SDS–PAGE as described in the text. An autoradiograph is shown.

morning, and cells are put on ice and lysed for 3 min in 0.3 ml per well of 50 mM HEPES, pH 7.2, 150 mM NaCl, 1.5 mM MgCl$_2$, 1 mM EGTA, 10% glycerol (v/v), 1% Triton X-100, 2 mM phenylmethylsulfonyl fluoride, and 200 U/ml aprotinin. The lysate is collected, incubated on ice for 5 min, and centrifuged at 12,000 g for 2 min, and the supernatant is used for immunoprecipitation with antireceptor antibodies and protein A-Sepharose beads (Pharmacia, Uppsala, Sweden). Immunoprecipitated receptors are washed 4 times in 500 μl of 20 mM HEPES, pH 7.2, 150 mM NaCl, 0.1% Triton X-100, and 10% glycerol, boiled for 5 min in Laemmli buffer, and subjected to analysis on SDS–PAGE (7% acrylamide, reducing conditions). The gel is then treated with 10% acetic acid and 25% 2-propanol for 30 min on a shaker at room temperature, for another 30 min with Amplify (The Radiochemical Centre, Amersham, UK), dried, and exposed to X-ray film (Fig. 4).

[22] Identification of Phosphorylation Sites: Use of the Epidermal Growth Factor Receptor

By GARY J. HEISERMANN and GORDON N. GILL

Introduction

Phosphorylation is basic to our understanding of the action of many hormones and growth factors. One well-studied example is the epidermal growth factor (EGF) receptor which contains an intrinsic protein tyrosine kinase activity[1] essential to its ability to stimulate cellular growth.[2] The transforming protein product of the v-erbB oncogene resulted from truncation of the receptor ligand-binding domain, suggesting that this domain normally acts to restrain the tyrosine kinase activity of the receptor in the absence of ligand.[3] The tyrosine kinase activity of the EGF receptor appears to also be inhibited by several self-phosphorylation sites located near its C terminus.[4] Kinetic analysis indicates that these sites in their nonphosphorylated state act as competitive inhibitors to block phosphorylation of exogenous substrates.[5] The EGF receptor is also regulated by heterologous phosphorylation following treatment of cells with phorbol esters. Phorbol esters activate protein kinase C resulting in phosphorylation of EGF receptor Thr-654 and inhibition of EGF binding and signal transduction.[6,7] The EGF receptor is phosphorylated *in vivo* on several additional threonine and serine residues,[6,8] which may also act to regulate receptor activity. Identification of the phosphorylation sites is an important step toward understanding their function in EGF receptor signaling. Other growth factor receptors and regulatory enzymes are similarly phosphorylated, and identification of sites of phosphorylation is a necessary prerequisite to investigation of the regulatory function of these sites.

---

[1] H. Ushiro and S. Cohen, *J. Biol. Chem.* **255**, 8363 (1980).
[2] W. C. Chen, C. S. Lazar, M. Poenie, R. Y. Tsien, G. N. Gill, and M. G. Rosenfeld, *Nature (London)* **328**, 820 (1987).
[3] A. Wells and J. M. Bishop, *Proc. Natl. Acad. Sci. U.S.A.* **85**, 7597 (1988).
[4] J. Downward, P. Parker, and M. D. Waterfield, *Nature (London)* **311**, 483 (1987).
[5] P. J. Bertics and G. N. Gill, *J. Biol. Chem.* **260**, 14642 (1985).
[6] C. Cochet, G. N. Gill, J. Meisenhelder, J. A. Cooper, and T. Hunter, *J. Biol. Chem.* **259**, 2553 (1984).
[7] T. Hunter, N. A. Ling, and J. A. Cooper, *Nature (London)* **311**, 480 (1985).
[8] T. Hunter and J. A. Cooper, *Cell (Cambridge, Mass.)* **24**, 741 (1981).

Principle

The location of phosphorylation sites in the EGF receptor sequence is based on radioactive labeling of the receptor either in intact cells or using purified protein kinases. Receptor radiolabeled *in vivo* is isolated under conditions which preserve its native phosphorylation state. Phosphopeptides are generated by protease treatment and are purified by successive chromatography on two reversed-phase HPLC columns in different buffers. The purified peptides are sequenced by automated Edman degradation, and the phosphorylated residues are identified by one of three methods: (1) localization of the radiolabel to a peptide containing a single potential phosphorylation site, (2) release of ^{32}P as inorganic phosphate during the sequencing cycle of the phosphorylated residue, and (3) loss of ^{32}P-labeled peptide and an increase in ^{32}PO$_4$ bound to the sequencing filter during Edman degradation of the phosphorylated residue. Site-directed mutagenesis of the identified phosphorylation sites permits assessment of their role in EGF receptor signal transduction.

Materials

Phosphopeptides are purified by HPLC on an LKB UltraChrom GTi Bioseparation System. The separations are performed on a Vydac C_{18} reversed-phase column (218TP54) and a Brownlee C_8 reversed-phase column (RP300). Mouse EGF and immune absorbent 528 IgG-agarose are prepared as previously described.[9]

Cells. Human epidermoid carcinoma A431 cells, clone 29I,[10] are grown in a 1:1 mixture of Dulbecco's modified Eagle's medium (DMEM) and Ham's F12 medium containing 5% calf serum. Mouse B82 L cells expressing transfected human EGF receptors are grown in DMEM containing 5% dialyzed fetal calf serum and 5 μM methotrexate.

Buffers

Buffer A: 200 mM potassium phosphate, pH 7.2, 150 mM NaCl, 2 mM EDTA, 1% (w/v) deoxycholic acid, 1% (v/v) Nonidet P-40 (NP40), 0.1% (w/v) sodium dodecyl sulfate (SDS), 4 mM benzamidine, 0.5 mM phenylmethylsulfonyl fluoride, 0.5 units/ml aprotinin, 0.04 mg/ml leupeptin, 6 mM 2-mercaptoethanol, 10 μM ammonium molybdate, 1 mM sodium vanadate, 50 mM sodium fluoride, 5 mM *p*-nitrophenyl phosphate

[9] G. N. Gill and W. Weber, this series, Vol. 146, p. 82.
[10] J. B. Santon, M. T. Cronin, C. L. MacLeod, J. Mendelsohn, H. Masui, and G. N. Gill, *Cancer Res.* **46**, 4701 (1986).

Buffer B: 20 mM HEPES, pH 7.4, 1 mM EDTA, 130 mM NaCl, 0.05% Triton X-100, 10% glycerol, 1 mM dithiothreitol (DTT)
Low-phosphate DMEM: 95% phosphate-free DMEM (Irvine, Santa Ana, CA), 5% regular DMEM

Methods

Labeling of Cells with $^{32}PO_4$. Cultured cells at 50% confluence are washed once in low-phosphate DMEM and incubated with 1 ml per 10-cm dish of low-phosphate DMEM containing 0.2–1.0 mCi $^{32}PO_4$. Cells are rocked for 12–14 hr in a Plexiglas box at 37° in a humidified 10% CO_2 atmosphere. Prior to harvesting, cells are transferred to an ice tray and are washed with 5 ml of ice-cold phosphate-buffered saline (PBS) containing 5 mM EGTA. Cells are scraped in PBS and collected by centrifugation (1 min, 800 g). The cell pellet is immediately resuspended in 3 volumes of buffer A, 10 µl per ml of 100 mM PMSF in 2-propanol is added, and extracts are frozen on dry ice and stored at −70°.

Purification of ^{32}P-Labeled EGF Receptor. EGF receptor is purified by a modification of the immunoaffinity method described earlier in this series.[9] Frozen extracts are thawed and homogenized with 3–4 strokes of a Teflon–glass homogenizer. The homogenate is transferred to a plastic centrifuge tube, and the homogenizer is rinsed with 0.25 volume of buffer B, which is pooled with the sample. Cellular debris is removed by centrifugation (10 min, 4°, 10,000 g), and the supernatant is filtered through glass wool into a tube containing 528 IgG-agarose beads. The beads and sample are mixed gently at 4° for 1–2 hr, and the beads are pelleted by a 5- to 10-sec spin in a microcentrifuge. The supernatant is removed with a disposable plastic Pasteur pipette, and the beads are washed with 5 volumes of buffer B (8 washes), buffer B containing 1 M NaCl (3 washes), buffer B (3 washes), again with buffer B containing 1 M NaCl (3 washes), buffer B (3 washes), and buffer B containing 1 M urea (3 washes). The purified EGF receptor is eluted by three incubations with equal volumes of buffer C containing 8 M urea for 30 min at 22°.

Phosphorylation of EGF Receptor in Vitro by Purified Protein Kinases. The EGF receptor may be phosphorylated *in vitro* by purified protein kinases, as has been reported using protein kinase C[6] and cAMP-dependent protein kinase.[11] Analysis of *in vitro* phosphorylation data is complicated by the additional phosphopeptides resulting from extensive autophosphorylation of active solubilized EGF receptor. This also creates a need to separate the two protein kinases prior to proteolysis. One approach which

---

[11] W. R. Rackoff, R. A. Rubin, and H. S. Earp, *Mol. Cell. Endocrinol.* **34**, 113 (1984).

has been used to circumvent this problem is to self-phosphorylate the EGF receptor immobilized on an affinity matrix, first with nonradioactive ATP, followed by incubation with [γ-^{32}P]ATP and the second protein kinase.[12] The components of the reaction mixture are readily washed away, and the phosphorylated EGF receptor is then eluted. Phosphorylation of inactive receptor preparations is problematic because denaturation may expose spurious phosphorylation sites unavailable in the native receptor structure.

Proteolytic Digestion and Peptide Purification. The purified ^{32}P-labeled EGF receptor is dialyzed against 100 mM NH$_4$HCO$_3$ containing 1 mM DTT and 0.01% Triton X-100. The purity of the dialyzed receptor is assessed by electrophoresis of an aliquot on a 10% SDS–polyacrylamide gel. The yield is determined by comparison of Coomassie blue or silver-stained EGF receptor with bovine serum albumin.[13] The dialyzed receptor is digested by addition of 0.5 mM CaCl$_2$ and 1% (w/v) trypsin and incubation overnight at 37°, followed by a second trypsin addition and incubation. The tryptic digests are concentrated under reduced pressure in a Savant Speed-Vac (Hicksville, NY). Purified receptor can alternatively be digested with V8 protease.[14] This protease recognizes different substrate sites than trypsin,[15] and it may be advantageous in the identification of phosphorylated residues located distant from sites of tryptic cleavage. Before protease treatment, the receptor may be reduced and its sulfhydryl groups alkylated[16] to prevent secondary structure from impeding the proteolytic digestion.

The concentrated tryptic peptides are mixed with an equal volume of freshly made 6 M guanidine hydrochloride and filtered through a Microfilterfuge tube (Rainin, Woburn, MA). The filtrate is applied to a C$_{18}$ reversed-phase HPLC column equilibrated in 10 mM potassium phosphate buffer (pH 6.0) and 3% acetonitrile. The column is washed for 10 min at a flow rate of 0.5 ml/min, and the peptides are eluted by a 3 to 53% acetonitrile gradient (0.5%/min). The absorbance at 220 nm is monitored to follow peptide elution, and phosphopeptides are detected by Cerenkov radiation. The phosphoamino acids present in the purified phosphopeptides are determined after hydrolysis in 0.1 ml of 6 N HCl for 1.5 hr at 110°. Samples are dried, washed twice in water, and resuspended in an aqueous solution of phosphoamino acid standards. The phosphoamino

[12] J. Downward, M. D. Waterfield, and P. J. Parker, *J. Biol. Chem.* **260**, 14538 (1985).
[13] W. Weber, P. J. Bertics, and G. N. Gill, *J. Biol. Chem.* **259**, 14631 (1984).
[14] G. M. Walton, P. J. Bertics, L. G. Hudson, T. S. Vedvick, and G. N. Gill, *Anal. Biochem.* **161**, 425 (1987).
[15] G. R. Drapeau, this series, Vol. 47, p. 189.
[16] A. Henschen, *in* "Advanced Methods in Protein Microsequence Analysis" (B. Wittmann-Liebold, ed.), Springer-Verlag, Heidelberg, 1986.

acids are separated by high-voltage electrophoresis at 1000 V for 2 hr at pH 3.5[17] and analyzed by standard procedures.[18]

Once the pattern of phosphopeptides eluting from the C_{18} column at pH 6.0 is established, a larger pool of EGF receptor is prepared to generate peptides for sequencing. EGF receptor which is phosphorylated but not radiolabeled is obtained by treating and harvesting cells as described above except that $^{32}PO_4$ is omitted from the low-phosphate DMEM. The inclusion of this nonradioactive EGF receptor increases the total amount of material available for sequencing and has the additional advantage of increasing the yield of ^{32}P-labeled phosphopeptides up to 2- to 3-fold. The nonradioactive and radioactive frozen cell extracts are pooled on thawing, and the EGF receptor is purified and digested with protease as described above. The peptides eluted from the C_{18} column in phosphate buffer are concentrated on the Speed-Vac to remove the acetonitrile. The peptides are then applied to a C_8 reversed-phase HPLC column equilibrated in 0.05% trifluoroacetic acid (TFA) and 3% acetonitrile. The column is washed for 10 min at 0.2 ml/min, and the peptides are eluted with a 3 to 53% acetonitrile gradient (0.5%/min). The absorbance and Cerenkov radiation are monitored as above. The phosphopeptides eluted from the C_8 column are concentrated on the Speed-Vac and are applied directly to a Polybrene-coated glass fiber filter for automated Edman degradation.[19]

The two-step peptide purification scheme described above is useful because of the large size of the EGF receptor and because of its multisite phosphorylation.[6,8] This two-step purification is effective with tryptic peptide digests, permitting the separation of EGF receptor phosphopeptides of similar size which differ in negative charge or phosphorylation state.[17] These phosphopeptides elute from the C_{18} column in phosphate buffer at unique positions (Table I). When rechromatographed in 0.05% TFA the phosphopeptides become more hydrophobic because their negatively charged amino acids are protonated, and consequently they elute at higher acetonitrile concentrations. Pairs of peptides which eluted in distinct positions from the C_{18} column in phosphate buffer now elute at similar acetonitrile concentrations in TFA (Table I), emphasizing the value of this two-step purification procedure relative to a single HPLC separation in TFA.

The relative elution positions of the positively charged peptides resulting from digestion with V8 protease would not be expected to vary as significantly between the phosphate buffer and TFA systems. A new strong

---

[17] G. J. Heisermann and G. N. Gill, *J. Biol. Chem.* **263**, 13152 (1988).
[18] J. A. Cooper, B. M. Sefton, and T. Hunter, this series, Vol. 99, p. 387.
[19] M. W. Hunkapiller, R. M. Hewick, W. J. Dreyer, and L. E. Hood, this series, Vol. 91, p. 399.

TABLE I
Epidermal Growth Factor Receptor Tryptic Phosphopeptide Sequences and Elution Behavior[a]

Peptide	Sequence	Acetonitrile (%)	
		pH 6.0	pH 2.3
P1	P P ELVEPLTPSGEAPNQALLR[681]	12.2	21.2
P2	P ELVEPLTPSGEAPNQALLR[681]	14.2	21.2
P3	* * MHLPSPTDSNFYR[975]	14.9	22.5
P4	** * * ALMDEEDMDDVVDADEYLIPQQGFFSSPSTSR[1007]	15.7	30.0
P5	PP YSSDPTGALTEDSIDDTFLPVPEYINQSVPKR[1076]	18.1	30.6/31.3

[a] EGF receptor tryptic phosphopeptides, purified by two sequential HPLC steps, were sequenced on an Applied Biosystems 470A gas-phase sequencer. The sequences of the EGF receptor tryptic peptides are shown in single-letter code. Superscript numbers refer to residues in the predicted EGF receptor sequence. A "P" above an amino acid designates a phosphorylation site identified by the methods described in the text. Potential phosphorylation sites as determined by phosphoamino acid analysis are indicated by asterisks. Also listed are the positions in the acetonitrile gradient where each peptide eluted in phosphate buffer (pH 6.0) and in TFA (pH 2.3). Peptide P5 has two adjacent tryptic cleavage sites at its C terminus and eluted as two closely spaced peaks when chromatographed in TFA.

cation-exchange (SCX) HPLC matrix has been reported which separates peptides on the basis of positive charge,[20] and SCX columns have been shown to efficiently separate peptides resulting from V8 digestion.[21] Further purification by reversed-phase HPLC in TFA should result in homogeneous peptides even from a very large protein.

Identification of Phosphorylated Residues

Localization by Inspection. If a purified phosphopeptide contains a single serine or threonine residue, then this residue must be the phosphorylation site. This was the case with EGF receptor phosphopeptides P1 and P2 (Table I), which represent the same peptide sequence phosphorylated on Thr-669 and Ser-671, or on Thr-669 alone, respectively. The major

[20] D. L. Crimmins, J. Gorka, R. S. Thoma, and B. D. Schwartz, *J. Chromatogr.* **443**, 63 (1988).
[21] D. L. Crimmins, R. S. Thoma, D. W. McCourt, and B. D. Schwartz, *Anal. Biochem.* **176**, 255 (1989).

limitation to identification of phosphorylation sites by this method is the rarity of candidate peptides containing a single serine or threonine. One caveat with the method is that the phosphopeptide must be pure; a small amount of phosphopeptide unresolved from a larger quantity of a nonphosphorylated peptide will make the latter appear to be phosphorylated. Although a sharp symmetrical peak on the final HPLC purification is suggestive of a pure peptide, the most stringent test of purity is the release of a single phenylthiohydantoin (PTH) amino acid during each sequencing cycle.

Identification by ^{32}P Release during Edman Degradation. The most common way to identify a phosphorylated residue is to measure the free ^{32}P released during Edman degradation. The ^{32}P is released from phosphoserine by β-elimination under the acid conditions employed in Edman degradation, and two PTH-derivatized serine breakdown products are produced.[22] Analogous breakdown products and ^{32}P release occur during the sequencing of peptides containing phosphothreonine and phosphotyrosine. Nearly all of the released ^{32}P binds to the glass fiber filter as inorganic phosphate because of the nonpolar solvents used in the gas-phase sequencer. Typically only 1–2% of the total radioactivity accompanies the released PTH-amino acid, but this is often adequate. This method has been used most often to identify phosphorylated residues after *in vitro* phosphorylation, and in such cases it is not difficult to prepare a very "hot" phosphopeptide. One major advantage of this technique is that it requires no extra sample manipulation. Prior to HPLC analysis of the PTH-amino acid derivative released during each sequencing cycle, an aliquot (typically 40–50%) is removed and counted for ^{32}P after addition of scintillation fluid. When aliquots were taken and counted during the sequencing of EGF receptor phosphopeptide P2, a peak of radioactivity was observed in the seventh sequencing cycle, establishing Thr-669 as the phosphorylated amino acid (Table I, Ref. 17).

Split-Filter Technique for Phosphorylation Site Identification. The split-filter method is based on the shift of sequencer filter-bound radioactivity from ^{32}P-labeled peptide to inorganic $^{32}PO_4$ following Edman degradation of a phosphorylated residue.[23] The method involves splitting a sequencer filter into multiple sections after sample application. The sample is sequenced normally, and individual filter pieces are removed from the gas-phase sequencer in the cycles surrounding suspected phosphorylation sites. This requires prior knowledge of the peptide sequence, making this

---

[22] G. Allen, "Sequencing of Proteins and Peptides." Elsevier, Amsterdam, 1989.
[23] C. J. Fiol, A. M. Mahrenholz, Y. Wang, R. W. Roeske, and P. J. Roach, *J. Biol. Chem.* **262**, 14042 (1987).

a convenient technique to use if a phosphorylation site is not precisely located by ^{32}P release. The pieces of sequencer filter removed in successive sequencing cycles are sonicated 3 times in 50% formic acid to extract both bound peptide and inorganic ^{32}PO$_4$. These components of the extracted samples are separated by reversed-phase HPLC or thin-layer chromatography. Edman degradation of a phosphorylated residue results in a loss of radioactivity from the peptide and an increase in inorganic ^{32}PO$_4$. This method is several times more sensitive than simply measuring ^{32}P release and is particularly useful in characterizing adjacent potential phosphorylation sites. The split-filter technique was employed to identify Ser-1046 and Ser-1047 as the phosphorylated residues in EGF receptor phosphopeptide P5 (Table I).

Comments

The ability to purify sufficient protein for characterization is a basic requirement for the identification of phosphorylation sites. This is easily accomplished with the EGF receptor because of the availability of cell lines such as A431 human epidermoid carcinoma cells, which contain an amplified EGF receptor gene and express the receptor at greatly elevated levels.[24] High level expression of the EGF receptor can also be achieved by selection after transfection of receptor cDNA into recipient cells.[25] An efficient purification method is required not only to generate pure EGF receptor but also to rapidly separate the phosphorylated receptor from cellular phosphatases. A variety of purification strategies for the EGF receptor have utilized the power of affinity chromatography, including use of immobilized EGF,[26] monoclonal anti-EGF receptor antibodies,[9] and monoclonal antiphosphotyrosine antibodies.[27] To retain the phosphorylation state of the EGF receptor during purification, cells are lysed in a denaturing buffer which contains several phosphatase inhibitors. Even with these precautions it is important to work quickly, and in certain situations it may be necessary to lyse cells on the plate rather than first scraping in PBS. A recently developed method promises to greatly increase the sensitivity of detection of phosphoserine residues.[28] This method entails modification of purified phosphopeptides with ethanethiol, which

[24] C. M. Stoscheck and G. Carpenter, *J. Cell. Biochem.* **23**, 191 (1983).
[25] C. R. Lin, W. S. Chen, C. S. Lazar, C. D. Carpenter, G. N. Gill, R. M. Evans, and M. G. Rosenfeld, *Cell (Cambridge, Mass.)* **44**, 839 (1986).
[26] S. Cohen, this series, Vol. 99, p. 379.
[27] P. B. Wedegaertner and G. N. Gill, *J. Biol. Chem.* **264**, 11346 (1989).
[28] H. E. Meyer, E. Hoffmann-Posorske, H. Korte, and M. G. Heilmeyer, Jr., *FEBS Lett.* **204**, 61 (1986).

reacts specifically with phosphoserine residues. On Edman degradation a unique PTH derivative is quantitatively produced. This circumvents problems of identification and yield of PTH-serine and avoids the need to follow radioactivity through the automated sequencer.

It is necessary to characterize a phosphorylation site *in vivo* to establish its physiological significance. In certain cases it may be easier to phosphorylate a protein *in vitro,* but the results should always be correlated with *in vivo* data. This is critical because of the tendency of protein kinases to phosphorylate many more proteins as well as additional sites on established substrates, when assayed *in vitro.* The range of *in vitro* studies is restricted because of the limited number of purified protein kinases available. Correlation of *in vitro* and *in vivo* phosphorylation data is not always possible because of the inability to specifically activate certain protein kinases in intact cells. It is nevertheless possible to qualitatively compare an identified *in vitro* phosphorylation site with *in vivo* phosphopeptide results, especially if the phosphopeptides do not contain multiple serine or threonine residues.

Once a phosphorylation site has been established, its role in regulation of protein function can be studied by specific mutation of the phosphorylated residue. The function of a mutant EGF receptor was studied after transfection into a cell line which does not normally express the receptor, mouse B82 L cells.[25] To create the mutant receptor, a fragment of the EGF receptor cDNA was inserted into M13 single-stranded phage to serve as a template for mutagenesis.[29] A mutagenic oligonucleotide encoding the desired amino acid alteration served as a primer for generating the replicative form of the phage containing the mutation. The EGF receptor cDNA fragment containing the desired mutation was excised and placed into an expression vector containing the remaining portions of the EGF receptor cDNA linked to the SV40 promoter. The expression vector was transfected into recipient cells, and high levels of EGF receptor expression were achieved by selection in methotrexate. Cells expressing the mutant EGF receptor were then used in studies of EGF-induced receptor internalization, tyrosine phosphorylation, and mitogenesis.[25,30]

Acknowledgments

We would like to thank Dr. Deborah L. Cadena for helpful discussions. Studies from the authors' laboratory were supported by a grant from The Council for Tobacco Research—U.S.A., Inc. (1622) and by The Markey Charitable Trust.

[29] M. J. Zoller and M. Smith, this series, Vol. 100, p. 468.
[30] G. J. Heisermann, H. S. Wiley, B. J. Walsh, H. A. Ingraham, C. J. Fiol, and G. N. Gill, *J. Biol. Chem.* **265,** 12820 (1990).

[23] In Vitro Transcription of Epidermal Growth Factor Receptor Gene

By RYOICHIRO KAGEYAMA and GLENN T. MERLINO

Introduction

The promoter of the epidermal growth factor (EGF)[1] receptor gene has a high GC content and lacks a typical TATA box and CAAT box.[2] Recently, transcriptional mechanisms have been intensively analyzed[3,4]; however, our understanding of the factors involved in transcription from promoters without TATA boxes is still rather limited because of the low efficiency of cell-free transcription from these promoters. We have developed an *in vitro* transcription system for the human EGF receptor gene using nuclear extracts from A431 human epidermoid carcinoma cells, which overproduce EGF receptors.[5,6] We found that transcription factors Sp1 and ETF (EGF receptor transcription factor) are required for the expression of EGF receptor.[5-8] Sp1 is a well-characterized factor that binds to GC boxes (GGGCGG) and stimulates transcription from various promoters containing GC boxes such as the SV40 promoter.[9,10] ETF is a novel trans-acting factor which also binds to GC-rich regions, but it specifically stimulates EGF receptor transcription and has little or no effect on transcription from the SV40 early promoter.[6,11]

In this chapter we describe methods for the preparation and fractionation of A431 nuclear extracts, a technique for purification of ETF, and the *in vitro* transcription assay used in the analysis of the EGF receptor

---

[1] AMV, Avian myeloblastosis virus; CAT, chloramphenicol acetyltransferase; DTT, dithiothreitol; EGF, epidermal growth factor; HEPES, N-2-hydroxyethylpiperazine-N'-2-ethanesulfonic acid; PBS, phosphate-buffered saline; PCPV, packed cell pellet volumes; SDS, sodium dodecyl sulfate; ETF, EGF receptor transcription factor.
[2] S. Ishii, Y.-h. Xu, R. H. Stratton, B. A. Roe, G. T. Merlino, and I. Pastan, *Proc. Natl. Acad. Sci. U.S.A.* **82**, 4920 (1985).
[3] P. F. Johnson and S. L. McKnight, *Annu. Rev. Biochem.* **58**, 799 (1989).
[4] P. J. Mitchell and R. Tjian, *Science* **245**, 371 (1989).
[5] R. Kageyama, G. T. Merlino, and I. Pastan, *J. Biol. Chem.* **263**, 6329 (1988).
[6] R. Kageyama, G. T. Merlino, and I. Pastan, *Proc. Natl. Acad. Sci. U.S.A.* **85**, 5016 (1988).
[7] A. C. Johnson, S. Ishii, Y. Jinno, I. Pastan, and G. T. Merlino, *J. Biol. Chem.* **263**, 5693 (1988).
[8] A. C. Johnson, Y. Jinno, and G. T. Merlino, *Mol. Cell. Biol.* **8**, 4174 (1988).
[9] W. S. Dynan and R. Tjian, *Nature (London)* **316**, 774 (1985).
[10] J. T. Kadonaga, K. A. Jones, and R. Tjian, *Trends Biochem.* **11**, 20 (1986).
[11] R. Kageyama, G. T. Merlino, and I. Pastan, *J. Biol. Chem.* **264**, 15508 (1989).

promoter. Nuclear extract preparation and fractionation are basically performed according to Wildeman et al.[12] and Dynan and Tjian,[13] respectively. Preparation of the sequence-specific affinity resin for ETF purification is according to Wu et al.[14]

Preparation of A431 Nuclear Extract

Materials

Dulbecco's modified Eagle's medium supplemented with 10% fetal bovine serum

Phosphate-buffered saline without calcium and magnesium (PBS)

Buffer 1: 10 mM HEPES, pH 7.9, 10 mM KCl, 1.5 mM MgCl$_2$, and 0.5 mM dithiothreitol (DTT)

Buffer 2: 20 mM HEPES, pH 7.9, 420 mM NaCl, 0.2 mM EDTA, 1.5 mM MgCl$_2$, 25% (v/v) glycerol, and 0.5 mM DTT

Buffer 3: 20 mM HEPES, pH 7.9, 20 mM KCl, 1 mM MgCl$_2$, 17% (v/v) glycerol, and 2 mM DTT

Methods

A431 cells (3 × 10^7 plated per roller bottle, 900 cm^2) are grown in 100 ml of Dulbecco's modified Eagle's medium supplemented with 10% fetal bovine serum. Typically, 20 roller bottles are prepared. On the third day the medium is changed, and on the fourth day the cells are harvested. After washing with PBS the cells are collected by a scraper with 15 ml of ice-cold PBS per roller bottle. From this point all procedures are done at 4°. The cell suspension is subjected to centrifugation at 4700 g for 5 min, and the cell pellets are suspended in the same volume of PBS. After centrifugation at 1800 g for 5 min, the volume of the cell pellet is measured. We usually obtain 30 ml of packed cells from 20 roller bottles.

The cells are resuspended in 5 packed cell pellet volumes (PCPV) of buffer 1 and placed on ice for 10 min. After centrifugation at 7300 g for 10 min, the cells are resuspended in 2 PCPV of buffer 1 and lysed by homogenization in a Dounce homogenizer (~20 strokes). Cell lysis should be checked by light microscopy; lysis should be greater than 90%. After centrifugation at 1200 g for 20 min, the homogenate separates into two

---

[12] A. G. Wildeman, P. Sassone-Corsi, T. Grundstroem, M. Zenke, and P. Chambon, *EMBO J.* **3**, 3129 (1984).

[13] W. S. Dynan and R. Tjian, *Cell (Cambridge, Mass.)* **32**, 669 (1983).

[14] C. Wu, S. Wilson, B. Walker, I. Dawid, T. Paisley, V. Zimarino, and H. Ueda, *Science* **238**, 1247 (1987).

phases. The top phase contains cytoplasmic proteins, and the bottom phase is a crude nuclear pellet. The pellet is suspended in 2 PCPV of buffer 2 by 10 strokes in a Dounce homogenizer. The homogenate is stirred gently on ice for 30 min, then subjected to centrifugation at 12,000 g for 20 min. The supernatant is collected, and 0.33 g of $(NH_4)_2SO_4$ per 1 ml of the supernatant is slowly added with gentle stirring. After this addition, the solution is stirred for an additional 20 min and then centrifuged at 12,000 g for 25 min. The pellet is suspended in buffer 3 at 1/12 the volume of the 12,000 g supernatant, and the resulting suspension is dialyzed against two changes of 100 volumes each of buffer 3 for a total of 5–8 hr. The dialyzate is centrifuged at 10,000 g for 10 min to remove insoluble material. The supernatant is divided into small aliquots, quickly frozen in dry ice, and stored at $-80°$. We usually obtain 100–150 mg of crude nuclear extract from 30 g of A431 cells.

Fractionation of A431 Nuclear Extract

Materials

Heparin-agarose (Bethesda Research Laboratories, Gaithersburg, MD)

DEAE-Sepharose CL-6B (Pharmacia, Piscataway, NJ)

HM buffer: 20 mM HEPES, pH 7.9, 1 mM $MgCl_2$, 2 mM DTT, 0.5 mM phenylmethylsulfonyl fluoride, and 17% (v/v) glycerol

Note: 0.1 M KCl/HM indicates 0.1 M KCl solution in HM buffer.

Methods

Fractionation is performed according to the scheme shown in Fig. 1A. First, 250 mg of nuclear extract is applied to a heparin-agarose column (30-ml bed volume, 2.5 × 6 cm) equilibrated in 0.1 M KCl/HM, and 1.5-ml fractions are collected. The column is washed with 3 bed volumes of 0.1 M KCl/HM (fraction A). In general, fraction A is not required for EGF receptor gene transcription. However, it may be necessary for optimal transcription of other promoters and should be tested in each case. The column is then eluted with 0.4 M KCl/HM (fraction B). The peak fractions are collected and dialyzed against 100 volumes of 0.1 M KCl/HM for 3 hr. Long dialysis should be avoided because the transcriptional activity is very labile.

The 0.4 M KCl B fraction (75 mg) is next applied to a DEAE-Sepharose CL-6B column (30-ml bed volume, 2.5 × 6 cm) equilibrated in 0.12 M KCl/HM. The column is step-eluted with 2.5 bed volumes of 0.12 M KCl/

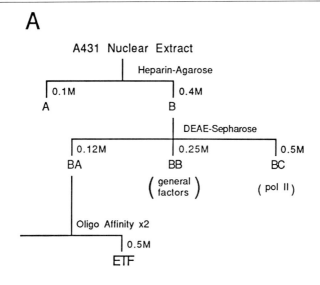

FIG. 1. (A) Fractionation of A431 nuclear extract. DEAE-Sepharose fraction BA contains ETF, which is further purified by sequence-specific oligonucleotide affinity chromatography. DEAE-Sepharose fractions BB and BC contain general transcription factors and RNA polymerase II, respectively. (B) Sequences of the two oligonucleotides used for the affinity resin. The ETF-binding region is indicated by brackets.

HM (fraction BA), 0.25 M KCl/HM (fraction BB), and then 0.5 M KCl/HM (fraction BC). The peak protein-containing fractions from each step are individually pooled, adjusted to 0.33 mg/ml $(NH_4)_2SO_4$ and stirred gently at 4° for 1 hr. After centrifugation at 10,000 g for 15 min, each pellet is dissolved in 500 μl of 0.12 M KCl/HM. The three fractions are individually dialyzed against 100 volumes of 0.12 M KCl/HM for 3 hr. The final yields of the fractions BA, BB, and BC are usually approximately 6, 15, and 5 mg, respectively. Fraction BB contains most general transcription factors, and fraction BC has RNA polymerase II. Transcription factor

ETF is present in fraction BA. For purification of ETF we subject pooled 0.12 M KCl fractions directly to sequence-specific oligonucleotide affinity chromatography without concentration by $(NH_4)_2SO_4$ precipitation.

Purification of Epidermal Growth Factor Receptor Transcription Factor

Materials. Two complementary oligonucleotides, a 46-mer and a 36-mer, and CNBr-activated Sepharose 4B (Pharmacia) are required to prepare the affinity column. The two oligonucleotides

5'-CCCGCGCGAGCTAGACGTCCGGGCAGCCCCCGGCGCAGCGCGGCCG-3'

and

5'-CGGCCGCGCTGCGCCGGGGGCTGCCCGGACGTCTAG-3'

are made using an Applied Biosystems (Foster City, CA) DNA synthesizer (Model 380A).

Methods

The sequence-specific oligonucleotide affinity resin is prepared as follows. We make two complementary oligonucleotides (46-mer and 36-mer) in order that the annealed oligonucleotides consist of a 36-base pair (bp) double-stranded portion and a 10-nucleotide single-stranded portion (Fig. 1B). The single-stranded part is necessary for the coupling reaction to CNBr-activated Sepharose. We use the oligonucleotides without purification. For annealing, 700 μg of the 46-mer and 540 μg of the 36-mer are mixed in a final volume of 400 μl of solution containing 10 mM potassium phosphate, pH 8.2, and 100 mM NaCl. This mixture is heated at 85° for 2 min and then gradually cooled to room temperature over several hours. The hybridized oligonucleotides can be stored at 4°. The annealing efficiency should be checked by an 8% polyacrylamide nondenaturing gel. Only a double-stranded band should be seen after staining with 1 μg/ml ethidium bromide. This hybridized DNA fragment is used for the coupling reaction without further purification.

The coupling reaction is carried out in a 4° cold room. CNBr-activated Sepharose 4B (1.2 g) is suspended in 40 ml of 1 mM HCl and poured into a 15-ml sintered glass funnel under slight vacuum. The Sepharose is washed sequentially with 150 ml of 1 mM HCl and 3 times with 15 ml of 10 mM potassium phosphate, pH 8.2, after which the vacuum is released. The Sepharose solution is resuspended in 2.5 ml of 10 mM potassium phosphate and transferred to a 15-ml polypropylene tube. The funnel is washed with an additional 2.5 ml of 10 mM potassium phosphate, and this solution is also transferred to the same tube. The hybridized olgionucleotide mixture (400 μl) is then added to the Sepharose solution, and the tube is rotated at room temperature for 18 hr. The coupling reaction is stopped by addition of Tris-HCl, pH 8.0, to a final concentration of 100 mM. This mixture is incubated at 4° for 24 hr and then transferred to a 15-ml sintered glass funnel at room temperature. The Sepharose is washed with 10 ml of 100 mM Tris-HCl, pH 8.0; 30 ml of 100 mM potassium phosphate, pH 8.2; 30 ml of a solution of 1.5 M NaCl and 10 mM Tris-HCl, pH 8.0; and 30 ml of a solution of 100 mM NaCl, 10 mM Tris-HCl, pH 8.0, and 1 mM EDTA in this order under slight vacuum. The vacuum is released, and the Sepharose is suspended in 5 ml of the latter buffer and stored at 4°. The efficiency of the coupling reaction can be checked by using ^{32}P-labeled oligonucleotides. More than 20% of the oligonucleotides should be coupled.

To purify the transcription factor ETF, fraction BA (10 ml, ~1.5 mg/ml) is applied to the sequence-specific oligonucleotide affinity column (1-ml bed volume in Econo-column, Bio-Rad, Richmond, CA) equilibrated in 0.12 M KCl/HM. The column is washed with 5 ml of 0.25 M KCl/HM and then eluted with 2 ml of 0.5 KCl/HM. The eluate (~2 ml) is diluted to 0.1 M KCl with HM buffer, and reapplied to a second affinity column (1-ml bed volume in Econo-column) equilibrated in 0.12 M KCl/HM. The column is washed with 5 ml of 0.25 M KCl/HM and eluted with 2 ml of 0.5 M KCl/HM. For the elution, 0.5-ml fractions are collected and analyzed by sodium dodecyl sulfate (SDS)/10% polyacrylamide gel electrophoresis. The peak fraction which contains the majority of the 120-kDa protein is used for the DNase I footprinting, gel mobility shift, and *in vitro* transcription assays.

In Vitro Transcription Assay

Materials. Supercoiled DNA templates containing the chloramphenicol acetyltransferase (CAT) gene, and a CAT-specific primer,

5'-TGCCATTGGGATATATCAACGGTG-3'

that is complementary to the region between nucleotides 4920 and 4943 of pSV2 cat were prepared.[15] Sephadex G-50, DNA grade (Pharmacia), T4 polynucleotide kinase (Bethesda Research Laboratories), avian myeloblastosis virus (AMV) reverse transcriptase (Life Sciences, Pharmacia, or Boehringer Mannheim, Indianapolis, IN), RNase-free DNase I (Bethesda Research Laboratories), and [γ-^{32}P]ATP, 5000 Ci/mmol (Amersham, Ar-

---

[15] C. Gorman, in "DNA Cloning, Volume II, A Practical Approach" (D. M. Glover, ed.), p. 143. IRL, Oxford and Washington, D.C., 1985.

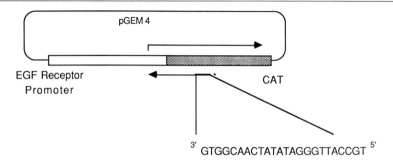

FIG. 2. Structure of the supercoiled DNA template. The EGF receptor promoter region and the truncated CAT gene (between nucleotides 4751 and 5001 of pSV2 cat[15]) are cloned into pGEM-4 (Promega, Madison, WI). The start site (-48) and direction of transcription are indicated by the upper arrow. The asterisk shows the 5′-end-labeled site of the CAT-specific primer (thick line), and the lower arrow indicates the primer-extended product. The sequence of the primer is also shown.

lington Heights, IL), were obtained from the indicated sources. Glycine buffer contains 170 mM glycine, 170 mM NaCl, and 32 mM NaOH.

Methods

There are several methods available to study *in vitro* transcription: run-off,[16,17] S1 nuclease mapping,[16] and primer extension assays. The clearest results are usually obtained using a primer extension assay. The templates we use are supercoiled plasmids containing the EGF receptor promoter and the CAT reporter gene (Fig. 2). In general, the use of a truncated CAT gene can substantially reduce the background. However, when using the highest quality extracts and fractions we get virtually the same signal from plasmids containing the whole CAT gene. A 5′-end-labeled CAT-specific primer is used to detect transcripts initiating within the EGF receptor promoter. The distance between the labeled site of the primer and the 5′ end of the CAT gene is 80 nucleotides. Therefore, the primer-extended product corresponding to the initiation site in the promoter should be greater than 80 nucleotides.

The primer is 5′-end-labeled as follows. First, the primer (5 μl, 70 ng) is heated at 90° for 30 sec and transferred to ice. Then, 1.75 μl of 100 mM MgCl$_2$, 1.75 μl of 100 mM DTT, 4.4 μl of glycine buffer, 10 μl of [γ-^{32}P]ATP, and 1 μl of T4 polynucleotide kinase (10 units/μl) are added, and the mixture is incubated at 37° for 30 min. This solution is applied

[16] J. L. Manley, A. Fire, M. Samuels, and P. A. Sharp, this series, Vol. 101, p. 568.
[17] J. D. Dignam, P. L. Martin, B. S. Shastry, and R. G. Roeder, this series, Vol. 101, p. 582.

to a Sephadex G-50 column (1.8-ml bed volume in an Econo-column) equilibrated in TE buffer (10 mM Tris-HCl, pH 7.5, and 1 mM EDTA) to separate the labeled primer from the free [γ-^{32}P]ATP. Fractions (3 drops, ~100 μl) are collected and counted. The two or three peak fractions are pooled and directly used for the primer extension assay.

In vitro transcription reactions are carried out as follows. When using a crude nuclear extract, 1.25 μl of a mixture of 10 mM each of ATP, GTP, CTP, and UTP (10 mM 4NTP), 1.2 μl of 40 mM spermidine, 1.5 μl of 16 mM MgCl$_2$, 2.3 μl of 500 mM KCl, and 1 μg of a DNA template are mixed with an appropriate amount of water to make the volume 17.5 μl. The crude nuclear extract (7.5 μl, ~12 mg/ml) is then added to start the reaction. When using the heparin-agarose fraction B, 2.5 μl of 10 mM 4NTP, 2.4 μl of 40 mM spermidine, 3 μl of 16 mM MgCl$_2$, and 1 μg of a DNA template are mixed with an appropriate amount of water to make the volumn 38 μl. Fraction B (12 μl, ~6 mg/ml) is then added to start the reaction. When reconstituting DEAE-Sepharose fractions, the same amounts of reagents and DNA are used as for the heparin-agarose fraction, but they are mixed with an appropriate amount of water to make the volume 31 μl. Then 7 μl of fraction BA, 7 μl of fraction BB, and 5 μl of fraction BC are added. When using purified ETF, 2 μl of the affinity column eluate and 5 μl of HM buffer are added instead of fraction BA.

After incubation at 30° for 1 hr, the reaction is stopped by the addition of 400 μl of a solution of 0.3 M sodium acetate, 0.4% SDS, and 1 mM EDTA and extracted with phenol/chloroform and then with chloroform. After the addition of 1 ml of ethanol, the mixture is incubated in a dry ice/ethanol bath for 5 min and centrifuged for 10 min. After washing with ethanol and drying, the pellet is dissolved in 200 μl of 2 M LiCl. Ethanol (400 μl) is added, and the solution is incubated on ice for 30 min. After centrifugation for 10 min and washing with ethanol, the pellet is suspended in 100 μl of a solution of 10 mM Tris-HCl, pH 7.5, 100 mM NaCl, and 10 mM MgCl$_2$ and treated with 13 μg of RNase-free DNase I at 37° for 10 min. The reaction is stopped by the addition of a solution of 100 μl of 10 mM EDTA, 0.2% SDS (w/v), and 150 mM NaCl, and the resulting mixture is extracted with phenol/chloroform.

The ^{32}P-end-labeled primer (5–10 μl, 0.1–0.2 pmol) and 500 μl of ethanol are added to the cell-free RNA. After incubation in a dry ice/ethanol bath for 5 min, the mixture is subjected to centrifugation for 10 min. At least 50% of the radioactivity should be in the pellet. After washing with ethanol, the pellet is dissolved in 10 μl of a solution of 40% (v/v) formamide, 0.4 M NaCl, 40 mM piperazine-N,N'-bis(2-ethanesulfonic acid), pH 6.4, and 1 mM EDTA and then heated at 75° for 5 min. After incubation at 42° for 2 hr, 80 μl of water, 10 μl of 3 M sodium acetate, and 250 μl of ethanol are added, and the mixture is placed in a dry ice/ethanol bath for 5 min. Nucleic acids are recovered by centrifugation in a microcentrifuge for 10 min. After ethanol washing, the pellet is dissolved in 49 μl of a solution of 50 mM Tris-HCl, pH 8.3, 8 mM MgCl$_2$, 30 mM KCl, 3 mM DTT, 1 mM each of dATP, dGTP, dCTP, and dTTP, and 1.6 μg/ml of actinomycin D. AMV reverse transcriptase (1 μl, 10–20 units) is added to start the reaction. After incubation at 37° for 1 hr, the reaction is stopped by sequentially adding 6 μl of 0.25 M EDTA and 6 μl of 2 N NaOH, then heating at 37° for 20 min. The reaction is neutralized by the addition of 12 μl of 1 N HCl. tRNA (1 μl, 5 mg/ml) is then introduced as a carrier. The solution is extracted with phenol/chloroform and precipitated with 200 μl of ethanol. The pellet is dissolved in 5 μl of a solution of 80% (v/v) formamide, 0.01 N NaOH, 1 mM EDTA, 1 mg/ml xylene cyanol, and 1 mg/ml bromphenol blue. The sample (2–3 μl) is heated at 90° for 2 min, transferred to ice, and applied to a 5% polyacrylamide/7 M urea denaturing gel. After electrophoresis, the gel is subjected to standard autoradiographic techniques.

Concluding Remarks

We have successfully used this sensitive assay system to identify novel trans-acting factors involved in EGF receptor gene transcription regulation.[5,6,11] It is likely that this system can be used to analyze transcription by related TATA box-minus promoters and that together these studies may help elucidate general mechanisms by which expression of this important class of genes is regulated.

Acknowledgments

The authors wish to thank Drs. Alfred Johnson, Pamela Marino, and Ira Pastan for useful discussions, and Althea Gaddis and Jennie Evans for editorial assistance.

[24] Isolation of Cell Membrane for Epidermal Growth Factor Receptor Studies

By PETER H. LIN, RICHARD SELINFREUND, and WALKER WHARTON

Introduction

We have developed a rapid procedure for the purification of cell membrane from cultured cells. Two versions of this purification method are included in this chapter. The large-scale version provides a convenient way to process either liter quantities of high-density spin cultures or several hundred large plates of cell cultures.[1] The small-scale version is routinely used to extract milligrams amounts of cell membrane from as few as four confluent plates (25 × 150 mm) using a 7 min centrifugation protocol.[2]

Traditional extraction of cell membranes from cultured systems relies heavily on lengthy ultracentrifugation cycles (i.e., 2 to 20 hr) to selectively sort out the plasmalemma from other organellar membranes.[3-5] The yield and lot-to-lot reproducibility is typically poor because a large number of physical parameters has to be carefully controlled. Small variations in key steps, such as the type and duration of cell homogenization, can significantly alter the size of different membrane fractions and decrease the purity of the final product.

The new purification method developed here is substantially faster (i.e., <1 hr of centrifugation time) and has eliminated the need for density gradient ultracentrifugation. Deviations during cell homogenization is also less critical since the process is taken to completion with all membranous components sheared into small vesicles. The new method takes advantage of a unique property of calcium ion, which at millimolar concentrations preferentially induces aggregation of nonplasmalemma membranes.[6-8] The membrane aggregate is easily removed from the homogenate by a

---

[1] P. H. Lin, R. Selinfreund, E. Wakshull, and W. Wharton, *Biochemistry* **26**, 731 (1987).
[2] P. H. Lin, R. Selinfreund, E. Wakshull, and W. Wharton, *Anal. Biochem.* **168**, 300 (1988).
[3] D. Thom, A. Powell, C. W. Lloyd, and D. A. Rees, *Biochem. J.* **168**, 187 (1977).
[4] S. Cohen, this series, Vol. 99, p. 379.
[5] T. D. Butters and R. C. Hughes, *Biochem. J.* **140**, 469 (1974).
[6] J. B. Shenkman and D. L. Cinti, this series, Vol. 52, p. 83.
[7] S. A. Kamath and E. Rubin, *Biochem. Biophys. Res. Commun.* **49**, 52 (1972).
[8] P. Malathi, H. Preiser, P. Fairclough, P. Mallett, and R. K. Crane, *Biochim. Biophys. Acta* **554**, 259 (1979).

simple centrifugation step. The plasma membrane is subsequently collected from the supernatant with a second centrifugation cycle.

Cell Membrane Purification Procedure

An outline of the cell membrane purification is shown in Fig. 1. All steps are performed at 4°.

Reagents. All reagents are stored at 4°.

Phosphate-buffered saline (PBS: 140 mM NaCl, 10 mM sodium phosphate, pH 7.4)

Hypotonic lysis buffer (50 mM mannitol, 5 mM HEPES, pH 7.4)

$CaCl_2$, 1 M in deionized water

Cell Harvest and Wet Weight Determination

Cells are grown in 5% CO_2 at 37°. The culture medium is changed every 4 days. On reaching confluency, cells are rinsed once with PBS and harvested from the plates with a wide blade rubber policeman. The wet weight of the cells is determined by pelleting the cells in preweighed centrifuge tubes (10 min, 2000 g). Approximately 0.1–0.2 g of cells can be obtained from a single confluent P150 (25 × 150 mm) culture plate. High-density spin cultured cells are directly harvested by centrifugation using preweighed 250-ml centrifugation bottles.

Large-scale Procedure. The large-scale procedure should be used for cell harvests that are greater than 1 g per harvest. Thirty volumes (1 : 30, w/v) of hypotonic lysis buffer is added to the harvested cells. The mixture is immediately homogenized in a Waring blendor at maximum speed for 5 min. The 1 M calcium chloride solution is added to the homogenate with stirring to a final concentration of 10 mM. The mixture is stirred vigorously for 10 min more to ensure even distribution of the calcium ion.

Calcium-induced membrane aggregates are sedimented by a 3000 g centrifugation (i.e., SS-34 rotor at 5000 rpm) for 15 min. The solid pellet (P1 fraction) containing nuclei, lysosomes, and endoplasmic reticulum is discarded. The slightly turbid supernatant is tranferred to a clean set of centrifuge tubes. The plasma membrane is pelleted with a 48,000 g spin (i.e., SS-34 rotor at 20,000 rpm in a RC-5C centrifuge) for 30 min. The whitish, translucent pellet (P2 fraction) is the purified cell membrane.

Small-scale Procedure. Cells from four confluent P150 plates are harvested. Ten volumes (1 : 10, w/v) of the hypotonic buffer is added to the harvested cells. The mixture is homogenized by aspiration (6 times) through two 25-gauge needles using the vacuum homogenization apparatus (Fig. 2). Alternatively, when very small amounts of cells are used, homoge-

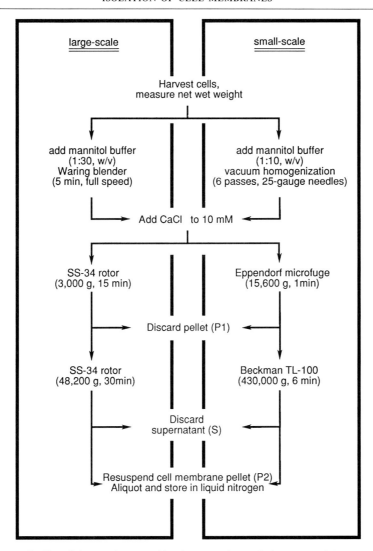

FIG. 1. Outline of the membrane purification procedure. Distinct steps of the large-scale and small-scale purifications are bracketed on the left-hand and right-hand sides, respectively. Common steps for the two protocols are positioned in the middle.

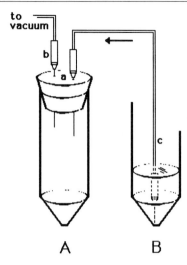

FIG. 2. The vacuum homogenization apparatus is constructed from sterile, disposable plasticware. A 50-ml conical centrifuge tube (A) is stopped with a silastic rubber stopper (a). Vacuum is applied to the tube through a 1.5-mm cannula and 18-gauge needles (b, shown with Leur adaptors). Cells in hypotonic lysis buffer (B) are homogenized by aspiration through the two 25-gauge needles (c) and is collected in tube A. Multiple units can be constructed for parallel sample processing. Needles are replaced upon clogging.

nization can be done manually by aspiration through a 1-ml syringe and a 25-gauge needle. Calcium chloride solution (1 M) is immediately added to the homogenate to a final concentration of 10 mM. The mixture is vortexed vigorously for 1 min.

Aliquots of the homogenate are placed into 1.5-ml microfuge tubes. Calcium-induced membrane aggregates (i.e., the P1 fraction) are sedimented by a 1-min 15,600 g centrifugation (14,000 rpm in a Eppendorf microfuge). The supernatant is transferred to 1-ml polycarbonate tubes, and the cell membranes (i.e., the P2 fraction) are collected with a 6-min centrifugation in a Beckman TL-100 tabletop ultracentrifuge (TLA100.2 fixed-angle rotor at 430,000 g).

Resuspension and Storage of Purified Cell Membrane

Resuspension of the purified membrane pellet (P2 fraction) from the 1-ml polycarbonate tubes is easily accomplished by aspiration through a 25-gauge needle using a 1-ml plastic syringe. For the large-scale preparation, membrane resuspension is done automatically using the vacuum homogenization unit. The membrane pellets are scraped away from the centrifugation tubes and repeatedly aspirated through 25-gauge needles as explained (Fig. 2).

The purified cell membranes can be resuspended in different buffers by repeating the 430,000 g centrifugation cycle as needed. Excess calcium in the membrane is removed, if necessary, with one wash of EDTA (2 mM) buffer. The protein concentration is determined using the bincinchoninic acid assay[9] (available through Pierce, Rockford, IL). Aliquots of the cell membrane preparation are stored frozen in liquid nitrogen. Under such storage conditions, the biological activity of the membranes is preserved with no loss of activity for 6 months.

Characterization of Purified Cell Membranes

Plasma membranes from human epidermoid carcinoma A431 cells or KB cells are purified as described above. The purity of the cell membranes from both large-scale and small-scale methods is determined by membrane marker enzyme assays. In addition, membrane samples from the large-scale purification are processed for transmission electron microscopy. The activity of the EGF receptor is determined by ^{125}I-labeled EGF binding and receptor phosphorylation assays.

Relative Specific Activity Analysis

The purified A431 membranes are enriched 7- to 12-fold over the starting homogenate, as indicated by the relative specific activity (RSA) analysis of alkaline phosphatase (Table I). The ratios are substantially better than the 1.8- to 3.1-fold purification of a 2-hr procedure[3] and are comparable to the purification obtained using a 20-hr gradient procedure.[5] Cell membranes purified from KB cells have a similar profile (RSA ratio of 9) and has been reported.[1] Consistent with the electron microscopy data, RSA analysis also demonstrates that contamination from endoplasmic reticulum or lysosomal membranes in the purified cell membrane preparation is minimal (Table I).

Transmission Electron Microscopy

The purified A431 cell membrane fraction consists of a population of small vesicles with an average diameter of 100 Å (Fig. 3). No contamination from nuclei, lysosomes, or endoplasmic reticulum is detected. Transmis-

---

[9] P. K. Smith, R. I. Krohn, G. R. Hermanson, A. K. Mallia, F. H. Gartner, M. D. Provenzano, E. K. Fujimoto, N. M. Goeke, B. J. Olson, and D. C. Klenk, *Anal. Biochem.* **150,** 76 (1985).

TABLE I
RELATIVE SPECIFIC ACTIVITY ANALYSIS OF PURIFIED MEMBRANE PELLET FROM A431 CELLS[a]

Enzyme	Marker of	RSA ratio (P2/H)[b]	
		Large-scale	Small-scale
Alkaline phosphatase	Cell membrane	12.0	6.7
5'-Nucleotidase	Cell membrane	5.2	3.6
Glucose-6-phosphatase	Endoplasmic reticulum	1.2	0.3
Acid phosphatase	Lysosome	2.3	1.3

[a] P2 fraction. From Refs. 1 and 2.
[b] RSA ratio equals the enzyme specific activity of purified cell membrane (P2) divided by the enzyme specific activity of homogenate (H).

sion electron microscopy of purified cell membranes from KB cells reveals a similar morphology.[1]

Epidermal Growth Factor Receptor Profile of Purified Cell Membranes

Binding Profile

^{125}I-Labeled EGF binding to purified cell membranes is rapid, saturable, and reaches equilibrium within 30 min.[1,2] The binding specificity of ^{125}I-labeled EGF to purified cell membranes of both A431 and KB cells is greater than 90% (Table II). Scatchard analysis shows that the equilibrium binding constants for both cell lines are in the nanomolar range (Table II). In particular, the K_d of 1.2 nM for purified A431 cells membranes agrees with a previously reported value of 1.5 nM.[10]

EGF-Dependent Receptor Autophosphorylation Using Purified A431 Cell Membranes

Using the purified cell membranes, distinct EGF-dependent receptor autophosphorylation can be demonstrated at 170 kDa without the need for prior immunoprecipitation (Fig. 4a).[11,12] The known divalent cation dependency of the EGF receptor kinase[13] is illustrated in Fig. 4b where

[10] J. A. Fernandez-Pol, *J. Biol. Chem.* **260,** 5003 (1985).
[11] G. Carpenter, L. King, and S. Cohen, *J. Biol. Chem.* **254,** 4884 (1979).
[12] P. H. Lin, R. Selinfreund, and W. Wharton, *Anal. Biochem.* **167,** 128 (1987).
[13] E. M. Wakshull and W. Wharton, *Proc. Natl. Acad. Sci. U.S.A.* **82,** 8513 (1985).

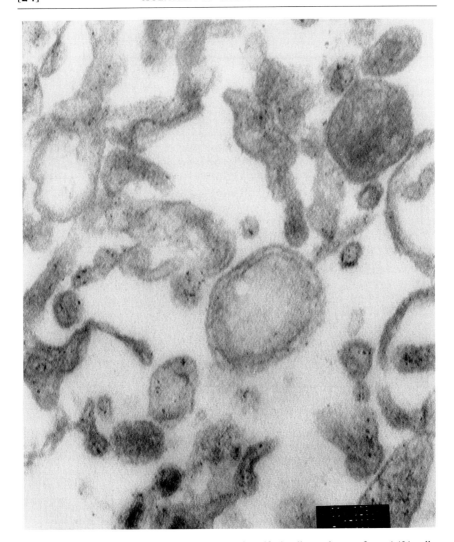

FIG. 3. Transmission electron microscopy of purified cell membranes from A431 cells. The membrane sample was fixed in 3% glutaraldehyde, 1% OsO$_3$ and stained with lead citrate and uranyl acetate as described.[1] Black Box: 100 Å.

the addition of 3 mM EGTA inhibits the EGF-dependent receptor autophosphorylation.

Western blotting of the purified cell membranes with ^{125}I-labeled EGF detects both forms of the EGF receptor at 150 and 170 kDa (Fig. 4c). The

TABLE II
125I-LABELED EGF-BINDING TO PURIFIED CELL MEMBRANES
FROM A431 AND KB CELL LINES[a]

Cell type	Specific[b] binding (%)	K_d (nM)	B_{max} (pm/mg protein)
A431	93	1.2	5.3
KB	92	0.14	0.1

[a] From Ref. 1.
[b] Specific binding equals total 125I-labeled EGF bound minus background binding (in the presence of excessive unlabeled EGF).

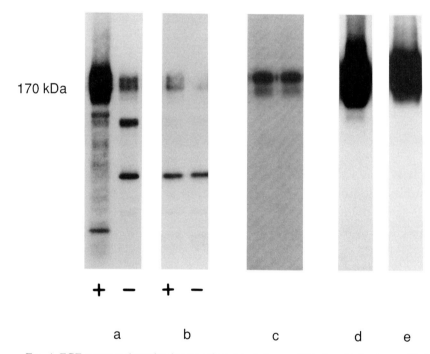

FIG. 4. EGF receptor detection by autophosphorylation and Western blot using purified A431 cell membranes. (a) A431 receptor autophosphorylation in the presence (+) and absence (−) of EGF. (b) EGF receptor autophosphorylation with (+) and without (−) EGF in the presence of 3 mM EGTA. (c) Western blot of EGF receptor using 125I-labeled EGF. (d) Immunoblot of EGF receptor using 125I-labeled anti-EGF receptor antibody (#528). (e) Immunoblot of EGF receptor using 125I-labeled anti-EGF receptor antibody (#225). EGF receptor autophosphorylation using [γ-32P]ATP was performed as described.[1,11] Western blot and immunoblots using 125I-labeled proteins were processed as reported.[12]

150 kDa form of the receptor has been suggested to be a proteolytic product of the 170 kDa receptor.[14-16] Last, immunoblot analysis using ^{125}I-labeled anti-EGF receptor antibody (#528 and #225) to the purified cell membrane shows a strong reactive band at 170 kDa indicating the presence of EGF receptor (Fig. 4d,e).

Summary

The cell membrane isolation procedure we developed here can be scaled up from four to several hundred plates of cultured cells. Transmission electron microscopy, membrane marker enzyme analysis, binding study, EGF-dependent receptor autophosphorylation, and Western blots all demonstrate the biological activity of the purified cell membranes. The membrane purification procedure has been adapted by others[17] in assessing EGF kinase activity and has been used for the purification of cell membranes from other types of cultured cells.[18]

Acknowledgments

The authors wish to thank Dr. D. Urquhart for helpful comments. This work was supported in part by the United States Department of Energy under Contract W-7405-ENG36 with the University of California, and in part by the United States Army Medical Research and Development Command, Fort Detrick, Maryland.

[14] S. A. Burow, S. Cohen, and J. V. Staros, *J. Biol. Chem.* **257,** 4019 (1982).
[15] R. Gates and L. E. King, *Biochemistry* **24,** 5209 (1985).
[16] P. Ghosh-Dastidar and C. F. Fox, *J. Biol. Chem.* **259,** 3864 (1984).
[17] S. E. Shoelson, M. F. White, and C. R. Kahn, *J. Biol. Chem.* **264,** 7831 (1989).
[18] R. Hattori, K. K. Hamilton, R. P. McEver, and P. J. Sims, *J. Biol. Chem.* **264,** 9053 (1989).

[25] Phosphorylation of Lipocortin-1 by the Epidermal Growth Factor Receptor

By R. BLAKE PEPINSKY

Introduction

The lipocortins are a recently characterized family of Ca^{2+}- and phospholipid-binding proteins which differ from conventional Ca^{2+}-binding proteins in that they lack an EF-hand type Ca^{2+}-binding loop and they inhibit phospholipase A_2 in *in vitro* assay systems. Eight related proteins have been identified to date, all sharing extensive sequence similarity (for references, see Refs. 1 and 2). Most conserved are a series of 70 amino acid repeat units that are organized into a core structure that binds Ca^{2+} and phospholipid. The 35- to 45-kDa proteins contain four copies of the repeat and the 70-kDa form, eight copies. Each protein also contains a short amino-terminal segment that is distinct from the core structure and unique to each protein. While the repeat units directly bind Ca^{2+} and phospholipid, the amino terminus regulates the binding affinities and thus may provide functional identity to the individual family members.[3-5] Lipocortin-like proteins have been implicated in diverse processes, ranging from intracellular events where they affect membrane–cytoskeleton interactions and signal transduction to extracellular events where they affect inflammation, the immune response, blood coagulation, growth, and differentiation (for references, see Ref. 6). Of the eight proteins lipocortins-1 and -2 have been of particular interest as major substrates for oncogene and receptor protein kinases.

In 1984, Fava and Cohen characterized a 39-kDa Ca^{2+}-binding protein from A431 cells that was a substrate of the epidermal growth factor (EGF) receptor/kinase *in vitro* using receptor-rich cell membranes as a source of

---

[1] R. B. Pepinsky, R. Tizard, R. J. Mattaliano, L. K. Sinclair, G. T. Miller, J. L. Browning, E. P. Chow, C. Burne, K.-S. Huang, D. Pratt, L. Wachter, C. Hession, A. Z. Frey, and B. P. Wallner, *J. Biol. Chem.* **263,** 10799 (1988).
[2] M. J. Crumpton and J. R. Dedman, *Nature (London)* **345,** 212 (1990).
[3] Y. Ando, S. Imamura, Y.-M. Hong, M. K. Owada, T. Kakunaga, and R. Kannag, *J. Biol. Chem.* **264,** 6948 (1989).
[4] D. S. Drust and C. E. Creutz, *Nature (London)* **331,** 88 (1988).
[5] J. Glenney and L. Zokas, *Biochemistry* **27,** 2069 (1988).
[6] R. B. Pepinsky, L. K. Sinclair, E. P. Chow, and B. Greco-O'brine, *Biochem. J.* **263,** 97 (1989).

the kinase.[7] The phosphorylation was EGF-dependent and occurred on tyrosine. Subsequent studies revealed that the 39-kDa protein was phosphorylated in intact cells and, excluding the receptor itself, represented the first known physiological substrate of the EGF receptor.[8] The extent of phosphorylation was growth cycle-dependent and accounted for as much as 25% of the protein. In studying lipocortin-like proteins, we observed that lipocortin-1 also was a substrate of the EGF receptor and demonstrated that it and the 39-kDa band were the same protein.[9] Because we had already cloned and expressed lipocortin-1,[10] this finding placed us in an unusual position to evaluate the activities and properties of the protein. The following sections describe some of these studies, focusing on those that pertain to the phosphorylation of lipocortin by the EGF receptor and characterization of the phosphorylated product.

Materials and Methods

Labeling and Immune Precipitation Procedures

The A431 cell line, derived from a human epidermal carcinoma, is obtained from the American Type Culture Collection (CRL 1555, Rockville, MD). The cells are grown as monolayers at 36° in a tissue culture incubator in Dulbecco's modified Eagle's medium supplemented with 10% calf serum, 50 units/ml penicillin, and 50 μg/ml streptomycin. Cells are maintained in T75 flasks and trypsinized and passaged at a 1 : 5 split ratio every third day. Cells for metabolic labeling studies are split at the same ratio into 6-well culture dishes. After 48 hr, the growth medium is discarded and the wells washed twice with 2.5 ml serum-free medium that is deficient either in methionine or phosphate (Flow Laboratories, McLean, VA). Parallel cultures are labeled for 4 hr with 100 μCi/well [^{35}S]methionine (900 Ci/mmol) or 1 mCi/well [^{32}P]phosphate (1200 Ci/mol) in either the presence or absence of 200 ng/ml EGF. The labeling medium is discarded, and the cells are scraped into 1 ml of ice-cold lysis buffer (50 mM NaCl, 20 mM Tris-HCl, pH 7.3, 0.5% Nonidet P-40, 0.5% sodium deoxycholate, 4 mM iodoacetic acid, 1 mM ammonium vanadate). The lysates are subjected to vortex mixing for 30 sec and clarified in an Eppendorf centrifuge (5 min, 10,000 rpm, 4°).

[7] R. A. Fava and S. Cohen, *J. Biol. Chem.* **259**, 2636 (1984).
[8] S. T. Sawyer and S. Cohen, *J. Biol. Chem.* **260**, 8233 (1985).
[9] R. B. Pepinsky and L. K. Sinclair, *Nature (London)* **321**, 81 (1986).
[10] B. P. Wallner, R. J. Mattaliano, C. Hession, R. L. Cate, R. Tizard, L. K. Sinclair, C. Foeller, E. P. Chow, J. L. Browning, K. L. Ramachandran, and R. B. Pepinsky, *Nature (London)* **320**, 77 (1986).

For immune precipitations the labeled lysates are split into two aliquots, one receiving 2 µl of preimmune serum and the other receiving 2 µl of antilipocortin-1 antiserum (lipocortin antiserum is developed in rabbits using recombinant protein produced in *Escherichia coli* as immunogen; see Ref. 11). The samples are incubated on ice for 2 hr, and then the immune complexes are collected by adsorption to protein A-Sepharose (Sigma Chemical Company, St. Louis, MO) for 1 hr at 4° with continuous mixing (25 µl of packed beads per sample). The adsorbed complexes are washed 4 times, each with 0.5 ml of lysis buffer, suspended in 50 µl of electrophoresis sample buffer [2% sodium dodecyl sulfate (SDS), 50 mM Tris-HCl, pH 6.8, 12.5% glycerol, 1.5% 2-mercaptoethanol, 0.1% bromophenol blue], and heated at 65° for 10 min. The protein A beads are pelleted in an Eppendorf centrifuge (1 min, 10,000 rpm, 4°). Supernatants are transferred to new tubes and stored at $-20°$ for subsequent analysis. Aliquots (10 µl) are subjected to SDS–PAGE in minigels ($7 \times 7 \times 0.14$ cm) using the Laemmli system. Separating gels contain 12.5% acrylamide, 0.1% methylene bisacrylamide and stacking gels, 7.6% acrylamide, 0.21% methylene bisacrylamide. Gels are dried under reduced pressure for 1.5 hr at 80° and radioactive bands visualized by autoradiography. Generally, an overnight exposure is needed to visualize the immune precipitates.

In Vitro Phosphorylation Reactions

A431 cells are split at a 1 : 5 ratio into T75 flasks. After 48 hr the cells are washed with serum-free growth medium and incubated for 1 hr at 37° in 7 ml of the same medium without or with 200 ng/ml EGF. Monolayers are rinsed with 2.5 ml of ice-cold hypotonic buffer (10 mM HEPES, pH 7.5, 0.5 mM MgCl$_2$) and the cells scraped into 1.5 ml of the same buffer. After 30 min on ice, the cells are lysed with a Dounce homogenizer (20 strokes, Type A pestle) and clarified of large debris by centrifugation for 5 min at 2000 g. Cell membranes are pelleted by centrifugation at 10,000 g for 10 min and suspended in 300 µl of phosphorylation buffer (20 mM HEPES, pH 7.5, 2 mM MgCl$_2$, 10 µM ammonium vanadate) plus 2 µg/ml EGF for EGF-treated samples. For *in vitro* phosphorylations, 60-µl reactions, each containing 30 µl of the membrane fraction plus 30 µCi of [γ-^{32}P]ATP and the indicated amount of lipocortin in 30 µl of the same buffer, are held on ice for 10 min. Reactions are stopped by adding 20 µl of 5× electrophoresis sample buffer. Samples are heated at 65° for 10 min,

---

[11] K.-S. Huang, B. P. Wallner, R. J. Mattaliano, R. Tizard, C. Burne, A. Frey, C. Hession, P. McGray, L. Sinclair, E. P. Chow, J. L. Browning, K. L. Ramachandran, J. Tang, J. E. Smart, and R. B. Pepinsky, *Cell (Cambridge, Mass.)* **46**, 191 (1986).

Epidermal Growth Factor-Dependent Phosphorylation of Lipocortin-1 in A431 Cells

and 5-μl aliquots are subjected to SDS gel analysis. Gels are dried under reduced pressure and exposed to X-ray film for 1 hr at 23°.

The mitogenic or in some instances inhibitory effects of EGF on target cells are mediated through its receptor, a 170-kDa transmembrane protein with an extracellular EGF binding site and an intracellular tyrosine kinase domain. Of multiple immediate changes in target cells that are initiated on binding of EGF to the receptor, the earliest is the activation of the kinase, producing a cascade of phosphorylation reactions presumed to be crucial for transduction of the EGF signal. Two prominent phosphorylation products of the kinase are the receptor itself and lipocortin-1.[7,9,12] Recent immunohistochemical studies with lung tissue have revealed a striking correlation between the cellular distribution of lipocortin-1 and the EGF receptor, supporting a role for the protein in receptor function.[13]

In A431 cells the EGF receptor is overproduced,[14] leading to the frequent use of the cell line as a source of the kinase. A431 cells also produce large amounts of lipocortin-1 (~0.5% of the total protein) and therefore provide a simple, accessible system for studying the phosphorylation of lipocortin-1 by the EGF receptor.[7-9,15] Figure 1 shows results from two sets of experiments where EGF-dependent phosphorylation of lipocortin-1 was studied in intact cells (Fig. 1, lanes i–p) and *in vitro* using A431 cell membranes as a source of the kinase (Fig. 1, lanes a–h). To assess phosphorylation in intact cells, A431 cells or EGF-treated A431 cells are metabolically labeled with [^{32}P]phosphate. Lipocortin-1 is selectively precipitated from the lysates with lipocortin-1-specific antiserum and analyzed for [^{32}P]phosphate by SDS-PAGE. From control cultures labeled with [^{35}S]methionine, lipocortin-1 is the only band detected by the gel analysis (Fig. 1, lanes i–l) with EGF having no effect on the labeling. In contrast, from [^{32}P]phosphate-labeled cells (lanes m–p, Fig. 1), lipocortin-1 is only observed in extracts from cells that are first treated with EGF. Typically, about a 5-fold increase in incorporation of phosphate into lipocortin-1 was observed in the EGF-treated samples. Phosphoamino acid

---

[12] B. K. De, K. S. Misono, T. J. Lukas, B. Mroczkowski, and S. Cohen, *J. Biol. Chem.* **261**, 13784 (1986).

[13] M. D. Johnson, M. E. Gray, G. Carpenter, R. B. Pepinsky, H. Sundell, and M. T. Stahlman, *Pediatr. Res.* **25**, 535 (1989).

[14] R. N. Fabricant, J. E. DeLarco, and G. J. Todaro, *Proc. Natl. Acad. Sci. U.S.A.* **74**, 565 (1977).

[15] D. D. Schlaepfer and H. T. Haigler, *J. Biol. Chem.* **262**, 6931 (1987).

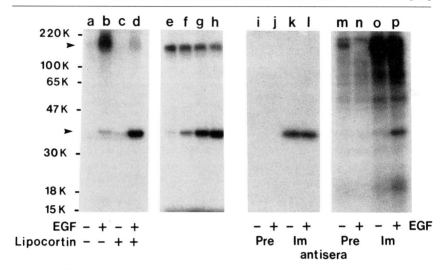

FIG. 1. EGF-dependent phosphorylation of lipocortin-1. *In vitro* phosphorylation reactions using isolated A431 cell membranes (lanes a–h) and immune precipitations from metabolically labeled cells ([^{35}S]methionine, lanes i–l; [^{32}P]phosphate, lanes m–p) were analyzed by SDS–PAGE and radioactive products visualized by autoradiography. Arrowheads denote the positions of the EGF receptor. (170 kDa) and of lipocortin-1 (39 kDa). Specific variables are indicated (Pre, preimmune serum; Im, immune serum). Lanes c and d each received 2 µg lipocortin. Lanes e–h show phosphorylation results obtained with membranes from EGF-treated cells that were incubated with 0.2, 2, 20, and 200 µg/ml of lipocortin, respectively. [Reprinted with permission from R. B. Pepinsky and L. K. Sinclair, *Nature (London)* **321**, 81 (1989). Copyright © Macmillan Magazines Ltd.]

analysis revealed that the increase in label was predominantly on tyrosine. A series of bands were detected in the samples from phosphate-labeled cells, but of these only lipocortin-1 and an 18K band presumed to be a fragment of lipocortin showed EGF-dependent phosphorylation. While some of the other proteins may be immunologically related to lipocortin,[6,16] many probably result from nonspecific binding to the protein A-Sepharose beads, since the appearance of the precipitations in subsequent analysis was improved by including SDS in the lysis buffer.[17] Perhaps the cleanest results to date were generated by Sawyer and Cohen, who partially purified the phosphorylated lipocortin prior to the immune precipitation step.[8] They obtained a 15-fold increase in labeled lipocortin from EGF-treated A431 cells. Phosphate incorporation increased steadily over the labeling

[16] A. Karasik, R. B. Pepinsky, and R. C. Kahn, *J. Biol. Chem.* **263**, 11862 (1988).
[17] A. Karasik, R. B. Pepinsky, S. E. Shoelson, and R. C. Kahn, *J. Biol. Chem.* **263**, 18558 (1988).

period (4 hr), indicating that the phosphorylated adduct was relatively long lived.

In working with A431 cells, we routinely use cells at around 48 hr after plating. At this time EGF induces a dramatic morphologic change, marked by a rapid movement of the cells from a confluent monolayer into a weblike network where the cells aggregate to form the cables of the web. If the cells are used too early, namely, within the first 24 hr after plating, the EGF treatment causes them to round up and detach, thus being lost to the analysis. By 72 hr growth is no longer logarithmic, and the cells are not as responsive to EGF.

Phosphorylation of Lipocortin-1 by Epidermal Growth Factor Receptor

The phosphorylation of lipocortin-1 by the EGF receptor is particularly dramatic in cell-free systems using A431 cell membranes as a source of the kinase (see Fig. 1, lanes a–d). The reconstituted preparation provides a simple system for evaluating the various parameters affecting phosphorylation and for testing structural variants of the protein as substrates or inhibitors of the kinase. Whereas lipocortin was barely detected in the absence of EGF (Fig. 1, lane c), it was the major labeled band in the EGF-treated preparations (lane d, Fig. 1). Phosphate incorporation varied as a function of the concentration of added protein (lanes e–h, Fig. 1) and coincided with a decrease in phosphorylation of the receptor. When similar reactions were performed without added protein (lanes a and b, Fig. 1), endogenous lipocortin-1 was phosphorylated in an EGF-dependent manner. The identity of this band as lipocortin was previously verified by peptide mapping.[9]

EGF-dependent phosphorylation of lipocortin-1 is a Ca^{2+}-dependent reaction. Since the membranes for most phosphorylation studies are isolated in the presence of EDTA, Ca^{2+} must be added to the reaction cocktails in order to obtain phosphorylated protein. However, since our protocol is run in the absence of EDTA, there is no Ca^{2+} requirement, relying on residual Ca^{2+} in the preparation to promote phosphorylation. While micromolar Ca^{2+} may be necessary for quantitative phosphorylations, Ca^{2+} induces proteolytic clipping of the amino-terminal segment, and thus its addition can be problematic. Care should be taken in monitoring this competing reaction, since the phosphorylated tail region is particularly sensitive to Ca^{2+}-dependent proteolysis.[18]

---

[18] H. T. Haigler, D. D. Schlaepfer, and W. H. Burgess, *J. Biol. Chem.* **262**, 6921 (1987).

Purification of Lipocortin-1

Most purification strategies used for lipocortin-like proteins rely on Ca^{2+} and/or phospholipid binding as an affinity step.[11,18–20] Typically, a particulate subcellular fraction composed either of membranes or detergent-insoluble complexes is produced in the presence of Ca^{2+}, washed, and the proteins of interest are selectively eluted from the insoluble fraction with EDTA. The Ca^{2+}-binding proteins, which are enriched in the EDTA extract, are then fractionated by a combination of cation- and anion-exchange chromatography steps and gel filtration. Using such a strategy, we can recover about 10 mg of lipocortin-1 from a single human placenta.[6] A similar approach was developed for purifying the recombinant protein.[9] Of the tissues that express lipocortin-1, lung and placenta are particularly rich sources of the protein and thus frequently used for its purification.

Since milligram quantities of lipocortin-1 can be purified from natural sources, many groups have purified the protein; however, in few instances has the protein been analyzed biochemically. Lack of rigor in characterizing the final product, particularly with respect to the state of the amino terminus, can lead to erroneous conclusions that result from differences in protein preparations. Routine gel analysis, while useful for evaluating purity, generally is insensitive to minor sequence differences and is particularly inadequate for lipocortin-1, which is susceptible to proteolytic clipping at multiple sites near the amino terminus. While results from binding studies have correlated specific clipping patterns with changes in Ca^{2+}-binding affinities,[3] experiments comparing the effects of clipped adducts on phosphorylation have not been performed. In the one set of experiments where phosphorylated lipocortin was directly sequenced, the protein was found to be a des-12 variant.[12]

In producing recombinant lipocortin, by chance we obtained a des-9 preparation of the protein. Although the truncated and full-length proteins both were phosphorylated in an EGF-dependent manner, the label in the des-9 protein was tyrosine-specific whereas the majority of the label in the full-length protein was on threonine. The nature of the serine(threonine) protein kinase is unknown but the finding suggests that a second kinase may be associated with the EGF receptor.[21] Because most lipocortin preparations are a mixture of full-length and clipped species, we decided

[19] P. J. Shadle, V. Gerke, and K. Weber, *J. Biol. Chem.* **260**, 16354 (1985).
[20] C. E. Creutz, W. J. Zaks, H. C. Hamman, S. Crane, W. H. Martin, K. L. Gould, K. M. Oddie, and S. J. Parsons, *J. Biol. Chem.* **262**, 1860 (1987).
[21] M. Abdel-Ghany, H. K. Kole, M. A. Saad, and E. Racker, *Proc. Natl. Acad. Sci. U.S.A.* **86**, 6072 (1989).

to mix the full-length and des-9 protein and test the effect of the heterogeneous mixture on phosphorylation. Surprisingly, phosphorylation was almost exclusively on tyrosine, indicating that the clipped moiety was a better substrate for the receptor.

Localization of Phosphorylation Site by Peptide Mapping

To identify a specific site of interest within a protein, it is useful to have at hand structural tools that allow the molecule to be rapidly dissected into smaller and smaller segments. While construction of defined peptide maps can be laborious, once established they serve as a blueprint for the protein, which can be referred to repeatedly. For analyzing lipocortin-1-specific modifications we have developed a two-step process, first relying on CNBr mapping to define the specificity of the modification and to produce a crude fractionation and then using tryptic digests to provide a more detailed analysis. This two-step process was particularly important when studying phosphorylated lipocortin since only a small fraction of the protein was phosphorylated and the perturbation due to label altered the chromatographic mobility of the phosphorylated fragments. While a description of the construction of the maps is beyond the limits of this chapter, the details have been laid out previously[22,23] and applied specifically to lipocortin-1 to identify sites of phosphorylation by pp60^{c-src},[24] insulin receptor kinase,[17] protein kinases A and C,[24] and to characterize a dimeric form of lipocortin-1 which we had purified from placenta.[6]

Cyanogen Bromide Mapping

Figure 2 summarizes the CNBr mapping data. CNBr cleaves lipocortin-1 into eight fragments with masses of 6.2, 7.8, 13.3, 4.2, 1.5, 0.9, 1.1, and 3.1 kDa from the amino to carboxy termini respectively. The positioning of the fragments with respect to the lipocortin-1 sequence is shown at the top of the panel. Under limiting conditions, digestion with CNBr should produce 36 (28 partial and 8 complete) cleavage fragments, where partials represent fragments that contain 2 or more cleavage products. Of the 36 products, 23 are detected on silver-stained SDS gels (land a, Fig.2). Of the missing cleavage products, 10 are simply too small to be detected by SDS–PAGE, and the other 3 were undetected because they comigrate

[22] R. B. Pepinsky, *J. Biol. Chem.* **258**, 11229 (1983).
[23] K-S. Huang, P. McGray, R. J. Mattaliano, C. Burne, E. P. Chow, L. K. Sinclair, and R. B. Pepinsky, *J. Biol. Chem.* **262**, 7639 (1987).
[24] L. Varticovski, S. B. Chahwala, M. Whitman, L. Cantley, D. Schindler, E. P. Chow, L. K. Sinclair, and R. B. Pepinsky, *Biochemistry* **27**, 3682 (1988).

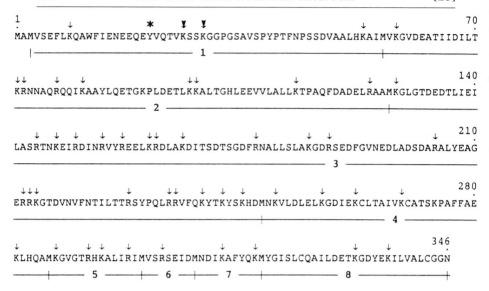

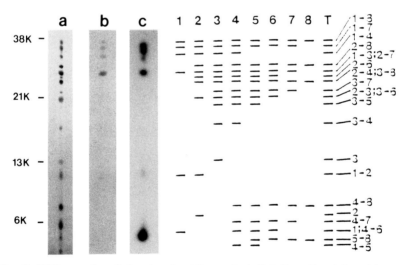

Fig. 2. CNBr mapping of phosphorylated lipocortin-1. Gel slices (2 × 1.5 × 1.4 mm) containing 0.5 μg of unlabeled or *in vitro* phosphorylated lipocortin were incubated at 23° for 1 hr with 21 mg/ml CNBr in 0.1 N HCl, 0.1% 2-mercaptoethanol. CNBr-treated gel slices were washed twice for 5 min each with water, once for 5 min with 0.25 M Tris-HCl, pH 6.8, and once for 10 min at 37° with electrophoresis sample buffer. The gel slices were loaded onto SDS slab gels and the cleavage products subjected to SDS-PAGE. (Top) Organization of CNBr fragments within the lipocortin-1 sequence. Small arrows denote trypsin cleavage sites. Large arrows denote plasmin cleavage sites. Tyrosine-21 is marked with an asterisk.

with other partials. All of the observed cleavage products have been characterized by CNBr mapping, and thus the pattern of bands can be used to distinguish between each of the CNBr fragments.[24]

The schematic shown at the bottom right of Fig. 2 summarizes the mapping results. Profiles representing the total digest (lane T, Fig. 2) and specific subsets of the digest (lanes 1–8, Fig. 2) are indicated. Each CNBr fragment is represented by a distinct subset of products, which in turn serves as a hallmark for its identification. Lane c (Fig. 2) shows CNBr mapping results for lipocortin that had been phosphorylated on tyrosine by the EGF receptor; six fragments were labeled. The pattern of labeled products indicates that CNBr fragment 1 is phosphorylated, since a modification in fragment 1 is the only type that would produce such a profile (compare lane c with lanes 1–8 in Fig. 2) and thus localizes the site of phosphorylation within the first 55 amino acids of lipocortin.

The CNBr mapping data were confirmed by epitope mapping, using a CNBr fragment 1-specific monoclonal antibody to directly monitor the fragment 1-containing cleavage products. When cleavage products were subjected to Western blotting and probed with the fragment 1-specific antibody (lane b, Fig. 2), the five fragment 1-containing partials were detected. Except for fragment 1 itself, which is not detected by the antibody on Western blots, the profiles of the fragment 1-containing cleavage products (lane b, Fig. 2) and phosphorylated fragments (lane c, Fig. 2) were identical.

Tryptic Peptide Mapping

To localize the phosphorylation site further, lipocortin-1 was subjected to tryptic peptide mapping, using reversed-phase HPLC to separate cleavage products. Previously, all of the peaks from the lipocortin-1 tryptic map were sequenced, which covered approximately 85% of the primary structure.[23] Most of the missing fragments were small and eluted in the flow-through. Figure 3 summarizes the sequence data, where fragment

---

(Bottom) CNBr mapping results. Lanes a–c show lipocortin-1 cleavage products that were visualized by silver staining, by Western blotting with CNBr fragment 1-specific antiserum, and by autoradiography, respectively. Lanes 1–8 and T show schematic diagrams summarizing mapping data for lipocortin-1. Numbers at right reflect fragment compositions, where 1 refers to the N-terminal CNBr fragment and 8 to the C-terminal fragment. Lane T represents a profile of the total digest. Apparent molecular weights of specific fragments are indicated at left. [Portions reprinted with permission from L. Varticovski, S. B. Chahwala, M. Whitman, L. Cantley, D. Schindler, E. P. Chow, L. K. Sinclair, and R. B. Pepinsky, *Biochemistry* **27**, 3682 (1988). Copyright © American Chemical Society.]

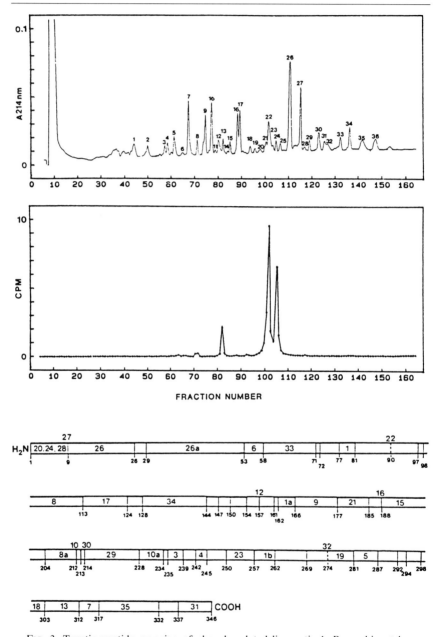

Fig. 3. Tryptic peptide mapping of phosphorylated lipocortin-1. Recombinant human lipocortin-1 (1 nmol) was digested with trypsin for 20 hr at 37° and the proteolytic fragments separated by reversed phase HPLC on a C_{18} column (SpectraPhysics, San Jose, CA, 0.46 × 25 cm). Bound components were eluted with a 95-min gradient (0–75% acetonitrile in 0.1%

designations correspond to peak numbers listed in the tryptic map shown at the top. All of the tryptic fragments from CNBr fragment 1 have been identified, and thus the map can be used directly for characterizing fragment 1-specific modifications. When [^{32}P]phosphate-labeled protein that had been phosphorylated by the EGF receptor was digested with trypsin and the labeled fragments analyzed by HPLC, the peptide map shown in the middle was obtained. Two major peaks at fractions 102 and 106 were observed. Although the fragments are shifted in their chromatographic mobility due to the added phosphate group, the pattern is consistent with phosphorylation within peptide 9–26, since this is the only peptide from within CNBr fragment 1 that would produce such a profile. Subsequent sequencing of the peptide verified its identity. The observed site, Tyr-21, is a canonical phosphorylation site that is rich in glutamic acids. Tyr-21 was confirmed as the only phosphorylation site for the EGF receptor by two independent approaches, first using site-directed mutagenesis to convert the tyrosine into phenylalanine and second using limited proteolysis with plasmin to release selectively the phosphorylated tail region from the remainder of the protein.[24]

The amino-terminal domain of lipocortin-like proteins, by regulating Ca^{2+}- and phospholipid-binding affinities, provides a mechanism for increasing or decreasing the affinity of the proteins for membranes. Tyr-21 falls within this region and through its phosphorylation has created a pathway whereby EGF can impact function. Although *in vitro* studies have demonstrated that tyrosine-specific phosphorylation of lipocortin alters the Ca^{2+}-/phospholipid-binding affinities and enhances the suscepti-

---

trifluoroacetic acid at 1.4 ml/min (0.5-min fractions were collected). The column eluate was monitored simultaneously at 280 and 214 nm. For mapping phosphorylated fragments, 1 nmol of lipocortin-1 was spiked with 400,000 counts/min (cpm) of labeled protein that had been gel purified and then processed as described above. Column fractions were analyzed for absorbance at 214 and 280 nm and by scintillation counting. (Top) Elution profile monitored at 214 nm. Numbered peaks were subjected to amino acid and sequence analysis. These results are summarized with respect to the lipocortin sequence shown below. (Middle) Profile of phosphorylated fragments (10 = 76,000 cpm). (Bottom) Summary of sequence data. Potential tryptic peptides are represented with boxes. Large numbers refer to peak numbers in the top graph. Peaks 1, 8, 10, and 26 produced multiple sequences. Sequences that were derived from partial cleavage products are indicated above the junction between boxes. In two instances indicated by dashed lines, partials were generated because of a trypsin-resistant Lys-Pro sequence. Several peaks contain structural variants of the amino-terminal fragment, consistent with the known amino-terminal heterogeneity of this particular batch of *E. coli*-derived protein. [Peptide sequences reprinted with permission from K.-S. Huang, P. McGray, R. J. Mattaliano, C. Burne, E. P. Chow, L. K. Sinclair, and R. B. Pepinsky, *J. Biol. Chem.* **262**, 7639 (1987). Copyright © American Society for Biochemistry and Molecular Biology.]

bility of the tail region to proteolysis,[15,18] the significance of the event awaits a clearer understanding of the physiological roles of the protein. The phosphorylation of lipocortin-1 in intact cells as a tag for activation of the EGF receptor provides a simple method for monitoring the activation of the transduction pathway.

[26] Cloning, Expression, and Biological Effects of erbB-2/neu Gene in Mammalian Cells

By PIER PAOLO DI FIORE, ORESTE SEGATTO, and
STUART A. AARONSON

Independent Approaches Identifying erbB-2/neu Oncogene

Human tumors often contain genes (referred to as oncogenes) able to confer a dominant transformed phenotype when transfected into murine fibroblasts.[1] The majority of oncogenes encode altered versions of one of three closely related members of the *ras* gene family.[1] Oncogenes different from *ras* have also been detected.[1] One, termed *neu*, was first identified in rat neuroblastomas induced with the chemical carcinogen ethylnitrosourea (ENU).[2] Sera obtained from mice bearing tumors induced with neuroblastoma transfectants reacted with a 185-kDa phosphoprotein specifically induced by the neuroblastoma transforming sequence,[3] thus identifying the *neu* oncogene as distinct from the *ras* family. Further studies revealed that *neu* was distinct from but related to the epidermal growth factor receptor (EGFR).[4]

Independent efforts aimed at detection of human oncogenes were based on evidence that the EGFR gene was amplified in certain human tumors. Using a v-*erbB* probe in moderate stringency conditions, King *et al.* analyzed alterations affecting EGFR-related genes in a series of human mammary tumors.[5] DNA prepared from a human mammary carcinoma, MAC117, showed a pattern of hybridization differing from that observed

[1] P. Kahn and T. Graf, (eds.), "Oncogenes and Growth Control." Springer-Verlag, New York, 1986.
[2] C. Shih, L. C. Padhy, M. Murray, and R. A. Weinberg, *Nature (London)* **290**, 261 (1981).
[3] L. C. Padhy, C. Shih, D. Cowing, R. Finkelstein, and R. A. Weinberg, *Cell (Cambridge, Mass.)* **28**, 865 (1982).
[4] A. L. Schechter, D. F. Stern, L. Vaidyanathan, S. J. Decker, J. A. Drebin, M. I. Greene, and R. A. Weinberg, *Nature (London)* **312**, 513 (1984).
[5] C. R. King, M. H. Kraus, and S. A. Aaronson, *Science* **229**, 974 (1985).

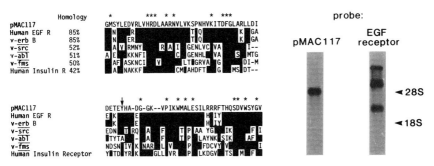

FIG. 1. (Left) Comparison of the putative encoded amino acid sequence of pMAC117 with known tyrosine kinase sequences. Black regions represent homologous amino acids. Asterisks denote amino acids conserved in all sequences. (Right) Detection of the erbB-2 transcript in mRNA extracted from MAC117 tumors in comparison to EGFR-specific transcripts from A431 cells.

both with normal human DNA and with DNA of a tumor with an amplified EGFR gene. These findings were consistent with the possibility that the MAC117 tumor contained an amplified DNA sequence related to, but distinct from, the cellular *erbB* gene (EGFR).

To define its structure, the specific DNA fragment was molecularly cloned and its nucleotide sequence determined. The sequence contained two regions of nucleotide sequence similarity to v-*erbB* separated by 122 nucleotides. The predicted amino acid sequence (Fig. 1) was 85% homologous to two regions that are contiguous in the EGFR sequence and showed similarity with other tyrosine kinases as well (Fig. 1). Findings that a cloned probe of the gene detected a novel single 5-kilobase (kb) transcript suggested that it was a functional gene[5] (Fig. 1). The gene was designated *erbB-2*, representing a new functional gene within the tyrosine kinase family, closely related to but distinct from the gene encoding the EGFR. Several other laboratories independently identified *erbB-2* as well.[6,7] The chromosomal localization of human *erbB-2* on chromosome 17 at q21 was shown to coincide with the localization of the *neu* gene on the human genome,[7] strongly suggesting identity of the two genes. This identity was further established by comparison of DNA restriction, cross-hybridization, and finally nucleotide sequence analysis.[7,8]

[6] K. Semba, N. Kamata, K. Toyoshima, and T. Yamamoto, *Proc. Natl. Acad. Sci. U.S.A.* **82**, 6497 (1985).
[7] L. Coussens, T. L. Yang-Feng, Y. C. Liao, E. Chen, A. Gray, J. McGrath, P. H. Seeburg, T. A. Libermann, J. Schlessinger, and U. Francke, *Science* **230**, 1132 (1985).
[8] C. I. Bargmann, M. C. Hung, and R. A. Weinberg, *Nature (London)* **319**, 226 (1986).

The gene has also been referred to as *HER-2* and *NGL*. Based on the designation *erbB-2* adopted by the Howard Hughes Medical Institute Human Gene Mapping Library, we utilize this nomenclature for the human gene and its product, gp185^{erbB-2}. However, we have retained the *neu* designation for the rat homolog to facilitate discussion of some apparent differences in biological properties between the rat and human homologs (see below). We refer to the rat *neu* oncogene as *neuT* (for *neu*-transforming) and to its normal rat counterpart as *neuN* (for *neu*-normal).

Role of erbB-2/neu in Experimentally Induced and Naturally Occurring Tumors

The identification of *neuT* as a dominant transforming gene led to the investigation of mechanisms involved in its activation. To address this question, Hung et al. isolated genomic clones of *neuN* and *neuT*.[9] The *neuN* genomic clone exhibited no detectable transforming activity and no obvious differences in its overall structure as compared to the transforming allele,[9] suggesting that the lesion activating the *neu* gene was a minor change in the DNA sequence. Later, Bargmann et al. identified the activating lesion as a single point mutation causing a change from valine to glutamic acid at position 664 in the transmembrane domain of the predicted protein.[10]

At variance with ENU-induced rat neuroblastomas, the *erbB-2* gene has not been detected as a transforming gene by NIH/3T3 transfection analysis.[11] In addition, *erbB-2* genes cloned directly from two human mammary carcinomas displaying *erbB-2* amplification showed no mutations in the predicted transmembrane domain.[12] These findings suggested that mutations similar to those observed in *neuT* were not frequently associated with *erbB-2* activation as an oncogene in human malignancies.

The initial identification of *erbB-2* gene amplification in a primary human tumor,[5] however, suggested the possibility that *erbB-2* overexpression might contribute to neoplastic growth. Indeed, studies have shown that the *erbB-2* gene is amplified and/or overexpressed in a variety of adenocarcinomas,[13] particularly in mammary and ovarian adenocarcino-

[9] M. C. Hung, A. L. Schechter, P. Y. Chevray, D. F. Stern, and R. A. Weinberg, *Proc. Natl. Acad. Sci. U.S.A.* **83,** 261 (1986).
[10] C. I. Bargmann, M. C. Hung, and R. A. Weinberg, *Cell (Cambridge, Mass.)* **A5,** 649 (1986).
[11] M. H. Kraus, N. C. Popescu, S. C. Amsbaugh, and C. R. King, *EMBO J.* **6,** 605 (1987).
[12] D. J. Slamon, W. Godolphin, L. A. Jones, J. A. Holt, S. G. Wong, D. E. Keith, W. J. Levin, S. G. Stuart, J. Udove, and A. Ullrich, *Science* **244,** 707 (1989).
[13] J. Yokota, T. Yamamoto, K. Toyoshima, M. Terada, T. Sugimura, H. Battifora, and M. J. Cline, *Lancet* **1,** 765 (1986).

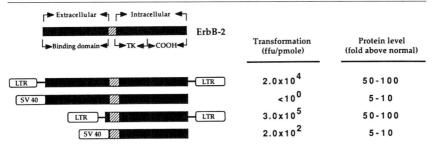

Fig. 2. Transforming activity and expression level of eukaryotic expression vectors for erbB-2 and its amino-truncated form. The LTR- and SV40-based eukaryotic expression vectors for erbB-2 and the amino-truncated counterpart (ΔN erbB-2) are depicted. The transforming activity of the constructs was evaluated as the ability to induce foci of morphological alteration on NIH/3T3 cells. The transforming efficiency was calculated in focus-forming units per picomole of DNA added, based on the relative molecular weights of the respective plasmids.[15] erbB-2 and ΔN erbB-2 expression was determined by immunoblot analysis using antibodies generated against synthetic peptides derived from the erbB-2 sequence.[11]

mas.[12,14] With reference to primary breast cancers, erbB-2 gene amplification has been detected at frequencies from 10 to 40%.[12-14] Evidence has also been provided that alterations of erbB-2 expression might correlate with a more aggressive disease course in breast cancer patients.[12,14]

Transforming Potential of Normal and Activated erbB-2/neu in Model Systems

Findings that the most frequent alteration affecting the erbB-2 gene in human cancers is overexpression in the absence of any other apparent genetic change in the coding sequence[5,6,11,12] prompted efforts to assess the effect of overexpression of gp185^{erbB-2} in model systems in vitro. erbB-2 eukaryotic expression vectors were engineered based on the transcriptional initiation sequences of either the Moloney murine leukemia virus long terminal repeat (Mo-MLV LTR) or the SV40 early promoter, in an attempt to express the erbB-2 cDNA at different levels in NIH/3T3 cells.[15] As shown in Fig. 2, an LTR-based erbB-2 expression vector induced transformed foci at high efficiency. In striking contrast, the SV40/erbB-2 construct failed to induce any detectable morphological alteration of NIH/3T3 cells (Fig. 2). Immunological analysis revealed that SV40/erbB-2

[14] D. J. Slamon, G. M. Clark, S. G. Wong, W. J. Levin, A. Ullrich, and W. L. McGuire, Science 235, 177 (1987).
[15] P. P. Di Fiore, J. H. Pierce, M. H. Kraus, O. Segatto, C. R. King, and S. A. Aaronson, Science 237, 178 (1987).

transfectants expressed gp185^{erbB-2} at 10-fold higher levels than control NIH/3T3 cells. A further 10-fold increase in gp185^{erbB-2} expression was detected in LTR/*erbB-2* transfectants (Fig. 2). These results demonstrated that the high levels of *erbB-2* expression under LTR influence correlated with its ability to exert transforming activity.[15]

The level of overexpression of gp185^{erbB-2} in human mammary tumor cell lines possessing amplified *erbB-2* genes was compared with that of NIH/3T3 cells transformed by the *erbB-2* coding sequence. We found that human mammary tumor cells which overexpressed the *erbB-2* gene demonstrated levels of the *erbB-2* gene product similar to those needed to induce malignant transformation in a model system.[15] These findings established a mechanistic basis for normal *erbB-2* gene amplification as a causal driving force in the clonal evolution of a tumor cell rather than being an incidental consequence of tumorigenesis.

While the transforming potential of the overexpressed normal human gp185^{erbB-2} has been independently confirmed,[16] contradictory results have been reported for its normal rat homolog. A number of studies have found no detectable transformation of murine fibroblasts resulting from overexpression of *neuN*.[9,10] However, there was no attempt to estimate the levels of overexpression of gp185^{neuN} obtained in the various model systems, as compared to the levels of gp185^{erbB-2} needed to induce transformation *in vitro*. To address this question, we have systematically compared the effects of different levels of expression of *erbB-2* and *neu* in NIH/3T3 cells. Our results show that the *neuN* gene also acts as a potent oncogene when expressed in NIH/3T3 cells, at levels similar to those needed for *erbB-2*-induced transformation.[17]

A number of other genetic alterations have been shown to confer transforming potential to *erbB-2/neu*. Bargmann and Weinberg engineered several point mutations in the region encoding the transmembrane domain of gp185^{neu}. Only a mutation which caused a substitution of glutamine for valine at position 664 was able to mimic the Val-664 → Glu-664 substitution in terms of oncogene activation.[18] Four other substitutions (Lys, Hys, Tyr, and Gly, respectively, for Val-664) were ineffective. A fifth change leading to an aspartic acid for Val-664 substitution caused a modest enhancement of transforming ability. Of note, changes leading to a substituted glutamic acid residue at position 663 or 665, immediately bordering Val-664, did not activate the *neu* gene.[18] It was concluded that a structural

---

[16] R. M. Hudziak, J. Schelssinger, and A. Ullrich, *Proc. Natl. Acad. Sci. U.S.A.* **84,** 7159 (1987).

[17] E. Di Marco, J. H. Pierce, C. L. Knicley, and P. P. Di Fiore, *Mol. Cell. Biol.* **10,** 3247 (1990).

[18] C. I. Bargmann and R. A. Weinberg, *EMBO J.* **7,** 2043 (1988).

requirement, independent of charge and stringently dependent on site and position of Val-664, was altered in the Val-664 → Glu-664 substitution. Segatto et al. engineered similar mutations in the human normal *erbB-2* cDNA.[19] At variance with data reported by Bargmann and Weinberg,[18] both glutamic acid for valine and aspartic acid for valine substitutions at position 659 (the position analogous to Val-664 in rat *neuN*) activated the oncogenic potential of *erbB-2* to a similar extent.[19]

Another alteration capable of activating *erbB-2/neu* is amino-terminal truncation (Fig. 2). Expression vectors encoding a protein, ΔN *erbB-2*, in which the extracellular ligand-binding domain was deleted showed increased transforming efficiency at both low and high levels of expression.[15] Similar results have been obtained in the rat system.[18] In the latter system it has also been reported that the effect of the amino-terminal truncation is not additive to that of the Val-664 → Glu-664 transmembrane substitution, suggesting that the two alterations might be influencing transforming activity through a common mechanism.[18] Activation of *erbB-2/neu* by amino-terminal truncation resembles activation of EGFR by a similar alteration in v-*erb*,[20] suggesting that the extracellular domains of both proteins exert a negative regulatory influence that can be abolished by deletion or modulated in response to ligand binding.

[19] O. Segatto, C. R. King, J. H. Pierce, P. P. Di Fiore, and S. A. Aaronson, *Mol. Cell. Biol.* **8**, 5570 (1988).
[20] J. Schlessinger, *Adv. Exp. Med. Biol.* **234**, 65 (1988).

[27] Biological Effects of Monoclonal Antireceptor Antibodies Reactive with *neu* Oncogene Product, p185^{neu}

By Jeffrey N. Myers, Jeffrey A. Drebin, Takuro Wada, and Mark I. Greene

Introduction

The product of the *neu* oncogene, p185^{neu}, is a 185-kilodalton (kDa) plasma membrane glycoprotein with a cytoplasmic tyrosine kinase domain. The similarity of p185^{neu} to the epidermal growth factor receptor (EGFR) in terms of size, sequence, domain structure, and subcellular localization suggest that p185^{neu} is a growth factor receptor. However, the ligand for this receptor-like molecule has not been isolated. Therefore, we and other investigators have had to devise alternative strategies for studying the role of p185^{neu} in normal and neoplastic cell growth. Several years ago we developed monoclonal antibodies to p185^{neu} that have been extremely useful in our studies of the role of p185^{neu} in neoplastic tranformation. We have also used these antibodies to treat tumors that arise in mice inoculated with *neu*-transformed cells. The results of this work demonstrate that the cell surface expression of p185^{neu} is required for maintenance of the transformed phenotype and indicate that *in vivo* treatment with anti-p185^{neu} antibodies leads to inhibition of tumor growth. Taken together, the data suggest that monoclonal antibodies to cell surface molecules essential for maintaining the transformed phenotype are potentially useful in the treatment of cancer.

Development of Monoclonal Antibodies to p185^{neu}

The *neu* oncogene was originally identified as a gene derived from ethylnitrosourea-induced rat neuroblastomas capable of transforming NIH 3T3 cells in DNA transfection assays.[1] During initial studies, it was found that mice inoculated with *neu*-transformed NIH 3T3 cells developed a serologic response to a 185-kDa phosphoprotein which was found only in the lysates of *neu*-transformed cells. These studies suggested that *neu* oncogene transfection induces the cell surface expression of a 185-kDa phosphoprotein antigen. This cell surface protein has subsequently been shown to be the product of the *neu* oncogene, and it has been called p185^{neu}.[2] In order to further characterize the p185^{neu} antigen and study its role in cellular transformation, we produced monoclonal antibodies which recognize this molecule.[3] Our strategy for producing anti-p185^{neu} antibodies is outlined in Fig. 1.

To prepare p185^{neu}-specific monoclonal antibodies, C3H female mice are immunized by intraperitoneal injection of 2×10^6 B104-1-1 cells, *neu* oncogene-transformed NIH 3T3 cells, emulsified in Freund's complete adjuvant (Difco, Detroit, MI). Booster injections of $1-5 \times 10^6$ B104-1-1 cells in Freund's incomplete adjuvant are given at 2- to 4-week intervals for a total of 5 injections over a 3-month period. Ascites is tapped from the abdominal cavity with an 18-gauge needle and spun at 2500 g to remove cellular debris.

The ascites fluid is screened for p185 immunoreactivity by immunoprecipitation and immunofluorescent flow cytometry, after absorption on NIH

---

[1] L. C. Padhy, C. Shih, D. Cowing, R. Finkelstein, and R. A. Weinberg, *Cell* (*Cambridge, Mass.*) **28**, 865 (1982).

[2] A. L. Schecter, D. F. Stern, L. Vaidyanathan, S. J. Decker, J. A. Drebin, M. I. Greene, and R. A. Weinber, *Nature* (*London*) **312**, 513 (1984).

[3] J. A. Drebin, D. F. Stern, V. C. Link, R. A. Weinberg, and M. I. Greene, *Nature* (*London*) **312**, 545 (1984).

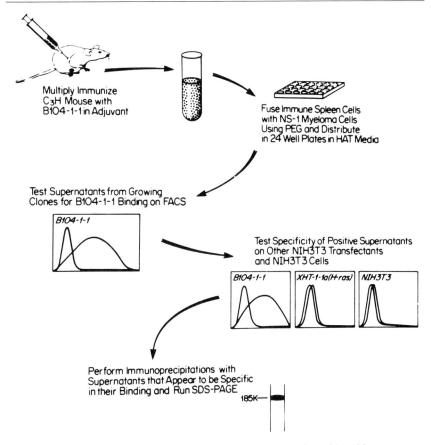

Fig. 1. Strategy for production of antibodies reactive with p185^{neu}.

3T3 cells to eliminate antibodies reactive to cell surface antigens that are not specific for *neu* oncogene transformation. To absorb the serum, 0.1 ml of ascites fluid is diluted 10-fold in phosphate-buffered saline (PBS) and incubated with 5×10^7 NIH 3T3 cells for 1 hr at 4° on a rotating wheel. The cells are then pelleted by centrifugation and the supernatant retained for analysis.

To screen the antiserum by immunofluorescence cell cytometry, B104-1-1 and NIH 3T3 cells are removed from tissue culture flasks with Versene, a buffered EDTA solution from M. A. Bioproducts (Walkersville, MD), and washed twice in fluorescence-activated cell sorter (FACS) buffer consisting of Hanks' balanced salt solution (GIBCO, Grand Island, NY) (HBSS) with 2% fetal calf serum, 0.1% sodium azide, and 10 mM HEPES.

Between 1×10^5 and 1×10^6 cells in 0.1 ml of FACS buffer are incubated with 0.1 ml antiserum dilution or control supernatant for 1 hr at 4°. The cells are then washed 3 times in 2.5 ml of FACS buffer by resuspension in fresh buffer followed by cell pelleting at 1000 g. The cells are then resuspended in 0.1 ml of fluorescein isothiocyanate (FITC)-conjugated rabbit anti-mouse IgG diluted 1:20 to 1:50 in FACS buffer at 4° for 1 hr. After 3 washes in 2.5 ml FACS buffer, the cells are resuspended and fixed in 0.5–1.0 ml of 2% paraformaldehyde–PBS. About 10,000 cells/sample are routinely analyzed, and the specific fluorescence is quantitated by subtracting the median fluorescence channel of cells stained with FITC-conjugated rabbit anti-mouse IgG alone from the median fluorescence channel of cells stained with specific antibody followed by FITC-conjugated rabbit anti-mouse IgG.

Mice whose sera proved positive for reactivity to cell surface determinants on *neu*-transformed cells by immunofluorescence and immunoprecipitation are sacrificed by cervical dislocation and their spleens removed. The spleens are minced with forceps to yield a single-cell suspension that is then filtered through a nylon mesh to remove clumped cells. Erythrocytes are lysed by treatment with Tris–ammonium chloride. Splenocytes are then pelleted with aminopterin-sensitive NS-1 myeloma cells at a 7:1 ratio by spinning at 1000 g and then washed 1 time in serum-free Dulbecco's modified Eagle's medium (DMEM). Three-tenths milliliter of 40% polyethylene glycol 1600 (ATCC, Rockville, MD) in PBS is added to the pellet over 1 min, after which time the suspension is warmed for 1 min at 37°. After addition of 15 ml of DMEM over a 5-min period, the cells are pelleted by centrifugation at 800 g.

The fused cells are then resuspended and mixed with 500 ml of a selection medium consisting of 20% zeta serum (AMF Biologicals), 1 mM sodium pyruvate (GIBCO), 0.1 mM essential amino acids (GIBCO), 0.1 mM nonessential amino acids (GIBCO), 2 mM L-glutamine (GIBCO), 1% penicillin–streptomycin–Fungizone mixture (M. A. Bioproducts), 10^{-4} M hypoxanthine (Sigma, St. Louis, MO), 4×10^{-7} M aminopterin (Sigma), and 1.6×10^{-5} M thymidine (Sigma), and 1 ml of medium with approximately 2×10^5 cells is added per well of 24-well plates for incubation at 37° in a humidified 10% CO_2 incubator. The cultures are fed with 1 ml/well hybrid selection medium on days 7 and 14 and were examined for the presence of hybridomas on day 14.

Hybridoma supernatants are tested for B104-1-1-specific binding using indirect immunofluorescence as described above, and cells positive for B-104-1-1, but nonreactive to NIH 3T3 or H-*ras*-transformed NIH 3T3 cells, are cloned by 3 cycles of limiting dilution. In order to prepare large quantities of purified antibodies the hybridomas are injected intraperitoneally into pristane-primed, 400-rad irradiated C_3D2F_1 mice to induce ascites fluid production. Cleared ascites fluid is precipitated by addition of ammonium sulfate to 50%. The resuspended precipitate is passed over a protein A-Sepharose column in order to purify IgG_2 and IgG_{2b} antibodies, or eluted from DEAE-Affi-Gel blue or Sephadex G-200 in order to purify IgG_1 and IgM monoclonal antibodies, respectively. Several antibodies including an IgG_{2a} antibody, MAb 7.16.4, are found to precipitate $p185^{neu}$ from the lysates of cells transfected with eukaryotic expression vectors bearing the *neu* oncogene.

Further characterization of the monoclonals includes competitive binding studies to distinguish whether the several anti-$p185^{neu}$ antibodies we produced were to the same or different antigenic determinants.[4] In these studies, antibodies are iodinated with ^{125}I to a specific activity of 500–2500 counts/min (cpm)/ng using the chloramine-T method. Then 1×10^6 B104-1-1 cells are incubated with labeled antibody at 75% of saturating concentration and varying amounts of unlabeled competing antibody in HBSS with 5% fetal calf serum and 0.2% sodium azide for 4 hr at 4°. The cells are washed 3 times in 2 ml of the same buffer before the bound radioactivity is measured by γ counting.

Competition assays using five different monoclonal antibodies identified three different binding classes. Antibodies of each class could compete for binding with members of the same class but could not inhibit the binding of antibodies of the other two classes. These data suggest that the three binding classes represent antibodies which bind to three distinct regions of $p185^{neu}$.

Down-Modulation of $p185^{neu}$ Expression by Monoclonal Antibodies

A number of studies with antibodies to cell surface antigens have demonstrated that polyvalent antibodies can induce the down-modulation of their cognate antigens through a cross-linking-induced internalization process.[5] With the antibodies we parepared to $p185^{neu}$, we were able to show that $p185^{neu}$ could be internalized by antibody treatment, and by preparing F(ab) fragments from the monoclonal antibodies, we demonstrated that antibody-mediated internalization of $p185^{neu}$ is most likely due to a cross-linking process.[6]

To assay for antibody-induced down-modulation of $p185^{neu}$ cell surface

---

[4] J. A. Drebin, V. C. Link, and M. I. Greene, *Oncogene* **2**, 387 (1988).
[5] G. M. Edelman, *Science* **192**, 218 (1976).
[6] J. A. Drebin, V. C. Link, D. F. Stern, R. A. Weinberg, and M. I. Greene, *Cell* (*Cambridge, Mass.*) **41**, 695 (1985).

expression, 2×10^5 B104-1-1 cells are plated in 60-mm culture dishes containing DMEM with 10% fetal bovine serum and cultured for 24 hr. Varying amounts of affinity-purified monoclonal antibody, 7.16.4, are added to the plates and incubated at 37° for different time periods, at which time the cells are removed from the plates with Versene, restained with saturating quantities of antibody, and processed as above for immunofluorescent cell cytometry. Preinternalization levels of expression are determined by preparing parallel wells of cells for flow cytometry without prior antibody incubation at 37°. The percentage surface expression of p185^{neu}, E, is determined using Eq. (1)[7]:

$$E = 10^{(X/512)} - 10^{(N/512)}/10^{(P/512)} - 10^{(N/512)} \quad (1)$$

where X represents the logarithmic median fluorescence channel of cells in the experimental group, N represents the logarithmic median fluorescence channel of cells stained with fluorescent second antibody alone (to correct for nonspecific binding), and P represents the logarithmic median fluorescence channel of untreated cells stained with saturating amounts of first and second antibody (postive control). p185^{neu} internalization has also been determined by direct binding studies in which cells pretreated with 7.16.4 at 37° are washed and incubated with a noncross-reactive radioiodinated p185^{neu} antibody.

Phenotypic Effects of Monoclonal Antibody-Induced p185^{neu} Down-Modulation

Since our studies with the monoclonal antibodies specific for p185^{neu} demonstrated that the antibodies will down-modulate the cell surface expression of p185^{neu} by more than 70% in 2 hr with 5 μg of antibody added per 60-mm dish, we were in a position to address questions about the effect of down-modulation of the cell surface expression of p185^{neu} on the phenotype of *neu* oncogene-transformed cells. To investigate these questions we added monoclonal antibodies to p185^{neu} or F(ab) fragments prepared from them to *neu*-transformed cells in transformation assays.[6]

The ability of transformed cells to grow without adherence to the extracellular matrix distinguishes them from normal cells grown *in vitro*. This property, termed anchorage-independent growth, is most conveniently assayed by suspending cells in soft agar and observing whether or not they can grow and divide to form colonies. Normal NIH 3T3 cells will

[7] J. A. Drebin, "Effects of Monoclonal Antibodies Reactive with the *neu*-Oncogene Product on the Neoplastic Properties of *neu*-Transformed Cells." Ph.D. Thesis, Harvard University Division of Medical Sciences, Cambridge, Massachusetts, 1987.

grow slowly and then stop dividing when plated in soft agar, whereas *neu*-transformed NIH 3T3 cells, such as the cell line B104-1-1, grow well in soft agar; 5–10% of input B104-1-1 cells give rise to colonies of greater than 0.5 mm diam.

To determine the effect of monoclonal antibodies to p185^{neu} on the anchorage-independent growth of B104-1-1 cells or a *ras*-transformed NIH 3T3 cell line, 1×10^3 cells are suspended in 1 ml of 0.18% agarose RPMI 1640 containing 10% fetal bovine serum, penicillin–streptomycin–Fungizone, and gentamicin with 0–10 μg MAb 7.16.4 and layered onto 60-mm plates containing 5 ml of 0.24% agarose in RPMI 1640 supplemented with 10% fetal calf serum, penicillin–streptomycin–Fungizone, and gentamicin. The cells are grown at 37° in a 5% CO_2 humidified incubator and are fed at weekly intervals by the addition of 1 ml DMEM with 10% fetal bovine serum, antibiotics, and the same quantity of antibody initially added to the top agarose layer. On the day before colonies are counted, the cultures are fed with 1 ml of HBSS containing 1 mg/ml *p*-iodonitrotetrazolium violet (INT, Sigma) to stain the colonies. On the following day (usually day 14) colonies of diameter 0.5 mm or greater are counted using a dissecting microscope and a calibrated template (see Fig. 2).

Studies of the effect of anti-p185^{neu} antibodies on the anchorage-independent growth of *neu*-transformed cells demonstrated an antibody dose-dependent inhibition of *neu*-transformed NIH 3T3 cell growth in soft agar. These effects were shown to be reversible and reproducible and were not attributable to a nonspecific antibody toxicity. Furthermore, the anti-p185^{neu} inhibition of colony growth was specific for *neu*-transformed cells, as the antibody had no effect on the anchorage-independent growth of *ras* transformants or *ras*/*neu* double transformants. Colony growth was partially inhibited with 10 ng/dish of MAb 7.16.4 and it was inhibited by greater than 90% with 1 μg/dish, whereas an equal concentration of isotype-matched antibody had no inhibitory effect.

Although these studies indicate that anti-p185^{neu} antibodies have a direct cytostatic effect on *neu*-transformed cells, they do not delineate the mechanisms through which the antibodies inhibit growth in soft agar. To determine whether these effects occur as a result of the processes of patching and capping in which multiple p185^{neu} molecules are cross-linked by divalent antibodies and internalized, or whether the binding of antibody by p185^{neu} has a direct cytostatic effect on cells bearing p185^{neu}, we tested the effects of monovalent F(ab) fragments prepared from the anti-p185^{neu} monoclonal antibody, 7.16.4, on the anchorage-independent growth of *neu*-transformed cell lines, and compared these results to those obtained with intact 7.16.4.[6]

To prepare the F(ab) fragments, purified antibody is resuspended to 2

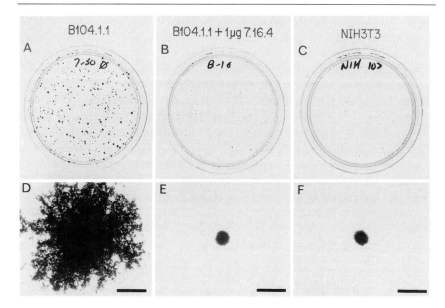

FIG. 2. Effects of anti-p185^{neu} monoclonal antibody on soft agar growth of *neu*-transformed cells. One thousand cells of the indicated cell types were plated in soft agar in the presence or absence of anti-p185^{neu} antibody, 7.16.4, as described in the text. After 14 days the entire culture dish (A–C) or representative colonies (magnification): ×40 (D–F) were photographed. (A, D) Untreated *neu*-transformed NIH 3T3 cells, B104-1-1; (B, E) B104-1-1 cells treated with 1 μg of 7.16.4; (C, F) untreated NIH 3T3 cells. Solid bars in D–F indicate 200 μm.

mg/ml in 4 mM ethylenediaminetetraacetic acid (EDTA, Sigma), 20 mM 2-mercaptoethanol (Sigma), and then mercuripapain (Sigma) is added to a final ratio of 2:100 (w/w) relative to the amount of antibody for a 16-hr incubation at 37°. The reaction is quenched by chilling the mixture to 4° and adding 0.3 M iodoacetamide (Sigma) in 1 M Tris, pH 8.0, to a final concentration of 15 mM. The antibody preparation is dialyzed against PBS overnight with 3 buffer changes before passing it over a protein A-Sepharose column. The column filtrate is asssayed for B104-1-1 reactivity by indirect immunofluorescence using a light-chain reactive second antibody, and proteolytic digestion is verified by reducing polyacrylamide gel electrophoresis followed by Coomassie blue staining. To assay the effects of the F(ab) fragments on the growth of *neu*-transformed cells, the fragments are added to the soft agar assay in the same manner as the intact antibody is added.

Monovalent F(ab) fragments prepared from 7.16.4 were found to have no direct effect on the anchorage-independent growth of *neu*-transformed cells, and, in fact, the monomeric antibody fragments partially antagonized the growth-inhibitory effects of the intact antibody. The data suggest that cross-linking of p185^{neu} molecules by multivalent antibodies is required for biological effects of antibody treatment. Further support for this hypothesis comes from experiments in which the F(ab) fragments were cross-linked by the addition of goat anti-mouse antibody to the cells incubated in soft agar with 7.16.4 F(ab) fragments. F(ab) monomers cross-linked with a second antibody were found to inhibit soft agar growth to the same degree as intact antibody.

In Vivo Effects of p185^{neu}-Specific Monoclonal Antibodies on Tumorigenic Growth of neu-Transformed Cells

In vitro studies demonstrated that antibodies to p185^{neu} were capable of down-modulating p185^{neu} and reverting the transformed phenotype, suggesting that cell surface expression of p185^{neu} is essential for maintaining the transformed phenotype of *neu*-transformed cells. Therefore, we were interested in the effects that these antibodies would have on the growth of *neu*-transformed cells *in vivo*.[4,8]

In order to assay for the effect on p185^{neu}-specific antibodies on the *in vivo* tumorigenesis of various transformed cell lines, NIH 3T3, B104-1-1, *ras*-transformed 3T3 cells, and *neu/ras* double transformants grown in DMEM with 10% fetal calf serum are washed with sterile PBS and removed from tissue culture flasks with Versene. After washing the cells 3 times with HBSS 1 × 10^6 cells are injected subcutaneously in the middorsum of nude mice. Mice are injected intraperitoneally with ascites fluid containing 7.16.4 or an isotype-matched control antibody on the day of tumor implantation and on the following day. The animals are checked on alternate days for palpable tumors and mortality. Vernier calipers are used to measure tumor length, width, and the product of the two measurements is calculated.

Anti-p185^{neu} antibodies effectively inhibit the growth of *neu*-transformed tumors in nude mice in a dose-dependent fashion, whereas ascites from mice bearing isotype-matched hybridomas do not. In addition, 7.16.4 treatment increases the survival time of the mice nearly 2-fold. These effects have been shown to be specific for *neu*-transformed cells, since MAb 7.16.4 does not inhibit the tumorgenic growth of *ras*-transformed cells. 7.16.4 also inhibits the growth of the neuroblastoma line B104 from

---

[8] J. A. Drebin, V. C. Link, R. A. Weinberg, and M. I. Greene, *Proc. Natl. Acad. Sci. U.S.A.* **83**, 9129 (1986).

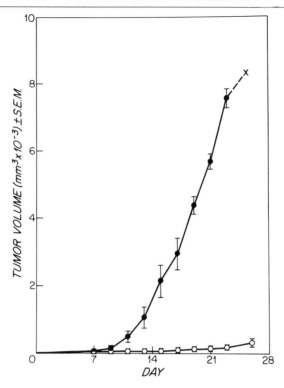

FIG. 3. Effects of immunotherapy with a mixture of anti-p185^{neu} antibodies on established tumors in nude mice. BALB/c nude mice were injected subcutaneously in the middorsum with 1×10^6 neu-transformed, B104-1-1 cells. After 7 days, when the tumors had reached a volume of 75 mm^3, animals were randomized into two groups of six mice each for monoclonal antibody treatment. One group (○) received twice weekly intravenous injections of a synergistic mixture of anti-p185^{neu} antibodies (500 μg antibody 7.9.5 plus 500 μg antibody 7.16.4) for 2 weeks, while the other group (●) received injections of saline. Antibody treatment was stopped on day 21. Tumors were measured on the days indicated with vernier calipers, and tumor volume was calculated as the product of tumor length, width, and height. Vertical bars represent the standard error of the mean.

which the *neu* oncogene was first isolated, in both nude mice and syngeneic BDIX rats.

In a more recent series of experiments, we used combinations of monoclonal antibodies from different binding classes, which do not compete for p185^{neu} binding, to treat tumors arising from *neu*-transformed cells, *in vivo*.[4] These studies revealed a therapeutic synergism wherein two anti-p185^{neu} antibodies of direct binding classes prevent the growth of established tumors in nude mice while treatment with a single antibody does not (Fig. 3).

Cytostatic and Cytotoxic Antitumor Effects of p185^{neu}-Specific Monoclonal Antibodies

The ability of monoclonal antibodies specific for p185^{neu} to inhibit the growth of *neu*-transformed cells in soft agar growth assays and in syngeneic animals suggests that this antibody exerts direct cytostatic effects on cells bearing p185^{neu}. To investigate whether anti-p185^{neu} monoclonal antibodies exert antitumor effects through additional mechanisms, we examined whether the antibodies could kill *neu*-transformed cells through complement- or antibody-dependent cell-mediated lysis.[7]

To determine whether anti-p185^{neu} antibodies could lyse various target cells, the target cells are first loaded with ^{51}Cr by incubating them in 0.5 ml with 200 μCi of Na^{51}CrO$_4$ at 37° for 60 min. The cells are then washed 3 times in HBSS, counted in a hemocytometer, and diluted to 10^5 cells/ml. One hundred microliters of cells is then added to wells of a 96-well microtiter plate. Fifty microliters of monoclonal antibody dilutions or detergent is added to the wells. After a 1-hr incubation at 37°, the cells are centrifuged at 1000 g for 10 min, and 100 μl of the supernatant is assayed for ^{51}Cr release in a γ counter. Specific release, R, is calculated using Eq. (2):

$$R = X - C/M - C \qquad (2)$$

where X is equal to the ^{51}Cr activity in counts per minute in the experimental well, C represents the ^{51}Cr activity in wells containing complement but no antibody, and M is the maximal ^{51}Cr release as determined by detergent lysis. All experimental groups represent the mean of triplicate samples.

Although the addition of monoclonal antibody 7.16.4 to ^{51}Cr-loaded *neu*-transformed cells does not lead to appreciable ^{51}Cr release, cells incubated with 7.16.4 in the presence of complement were lysed by as little as 5 ng/μl of 7.16.4. However, the XHT-1-1a *ras*-transformed cell line was not lysed by the combination of 7.16.4 and complement. The data indicate that whereas 7.16.4 is not directly cytotoxic it can led to the *in vitro* lysis of cells bearing p185^{neu} through complement fixation.

Antibody-dependent cell-mediated cytotoxicity (ADCC) is assayed in the following manner. The cells are loaded with ^{51}Cr as above, and 1 $\times$ 10^4 cells are aliquoted per well with 20 μg 7.16.4 in DMEM/10% fetal calf serum with various numbers of effector spleen cells, Freund's complete adjuvant-elicited macrophages, or thioglycolate-elicited macrophages. After 24 hr at 37° in a 5% CO$_2$ incubator, the cells are centrifuged at 1000 g, and 100 μl of the supernatant is counted in a γ counter. Specific release is calculated as above for complement-mediated lysis.

Treatment of cells with 7.16.4 in ADCC assays showed that this antibody can stimulate only low levels (<15%) of cell-mediated lysis regardless of the effector cell type. Thus, whereas 7.16.4 is directly cytostatic and can lyse cells through complement fixation *in vitro*, it is not very effective in mediating *in vitro* cell-dependent lysis. In order to test whether MAb 7.16.4 ADCC or complement-mediated cell lysis is important for inhibiting the growth of *neu*-transformed cells implanted in nude mice, we depleted the complement and macrophage functions of the tumor-bearing host. Complement is depleted by injection of cobra venom toxin factor (Sigma). One unit of toxin is injected in HBSS every 8 hr for a total of four doses 48 hr prior to tumor cell implantation and four doses 48 hr after implantation. In order to inhibit macrophage function, mice are injected with 0.5 mg of carrageenan in HBSS 48 hr preceding and following tumor cell implantation.

The *in vivo* studies of the effects of 7.16.4 administration on the growth of *neu*-transformed cells in complement- and macrophage-depleted animals revealed that depletion of complement and macrophages has little effect on tumor growth. These results suggest that monoclonal antibody 7.16.4 works directly on *neu*-transformed cells to inhibit their growth *in vivo*.

Discussion

In our studies of transformed cells expressing the product of the *neu* oncogene, $p185^{neu}$, we raised monoclonal antibodies specific for $p185^{neu}$ by immunization of mice with *neu*-transformed cells, preabsorption of immune sera on parent cell lines, and screening for binding to *neu*-transformed cells using immunofluorescence cytometry. In this way we produced monoclonal antibodies that are capable of down-modulating the surface expression of a growth factor receptor-like glycoprotein, and we have found that a loss of cell surface expression of $p185^{neu}$ correlates with a reversion of the transformed phenotype as assayed by focus formation, growth in soft agar, and tumorigenesis in nude mice. We believe that the *in vivo* effects of $p185^{neu}$-specific antibodies on the growth of *neu*-transformed cells is most likely due to direct cytostatic effects of the antibody since depletion of complement or macrophages in treated animals has little effect on the inhibition of cell growth by anti-$p185^{neu}$. This is further supported by the reversal of growth-inhibition effects on antibody removal.

Some more recent work in our laboratory indicates that the effects of antireceptor antibodies on *in vivo* tumor growth may eventually be applicable to human tumors. We have prepared cell lines which overexpress the EGFR as well as the product of the c-*neu* protooncogene, termed p185^{neu}.[9] When either of the receptor molecules are expressed independently in NIH 3T3 cells, the cells do not become completely transformed. However, the co-overexpression of these two receptors leads to complete oncogenic transformation. This is the first example that we know of where the expression of two normal cellular receptors can lead to cellular transformation. These tumors may bear more similarity to human tumors than cell lines transformed by transfection with the oncogenic form of the *neu* gene, since specific human tumors have been found to overexpress normal cellular receptors. In particular, the overexpression of p185$^{c\text{-}erbB\text{-}2}$, the human homolog of p185$^{c\text{-}neu}$, has been shown to correlate with a poor clinical prognosis in adenocarcinoma of the breast and ovary.[10]

When the cell lines overexpressing p185$^{c\text{-}neu}$ and the EGFR are implanted in nude mice, they form tumors that dramatically decrease the survival time of the animals. Treatment of animals bearing the doubly transfected cells with monoclonal antibodies to the external domain of p185$^{c\text{-}neu}$ (MAb 7.16.4) or the human EGFR (MAb 425) leads to inhibition of tumor growth by 60–90%.[11] Treatment with antibodies to both receptors has an apparent synergistic antitumor effect, completely inhibiting tumor growth in 20% of animals, with marked reduction in tumor growth in the others. Other recent work suggests that monoclonal antibody treatment may be effective for inhibiting the growth of human tumors. Using monoclonal antibodies to the extracellular domain of the c-*erbB-2* gene product, p185$^{c\text{-}erbB\text{-}2}$, one group has demonstrated inhibition of soft agar growth of human breast cancer cell lines that overexpress p185$^{c\text{-}erbB\text{-}2}$.[12] In addition, this group has found that anti-p185$^{c\text{-}erbB\text{-}2}$ antibodies can enhance the sensitivity of an adenocarcinoma line to the cytotoxic effects of tumor necrosis factor α (TNF-α).

Our findings and those of others indicate that monoclonal antibodies to cell surface molecules essential for tumor progression are effective agents for inhibiting tumor growth in *in vitro* and *in vivo* models of tumorigenesis. It appears that the most effective means of inhibiting tumor growth *in vivo* with antibodies is to use a combination of antibodies to distinct domains of a single molecule, antibodies to multiple cell surface molecules needed for cell growth, or antibodies and a cytokine such as tumor necrosis

---

[9] Y. Kokai, J. N. Myers, T. Wada, V. I. Brown, C. M. LeVea, J. G. Davis, and K. Dobashi, *Cell (Cambridge, Mass.)* **58**, 287 (1989).

[10] D. J. Slamon, W. Godolphin, L. A. Jones, J. A. Holt, S. G. Wong, D. E. Keith, W. J. Levin, S. G. Stuart, J. Udove, A. Ullrich, and M. F. Press, *Science* **244**, 707 (1989).

[11] T. Wada, J. N. Myers, Y. Kokai, V. I. Brown, J. Hamuro, C. M. LeVea, and M. I. Greene, *Oncogene* **5**, 489 (1990).

[12] R. M. Hudziak, G. D. Lesis, M. Winget, B. M. Fendly, H. M. Shepard, and A. Ullrich, *Mol. Cell. Biol.* **9**, 1165 (1989).

factor. Clearly, further understanding of the role of growth factor receptor overexpression in the etiology and progression of human neoplasms and pharmacokinetics and the toxicity of monoclonal antibody therapy is needed before the antibodies could be used in the treatment of cancer. While the full potential of the antireceptor monoclonal antibodies for diagnosing and treating human tumors remains to be tapped, the antibodies continue to be useful reagents for studying the cellular and biochemical mechanisms of the role of p185^{neu} in neoplastic transformation.

[28] Quantification of *erbB-2/neu* Levels in Tissue

By SOONMYOUNG PAIK, C. RICHTER KING, SUSAN SIMPSON, and MARC E. LIPPMAN

Introduction

Abnormalities of expression of the *erbB-2* (or *HER-2* or *neu*) gene have been identified frequently in cancers of the breast, ovary, stomach, and salivary gland.[1-6] For breast and ovarian cancers, *erbB-2* gene amplification and overexpression are associated with increased patient mortality.[1-4] Such abnormalities have been detected as gene amplification, mRNA overexpression, and protein overexpression.[1-4] Detection of gene amplification gave the first indication of abnormalities of the *erbB-2* gene and of the associated prognostic value.[1] Levels of gene amplification vary from 2-fold to greater than 25-fold. mRNA and protein overexpression occur at levels from 4- to 128-fold.[1] Although the greatest degree of mRNA and protein overexpression occurs in tumors with gene amplification, overexpression can occur in tumors with an apparently normal complement of *erbB-2* genes.[1,5] Therefore, measurement of mRNA or protein overexpres-

---

[1] D. J. Slamon, W. Godolphin, L. A. Jones, J. A. Holt, S. G. Wong, D. E. Keith, W. J. Levin, S. G. Stuart, S. Udove, A. Ulrich, and M. F. Press, *Science* **244**, 707 (1989).

[2] J. Yokota, K. Toyoshima, T. Sugimura, T. Yamamoto, M. Terada, H. Battifora, and M. J. Cline, *Lancet* **1**, 765 (1986).

[3] A. K. Tandon, G. M. Clark, G. C. Chamness, A. Ulrich, and W. L. McGuire, *J. Clin. Oncol.* **6**, 1076 (1988).

[4] M. H. Kraus, N. C. Popescu, C. Amsbaugh, and C. R. King, *EMBO J.* **6**, 605 (1987).

[5] C. R. King, S. M. Swain, L. P. Porter, S. M. Steinberg, M. E. Lippman, and E. P. Gelmann, *Cancer Res.* **49**, 4185 (1989).

[6] S. Paik, R. Hazan, E. R. Fisher, R. E. Sass, B. Fisher, C. Redmond, J. Schlessinger, M. E. Lippman, and C. R. King, *J. Clin. Oncol.* **8**, 103 (1990).

sion may represent more sensitive techniques for the detection of *erbB-2* abnormalities. Detection of *erbB-2* mRNA or *erb-2* protein by biochemical techniques allows quantitative estimates of the level of protein overexpression.

Immunohistochemical analysis facilitates the examination of archival material necessary to determine the clinical significance of *erbB-2* overexpression[6] but suffers from some difficulty in quantitation and some technical variability in results.[1] However, simple positive and negative or semiquantitative classification schemes can be used to derive useful clinical information.[6]

In this chapter, we describe a methodological framework suitable for the analysis of *erbB-2* expression in clinical samples. A simple method for the detection of *erbB-2* gene amplification and mRNA overexpression in small samples provides rapid, quantitative, and unequivocal results. Control samples standardized in this way can then be used to evaluate the consistency of immunohistochemical detection. Immunohistology of formalin-fixed paraffin-embedded sections can be used for routine testing for the determination of prognostic information.

Choice of Assay System

When over 5 to 50 mg of fresh frozen sample is available, one can choose from Southern, Northern, and Western blot analyses, flow cytometry, and immunohistochemistry. Among these, Southern blot analysis,[1] Western blot analysis,[3] and immunohistochemistry[6] were found to give prognostically valuable information. But Slamon *et al.* showed that Western blot analysis suffers from dilution by other proteins in the tissue, thus having the least correlation with other assay results.[1] When only archival materials or small amounts of tissue samples are available, the obvious choice is immunohistochemistry.

DNA/RNA Blot Analysis

Sample Preparation

Other investigators have used standard techniques for the analysis of either DNA or RNA from breast cancer specimens. In the case of analysis of *erbB-2* abnormalities, it can be advantageous to determine the extent of RNA overexpression and gene amplification in the same tissue sample. For this purpose, it is possible to divide the sample and use separate methods for RNA and DNA extraction. However, we have used a modification of the method of Chirgwin and co-workers[7] for the isolation of both RNA and DNA from a single procedure.

1. The tissue sample (0.5–1 g) is frozen in liquid nitrogen and lyophilized overnight.
2. The sample is ground to a dust in a mortar in liquid nitrogen.
3. Two milliliters of GTC (4 M guanidine thiocyanate, 5 mM sodium citrate, pH 7.0, 5% sodium sarcosyl, and 0.1 M 2-mercaptoethanol) solution is added, and grinding is continued to dissolve the sample at room temperature. The mortar is rinsed with an additional 1 ml of GTC.
4. The sample is allowed to mix for 30 min by rotation on a wheel.
5. Solid cesium chloride (1.6 g) is added and allowed to dissolve by gentle mixing by rotation on a wheel for 10 min.
6. To a SW 50.1 tube, add a cushion of 1.2 ml of 5.7 M CsCl$_2$, 0.1 M EDTA solution and layer the solubilized tissue sample on top. Fill the tube with GTC solution.
7. Centrifuge for at least 12 hr at 35,000 rpm.
8. There will be an opalescent band above the CsCl$_2$ cushion containing the DNA. Remove about 2–3 ml of this viscous material (DNA) to a 50-ml tube. Add 3 ml of GTC and gently mix for about 10 min while preparing RNA.

Preparation of RNA

1. Remove the CsCl$_2$ cushion with a Pasteur pipette. The RNA pellet will probably be slightly cloudy. Remove excess liquid with a cotton swab.
2. Resuspend the pellet in 0.2 ml of sterile TES [10 mM Tris, pH 7.5, 0.5 mM EDTA, and 1% sodium dodecyl sulfate (SDS)].
3. Extract with 0.2 ml chloroform/isobutanol (4 : 1) by vortexing.
4. Reextract the organic phase with 0.2 ml TES.
5. Repeat Step 4.
6. Reextract the pooled aqueous phases with chloroform.
7. Add 40 μl of 3 M soidum acetate, pH 5.2.
8. Ethanol-precipitate with 2 volumes absolute ethanol.
9. Resuspend in TES. (Expect yields of 50–200 μg.)

Preparation of DNA

1. Layer the DNA solution on top of 1.5 volumes of ethanol. Allow a precipitate to form by gentle mixing, taking care not to allow a tight spool of DNA to form. Remove precipitated DNA using the end of a pipette and place directly into TES.

[7] R. J. Macdonald, G. H. Swift, A. E. Przybyla, and J. M. Chirgwin, this series, Vol. 152, p. 219.

2. Allow resuspension by gentle inversion mixing overnight.
3. Add NaCl to 0.5 M.
4. Phenol-extract with distilled phenol by gently shaking for 30 min.
5. Centrifuge at 8000 rpm in an SS-34 rotor.
6. Extract the aqueous phase with chloroform/isoamyl alcohol (24:1, v/v) and reextract.
7. Gently ethanol-precipitate the DNA as in Step 1 and remove directly into a tube containing 2 ml of Tris, pH 7.5, 1 mM EDTA. (Expect 50–200 μg DNA.)

Analysis of RNA by Northern Blots

Standard techniques can be used to detect the *erbB-2* mRNA involving agarose gel electrophoresis with formaldehyde as a denaturant.[8] We have used loadings of 5 μg of total RNA into gel slots of 2 × 0.5 mm. The 5-kilobase (kb) *erbB-2* mRNA comigrates with the 28 S ribosomal RNA. Specific detection is accomplished using ^{32}P-labeled *erbB-2* cDNA probes [1 × 10^6 disintegrations/min (dpm)/ml]. Overexpressing samples are detectable following 1–2 hr of autoradiography and the normal level of *erbB-2* mRNA detected following overnight exposure. As controls, RNA from the cell line SK-BR-3 (ATCC, Rockville, MD) shows 128-fold overexpression, and cell line MCF-7 (ATCC) shows the normal level of expression.

To quantify the level of *erbB-2* mRNA it is necessary to perform dilution analysis of the RNA samples.[4] For this purpose RNA is diluted into 1 M ammonium acetate and 5 μg/ml tRNA and applied using gentle suction to a dot or slot blot. This is then baked to fix the RNA and hybridized to radioactive cDNA probes. Following autoradiography, the blots can be stripped by heating to 90° in 0.1× SSC. The same dot blot can then be probed with control probes such as β-actin or β_2-microglobulin. Incorporation of standards such as MCF-7 mRNA allows the extent of overexpression to be estimated.

Analysis of DNA by Southern Blots

DNA can be analyzed using standard Southern blotting techniques. We have successfully examined 2 μg samples using the small 2 × 0.5 mm gel slots. Again, incorporation of standards such as DNA from the cell line SK-BR-3 (4- to 16-fold gene amplification) and from MCF-7 (normal gene levels) provides standards for the detection of gene amplification. To ensure that equivalent amounts of DNA have been loaded for each sample,

[8] H. Lerach, D. Diamond, J. M. Wozney, and H. Boedtker, *Biochemistry* **16**, 4743 (1977).

blots should be stripped of radioactive *erbB-2* cDNA probe and reanalyzed using a β-actin probe. Hybridization with a p53 probe provides an interesting control in that this gene is located on the same chromosome as *erbB-2* but on the opposite arm.[1] This location assures that *erbB-2* gene amplification does not occur by chromosomal duplication.

Caution must be exercised in interpreting minor changes in *erbB-2* hybridization intensity for several reasons. First, comparisons with p53 hybridization may be subject to error as some tumors clearly show allelic elimination for this gene. Second, the level of *erbB-2* amplification is relatively modest. As a result, the detection of some samples with low level gene amplification may be obscured by the presence of DNA derived from normal cells (epithelial and stromal) present in the tumor sample.

Immunohistochemical Staining of *erbB-2/neu*

Choice of Antibodies against erbB-2/neu Protein

Many polyclonal and monoclonal antibodies to the *erbB-2/neu* protein have been described, and some of them are available commercially. We have generated our own monoclonal antibodies utilizing the membrane fraction of NIH 3T3 cells transfected with an *erbB-2/neu* cDNA expression vector. Antibody Hy83 was selected for giving the best staining for the tissue samples of known overexpression.[6] Hy83 detects an intracellular epitope, since it does not stain NIH 3T3 cells expressing an extracellular *erbB-2*–intracellular epidermal growth factor (EGF) receptor hybrid protein. We have compared several commercial antibodies with Hy83. One commercial polyclonal antibody (PAb-1, Triton Biosciences, Alameda, CA) raised against an intracellular epitope could detect only 30% of the cases which gave positive membrane staining with Hy83. The section shown in Fig. 1a stained positive with Hy83 but did not stain with PAb-1 (Fig. 1b). In addition, even in positive cases staining was focal in nature, marking only a portion of the section. Thus, the epitope for PAb-1 may be affected more by formalin fixation and paraffin embedding than the epitope for Hy83. Frozen sections should therefore be used for PAb-1 antibody. PAb-1 also has a disadvantage of cross-reacting with erythrocytes. A monoclonal antibody MAb-1 (Triton Biosciences) gave similar staining results as for Hy83, but staining was heterogeneous in some cases (Fig. 2a). Addition of PAb-1 to MAb-1 enhanced the staining, resulting in homogeneous staining of all tumor cells in the section without an increase in background (Fig. 2b). Many commercially available antibodies are raised against the extracellular domain and are not ideal for paraffin-embedded sections. These antibodies give heterogeneous staining of the

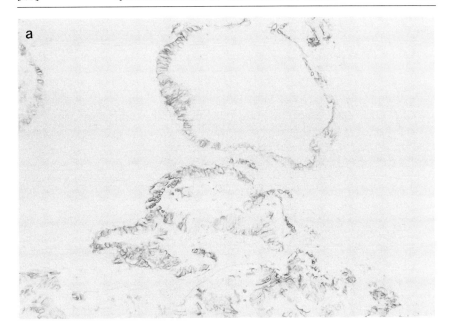

FIG. 1. Comparison of immunostaining by Hy83 and PAB-1. (a) Staining by Hy83 shows membrane staining of all tumor cells in the section. Magnification: ×40. (b) Staining of a serial section of the same tumor by PAb-1 shows no staining. Magnification: ×100.

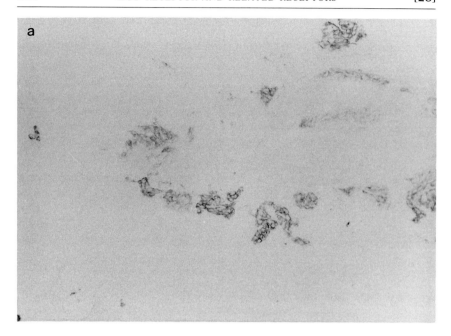
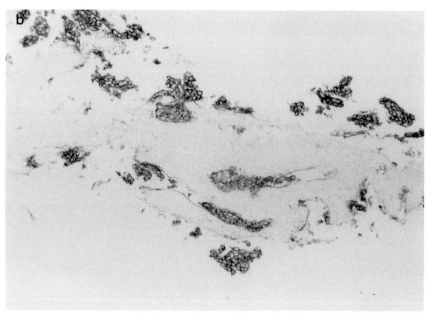

FIG. 2. Enhancement of immunostaining by MAb-1 with addition of PAb-1. (a) MAb-1 fails to stain the center of the section. (b) Addition of PAb-1 to MAb-1 enhances the staining, especially in the center of the section. Magnification: ×40.

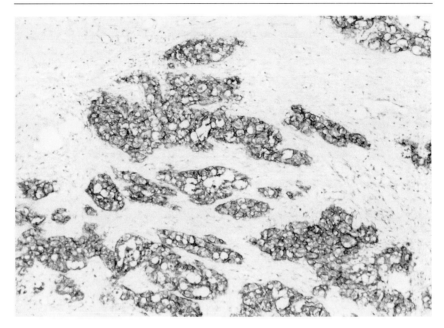

FIG. 3. Example of an ideal immunostaining for the *erbB-2* protein. All tumor cells in the section show strong membrane staining. Magnification: ×100.

tumor cells within the section. Attempts at quantification by calculating the percentage of stained cells using these antibodies may thus be quite misleading. A good antibody should result in homogeneous staining of all tumor cells in most of the cases (Fig. 3).

Preparation of Tissue

Freshly frozen sections are ideal, since paraffin embedding seems to affect staining results, especially for cases with only moderate overexpression.[1] However, in a large clinical trial, information gathered by immunohistochemical staining of formalin-fixed paraffin-embedded sections gave prognostically valuable information.[6] Fixation in formalin for 3 weeks and embedding in paraffin did not reduce the staining with Hy83 significantly in our hands. Thus, results from analysis of paraffin sections may be used in order to give prognostic information. Omnifix (Xenetics, Irvine, CA) decreases but does not eliminate the intensity of the staining with Hy83 compared to formalin fixation. However, Omnifix gave excellent staining with PAb-1. In formalin-fixed and paraffin-embedded sections, staining is localized to the tumor cell membrane. However, in frozen sections the staining is over the entire cell, probably due to the thickness of the section. One precaution in staining paraffin sections is that sections have to be cut freshly before staining. We found a marked decrease in staining when sections were cut and left at room temperature for more than 1 month before staining, resulting in heterogeneous staining of the sections, although this could be overcome by using cocktail preparations of antibodies.

Protocol for Immunohistochemistry

Since clinical use involves the staining of many cases together, we have introduced a simple modification of existing procedures which eliminates the possible source of variation of assay conditions between the slides. After deparaffinization and dehydration, sections are air-dried and circled with a PAP pen (Research Products International, Mount Prospect, IL). This gives an excellent water barrier and facilitates the removal of excess buffer after the washing step. A convenient humid chamber is commercially available (Lipshaw, Pittsburgh, PA).

Reagents

Buffer: TBS (50 mM Tris-buffered saline, pH 7.5) with Brij 35 (Sigma, St. Louis, MO). TBS can be made as a $10\times$ concentrated stock solution and is stable at room temperature for up to 1 year. Brij 35 is available as a convenient 30% solution (Fisher, Fairlawn, NJ). Simply add 1:400 into final solution. A premade $10\times$ buffer is available as Automation Buffer (Fisher). Phosphate-buffered saline (PBS) should be used for PAb-1 and MAb-1, since antibodies are provided in PBS. Using the Vector ABC Elite kit (Vector, Burlingame, CA), overnight incubation at 4° was successful with 1:40 dilutions. For other detection systems, titrations should be done to get the best results.

Detection system: We use the Vector ABC Elite peroxidase kit (Vector). Primary antibody is diluted in the blocking horse serum supplied in the kit.

DAB: A $100\times$ concentrate (5 g/100 ml) of diaminobenzidine tetrahydrochloride (Sigma) is made in 50 mM Tris, and 2.5-ml aliquots are kept at $-20°$.

30% H_2O_2 (v/v): Keep in the refrigerator and use within 2 weeks after opening.

Staining

1. Deparaffinization:
    a. Bake the sections in an oven at 58° for 1 hr.

b. Immediately transfer the sections into fresh xylene (ACS grade) and incubate for 10 min. Repeat with fresh xylene.
2. Application of PAP pen and rehydration:
   a. Incubate in reagent-grade alcohol for 5 min.
   b. Air-dry for 5 min.
   c. Circle the sections with a PAP pen.
   d. Air-dry for 5 min.
   e. Incubate in reagent-grade alcohol for 5 min.
   f. Incubate in TBS for 5 min.
3. Blocking of endogenous peroxidase activity:
   a. Incubate sections in 0.6% H_2O_2 (v/v) in methanol for 20 min with continuous agitation.
   b. Incubate in TBS for 5 min twice.
4. Incubate with blocking serum in humid chamber for 10 min.
5. Aspirate the blocking serum under reduced pressure and incubate with primary antibody diluted in blocking serum for 30 min to overnight at room temperature.
6. Perform TBS washing for 5 min twice.
7. Use reduced pressure to aspirate TBS and incubate with biotinylated secondary antibody for 30 min.
8. Perform TBS washing for 5 min.
9. Aspirate the TBS and incubate with the avidin–biotin complex for 30 min.
10. Perform TBS washing for 5 min. (During this washing, thaw out the DAB stock solution by vortexing with TBS at room temperature.)
11. Incubate the sections in 250 ml DAB solution in TBS for 5 min.
12. Add 87 μl fresh 30% H_2O_2 and incubate for 15 min.
13. Wash in running tap water for 5 min.
14. Counterstain:
    a. Incubate in Meyer's hematoxylin (Fisher) for 3 min.
    b. Wash in double-distilled water for 3 min.
    c. Wash in TBS for 1 min.
    d. Air-dry.
    e. Apply Crystal Mount (Fisher) and incubate in 57° in an oven for 1 hr.

Controls

To minimize batch-to-batch differences, it is essential to have necessary controls included within each batch of staining. As a positive control, either a formalin-fixed paraffin-embedded section of a breast cancer cell line overexpressing *erbB-2/neu* (SK-BR-3 or MDA-453) or a tissue sample known to overexpress *erbB-2/neu* can be used. As a negative control, a paraffin-embedded section of the MDA-468 cell line (ATCC) can be used. This cell line serves as an additional control for cross-reactivity to EGF receptor protein, which shares considerable sequence similarity to the *erbB-2/neu* protein.

Section V

Transforming Growth Factor β and Related Factors

[29] Generation of Antibodies and Assays for Transforming Growth Factor β

By CATHERINE LUCAS, BRIAN M. FENDLY, VENKAT R. MUKKU, WAI LEE WONG, and MICHAEL A. PALLADINO

Introduction

Transforming growth factor β (TGF-β) is a growth factor which, like transforming growth factor α (TGF-α), was discovered in an assay evaluating cellular transformation of immortalized nonneoplastic fibroblasts.[1,2] TGF-β, however, is structurally unrelated to TGF-α and binds to distinct receptors. It is a disulfide-linked dimer of two identical 112 amino acid polypeptides, each derived from a larger secreted polypeptide. Protein biochemical analysis and cDNA cloning have established the existence of at least three human TGF-β species, which are structurally closely related to each other.[3-6] TGF-β stimulates or inhibits the proliferation of cells, depending on the cell type and the physiological conditions. A major activity of TGF-β resides in its ability to induce extracellular matrix formation. These various biological activities strongly suggest that TGF-β plays a crucial role in cell proliferation and differentiation *in vivo*. The generation of antibodies to TGF-β was of critical importance in studying the functionality of TGF-β in normal physiology and malignant transformation, as well as in developing precise and specific assays. The development of such antibodies and assays are described below.

The assays which initially led to the discovery of TGF-β are based on the ability of TGF-β to induce anchorage independence in some immortalized fibroblasts, either in the presence or the absence of epidermal growth

[1] A. B. Roberts and M. B. Sporn, *in* "Peptide Growth Factors and Their Receptors" (M. B. Sporn and A. B. Roberts, eds.), Handbook of Experimental Pharmacology, Vol. 95/1, p. 3. Springer-Verlag, Heidelberg, 1990.
[2] M. B. Sporn, A. B. Roberts, L. M. Wakefield, and B. de Crombrugghe, *J. Cell Biol.* **105**, 1039 (1987).
[3] R. Derynck, P. B. Lindquist, A. Lee, D. Wen, J. Tamm, J. L. Graycar, L. Rhee, A. J. Mason, D. A. Miller, R. J. Coffey, H. L. Moses, and E. Y. Chen, *EMBO J.* **7**, 3737 (1988).
[4] S. B. Jakewlew, P. J. Dillard, P. Kondaiah, M. B. Sporn, and A. B. Roberts, *Mol. Endocrinol.* **2**, 747 (1988).
[5] P. van Dyke, T. P. Hansen, K. K. Iwata, C. Pieler, and J. G. Foulkes, *Proc. Natl. Acad. Sci. U.S.A.* **85**, 4715 (1988).
[6] R. de Martin, B. Haendler, R. Hofer-Warbinek, H. Gaugitsch, M. Wrann, H. Schlusener, J. M. Seifert, S. Bodmer, A. Fontana, and E. Hofer, *EMBO J.* **6**, 3673 (1987).

factor (EGF). These assays, however, are not specific for this growth factor and are subject to interference or synergism by other factors. Thus, they do not allow reliable quantitation of TGF-β in crude extracts or conditioned media. In addition, as discussed above, there are multiple TGF-β species which are structurally closely related and have similar activities,[3-6] and TGF-β is biosynthesized as a protein-bound, latent, inactive form.[7-10] An ideal assay for TGF-β should be able to quantitate the factor in impure preparations and discriminate between the different TGF-β species as well as between the active and latent complexed forms. Several laboratories have recently developed receptor-binding assays based on a competition of the TGF-β in the sample with ^{125}I-labeled TGF-β for binding to the TGF-β receptors.[11-13] These assays allow reliable quantitation of active TGF-β but do not discriminate between the different TGF-β species. The total amount of TGF-β in test samples can be quantitated after activation of TGF-β with acidic pH, followed by neutralization. We describe the production of monoclonal antibodies specific for TGF-β1 and the radiolabeling of this protein. These reagents together with human TGF-β1 from recombinant sources[14] were used for the development of a double-antibody enzyme immunoassay and a radioimmunoassay specific for TGF-β1. We also describe a radioreceptor assay and a bioactivity assay for TGF-β.

Radioiodination of Transforming Growth Factor β1

Examination of the literature indicates that most laboratories prepare the ^{125}I-labeled TGF-β needed for assays using the chloramine-T method.[15,16] We have compared several methods for iodination of TGF-β

[7] D. A. Lawrence, R. Pircher, and P. Jullien, *Biochem. Biophys. Res. Commun.* **133**, 1026 (1985).
[8] I. A. Silver, R. J. Murrills, and D. J. Etherington, *Exp. Cell Res.* **175**, 266 (1988).
[9] R. O. C. Oreffo, G. R. Mundy, S. M. Seyedin, and L. F. Bonewald, *Biochem. Biophys. Res. Commun.* **158**, 817 (1989).
[10] Y. Pilatte, J. Bignon, and C. R. Lambre, *Biochim. Biophys. Acta* **923**, 150 (1987).
[11] B. O. Fanger and M. B. Sporn, *Anal. Biochem.* **156**, 444 (1986).
[12] P. R. Segarini, A. B. Roberts, D. M. Rosen, and S. M. Seyedin, *J. Biol. Chem.* **262**, 14655 (1987).
[13] L. M. Wakefield, L. M. Smith, T. Masui, C. C. Harris, and M. B. Sporn, *J. Cell Biol.* **105**, 965 (1987).
[14] R. Derynck, J. A. Jarrett, E. Y. Chen, D. H. Eaton, J. R. Bell, R. K. Assoian, A. B. Roberts, M. B. Sporn, and D. V. Goeddel, *Nature (London)* **316**, 701 (1985).
[15] W. M. Hunter and F. C. Greenwood, *Nature (London)* **194**, 495 (1962).
[16] C. A. Frolik, L. M. Wakefield, D. M. Smith, and M. B. Sporn, *J. Biol. Chem.* **259**, 10995 (1984).

and found that a high specific radioactivity was obtained with both the chloramine-T[15] and lactoperoxidase[17] methods. Using these methods the ^{125}I-labeled TGF-β remained biologically active. The lactoperoxidase iodinations generally resulted in a higher specific radioactivity than the chloramine-T method, but the latter method generally resulted in a better purity as assessed by sodium dodecyl sulfate (SDS)–polyacrylamide gel electrophoresis. Both methods of iodination are described below. Separation of iodinated TGF-β from free sodium iodide was best performed by HPLC, although separation by Sephadex G-100 filtration was adequate. The TGF-β used in these preparations was recombinant human TGF-β1, derived from mammalian cells transfected with an expression vector.

Chloramine-T Iodination Procedure

Chloramine-T is a strong oxidant which is able to covalently link iodine to tyrosine residues of proteins. The iodination procedure is performed on ice at neutral pH and by three sequential additions of chloramine-T. Approximately 10 μl (1 mCi) of Na^{125}I (Amersham, Arlington Heights, IL; 13–17 mCi/μg) is added to 10 μg of TGF-β in 50 μl of iodination buffer (1.5 M potassium phosphate, pH 7.4). TGF-β is iodinated by 3 sequential additions of 20 μl of the chloramine-T solution [0.1 mg/ml chloramine-T (Kodak, Rochester, NY) in 50 mM sodium phosphate, pH 7.4]. Between each addition, the reaction mixture is incubated for 1.5–2 min, with occasional mixing. The reaction is stopped by adding 20 μl of 50 mM N-acetyl-L-tyrosine for 1 min, then 20 μl of 1 M KI for 1 min, and finally 200 μl of 8 M urea, pH 3.2. The tracer is purified by HPLC or by gel-filtration chromatography. The HPLC purification generally resulted in a better recovery (>90%) and a higher purity.

Lactoperoxidase Iodination

Lactoperoxidase is an oxidative enzyme which catalyzes the enzymatic iodination of tyrosine-containing proteins. This reaction is a mild procedure which can be carried out at room temperature over a wide pH range (4.0 to 8.5), an advantage for proteins with unusually high or low isoelectric points and limited solubility at neutral pH. We found purified TGF-β1 to be poorly soluble in aqueous solutions above pH 5.0 and therefore chose to perform the lactoperoxidase iodination at pH 4.5.

Ten micrograms of purified TGF-β1 (~0.5 nmol) in 35 μl of 20 mM acetic acid, pH 4.0, is placed in a microcentrifuge tube, and 5 μl of 0.5 M sodium phosphate buffer, pH 7.4, is added to adjust the pH of the solution.

[17] J. I. Thorell and B. G. Johansson, *Biochim. Biophys. Acta* **251**, 363 (1971).

Approximately 10 μl (1 mCi) of Na^{125}I (Amersham; 13–17 mCi/μg) is added to the protein, then 5 μl (7 μg) of a freshly made solution of 1.4 mg lactoperoxidase (Sigma, St. Louis, MO) per milliliter phosphate-buffered saline (PBS), followed by 5 μl of 20 μM H$_2$O$_2$. The tube is capped, gently agitated, and incubated at room temperature for 10 min with occasional gentle mixing. The reaction is terminated by the addition of 20 μl of a fresh solution of 0.5 M N-acetyl-L-tyrosine (Sigma), followed 2 min later by 20 μl of 0.5 M KI and 200 μl of 8 M urea to retrieve the tracer from the reaction tube. The ^{125}I-labeled TGF-β is then purified by HPLC or by size-exclusion chromatography as described below.

Chromatographic Purification of ^{125}I-Labeled Transforming Growth Factor β

Purification of the radiolabeled TGF-β took place on an LKB (Uppsala, Sweden) HPLC system (LKB 2150 pump and 2152 LC controller) using a reversed-phase HPLC column C$_{18}$-300 (Vydac, Alltech Associates, Deerfield, IL). After equilibrating the column with solution A (0.1% trifluoroacetic acid in water), 0.5 ml of the iodination mixture is injected onto the column, and 60 fractions of 0.5 ml each are collected into borosilicate tubes, each containing 50 μl of 10% bovine serum albumin (BSA). A stepwise gradient is then started by adding solution B (0.08% trifluoroacetic acid in acetonitrile) to 25% over 5 min, then to 50% over the following 25 min, then to 100% over the following 5 min. The column is equilibrated and subsequently stored in 100% solution B. Five-microliter fractions are taken from each tube and counted in a γ counter to identify the iodinated protein peak (which elutes at 35–40% acetonitrile). Peak fractions are pooled, diluted to 5 ml with 1% BSA in 4 mM HCl, and stored in aliquots at −60° for approximately 1 month.

Sephadex G-100 Chromatography

A 1 × 30 cm glass Econo-column (Bio-Rad, Richmond, CA) is packed with Sephadex G-100, hydrated and degassed according to the manufacturer's recommendations. The column is equilibrated with approximately 50 ml running buffer (0.1% gelatin in 4 mM HCl). The iodination reaction is then loaded; the gel filtration takes place in running buffer, collecting 60 0.6-ml fractions. Five microliters is taken from each tube and counted in the γ counter to identify the iodinated protein peak in the void volume. The peak fractions are pooled and diluted with running buffer to a final volume of 5 ml. The iodinated TGF-β is stored in aliquots at −60° for approximately 1 month.

Trichloroacetic Acid Precipitation

The trichloroacetic acid (TCA) precipitability of radioactive counts gives a measure of the proportion of radiolabeled TGF-β relative to the unincorporated ^{125}I label remaining in the tracer preparation. A small amount of the ^{125}I-labeled TGF-β is diluted in tracer diluent (PBS, pH 7.4, containing 0.1% gelatin and 0.01% thimerosal or 0.5% BSA and 0.01% thimerosal) to obtain a minimum of 0.5 ml at 10^5-10^6 cpm/ml. The radioactivity in counts per minute (cpm)/100 μl is determined using an automatic γ counter. Two hundred microliters of the diluted radiolabeled preparation is added to 50 μl of 50% TCA in a microcentrifuge tube, mixed, and incubated for 30 min on ice. The precipitated sample is then centrifuged for 5 min at 13,000 g in a microcentrifuge at room temperature, and the amount of soluble counts in 100 μl of the supernatant is determined. The calculated percentage of TCA precipitability of the diluted radiolabeled TGF-β preparation should be more than 90%.

Using both iodination procedures described above, we have obtained ^{125}I-labeled TGF-β tracer preparations which typically had specific activities of 25 to 40 μCi/μg and were more than 90% TCA precipitable. The radiolabeled TGF-β preparations usually retained more than 90% of the bioactivity of the starting material, when tested in the bioassay described below. We found it to be very important in iodinating TGF-β to keep the protein at acidic pH, since TGF-β is not very soluble above pH 5.0. When the iodinations were initially less successful, we observed that the apparent loss of bioactivity was not so much due to iodination damage but rather attributable to loss of solubility of the protein. For this reason, we discontinued an initial evaluation of the Bolton–Hunter method of iodination and do not recommend it for TGF-β, because this method is best performed at basic pH[18] and results in considerable aggregation of the purified protein and very poor incorporation of the label.

Radioimmunoassay for Transforming Growth Factor β1

The radioimmunoassay (RIA) for TGF-β measures the displacement of the binding of radiolabeled TGF-β to a specific anti-TGF-β antibody by increasing concentrations of unlabeled TGF-β. Establishment of a standard curve using purified TGF-β1 shows that the signal in counts per minute in this competitive assay is inversely proportional to the concentration of unlabeled TGF-β1 present in the sample. The useful range of this assay is 6.25–200 ng/ml, with a sensitivity of 2.4 ng/ml.

---

[18] A. E. Bolton and W. M. Hunter, *Biochem. J.* **133**, 529 (1973).

One hundred microliters of each standard dilution (6.25–200 ng/ml TGF-β1), control, or test sample is added in duplicate into conical test tubes [four additional tubes contain only 100 μl of assay diluent (PBS containing 50 mg/liter BSA, 0.5 ml/liter Tween 20, 1 M NaCl, and 0.2 g/liter sodium azide) for nonspecific binding and zero reference]. One hundred microliters of ^{125}I-labeled TGF-β tracer is added to each tube, including two additional tubes for total counts. These two tubes are set aside until the end of the procedure. Then 100 μl of goat anti-TGF-β antibody appropriately diluted in assay diluent is added to all the tubes, except the two nonspecific binding tubes (which receive instead 100 μl of 2% normal goat serum). All tubes are mixed and incubated overnight at 4°. Immune complexes are precipitated with the addition of 1 ml of precipitation mixture [20 ml/liter donkey anti-goat IgG antibody (Pel-Freez, Rogers, AR) in PBS, pH 7.4, containing 40 g/liter polyethylene glycol (PEG) 8000 (Sigma) and 2 g/liter sodium azide] added to each tube for 1 hr at ambient temperature. The tubes are centrifuged for 20 min at 2000 g at 4° in a Beckman J-6B centrifuge, the supernatants are decanted, and the pellets are counted in a γ counter for 2 min each. The data are reduced with a four-parameter curve-fitting program based on an algorithm for least-squares estimation of nonlinear parameters.[19]

To determine the appropriate dilution of anti-TGF-β antibody for use in the assay, the antibody is first titrated as a series of 2-fold dilutions against the ^{125}I-labeled TGF-β tracer. The immune complexes are then precipitated as described in the procedure for the RIA. The three dilutions which precipitate 35 (±5)% of the counts are used to generate standard curves. The dilution which gives the most sensitivity is selected for the assay.

Radioreceptor Assay for Transforming Growth Factor β

The radioreceptor assay measures the ability of serial dilutions of TGF-β-containing samples to compete with ^{125}I-labeled TGF-β for the binding to the specific cell surface receptors. This assay is specific for biologically active TGF-β and does not detect the latent, protein-associated form of TGF-β,[7–10] unless previously acid-activated. The radioreceptor assay measures total binding to the different types of cell surface receptors and does not discriminate between the different TGF-β species. It is likely, however, that there are differences between the binding of the different TGF-β species to the three types of receptors, suggesting that some caution may be needed for the absolute quantitation of TGF-β. Our radioreceptor assay is done essentially as described by Wakefield et al.[13]

[19] D. W. Marquardt, *J. Soc. Ind. Appl. Math.* **11**, 431 (1963).

using A549 human lung carcinoma cells grown in monolayers. The range of the assay is 25–0.19 ng/ml.

A549 cells are grown at 37° with 5% CO_2 in high glucose Dulbecco's modified Eagle's medium (DMEM) (GIBCO, Grand Island, NY) supplemented with 5% calf serum, 2 mM L-glutamine, and 10 mM sodium pyruvate. The cells are seeded at 2×10^5 cells/ml/well into the wells of 24-well cluster plates (Nunc, Naperville, IL) and grown in assay medium (i.e., Ham's F12 mixed 1:1 with low glucose DMEM and without $NaHCO_3$) for 24 hr until 90–95% confluency is reached. Four wells containing growth medium only are incubated under the same conditions and are the blanks in the assay. Monolayers are washed twice with 1 ml assay diluent (assay medium containing 1 g/liter BSA, 25 mM HEPES, pH 7.4, and 42.5 mM NaCl) immediately prior to sample addition. The TGF-β in the samples (100–200 μl each) is acid-activated by adding 10 μl of 1.2 N HCl for 30 min at ambient temperature and subsequently neutralized with 10 μl of a solution containing 1 M HEPES, pH 7.4, and 1.44 M NaOH. The activated samples are then diluted with assay diluent (final dilution 1:2.5 or higher). The standards (25–0.19 ng/ml in assay diluent), blanks (assay diluent), and activated test samples (0.5 ml each) are mixed in dilution tubes with 0.5 ml of ^{125}I-labeled TGF-β tracer diluted to 20,000 cpm/100 μl in assay diluent. In addition, four counting tubes containing 100 μl of diluted tracer alone are set aside for the determination of the total counts (from which percent binding will be determined for standards and samples).

The assay diluent is aspirated from the cell plates, and 200 μl of the activated test samples or control samples from the dilution tubes are added in quadruplicate to appropriate wells of the cell plates. The plates are covered and incubated for 2 hr at ambient temperature, rocking gently on a rocking platform, after which all wells are washed 4 times with 1 ml of tracer wash buffer (1 g/liter BSA in PBS, pH 7.4). Solubilization buffer (10% glycerol, 1% Triton X-100 in 25 mM HEPES, pH 7.4) is then added to each well (0.75 ml/well). Following incubation at 37° for 30 min on a plate rotator, the contents of each well are transferred to counting tubes and counted for 5 min in a γ counter. The concentration of TGF-β in the test samples is determined by comparison with the standard curve data in a four-parameter logistic curve-fitting program.

Production of Polyclonal Antibodies to Transforming Growth Factor β

The production of antibodies to TGF-β has been difficult, presumably because of its immunosuppressive activities and its very conserved polypeptide sequence, but successful attempts have been reported.[20] In order

---

[20] D. Danielpour, L. L. Dart, K. C. Flanders, A. B. Roberts, and M. B. Sporn, *J. Cell. Physiol.* **138,** 79 (1989).

to obtain polyclonal antibodies to TGF-β1, we have immunized a variety of animals (rabbits, goats, turkeys, and guinea pigs) by several immunization procedures, including alum-precipitated TGF-β1 or TGF-β1 in detox adjuvant (Ribi, Hamilton, MT) or Freund's adjuvant. Most animals developed titers to TGF-β1, but the titers were not very high nor sustained for long. Most striking was the finding that antibodies generated were directed to similar or competing epitopes. We were never able to develop a double-antibody ELISA with any combination of these antibodies. Since antibodies generally do not need to be purified for use in radioimmunoassays, we used unfractionated antiserum to TGF-β1 for this assay.

However, we also examined the affinity purification of rabbit anti-TGF-β1 antibodies for potential use in enzyme immunoassays. Before affinity chromatography, we isolated the γ-globulin fraction by two precipitations with 40% ammonium sulfate. In order to prepare an affinity column with TGF-β1, coupling conditions must be chosen which will be feasible at low pH, since TGF-β has only limited solubility at neutral or basic pH. In contrast, most coupling reactions of proteins to resins are best performed at alkaline pH (pH 8.0 to 9.0). However, we were successful with two procedures described below, with no obvious advantage of one over the other.

In the first procedure, we coupled TGF-β1 to aldehyde-activated polyethylene glycol-coated silica (Chromatochem, Merck, Darmstadt, Germany), which can be used in 100 mM sodium citrate, pH 3.5. Two milligrams of TGF-β1 in sodium citrate buffer is added to 0.62 g of resin rehydrated for 15 min with sodium citrate buffer and degassed. Ten milligrams of sodium cyanoborohydride (NaCNBH$_3$) in 0.2 ml citrate buffer is then immediately added to the slurry. Following an incubation of 30 min at 4°, the resin is centrifuged at room temperature for 5 min at 2000 g in a Micro-Centrifuge (Fisher Scientific, Fairlawn, NJ), washed 3 times with 5 ml each of citrate buffer, and blocked for 1 hr at 4° with 1 g glucosamine (Sigma) in 5 ml citrate buffer containing 10 mg NaCNBH$_3$.

Alternatively, TGF-β can be coupled to AH- or CH-Sepharose 4B (Pharmacia, Piscataway, NJ) or to Affi-Gel 102 aminoalkyl-agarose (Bio-Rad) via carbodiimide coupling. As coupling reagent we use diethylaminopropylcarbodiimide hydrochloride (EDAC, Bio-Rad) at 40 mg/ml in water, acidified to pH 4.5 with HCl. The coupling reagent (10 mg/ml resin) is added to 1 g of the resin (~4 ml) and 2 mg TGF-β1 in acidified water for an overnight incubation at 4°. After centrifugation for 5 min at 2000 g (Micro-Centrifuge, Micromedics Systems) and 3 washes with acidified water, the reactive resin is blocked for 1 hr at ambient temperature with 5 ml of 0.1 M sodium acetate, 0.5 M NaCl at pH 4.0.

Production of Monoclonal Antibodies to Transforming Growth Factor β1

As in the case of the polyclonal antisera, the production of monoclonal antibodies (MAb) to TGF-β1 was difficult, possibly owing to the immunosuppressive properties and the highly conserved amino acid sequence of this factor. We were not successful using intraperitoneal or subcutaneous immunization and conventional fusion protocols, whereas Dasch et al.[21] have succeeded using these methods. In our successful experiments, we immunized BALB/c mice with TGF-β1 using footpad injections, and their draining inguinal and popliteal lymph nodes were fused. This protocol resulted in a large percentage of positive clones with a specific fusion efficiency of 88%, whereas conventional fusion of splenocytes from mice immunized intraperitoneally or subcutaneously yielded low specific fusion efficiency, with predominantly antibodies of the IgM isotype.

Ten BALB/c mice are injected with 5 μg/dose of purified TGF-β1 in 100 μl detox adjuvant (Ribi) in the hind footpads on days 0, 3, 7, 10, and 14. On day 17 the animals are sacrificed, their draining inguinal and popliteal nodes removed, and the lymphocytes dissociated from the nodal stroma using stainless steel 200 mesh (Tylinter, Inc., Mentor, OH). The lymphocyte suspensions from all 10 mice are pooled and fused with the mouse myeloma line X63-Ag8.653,[22] using 50% PEG 4000 according to an established procedure.[23] The fused cells are inoculated in the wells of 96-well tissue culture plates (Falcon, Oxnard, CA) at a density of 2×10^5 cells/well followed by HAT (hypoxanthine, aminopterin, and thymidine) selection[24] on day 1 after fusion.

Parental hybridoma cultures are screened for production of antibody to TGF-β1 by indirect ELISA, using 96-well polystyrene microtiter plates (Immulon, Nunc) coated overnight at 4° with 100 μl/well of 1 μg/ml TGF-β1 in coating buffer (50 mM carbonate/bicarbonate, pH 9.6). Coated plates are washed 3 times with wash buffer (PBS, 0.05% Tween 20) and blocked for 1 hr at ambient temperature with ELISA diluent (PBS, pH 7.4, 0.5% BSA, 0.05% Tween 20, and 0.01% thimerosal). After 3 additional washes with wash buffer, hybridoma supernatants (100 μl) are added for 1 hr at ambient temperature. The plates are washed again 3 times with wash buffer, and bound antibodies are detected with 100 μl/well of goat anti-mouse IgG conjugated with horseradish peroxidase (Tago Inc., Bur-

---

[21] J. R. Dasch, D. R. Pace, W. Waegell, D. Inenaga, and L. Ellingsworth, *J. Immunol.* **142**, 1536 (1989).

[22] J. F. Kearney, A. Radbruch, B. Liesegang, and K. Rajewski, *J. Immunol.* **123**, 1548 (1979).

[23] V. Oi and L. Herzenberg, in "Selected Methods in Cellular Immunology" (B. Mishel and S. Shiigi, eds.), p. 351. Freeman, San Francisco, California, 1980.

[24] J. W. Littlefield, *Science* **145**, 709 (1964).

lingame, CA), freshly diluted 1:5000 in ELISA diluent. The plates are washed with wash buffer and developed with 100 μl/well of 2.2 mM o-phenylenediamine substrate in substrate buffer (50 mM sodium phosphate, 0.1 M sodium citrate, pH 5.0) containing 0.01% (v/v) H_2O_2. The reaction is stopped after 15 min with 100 μl/well of 4.5 M sulfuric acid, and plates are read at 492 nm with an automatic plate reader (Molecular Devices, Menlo Park, CA). Hybridoma cultures positive for antibody to TGF-β1 are cloned twice by limiting dilution and grown as ascites in pristane-primed BALB/c mice.[25] The MAbs are purified from ascites fluid by protein A-Sepharose affinity chromatography and eluted in 0.1 M acetic acid, 0.5 M NaCl, pH 2.4, according to established procedures.[26]

Three hybridomas were cloned and their monoclonal antibodies characterized in some detail. Each of the MAbs recognized the dimer form of TGF-β1 in immunoblots. MAb 2G7, an antibody of the $IgG_{1,\kappa}$ isotype, immunoprecipitated TGF-β1, -β2, and -β3 and neutralized all three species in a [^{3}H]thymidine uptake inhibition assay using a mink lung fibroblast cell line (described below). It had an affinity of 1.2×10^8 liters/mol for TGF-β1 as determined by Scatchard analysis using ^{125}I-labeled TGF-β1. MAb 4A11, also an $IgG_{1,\kappa}$ antibody, neutralized and immunoprecipitated only TGF-β1 and had a similar affinity constant. Finally MAb 12H5, an $IgG_{2b,\kappa}$ isotype, immunoprecipitated but was not neutralizing for TGF-β1, and it had an affinity constant of 5×10^7 liters/mol for this molecule.[27]

Enzyme-Linked Immunosorbent Assay for Quantitation of Transforming Growth Factor β1

Double-antibody enzyme-linked immunosorbent assays have been extensively described. The immunoassay we have developed for the quantitation of TGF-β1 is based on the classic procedure of Engvall and Perlmann[28] and uses two MAbs specific for TGF-β1, 12H5 and 4A11, for capture and detection, respectively. MAb 4A11, which neutralizes the activity of TGF-β1 *in vitro* (see bioassay procedure below), was conjugated to horseradish peroxidase according to established procedures.[29] The assay has a useful range of 40 to 0.63 ng/ml and a sensitivity of 0.63 ng/ml

[25] M. Potter, J. G. Humphrey, and J. L. Walter, *J. Natl. Cancer Inst.* **49**, 305 (1972).
[26] J. W. Goding, *J. Immunol. Methods* **20**, 241 (1978).
[27] C. Lucas, L. N. Bald, B. M. Fendly, M. Mora-Worms, I. S. Figari, E. J. Patzer, and M. A. Palladino, *J. Immunol.* **145**, 1415 (1990).
[28] E. Engvall and P. Perlmann, *Immunochemistry* **8**, 871 (1971).
[29] P. K. Nakane and G. B. Pierce, *J. Cell Biol.* **33**, 307 (1967).

as determined by the unpaired, one-tailed t-test method of Rodbard et al.[30] TGF-β2, activin, and inhibin were unreactive in the assay when tested at concentrations of 500–1000 ng/ml.

In order to develop such an assay, which depends on optimal concentrations of antibodies, there is first a need to titer the antibodies in a checkerboard fashion, as described.[28] Several concentrations of coat antibody are titered against several dilutions of the enzyme-conjugated antibody, using a standard curve range (typically 200–0.56 ng/ml). Maximal sensitivity is usually achieved by decreasing the amount of antibody used for coating while increasing the concentration of enzyme-conjugated antibody. Many double-antibody ELISAs are linear between 1 and 50 ng/ml of the peptide to be analyzed with antibody concentrations of 0.5 to 2 μg/ml used to coat the microtiter plates. Optimal capture antibody and conjugate dilutions are those which after a fixed reaction time (20–30 min) will allow maximal sensitivity and the highest absorbance within the linear range of the spectrophotometer (an absorbance of 2.0 is the upper limit of linearity of most instruments). ELISAs such as these are easily automated and provide high throughput analysis of large number of samples assayed in duplicate.

Conjugation of MAb 4A11 to horseradish peroxidase (HRP; Boehringer Mannheim, Indianapolis, IN) is performed essentially according to the method of Nakane and Pierce.[29] This method is based on the periodate oxidation of hydroxyl groups in the carbohydrates of the enzyme to aldehyde groups, followed by the formation of Schiff bonds between the newly formed aldehyde groups and the amino groups of the antibody. The crosslink is subsequently stabilized by reduction with $NaBH_4$. HRP is activated by adding 200 μl of freshly prepared 0.1 M $NaIO_4$ solution to 4 mg HRP in 1 ml water for 20 min at ambient temperature. The activated HRP is then chromatographed on Sephadex G-50 (Pharmacia), preequilibrated with 1 mM sodium acetate, pH 4.0, to avoid self-coupling of HRP (which would occur if the solution were kept at alkaline pH). The colored HRP peak collected from the chromatography column is adjusted to pH 9.5 with 0.2 M sodium carbonate just prior to adding it to the antibody. The antibody (MAb 4A11) at 2–5 mg/ml in coupling buffer (10 mM sodium carbonate, pH 9.5) is added to the activated HRP in a weight ratio of antibody to HRP of 1–2:1. The coupling reaction is performed at ambient temperature

---

[30] D. Rodbard, P. J. Munson, and A. DeLean, in "Radioimmunoassay and Related Procedures in Medicine," Vol. 1, p. 469. Proceedings of an International Symposium on Radioimmunoassay and Related Procedures in Medicine, International Atomic Energy Agency, Vienna, Austria, 1978.

for 2 hr, then stopped by adding 100 μl of freshly prepared 4 mg/ml NaBH$_4$ for 2 hr at 4°.

For the quantitation of TGF-β1, MAb 12H5 is added at 0.5 μg/ml (100 μl/well) in coating buffer (50 mM carbonate/bicarbonate, pH 9.6) to 96-well microtiter plates for an overnight incubation at 4°. The plates are then washed 3 times with wash buffer (PBS, 0.05% Tween 20) and blocked for 1 hr at ambient temperature with ELISA diluent (PBS, pH 7.4, 0.5% BSA, 0.05% Tween 20, and 0.01% thimerosal). They are then washed again 3 times with wash buffer, and 100 μl of each TGF-β1 standard (40 to 0.63 ng/ml), control, or sample in ELISA diluent is added to duplicate wells for a 2-hr incubation at ambient temperature. The plates are subsequently washed 3 times with wash buffer and incubated for 1 hr at ambient temperature with HRP-conjugated MAb 4A11, freshly diluted to its optimal concentration in ELISA diluent. After 3 subsequent washes with wash buffer, substrate solution (2.2 mM o-phenylenediamine substrate in 50 mM sodium phosphate, 0.1 M sodium citrate, pH 5.0, containing 0.01% H$_2$O$_2$) is added for 30 min at ambient temperature (100 μl/well), and the reaction is stopped with 100 μl/well of 4.5 M H$_2$SO$_4$. Optical densities at 492 nm are then read with an automatic plate reader (Molecular Devices, Menlo Park, CA). Since this is a noncompetitive assay, the optical density is directly proportional to the antigen concentration. Data are reduced using a four-parameter curve-fitting program based on an algorithm for least-squares estimation of nonlinear parameters[19] to provide antigen concentrations calculated from the standard curve.

The interassay precision was good, with percent coefficients of variation of 8–10% for TGF-β samples assayed in nine different plates. The sensitivity and precision of this assay compared well with values obtained with the radioimmunoassay and the radioreceptor assay, with the obvious advantage of not requiring the use of radioactive reagents. This assay was only possible with the availability of monoclonal antibodies to TGF-β1. Many combinations of polyclonal antibodies were previously tried with complete lack of success, in contrast to the recently reported ELISA based on polyclonal antibodies to TGF-β.[20] The assay is adaptable to the measurement of TGF-β in cell culture fluid or in serum or plasma, although these fluids may contain interfering substances or binding proteins. Treatments such as acidification, partial denaturation, and/or extraction may therefore be needed prior to quantitation of the TGF-β.

Bioassay for Transforming Growth Factor β

The multifunctionality of TGF-β has resulted in the development of many different types of cell-based assays to measure its activity.[2] Since

TGF-β can either inhibit or enhance cellular proliferation, depending on the cell type, assays can be developed which measure either property. To measure the growth-promoting or mitogenic effects of TGF-β, the assay system most generally used is the stimulation of proliferation of fibroblastic cell types such as AKR-2B (mouse embryo fibroblasts) or NRK cells (normal rat kidney fibroblasts) in soft agar. On the other hand, the growth-inhibitory effects of TGF-β can be measured on a variety of both normal and transformed cell types grown either as monolayers or in soft agar. These include hepatocytes, endothelial cells, lymphocytes, fibroblasts, and epithelial cells.

One of the most sensitive *in vitro* assays to detect TGF-β bioactivity is based on the antiproliferative effects of TGF-β on mink lung epithelial-like cells (Mv1Lu).[20] This cell line is available from the American Type Culture Collection (Rockville, MD; ATCC CCL64). The parental cell line was cloned by limiting dilution at 0.3 cells/well. Clone 3D9 (designated Mv3D9) was chosen from 78 individual clones because of its greater sensitivity to the growth-inhibitory effects of TGF-β1 when compared to the parental cell line. Mv3D9 cells are maintained *in vitro* by weekly passage in CMEM consisting of Eagle's minimum essential medium supplemented with 0.1 mM nonessential amino acids, 2 mM L-glutamine, 1 mM sodium pyruvate, 100 U/ml penicillin, 100 μg/ml streptomycin (GIBCO), and 10% heat-inactivated fetal bovine serum (FBS, HyClone Laboratories, Logan, UT).

The TGF-β assay utilizing these cells is based on the ability of TGF-β to inhibit [^{3}H]thymidine uptake by Mv3D9 cells. The assay is performed as follows: samples to be assayed (100 μl/well) are serially diluted in assay medium (CMEM supplemented with 0.1% FBS) into 96-well flat-bottomed microtiter plates, followed by the addition of 10^4 Mv3D9 cells/well in 100 μl of assay medium. Prior to addition to the wells, assay samples which contain inactive forms of TGF-β are acid-activated by incubating each 100-μl sample with 10 μl of 1.2 N HCl for 15 min at ambient temperature and neutralized with 20 μl of 0.5 M HEPES buffer containing 0.72 M NaOH. The plates are then incubated at 37° with 5% CO_2 for 24 hr. For the last 4 hr of culture, 20 μl/well (1 μCi) of [^{3}H]thymidine in CMEM (6.7 Ci/mmol, Amersham) is added. After incorporation of the label, 100 μl of 0.5% trypsin/5.3 mM EDTA solution (GIBCO) is added to each well, and the plates are incubated for an additional 15 min at 37°. Cells are then harvested onto glass fiber filters and counted in a liquid scintillation counter. Results (pg/ml of TGF-β) are calculated based on percent inhibition of thymidine incorporation compared with a recombinant human TGF-β1 laboratory standard. A typical dose–response is in the range of 2–250 pg/ml TGF-β.

Monoclonal Antibody Neutralization of Transforming Growth Factor β In Vitro

To test whether antibodies (polyclonal or monoclonal) to TGF-β have neutralizing activity *in vitro*, TGF-β at a constant concentration (1000–2000 pg, final concentration) in CMEM is preincubated overnight at 4° with a dose titration of 0.39–100 μg/ml (final concentration) of the antibodies before being tested in the bioassay described above. Mv3D9 cells (10^4/100 μl/well, see above) are then added to each 100-μl sample for 20 hr at 37°. The cells are pulsed for 4 hr at 37° with 1 μCi/20 μl/well of radiolabeled thymidine, harvested as described above, and counted in a scintillation counter. The reversal of TGF-β inhibition of thymidine uptake by the test antibody is compared to that of a negative control MAb preincubated with TGF-β (we used anti-human growth hormone MAb 6G12 from Genentech Inc., South San Francisco, CA).

Conclusion

The development of antibodies and assays for transforming growth factor β has been a difficult endeavor. Although platelets are an abundant source of material for purified TGF-β,[1,31,32] antibodies of high titer and neutralizing antibodies were difficult to obtain, presumably owing to the small size and conserved amino acid sequence of this protein. Some authors successfully produced monoclonal antibodies to TGF-β with conventional immunization and fusion protocols[21]; however, we had to immunize mice in the footpads and fuse cells from their lymph nodes to obtain these reagents. In the absence of monoclonal antibodies, a conventional radioimmunoassay was developed with polyclonal antibodies to TGF-β, and we found that the radioiodination of TGF-β was a procedure which required careful handling of the molecule. Provided attention was paid to the particular behavior and properties of TGF-β, we were able to produce good reagents and reliable procedures for the measurement of this growth factor.

---

[31] C. B. Childs, J. A. Proper, R. F. Tucker, and H. L. Moses, *Proc. Natl. Acad. Sci. U.S.A.* **79,** 5312 (1982).

[32] R. K. Assoian and M. B. Sporn, *J. Cell Biol.* **102,** 1217 (1986).

[30] Purification of Transforming Growth Factors β1 and β2 from Bovine Bone and Cell Culture Assays

By YASUSHI OGAWA and SAEID M. SEYEDIN

Introduction

Two related proteins, cartilage-inducing factors A and B (CIF-A and CIF-B), which promote expression of the chondrocyte phenotype in rat fetal mesenchymal cells were isolated from bovine bone extract by Seyedin et al.[1] CIF-A was later shown to be identical to transforming growth factor β1 (TGF-β1) isolated from human platelets.[2,3] CIF-B, now known as TGF-β2, is a unique protein with about 71% sequence identity to TGF-β1.[4] TGF-β2 has also been isolated from BSC-1 African green monkey kidney epithelial cells,[5,6] porcine platelets,[7] human glioblastoma cells,[8] and a human prostatic adenocarcinoma cell line.[9]

TGF-β1 and TGF-β2 are homodimeric proteins with molecular weights of 25,000. The proteins are highly conserved among different species. Only one residue difference exists between human and mouse TGF-β1.[10] TGF-β1 and TGF-β2 are multifunctional regulators of cell proliferation and differentiation and play important roles during development and tissue repair (for reviews, see Refs. 11 and 12). Immunohistochemical staining

---

[1] S. M. Seyedin, T. C. Thomas, A. Y. Thompson, D. M. Rosen, and K. A. Piez, *Proc. Natl. Acad. Sci. U.S.A.* **82**, 2267 (1985).
[2] R. K. Assoian, A. Komoriya, C. A. Meyers, D. M. Miller, and M. B. Sporn, *J. Biol. Chem.* **258**, 7155 (1983).
[3] S. M. Seyedin, A. Y. Thompson, H. Bentz, D. M. Rosen, J. M. McPherson, A. Conti, N. R. Siegel, G. R. Galluppi, and K. A. Piez, *J. Biol. Chem.* **261**, 5693 (1986).
[4] S. M. Seyedin, P. R. Segarini, D. M. Rosen, A. Y. Thompson, H. Bentz, and J. Graycar, *J. Biol. Chem.* **262**, 1946 (1987).
[5] S. K. Hanks, R. Armour, J. H. Baldwin, F. Maldonado, J. Spiess, and R. W. Holley, *Proc. Natl. Acad. Sci. U.S.A.* **85**, 79 (1988).
[6] J. M. McPherson, S. J. Sawamura, Y. Ogawa, K. Dineley, P. Carrillo, and K. A. Piez, *Biochemistry* **28**, 3442 (1989).
[7] S. Cheifetz, J. A. Weatherbee, M. L.-S. Tsang, J. K. Anderson, J. E. Mole, R. Lucas, and J. Massague, *Cell (Cambridge, Mass.)* **48**, 409 (1987).
[8] M. Wrann, S. Bodmer, R. de Martin, C. Siepl, R. Hofer-Warbinek, K. Frei, E. Hofer, and A. Fontana, *EMBO J.* **6**, 1633 (1987).
[9] T. Ikeda, M. N. Lioubin, and H. Marquardt, *Biochemistry* **26**, 2406 (1987).
[10] R. Derynck, J. A. Jarrett, E. Y. Chen, and D. V. Goeddel, *J. Biol. Chem.* **261**, 4377 (1986).
[11] M. B. Sporn, A. B. Roberts, L. M. Wakefield, and B. de Crombrugghe, *J. Cell Biol.* **105**, 1039 (1987).
[12] M. B. Sporn, A. B. Roberts, L. M. Wakefield, and R. K. Assoian, *Science* **233**, 532 (1986).

of mouse embryos has shown that expression of TGF-β1 is elevated in bone during development.[13] Expression of osteoblast and chondrocyte phenotypes and the proliferation of these cells are regulated by TGF-β1 and TGF-β2.[14,15] TGF-β1 and TGF-β2 synergize with osteoinductive factor in inducing ectopic bone formation *in vivo*.[16] The biological activities of TGF-β1 and TGF-β2 suggest that the factors play an important role during bone development. This chapter describes the purification of TGF-β1 and TGF-β2 from bovine bone and cell culture assays of the growth-inhibitory and chondrogenic activities of the factors.

Purification of Transforming Growth Factors β1 and β2

Step 1: Bovine Bones. Cattle feet (100 pieces) are packed in ice at the slaughterhouse within 45 min of slaughter and transported back to the laboratory. Skin and soft tissues are removed, and the hoofs are detached from the metatarsal and metacarpal bones and discarded. Bones are frozen on dry ice and scraped with a knife to remove remaining soft tissue. Cartilagenous tissues are removed by sawing off the both ends of each bone. Frozen bones are broken into fragments, and the marrow is scraped off. Bone fragments are crushed into chips, pulverized into powder in a liquid nitrogen-cooled mill, and stored at $-80°$. TGF-β activity in the bone powder is stable for at least 6 months if the bone powder is stored at $-80°$. The yield of bone powder from 100 bones is approximately 13 kg.

Step 2: Demineralization and Extraction. Bone powder (4–4.5 kg) is washed 5 times, 20 min each, with deionized water (8 liters/kg) at 4° by keeping the bone powder suspended with a motorized propeller in a 15-gallon high-density polyethylene tank. Between successive washes, allow the bone powder to settle and then pour off the wash. The washed bone powder is collected over a sheet of filter membrane [Millipore (Bedford, MA) AP 25] placed over a Büchner funnel (25.3 cm diameter). Lipids are extracted by washing the bone powder 7 times, 20 min each, with acetone (3 liters/kg) under a hood at room temperature. Owing to fire hazard, an air-driven motor, rather than an electrical powered motor, is used to keep the bone powder suspended in acetone. Bone powder is air-dried under a

---

[13] U. I. Heine, E. F. Munoz, K. C. Flanders, L. R. Ellingsworth, H.-Y. P. Lam, N. L. Thompson, A. B. Roberts, and M. B. Sporn, *J. Cell Biol.* **105**, 2861 (1987).

[14] D. M. Rosen, S. A. Stempien, A. Y. Thompson, and S. M. Seyedin, *J. Cell. Physiol.* **134**, 337 (1988).

[15] M. Centrella, T. L. McCarthy, and E. Canalis, *J. Biol. Chem.* **262**, 2869 (1987).

[16] H. Bentz, R. M. Nathan, D. M. Rosen, R. M. Armstrong, A. Y. Thompson, P. R. Segarini, M. C. Mathews, J. R. Dasch, K. A. Piez, and S. M. Seyedin, *J. Biol. Chem.* **264**, 20805 (1989).

hood. Delipidated bone powder is demineralized by stirring in 0.5 N HCl (22 liters/kg) at 4° for 18 hr in a 30-gallon tank. The demineralized bone powder is washed 6 times, 20 min each, with deionized water (10 liters/kg) at 4° and collected over a filter membrane. Proteins are extracted with 4 M guanidine hydrochloride, 10 mM disodium EDTA, pH 6.8 (2 liters/kg), by stirring the demineralized bone powder in the guanidine buffer for 6 hr at 4°. The extract is filtered through a filter membrane, and the bone powder is extracted with fresh guanidine buffer (2 liters/kg) for 16 hr. Two extracts are combined and stored at 4°. TGF-β1 and TGF-β2 are stable in the guanidine buffer at 4° for at least 6 weeks.

Step 3: Gel-Filtration Chromatography. To concentrate the extract, the extract is dialyzed against deionized water in a 55-gallon tank at 4°, lyophilized, and dissolved in 300–400 ml of 4 M guanidine hydrochloride, 10 mM disodium EDTA, pH 6.8. The concentrated extract is clarified by centrifugation. A column (25.2 × 74 cm; BioProcess column 252, Pharmacia, Piscataway, NJ) of Sephacryl S-200 (super fine, Pharmacia) is equilibrated in 4 M guanidine hydrochloride, 10 mM disodium EDTA, pH 6.8, at room temperature. The flow through the column is in the upward direction. The extract is fractionated through the column at a flow rate of 30–35 ml/min. Fractions of 400–500 ml are collected. TGF-β1 and TGF-β2 are present in the fractions containing mostly proteins with molecular weights less than 50,000, which elute shortly after the major peak of absorbance at 280 nm. These fractions are pooled and stored at 4°.

Step 4: Carboxymethyl-Cellulose Cation-Exchange Chromatography. The pool of fractions containing TGF-β1 and TGF-β2 is dialyzed against deionized water in a 55-gallon tank at 4° and lyophilized. The lyophilized residue is dissolved in 6 M urea, 50 mM sodium acetate, 10 mM NaCl, 1% 2-propanol, 1 mM N-ethylmaleimide, pH 4.6, at the final concentration of 5 mg of dry residue/ml. Any undissolved material is removed by centrifugation. The pool of TGF-β1 and TGF-β2 fractions is chromatographed at room temperature on a column (2.5 × 36 cm) of carboxymethyl-cellulose (CM52, Whatman, Clifton, NJ) equilibrated in the urea–sodium acetate buffer. TGF-β1 and TGF-β2 are eluted with a linear NaCl gradient of 10 to 600 mM NaCl in a total volume of 1200 ml at a flow rate of 30 ml/hr. TGF-β1 and TGF-β2 elute between 120 and 150 mM NaCl (Fig. 1). TGF-β2 elutes somewhat earlier than TGF-β1. Bands of TGF-β1 and TGF-β2 are clearly visible on silver-stained sodium dodecyl sulfate (SDS)–polyacrylamide gels.[17,18] The fractions that contain TGF-β1 and TGF-β2 are pooled, filtered through a 0.2-μm filter, and stored at 4°.

[17] U. K. Laemmli, *Nature (London)* **227,** 680 (1970).
[18] J. H. Morrissey, *Anal. Biochem.* **117,** 307 (1981).

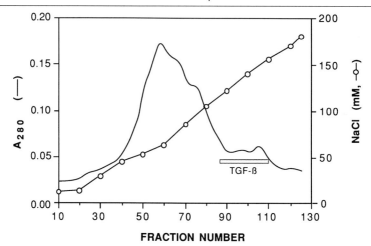

FIG. 1. Cation-exchange column chromatography. Concentrated low molecular weight fractions from the gel-filtration column chromatography step were applied onto a Whatman CM-52 column equilibrated in 50 mM sodium acetate, 6 M urea, 10 mM NaCl, 1% 2-propanol, 1 mM N-ethylmaleimide, pH 4.6, and TGF-β1 and TGF-β2 were eluted with a linear NaCl gradient. The position of the fractions that contained TGF-β1 and TGF-β2 is marked with a bar.

Step 5: C_{18} Reversed-Phase Chromatography. The pool of TGF-β1 and TGF-β2 is applied onto a C_{18} reversed-phase HPLC column (Vydac, 218TP54, 0.46 × 25 cm) equilibrated in 0.1% TFA (buffer A) at a flow rate of 1 ml/min. After the absorbance at 210 nm returns to baseline, a solvent mixture of 32% buffer B (0.1% TFA, 90% acetonitrile) and 68% buffer A is pumped through the column until a new baseline is established. TGF-β1 and TGF-β2 are eluted with a linear acetonitrile gradient formed with 32 to 62% buffer B over 30 min (Fig. 2).

TGF-β1 and TGF-β2 are major bands on the silver-stained SDS–polyacrylamide gels. TGF-β1 elutes from the column at 42–44% buffer B, while TGF-β2 elutes at 47–49%, both in distinct peaks. TGF-β1 and TGF-β2 fractions are diluted 3-fold with buffer A and separately rechromatographed on the C_{18} column.

Typical yields from 4.5 kg of bone powder is about 1.2 mg of TGF-β1 and 0.4 mg of TGF-β2. The purified TGF-β1 appears to be 95–99% pure, whereas TGF-β2 appears to be about 80% pure when analyzed by SDS–PAGE. Recently, we have been able to purify TGF-β1 to very high levels of purity (>99%) by chromatographing the TGF-β1 and TGF-β2 pool from the carboxymethyl-cellulose chromatography step on a S-Sepha-

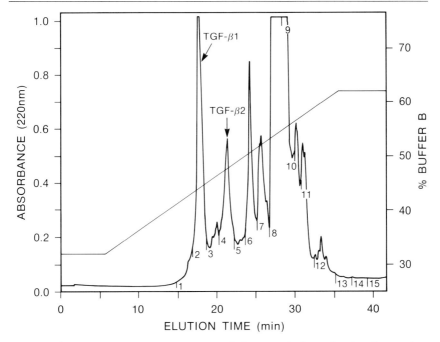

FIG. 2. C_{18} reversed-phase HPLC. The pool of TGF-β1 and TGF-β2 fractions from cation-exchange column chromatography was chromatographed on a C_{18} reversed-phase HPLC column. The bound proteins were eluted with a linear acetonitrile gradient in 0.1% TFA (buffer A: 0.1% TFA; buffer B: 90% acetonitrile, 0.1% TFA). TGF-β1 and TGF-β2 eluted in distinct peaks.

rose column (Pharmacia, 1 × 15 cm) in 25 mM HEPES, 8 M urea, pH 9, and eluting the bound TGF-β1 with a linear 10 to 400 mM NaCl gradient. TGF-β1 is further purified by C_{18} reversed-phase HPLC as already described.

The protein concentration is calculated by amino acid analysis or by differences in the absorbance at 215 and 225 nm.[19,20] Both TGF-β1 and TGF-β2 demonstrate ED$_{50}$ values in the range of 0.01 to 0.02 ng/ml when assayed by the mink lung epithelial cell culture assay described below. TGF-β1 and TGF-β2 are stable for at least 6 months if stored in the HPLC solvents at −80°. Bovine bone-derived TGF-β1 and TGF-β2 have been

[19] W. J. Waddell, *J. Lab. Clin. Med.* **48,** 311 (1956).
[20] J. B. Murphy and M. W. Kies, *Biochim. Biophys. Acta* **45,** 382 (1960).

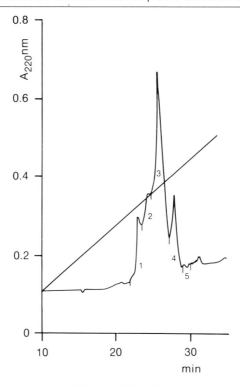

FIG. 3. Microheterogeneity of TGF-β2. TGF-β2 from the second C_{18} reversed-phase HPLC step was chromatographed on a C_4 reversed-phase HPLC column. TGF-β2 eluted in multiple peaks during a linear gradient of n-propanol in 0.1% TFA. Silver staining following SDS–polyacrylamide gel electrophoresis of the fractions revealed that all the peaks contained TGF-β2. The last peak (fraction 4) contained the majority of the contaminants.

completely sequenced, and the sequences have been determined to be identical at every residue to those of human TGF-β1[21] and TGF-β2.[22]

Microheterogeneity of Transforming Growth Factor β2 and Further Purifications

Chromatography of bovine bone-derived TGF-β2 preparations on C_4 reversed-phase HPLC columns reveals the existence of microheterogeneity, as demonstrated by the multiple peak elution profile (Fig. 3). The

[21] R. Derynck, J. A. Jarrett, E. Y. Chen, D. H. Eaton, J. R. Bell, R. K. Assoian, A. B. Roberts, M. B. Sporn, and D. V. Goeddel, *Nature (London)* **316**, 701 (1985).

[22] H. Marquardt, M. N. Lioubin, and T. Ikeda, *J. Biol. Chem.* **262**, 12127 (1987).

elution pattern is highly reproducible from one preparation of TGF-β2 to the next. TGF-β2 from the C_{18} reversed-phase HPLC step is applied onto a C_4 reversed-phase HPLC column (Vydac, 0.46 × 25 cm, 214TP54) equilibrated in 0.1% TFA (buffer A). The column is then equilibrated in 15% buffer B (0.1% TFA, 90% n-propanol) and 85% buffer A. TGF-β2 is eluted with a linear n-propanol gradient formed with 15–45% buffer B over 30 min at flow rate of 0.3–0.5 ml/min.

SDS–PAGE revealed that TGF-β2 was present in all of the peaks, except the last peak (fraction 4) which contained the majority of the contaminants. TGF-β2 in these peaks had a similar, if not identical, specific biological activity. Their identity as TGF-β2 was confirmed by immunoblotting with a TGF-β2-specific monoclonal antibody and by partial sequencing. The purity of TGF-β2 preparations, consisting of the pool of peaks of TGF-β2 from the C_4 reversed-phase HPLC, is approximately 99%. The nature of microheterogeneity of bovine bone-derived TGF-β2 is not known. In contrast to TGF-β2, TGF-β1 eluted as a single broad peak under identical chromatographic conditions.

Cell Culture Assays

Transforming Growth Factor β Assay

TGF-β1 and TGF-β2 stimulate dose-dependent formation of colonies of normal rat kidney fibroblasts cultured in soft agar in the presence of epidermal growth factor.[23] In contrast, TGF-β1 and TGF-β2 inhibit proliferation of a variety of cells, including fibroblasts, epithelial cells, keratinocytes, and endothelial cells, when grown in monolayer cultures.[11,12,24] Described below is a cell culture assay developed with mink lung epithelial cells. In this assay, the activity of a constitutively expressed enzyme, acid phosphatase, is assayed as a measure of cell number as described by Connolly et al.[25] In contrast to the soft agar assay and to other assays which measure incorporation of radiolabel into DNA,[9,26] the assay described below is very simple to perform, suited for assaying large numbers of samples, and does not involve handling of radiochemicals.

[23] A. B. Roberts, C. A. Frolik, M. A. Anzano, R. K. Assoian, and M. B. Sporn, in "Cell Culture Methods for Molecular and Cell Biology," (D. W. Barnes, D. A. Sirbasku, and G. H. Sato, eds.), Vol. 1, p. 181. Alan R. Liss, New York, 1984.
[24] A. B. Roberts and M. B. Sporn, in "Advances in Cancer Research," (G. Klein and S. Weinhouse, eds.), Vol. 51, p. 107. Academic Press, Orlando, Florida, 1988.
[25] D. T. Connolly, M. B. Knight, N. K. Harakas, A. J. Wittwer, and J. Feder, *Anal. Biochem.* **152**, 136 (1986).
[26] L. R. Ellingsworth, D. Nakayama, P. Segarini, J. Dasch, P. Carrillo, and W. Waegell, *Cell. Immunol.* **114**, 41 (1988).

Mink lung epithelial cells (ATCC, Rockville, MD, Mv1Lu CCL64) are plated at 0.5 to 1×10^6 cells/plate in 100-mm culture plates and grown in Eagle's minimum essential medium (MEM), supplemented with 50 units/ml penicillin, 50 µg/ml streptomycin, nonessential amino acids, L-glutamine, and 10% fetal bovine serum (FBS) until near confluency, which is usually reached in 3–4 days. Cells are detached with trypsin, collected by centrifugation at 800 g for 2 min, and suspended in the culture medium at 20,000 cells/ml. The cells are plated in 96-well microtiter plates at 1000 cells (50 µl)/well and allowed to attach for 30 min. Samples of TGF-β1 and TGF-β2 are diluted in the culture medium to cover the concentrations from 0.002 to 10 ng/ml. Aliquots (50 µl) of the sample are transferred into the wells in triplicate, and the plates are incubated under a 5% CO_2, 95% (v/v) air atmosphere at 37° for 4 days. The wells are rinsed with phosphate-buffered saline, then filled with 100 µl of 0.1 M sodium acetate, pH 5.5, 0.1% Triton X-100, 100 mM p-nitrophenyl phosphate (prepared within 1 hr of use), and the plates are incubated at 37° for 2 hr. Color development takes place when 10 µl of 1.0 N NaOH is added. After 20 min at room temperature, the absorbance at 405 nm is measured with a microtiter plate reader. The absorbance at 405 nm, which is proportional to the number of cells, decreases as a dose-dependent function of concentrations of TGF-β1 or TGF-β2. TGF-β1 and TGF-β2 typically demonstrate an ED_{50} values of 0.01–0.02 ng/ml.

Chondrogenesis Assays

Two *in vitro* chondrogenesis assays[27,28] are described below. In both assays, the expression of cartilage-specific proteoglycan and type II collagen by rat embryonic mesenchymal cells is assayed by ELISA. In the first assay, the cells are cultured in an agarose gel; in the second assay, the cells are grown in monolayer culture in the presence of dihydrocytochalasin B (DHCB), which is known to be an inhibitor of cytoskeletal structure organization, particularly the actin-containing cytoskeletal filaments.[29]

Agarose Gel Chondrogenesis Assay. The limbs of 19-day-old Sprague-Dawley rat fetuses are removed surgically. Muscle is dissected from the limbs and minced into small pieces. The fragments are plated onto 100-mm tissue culture plates in Eagle's MEM, 10% FBS, 50 units/ml penicillin, 50 µg/ml streptomycin. After about 1 week, the cells are harvested by

---

[27] A. Y. Thompson, K. A. Piez, and S. M. Seyedin, *Exp. Cell Res.* **157**, 483 (1985).
[28] D. M. Rosen, S. A. Stempien, A. Y. Thompson, J. E. Brennan, L. R. Ellingsworth, and S. M. Seyedin, *Exp. Cell Res.* **165**, 127 (1986).
[29] S. S. Brown and J. A. Spudich, *J. Cell Biol.* **88**, 487 (1981).

trypsinization and replated into two plates. The cells are used for studies within the first three passages.

A sample of TGF-β1 or TGF-β2 is dissolved in the culture medium (Ham's F12 medium supplemented with 10% FBS, 50 units/ml penicillin, and 50 μg/ml streptomycin). The cells are harvested by trypsinization and suspended in the culture medium. Each sample is assayed in triplicate. TGF-β1 or TGF-β2 is added to the cell suspension, such that the concentration is 2 times the final desired concentration and the concentration of the cells is 2.5×10^6 cells/ml. The cell suspension with TGF-β1 or TGF-β2 is diluted with an equal volume of 1% low melting temperature agarose (Bio-Rad, Richmond, CA) in the culture medium, and 0.2-ml aliquots are plated onto 17-mm wells (Falcon, Oxnard, CA) precoated with 0.15 ml of 1% high melting agarose (Bio-Rad) in the culture medium. The cultures are incubated at 37° for 5 min, chilled at 4° for 10 min, and overlaid with 1.0 ml of the culture medium containing the desired concentrations of TGF-β1 or TGF-β2. Plates are placed in a humidified atmosphere of 5% CO_2, 95% air at 37°, and the culture medium is replaced every 3–4 days with 1.0 ml of the culture medium without TGF-β1 or TGF-β2. At 7 to 14 days, cultures are frozen and stored at −80° until assayed by ELISA.

Seven days of culture is adequate to observe significantly higher levels of chondrocyte-specific proteoglycans, but the cultures should be incubated for 14 days to observe significant increases in the type II collagen levels. Although there is very little change in cell morphology after 7 days, striking changes in the culture are apparent after 14 to 21 days. Large colonies with the accumulation of extracellular matrix, which stains metachromatically with safranin O, are predominant in the treated cultures.

Each culture well is homogenized at 4° with 3 ml of 6 M guanidine hydrochloride, 75 mM sodium acetate, 20 mM EDTA, 10 mM N-ethylmaleimide, 5 mM phenylmethylsulfonyl fluoride, pH 5.8, and extracted by shaking overnight at 4°. Agarose is removed by centrifugation at 25,000 g for 45 min at 4°, and the supernatant is dialyzed overnight at 4° against 0.2 M NaCl, 50 mM Tris-HCl, pH 7.4.

Proteoglycan and type II collagen standards for ELISA are purified from Swarm rat chondrosarcoma tissue.[30,31] Antisera to rat cartilage proteoglycan and type II collagen are raised in rabbits. Antiserum to rat type II collagen is affinity-purified with purified type II collagen[32] and shows no immunocross-reactivity to type I collagen.

---

[30] E. J. Miller, *Biochemistry* **10**, 1652 (1971).
[31] S. L. Lee and K. A. Piez, *Collagen Rel. Res.* **3**, 89 (1983).
[32] R. Timpl, this series, Vol. 82, p. 472.

Inhibition ELISA is performed as described by Rennard et al.[33] and Schuurs and Van Weemen.[34] Briefly, the dialyzed samples are diluted with an equal volume of antibody solution [diluted 2500-fold in phosphate-buffered saline, 0.05% Tween 20, 1 mg/ml bovine serum albumin (BSA), pH 7.4] and incubated overnight at 4°. Microtiter plates (96 wells) are coated overnight at 4° with 2 µg/ml of proteoglycan or type II collagen standard in 50 mM sodium carbonate, pH 9.6. The plates are then blocked with 5 mg/ml BSA in phosphate-buffered saline for 30 min at room temperature. Samples (200 µl) are added to the wells and incubated for 1 hr at room temperature. The wells are rinsed with phosphate-buffered saline, 0.05% Tween 20, 1 mg/ml BSA, pH 7.4, and the bound antibodies are detected with horseradish peroxidase-conjugated goat anti-rabbit IgG and o-phenylenediamine (Sigma, St. Louis, MO, prepared in 0.1 M citrate, 0.03% (w/v) H_2O_2, pH 4.6) as the substrate of the peroxidase reaction. TGF-β1 and TGF-β2 stimulate synthesis of chondrocyte-specific proteoglycan and type II collagen with ED_{50} values of approximately 2–5 ng/ml.

Chondrogenesis in the Presence of Dihydrocytochalasin B. As described above, mesenchymal cells derived from rat embryonic tissues undergo chondrogenic differentiation in the presence of TGF-β1 or TGF-β2. However, in monolayer cultures, the cells do not differentiate into chondrocytes, nor do they express chondrocyte-specific proteoglycans in response to TGF-β1 or TGF-β2 unless DHCB is present. The chondrogenesis assay performed in the presence of DHCB is described below.

Embryonic rat mesenchymal cells are isolated from muscle tissues of 19-day Sprague-Dawley rat fetuses and cultured in Eagle's MEM containing 10% FBS as described for the agarose gel culture above. Cells are used within the first three passages. Cell suspensions are prepared at a concentration of 2×10^4 cells/ml in Ham's F12 containing 10% fetal bovine serum, 50 units/ml penicillin, 50 µg/ml streptomycin, 3 µM DHCB (Sigma), and desired concentrations of TGF-β1 or TGF-β2. One-milliliter aliquots of the cell suspension are transferred to 16-mm wells, and the plates are incubated in 5% CO_2, 95% air atmosphere at 37° for 5 days. The medium is collected and clarified by centrifugation. Chondrocyte-specific proteoglycan and type II collagen are assayed by ELISA as described for the agarose gel culture. TGF-β1 and TGF-β2 typically stimulate the

---

[33] S. I. Rennard, K. Kimata, B. Dusemund, H. J. Barrach, J. Wilczek, J. H. Kimura, and V. C. Hascall, *Arch. Biochem. Biophys.* **207**, 399 (1981).
[34] A. H. W. M. Schuurs and B. K. Van Weemen, *Clin. Chim. Acta* **81**, 1 (1977).

production of chrondrocyte-specific proteoglycan with ED_{50} values of approximately 0.5–1 ng/ml in the DHCB assay.

Acknowledgments

We thank Dr. Hans Marquardt for sequencing bovine TGF-β2, David Schmidt for purification and characterization of TGF-β1 and TGF-β2 from bovine bones, and Andrea Thompson for comments on the description of the chondrogenesis assays.

[31] Identification and Activation of Latent Transforming Growth Factor β

By DAVID A. LAWRENCE

Introduction

Transforming growth factor β (TGF-β) has been the subject of several recent reviews,[1,2] so here only some of its salient properties will be recalled as an anchor point for this chapter, which is specifically concerned with latent TGF-β (L-TGF-β). In its active form TGF-β is multifunctional and can stimulate or inhibit cell proliferation, cell differentiation, and cellular function depending on cell type (e.g., fibroblast or epithelial cell) and on what other growth factors are present in the local environment. Most of our knowledge about the biological effects of TGF-β concern TGF-β1, simply because the other isoforms TGF-β2, -β3, and -β4 were discovered more recently.[3–7] TGF-β1 is produced by most cell types[8] and, unlike

---

[1] M. B. Sporn, A. B. Roberts, L. M. Wakefield, and B. de Crombrugge, *J. Cell Biol.* **105**, 1039 (1987).
[2] D. A. Lawrence, "Malignant Cell Secretion" (V. Krsmanovic, ed.), p. 215. CRC Press, Boca Raton, Florida, 1990.
[3] S. Cheifetz, J. A. Weatherbee, M. L. S. Tsang, J. K. Anderson, J. E. Mole, R. Lucas, and J. Massagué, *Cell (Cambridge, Mass.)* **48**, 409 (1987).
[4] T. Ikeda, M. N. Liobin, and H. Marquardt, *Biochemistry* **26**, 2406 (1987).
[5] P. Ten Dijke, P. Hansen, K. K. Iwata, C. Pieler, and J. G. Foulkes, *Proc. Natl. Acad. Sci. U.S.A.* **85**, 4715 (1988).
[6] S. Jakowlew, P. J. Dillard, P. Kondaiah, M. B. Sporn, and A. B. Roberts, *Mol. Endocrinol.* **28**, 747 (1988).
[7] R. Derynck, P. B. Lindquist, A. Lee, D. Wen, J. Tamm, J. L. Graycar, L. Rhee, A. J. Mason, D. A. Miller, R. J. Coffey, H. L. Moses, and E. Y. Chen, *EMBO J.* **7**, 3737 (1988).
[8] R. Derynck, J. A. Jarrett, E. Y. Chen, D. H. Eaton, J. R. Bell, R. K. Assoian, A. B. Roberts, M. B. Sporn, and D. V. Goeddel, *Nature (London)* **316**, 701 (1985).

classic hormones, acts locally via paracrine and autocrine modes.[9,10] However, before such local action can occur, TGF-β1 must first be activated, as most of this growth regulatory molecule is released from producer cells in an inactive, latent form.[11-14]

Because TGF-β is resistant to acid pH (pH 2.0),[15,16] purification schemes developed for TGF-β made full use of acetic acid for dialysis of cell-conditioned media and cell extracts and as eluant for gel-filtration columns.[17] Purification under acid conditions always yielded the biologically active 25-kDa dimer, by dissociation of the latent complex (see later). Thus, the latency phenomenon had been completely overlooked until this laboratory reported that normal chicken, mouse, and human embryo fibroblasts all released a TGF-β activity in a latent form under neutral conditions.[11] Quite soon after we showed that other fibroblasts, normal and transformed by various RNA and DNA viruses, also secreted L-TGF-β.[12,18] Perhaps more important was the demonstration[14] that human blood platelets (with bone one of the richest sources of TGF-β[19]) contained a latent form of TGF-β. Gel-filtration runs performed during the initial studies indicated that L-TGF-β was of high molecular weight (>400K).[11,13,14] Considering the activation treatments (see later) it was proposed that L-TGF-β was a complex of 25-kDa TGF-β associated noncovalently with a carrier protein.[13] Moreover, it was suggested that activation of L-TGF-β could be a critical step in the regulation of biological activity.[14] As a general phenomenon, the latency of TGF-β1 has been confirmed by several laboratories.[20,21]

[9] M. B. Sporn, A. B. Roberts, L. M. Wakefield, and R. K. Assoian, *Science* **233**, 532 (1986).
[10] A. S. Goustin, E. B. Leof, G. D. Shipley, and H. L. Moses, *Cancer Res.* **46**, 1015 (1986).
[11] D. A. Lawrence, R. Pircher, C. Krycève-Martinerie, and P. Jullien, *J. Cell. Physiol.* **121**, 184 (1984).
[12] R. Pircher, D. A. Lawrence, and P. Jullien, *Cancer Res.* **44**, 5538 (1984).
[13] D. A. Lawrence, R. Pircher, and P. Jullien, *Biochem. Biophys. Res. Commun.* **133**, 1026 (1985).
[14] R. Pircher, P. Jullien, and D. A. Lawrence, *Biochem. Biophys. Res. Commun.* **136**, 30 (1986).
[15] H. L. Moses, E. L. Branum, J. A. Proper, and R. A. Robinson, *Cancer Res.* **41**, 2842 (1981).
[16] A. B. Roberts, M. A. Anzano, L. C. Lamb, J. M. Smith, and M. B. Sporn, *Proc. Natl. Acad. Sci. U.S.A.* **78**, 5339 (1981).
[17] M. B. Sporn, M. A. Anzano, R. K. Assoian, J. E. De Larco, C. A. Frolik, C. A. Meyers, and A. B. Roberts, *Cancer Cells* **1**, 1 (1984).
[18] C. Krycève-Martinerie, D. A. Lawrence, J. Crochet, P. Jullien, and P. Vigier, *Int. J. Cancer* **35**, 553 (1985).
[19] J. L. Carrington, A. B. Roberts, K. C. Flanders, N. S. Roche, and A. H. Reddi, *J. Cell Biol.* **107**, 1969 (1988).
[20] R. J. Coffey, G. D. Shipley, and H. L. Moses, *Cancer Res.* **46**, 1164 (1986).
[21] L. M. Wakefield, D. M. Smith, T. Masui, C. C. Harris, and M. B. Sporn, *J. Cell. Biol.* **105**, 965 (1987).

Sample Preparation

Latent Transforming Growth Factor β

Many cell types are still cultured in the presence of serum, despite the advancing knowledge regarding growth factor supplements[22] to minimal media containing salts, amino acids, and vitamins. *In vivo*, however, cells only come into contact with serum at wound sites.[22] A further disadvantage of using serum in the present context is that it contains numerous growth factors in variable quantities,[22] including TGF-β essentially in a latent form.[23,24] Where no satisfactory serum-free medium has been found in which the desired cells grow, one can often avoid the problem by first growing the cells in medium with serum and then washing the cells, at least twice, with serum-free medium and reincubating them for 12 to 48 hr in the serum-free medium, which is then harvested to provide a virtually serum-free cell-conditioned medium. During the washing procedure it is important to incline the culture dish and drain it to remove as much of the initial serum-containing growth medium as possible. Some investigators reincubate the washed cultures with serum-free medium for 6–24 hr and then discard the latter before further incubation in serum-free medium, which will constitute the sample for testing.[20,21] This additional step removes further serum components. Clearly, each investigator must determine the culture requirements of cells used, but where serum must be used this does not necessarily preclude identifying TGF-β in such samples. Many growth factors, including TGF-β, can be expected to accumulate in culture medium with time; harvesting after a 24-hr period is common, and, where cell viability is maintained, repeated harvests over 4–5 days can be made.

It is possible to identify activatable TGF-β from only 1.0 ml of a crude, unconcentrated sample, but using 5.0 ml or more is simpler at the bench level. After harvesting, the sample should be centrifuged (10 min, 4000 g) to remove whole cells and most debris. Depending on the volume available, the clarified conditioned medium can be divided into 2.0- to 5.0-ml aliquots, some of which can be stocked at $-20°$ or below for later use. One should avoid defrosting a sample more than once, as it has been reported that repeated freezing and thawing tend to activate L-TGF-β.[25] In contrast, purified TGF-β (25 kDa) tends to be inactivated by this treatment.

[22] D. Barnes and G. Sato, *Cell (Cambridge, Mass.)* **22**, 649 (1980).
[23] C. Bjornson-Childs, J. A. Proper, R. F. Tucker, and H. L. Moses, *Proc. Natl. Acad. Sci. U.S.A.* **79**, 5312 (1982).
[24] M. O'Connor-McCourt and L. M. Wakefield, *J. Biol. Chem.* **262**, 14090 (1987).
[25] R. M. Lyons, J. Keski-Oja, and H. L. Moses, *J. Cell Biol.* **106**, 1659 (1988).

Latent Transforming Growth Factor β

Neutral, nondissociating conditions must be used to prepare samples, otherwise some activation will occur. Sonication and/or homogenization (e.g., with a Dounce-type homogenizer) are convenient for preparing extracts of most cells. We have used sonication (30-sec duration at 8 μm peak to peak with a 1.0 cm diameter probe) to prepare L-TGF-β from human blood platelets.[14] Washed platelets are taken up in phosphate-buffered saline (PBS), or other neutral buffer (1 mg wet weight/20 ml). After sonication the homogenate is centrifuged (100,000 g for 1 hr) and the resulting supernatant taken as the neutral extract. For tissue extraction the reader is referred to standard procedures for tissue maceration, bearing in mind the necessity of using neutral buffers without detergents.

Activation of Latent Transforming Growth Factor β1

The most frequently used activation method is acidification, usually followed by reneutralization to render the sample biocompatible again.[11] It is convenient to prepare the following solutions: 1.0, 0.5, and 0.1 M HCl and NaOH in culture-grade, distilled water. Depending on the sample volume and the buffering capacity of its contents (e.g., serum or HEPES buffer will provide extra buffering strength), HCl is added dropwise with rapid mixing until pH 2.0 is obtained. Approximately 100 μl of 1.0 M HCl added to 2.0 ml Dulbecco's medium leads to pH 2.0. At this low pH all L-TGF-β1 is activated (e.g., a 20-fold increase in activity over an original sample of conditioned medium[13] and a 300-fold increase for an acidified human platelet extract compared to the neutral extract[14]), but considerable activation occurs (~7-fold) when chicken embryo fibroblast-conditioned medium, which contains latent TGF-β, is brought to pH 5.0.[13] To avoid excessive dilution during acidification, sample volumes greater than 10.0 ml should be treated with concentrated HCl (~10.0 M). The acidified sample is left for 1 hr at 4°. (These were the original conditions for standardizing the procedure. Because activation appears to be due to pH shock dissociating the latent complex, the above time might well be shortened to about 10 min as long as this practice is followed throughout.)

Reneutralization of the acidified sample is carried out by the dropwise addition of NaOH to return to the initial pH. Because pH can rise above pH 8.0, care must be exercised and use made of lower NaOH molarities to obtain the final pH. The alkaline overshoots are probably of little consequence as alkalinization is another activation treatment for L-TGF-β.[13]

Comments. The protein concentration of serum-free conditioned medium can be quite variable depending on cell type and culture conditions, but it rarely exceeds 100 µg/ml. Acidification of such media does not normally lead to visible turbidity unless serum is present. Such precipitated material can distort assays of activated samples and should be removed by centrifugation (10 min, 4000 g). Some microbial contamination of samples undergoing activation will inevitably occur during pH meter manipulations, but this can be cut to a minimum by using sterile pipettes, micropipette cones, and solutions of acid and base. To avoid potential losses of protein by filtration (e.g., 0.45-µm cartridges) one may sterilize the samples for bioassay by ^{60}Co-irradiation, although such facilities are not generally available, leaving filtration as the more common method for sample asepsis.

With an activated and sterile sample in hand, one can proceed to the actual assay of TGF-β. The investigator can choose one or more of these standard assays: stimulation of colony formation of NRK-49F fibroblasts in soft agar in the presence of epidermal growth factor (EGF)[26]; inhibition of [^{3}H]thymidine incorporation in certain sensitive cell lines, like the mink epithelial cell line CCL64[27]; radioreceptor competition assay[28,29]; and neutralization or Western blotting with specific anti-TGF-β antiserum.[30,31] The first three assays can be used with appropriate standards of known TGF-β concentrations to determine less than 1.0 ng/ml TGF-β. As a control for activated samples of TGF-β one should check, in addition to the original sample, an aliquot which received premixed amounts of acid and base (this sample ought to give the same result as the original unactivated sample). For assays on conditioned media, one should test an aliquot of culture medium that has not been in contact with the cells. One-half of the aliquot should be tested directly and the other half should receive the acid/base treatment. Secretion of spontaneously active L-TGF-β1 appears to be relatively uncommon.

For large samples needing further purification, another convenient acid-activating treatment, is dialysis in acetic acid. Both 1% and 1 M

[26] J. E. De Larco and G. J. Todaro, *Proc. Natl. Acad. Sci. U.S.A.* **75**, 4001 (1978).
[27] R. F. Tucker, G. D. Shipley, H. L. Moses, and R. W. Holley, *Science* **226**, 705 (1984).
[28] C. A. Frolik, L. M. Wakefield, D. M. Smith, and M. B. Sporn, *J. Biol. Chem.* **259**, 10995 (1984).
[29] R. F. Tucker, E. L. Branum, G. D. Shipley, R. J. Ryan, and H. L. Moses, *Proc. Natl. Acad. Sci. U.S.A.* **81**, 6757 (1984).
[30] T. Masui, L. M. Wakefield, J. F. Lechner, M. A. LaVeck, M. B. Sporn, and C. C. Harris, *Proc. Natl. Acad. Sci. U.S.A.* **83**, 2438 (1986).
[31] L. M. Wakefield, D. M. Smith, K. C. Flanders, and M. B. Sporn, *J. Biol. Chem.* **263**, 7646 (1988).

(5.7%) acetic acid solutions can be employed. Acetic acid (1 M) gives a pH of 2.4, low enough to activate all L-TGF-β. As acetic acid is evaporated during lyophilization, samples thus treated can be reconstituted in the desired buffer or culture medium and also concentrated at the same time by appropriate volume reduction.

Other Methods of Activating Latent Transforming Growth Factor β1

Boiling

Boiling is a convenient procedure for 20 or more samples (e.g., column fractions or conditioned media from several cell types or from various culture conditions) where acidification/neutralization becomes tedious. The sample is kept in boiling water for 3 min. This heat treatment is about 50% less effective than acidification.[13] Human platelet L-TGF-β1 cannot be activated by boiling,[32] whereas that of rat platelets can be activated in this way.[33]

Alkalinization

Alkalinization is simply the reverse of the acidification procedure. The pH of the sample is brought to pH 11.0 or more by dropwise addition of NaOH before reneutralization with HCl.

Urea

Exposure of the sample to 8 M urea at room temperature activates L-TGF-β1 but tends to be a less effective method for L-TGF-β1 in conditioned media[13] and platelet extracts.[14]

Enzymatic Activation

By Plasmin. A 2-hr incubation at 37° with 0.2 U plasmin per milliliter conditioned medium gives slightly less than one-half the activation obtained by acidifying to pH 1.5 (respectively 450 and 1000 colonies in the soft agar assay).[25] The plasmin treatment is stopped by adding phenylmethylsulfonyl fluoride and aprotonin at 2.0 mM and 0.55 TIU/ml final concentrations followed by 1.0 mg bovine serum albumin (as a carrier) per milliliter of reaction mixture. In the TGF-β radioreceptor assay, maximum inhibition is obtained following treatment with 0.5 U/ml plasmin. Evidence

---

[32] K. Miyazano, U. Hellman, C. Wernstedt, and C. H. Heldin, *J. Biol. Chem.* **263**, 6407 (1988).

[33] T. Nakamura, T. Kitazawa, and A. Ichihara, *Biochem. Biophys. Res. Commun.* **141**, 176 (1986).

is provided[25] for the existence of two pools of L-TGF-β, one being activated by plasmin (or mild acid pH 4.5) and the other by strong acid at pH 1.5. Further evidence for plasmin activation comes from coculture experiments.[34] Here, the migration of confluent bovine aortic endothelial cells into a "wounded area" (obtained by scraping across an adherent cell monolayer in a petri dish with a razor blade) is blocked by cocultivating bovine pericytes or smooth muscle cells in the denuded area. This blockage appears to be due to plasmin-mediated activation of L-TGF-β *in situ*.

By Endoglycosidase F. Endoglycosidase F (endo F) removes complex and oligomannose (high mannose) N-linked carbohydrates.[35–37] L-TGF-β1 purified from human platelets (see below) is incubated at 37° for 16 hr in 50.0 mM EDTA and 100.0 mM sodium phosphate, pH 7.4, containing 10 or 25 U of endo F from *Flavobacterium meningosepticum* (Boehringer, Mannheim, Germany) per milliliter of reaction mixture.[38] In a radioreceptor assay, 0.5 nM (purified) L-TGF-β1 pretreated with 10 or 25 U/ml of endo F gives the same degree of inhibition as does 0.1 nM L-TGF-β1 transiently acidified to pH 2.0.

By Sialidase (N-Acetylneuraminidase). Several enzymes have the general name sialidase; depending on their specificity, they break particular linkages to remove N-acetylneuraminic acid from various polysaccharides, glycoproteins, and gangliosides.[36,37] The sialidase used in the report[38] cited here is type V from *Clostridium perfringens* (Sigma, St. Louis, MO). Purified L-TGF-β1 at 320 pM is incubated with 2 U of sialidase per milliliter of reaction mixture in 100 mM sodium phosphate buffer, pH 5.6, for 18 hr at 37°. The effect of the treatment is assayed by the inhibition of [^{3}H]thymidine incorporation; the extent of inhibition is about 80%. Even with saturating concentrations of sialidase (15 U/ml) activation by this method is 8-fold less efficient than transient acidification to pH 2.5, when the L-TGF-β1 concentration is 0.4 nM. At higher concentrations of L-TGF-β1 (4 nM) sialidase treatment (15 U/ml) is only 10% less effective than transient acidification, as measured by inhibition of [^{3}H]thymidine incorporation.[38]

Monosaccharides

Carbohydrate structures present in the precursor remnant of L-TGF-β1 (see below) appear to be important in maintaining latency.[38] Several small monosaccharides have been tested to see if they can compete with these structures. In the presence of 15 mM sialic acid or 50 mM mannose

---

[34] Y. Sato and D. B. Rifkin, *J. Cell Biol.* **109**, 309 (1989).
[35] J. H. Elder and S. Alexander, *Proc. Natl. Acad. Sci. U.S.A.* **79**, 4540 (1982).
[36] N. T. Thotakura and O. P. Bahl, this series, Vol. 138, p. 350.
[37] M. Saito, K. Sugano, and Y. Nagai, *J. Biol. Chem.* **254**, 7845 (1979).
[38] K. Miyazono and C. H. Heldin, *Nature (London)* **338**, 158 (1989).

6-phosphate, activation of an unspecified amount of L-TGF-β1 occurs *in situ* and leads to inhibition of [^{3}H]thymidine incorporation (42 and 58%, respectively.[38]

Heparin

The sulfated mucopolysaccharide heparin has an antiproliferative action on several cell types that is distinct from its more well-known anticoagulant activity.[39-42] Recent evidence[43] suggests that the heparin-induced inhibition of smooth muscle cell proliferation in serum-containing medium could be due to heparin-mediated dissociation of TGF-β (the effective inhibitor) from α_2-macroglobulin (α_2-m). It was previously shown[24] that L-TGF-β in serum is a complex of 25-kDa TGF-β associated with α_2-m. Responsive cells are grown for 18 hr in 10% fetal bovine serum, which is a source of L-TGF-β (TGF-β plus α_2-m), and with or without 100 μg/ml heparin. These cultures are then pulsed for 2 hr with [^{3}H]thymidine to ascertain the inhibitory effect caused by heparin activation of serum L-TGF-β.[43]

Purification of Latent Transforming Growth Factor β1 from Human Blood Platelets

The starting point of the purification procedure[32] is the nonadsorbed fraction from the CM-Sephadex (Pharmacia-LKB, Uppsala, Sweden) chromatography step of the platelet-derived growth factor purification.[44] Fifteen liters of this fraction is mixed with dry QAE-Sephadex (Pharmacia P-L Biochemicals), 0.7 mg/liter, A-50, and left on a shaker overnight. After allowing the gel to sediment, the supernatant is discarded and the gel poured into a column (5 × 60 cm). The column is washed with 4 liters of 10 mM phosphate buffer, pH 7.4, containing 75 mM NaCl; then, by increasing the salt gradient, L-TGF-β1 is eluted between 250 and 800 mM NaCl. Precipitation of the proteins in the latter eluate is accomplished by addition of ammonium sulfate to 35% saturation (209 mg/liter). Following a 2-hr equilibration at 4°, the precipitated proteins are recovered by centrif-

[39] M. M. Lippman and M. B. Mathews, *Fed. Proc., Fed. Am. Soc. Exp. Biol.* **36**, 55 (1977).
[40] J. R. Guyton, R. Rosenberg, A. Clowes, and M. Karnovsky, *Circ. Res.* **46**, 625 (1980).
[41] W. E. Benitz, J. D. Lessler, J. D. Coulson, and M. Bernfield, *J. Cell. Physiol.* **127**, 1 (1986).
[42] C. F. Reilly, L. Fritze, and R. D. Rosenberg, *J. Cell. Physiol.* **129**, 11 (1986).
[43] T. A. McCaffrey, D. J. Falcone, C. F. Brayton, L. A. Agarwal, F. G. P. Welt, and B. B. Weksler, *J. Cell Biol.* **109**, 441 (1989).
[44] C.-H. Heldin, A. Johnsson, B. Ek, S. Wennergren, L. Rönnstrand, A. Hammacher, B. Faulders, Å. Wasteson, and B. Westermark, this series, Vol. 147, p. 3.

ugation at 2075 g for 15 min and resuspended in about 100 ml of 10 mM Tris-HCl, pH 7.4, with 150 mM NaCl. This fraction is then extensively dialyzed against the same buffer, after which an equal volume of 2 M ammonium sulfate in 10 mM Tris-HCl, pH 7.4, is added.

The sample is centrifuged at 2075 g for 15 min and the supernatant layered on a 20-ml column of octyl-Sepharose (Pharmacia-LKB) preequilibrated with 1 M ammonium sulfate in 10 mM Tris-HCl, pH 7.4. After washing the column with the same buffer, L-TGF-β1-containing material is eluted with 10 mM Tris-HCl, pH 7.4. The eluate is dialyzed against 10 mM phosphate buffer, pH 6.8, containing 10 μM CaCl$_2$ and injected on an HPLC hydroxyapatite column (7.8 × 100 mm, Bio-Rad, Richmond, CA), equipped with a guard column (4.0 × 50 mm, Bio-Rad). The hydroxyapatite column is equilibrated in 10 mM phosphate, pH 6.8, containing 10 μM CaCl$_2$, at a flow rate of 0.5 ml/min. Fractions rich in L-TGF-β1, which elutes at an elution volume of around 20 ml, are pooled and concentrated to 100 μl with a Centricon microconcentrator (Amicon Corp., Danvers, MA). The concentrate is injected onto a Superose 6 column (HR10/30, Pharmacia-LKB), equilibrated with 500 mM NaCl, 10 mM Tris-HCl, pH 7.4, and eluted with the same buffer at a flow rate of 0.5 ml/min. L-TGF-β1 elutes between thyroglobulin and ferritin at an elution volume around 14 ml; fractions containing L-TGF-β1 are pooled, mixed with an equal volume of 2.8 M ammonium sulfate (HPLC-grade, Bio-Rad), 100 mM phosphate, pH 6.8, and injected onto an alkyl-Sepharose HR 5/5 column (Pharmacia-LKB) equilibrated with 1.4 M ammonium sulfate, 100 mM phosphate, pH 6.8. The column is eluted with a gradient of 1.4–0 M ammonium sulfate in 100 mM phosphate, pH 6.8, at a flow rate of 0.5 ml/min. L-TGF-β1 elutes from the last column between 16 and 19 ml elution volume. As performed by Miyazono et al.,[32] this protocol gave a 2000-fold purification. Starting with 800–1000 liters of human blood the initial nonadsorbed fraction contained 300 g of platelet protein at the beginning, and after the final step 40 μg protein remained. In [^{3}H]thymidine incorporation assays the latter material gave maximal inhibition at 5 ng/ml.

Comments. Analysis by chromatography and electrophoretic methods[32] and Western blotting[31] revealed that L-TGF-β1 from human platelets is composed of an N-terminal precursor remnant present as a disulfide-bonded dimer bound with disulfide bonds to an approximately 125 kDa binding protein, with the activatable TGF-β1 dimer (25 kDa) being noncovalently attached to the precursor remnant. Only the N-terminal precursor dimer remnant appears to be essential for conferring latency.[45] The latter

---

[45] L. E. Gentry, N. R. Webb, G. J. Lim, A. M. Brunner, J. E. Ranchalis, D. R. Twardzik, M. N. Lioubin, H. Marquardt, and A. F. Purchio, *Mol. Cell. Biol.* **7,** 3418 (1987).

has been shown to contain mannose 6-phosphate residues,[46] which no doubt explains why L-TGF-β1 can bind to the cation-independent mannose 6-phosphate/insulin-like growth factor II (IGF-II) receptor.[47] Platelet L-TGF-β1 appears to be structurally similar, if not identical, to the latent form secreted by other cells.[31] However, one should note that this latent form is different from that in serum, which is a complex of 25-kDa TGF-β1 associated with α_2-m.[24]

Final Remarks

In the event that the other isoforms of TGF-β (i.e., β2, β3, and β4) exist as latent forms, they could well be different, not only from each other, but also from L-TGF-β1, given the dissimilarities in their respective precursor sequences.[7] It is possible that, for certain bioassays, activation of L-TGF-β1 occurs *in situ*, that is, would not require prior, exogenous activation. Indeed, the usual bioassays like soft agar colony formation or inhibition of [^{3}H]thymidine incorporation are clearly artificial laboratory expedients. It is equally clear that, *in vivo*, the organism must have acquired one or more ways to activate the latent form, since the latter does not bind to TGF-β receptors on the surface membrane,[21] though it can bind to the mannose 6-phosphate/IGF-II receptor[47] (the physiological significance of the latter binding is not yet known, and most of these receptors are intracellular). The possible activation of L-TGF-β1 in acidic cellular microenvironments (tumoral masses, osteoclasts), perhaps in cooperation with other activating mechanisms, has been proposed.[48] It could be that activating treatments of purified L-TGF-β1 do not fully reflect those occurring *in vivo*, where the complex is present with many other factors which might modulate natural, activating processes.

Acknowledgments

I thank the INSERM for financial support (Contract 884015), Drs. Thérèse Heyman and Philippe Vigier for helpful comments on the manuscript, and Françoise Arnouilh for typographical assistance.

---

[46] A. F. Purchio, J. A. Cooper, A. M. Brunner, M. N. Lioubin, L. E. Gentry, K. S. Kovacina, R. A. Roth, and H. Marquardt, *J. Biol. Chem.* **263**, 14211 (1988).

[47] K. S. Kovacina, G. Steele-Perkins, A. F. Purchio, M. N. Lioubin, K. Miyazono, C. H. Heldin, and R. A. Roth, *Biochem. Biophys. Res. Commun.* **160**, 393 (1989).

[48] P. Jullien, T. M. Berg, and D. A. Lawrence, *Int. J. Cancer* **43**, 886 (1989).

[32] Assay of Astrocyte Differentiation-Inducing Activity of Serum and Transforming Growth Factor β

By YOSHIO SAKAI and DAVID BARNES

Introduction

Mouse embryo cells derived in serum-supplemented media undergo a growth crisis accompanied by gross chromosomal aberration. We have derived lines of mouse embryo cells in serum-free media (serum-free mouse embryo cells, SFME). These cells do not exhibit growth crisis or chromosomal aberration detected by Giemsa banding, are dependent on epidermal growth factor (EGF) for survival, and are reversibly growth inhibited by serum.[1-4] Although originally established from trypsinized whole mouse embryos, cells with properties of SFME cultures can be established from embryonic, neonatal, or adult brain, and SFME cells exposed for 48 hr to 10% (v/v) calf serum (CS) or 10 ng/ml transforming growth factor β (TGF-β) express glial fibrillary acidic protein (GFAP), an astrocyte marker.[5] Experiments indicate that the major portion of the activity of serum on SFME cells in this regard is due to TGF-β present in the serum. The appearance of GFAP is reversible and not associated with inhibition of cell proliferation. In this chapter we detail procedures for the isolation and propagation of serum-free mouse embryo cells and describe a quantitative immunoassay for GFAP expression in SFME cells.

Serum-Free Cell Culture

SFME cells are available from the American Type Culture Collection (Rockville, MD). Alternatively, cultures can be initiated from whole bodies or heads of 16-day embryos. Pregnant mice are sacrificed by CO_2 asphyxiation or cervical dislocation. The embryos are surgically removed under sterile conditions. At this point, heads can be removed from bodies. The heads are washed with 10 ml of phosphate-buffered saline (PBS) without calcium or magnesium, minced, and incubated at 37° with trypsin

---

[1] D. T. Loo, J. I. Fuquay, C. L. Rawson, and D. W. Barnes, *Science* **236,** 200 (1987).
[2] D. T. Loo, C. L. Rawson, T. Ernst, S. Shirahata, and D. Barnes, in "Cell Growth and Cell Division: A Practical Approach" (R. Baserga, ed.), p. 17. IRL, Oxford, 1989.
[3] D. Loo, C. Rawson, A. Helmrich, and D. Barnes, *J. Cell. Physiol.* **139,** 484 (1989).
[4] Y. Sakai, C. Rawson, K. Lindburg, and D. Barnes, *Proc. Natl. Acad. Sci. U.S.A.* **87,** 8378 (1990).
[5] E. Lazarides, *Annu. Rev. Biochem.* **51,** 219 (1982).

solution: 0.25% crude trypsin with 1 mM ethylenediaminetetraacetic acid (EDTA) in PBS without calcium or magnesium. The incubation is terminated when the material is easily pipetted. It is not necessary to trypsinize until the entire suspension is single cells. The cells are centrifuged from the trypsin solution, resuspended in 10 ml of serum-free medium containing 1 mg/ml soybean trypsin inhibitor, and then recentrifuged and resuspended in fresh serum-free medium for counting.

The basal nutrient medium used for serum-free culture is a one to one mixture of Dulbecco's modified Eagle's medium (DMEM) containing 4.5 g/liter glucose and Ham's F12 (DMEM : F12)[6,7] supplemented with 15 mM 4-(2-hydroxyethyl)-1-piperazineethanesulfonic acid (HEPES), pH 7.4, 1.2 g/liter sodium bicarbonate, penicillin (200 U/ml), streptomycin (200 μg/ml), and ampicillin (25 μg/ml). Cells are cultured in a 5% CO_2–95% (v/v) air atmosphere at 37°. Powdered formulations are available commercially. Liquid medium may be stored for a maximum period of 3 weeks at $-20°$ after preparation from the powder. Antibiotics and HEPES are stored frozen as 100× concentrated stocks. Culture dishes are pretreated with fibronectin, and the basal nutrient medium is further supplemented with growth-stimulatory factors. For the preparation of concentrated stocks and the medium, water must be purified by triple glass distillation or passage through a Milli-Q (Millipore, Bedford, MA) water purification system immediately prior to use. Other details of serum-free cell culture have been published.[2,3]

Culture the cells in the basal serum-free medium supplemented with bovine insulin (10 μg/ml), human transferrin (25 μg/ml), human high density lipoprotein (HDL) (20 μg/ml), mouse EGF (50–100 ng/ml), and sodium selenite (10 nM) on dishes precoated with human fibronectin (20 μg/ml). Insulin, transferrin, EGF, and HDL are added directly to medium in individual plates or flasks as small aliquots from concentrated stocks immediately after plating cells. Insulin is stored at 1 mg/ml in 20 mM HCl and transferrin as a 5 mg/ml stock in PBS or culture medium (filter-sterilized). Insulin, transferrin, and peptide growth factors may be obtained commercially from a number of sources. Methods for the preparation of HDL (density 1.068–1.21 g/cm^3) and fibronectin have been published.[2]

To passage cultures, cells are exposed for a few seconds to the trypsin solution at room temperature, followed by dilution into an equal volume of soybean trypsin inhibitor solution (1 mg/ml in serum-free DMEM : F12) and centrifuged from suspension. Initial cultures are a mixture of cell

---

[6] J. P. Mather and G. H. Sato, *Exp. Cell Res.* **124**, 215 (1979).
[7] R. G. Ham and W. L. McKeehan, this series, Vol. 58, p. 44.

types. Homogeneous cultures with the properties of SFME are evident by about the third passage.

For the culture of serum-free adult mouse brain cells, 6-week-old BALB/c mice are sacrificed by cervical dislocation and brains dissected immediately under sterile conditions and dissociated by a modification of the procedure of Brunner et al.[8] Brains are washed twice in 10 ml sterile basal nutrient medium, then transferred to a sterile 10 cm diameter tissue culture dish and cut into pieces of 1–3 mm. Trypsin is added (10 ml of 0.5 mg/ml crude porcine trypsin in medium) and incubated at 37° for 10 min. Tissue fragments are resuspended by pipetting, and 3 ml of a RNase/DNase solution (1 mg/ml each in PBS) is added and incubated at 37° for 5 min.

Five milliliters of a collagenase/hyaluronidase solution (4 mg/ml each in PBS) is added and incubated at 37° for 30 min, resuspending the tissue every 5 min. This is followed by the addition of 3 ml RNase/DNase, 5 ml trypsin inhibitor (1 mg/ml in medium), and an additional 3 ml of RNase/DNase with pipetting between each addition. The tissue digest is filtered through a 30 μm pore nylon filter into a conical polypropylene tissue culture tube and centrifuged at 600 g for 5 min at room temperature. The cell pellet is washed once with medium, and cells are plated in tissue culture flasks and cultured as described for SFME cells, with additional supplementation of 10 ng/ml basic fibroblast growth factor (FGF) and 1 μg/ml heparin.

Immunoassay of Glial Fibrillary Acidic Protein in Culture

SFME cells are plated at 5×10^5 cells per 35 mm diameter dish. Maximal expression of GFAP occurs about 48 hr after addition of CS or TGF-β. At the appropriate time, cells are fixed by aspirating the medium and slowly adding 1 ml of 70% (v/v) ethanol followed by incubation for 5 min at room temperature, followed by an identical treatment with 95% ethanol. The ethanol is then aspirated, and plates are air-dried. Fixed cells are incubated with a solution containing 15 mM sodium phosphate, pH 7.1, 1% (v/v) Triton X-100, 0.6 M KCl, 10 mM MgCl$_2$, 2 mM EDTA, and 1 mM EGTA for 5 min at room temperature and then washed with 1 ml PBS for 5 min at room temperature. One milliliter of PBS with 10% CS is added and incubated for 1 hr at room temperature. The 10% CS solution is aspirated, and 1 ml of commercially available anti-GFAP antiserum (1 : 300 in PBS with 10% CS) is added and incubated at room temperature for 1 hr followed by two washes with 2 ml PBS, each 5 min at room temperature.

[8] G. Brunner, K. Lang, R. Wolfe, D. McClure, and G. Sato, *Dev. Brain Res.* **2**, 563 (1982).

Affinity-purified goat anti-rabbit IgG conjugated with horseradish peroxidase (1 ml of a 1:1000 dilution in PBS with 10% CS) is added and incubated for 1 hr at room temperature. This is aspirated, and plates are washed twice, 5 min each, in 2 ml PBS at room temperature. One milliliter of 2 mM 2,2'-azinodi(3-ethylbenzothiazoline sulfonate) in 0.1 M sodium acetate, 50 mM sodium phosphate, and 2.5 mM H_2O_2 is added, and the plates are incubated for 30 min in the dark at room temperature. The absorbance at 405 nm is determined for each sample. The cell number per dish for each condition is determined by trypsinizing replicate dishes cultured identically to those used for the GFAP assay and counting in a Coulter particle counter. Variation in this assay in individual plates of GFAP content per cell, expressed as $(A_{405})/10^6$ cells, is less than 10% from the average of duplicate dishes.

Plates of SFME treated with serum or TGF-β and assayed with anti-GFAP antiserum give values of 1.0 to 2.0 absorbance units per plate. Plates not receiving TGF-β or CS assayed with anti-GFAP give values of about 0.5 absorbance units. Controls with normal rabbit serum in place of rabbit anti-GFAP give values of 0.3 to 0.6 absorbance units per plate, for both untreated and TGF-β- or CS-treated plates.

Acknowledgments

Research was supported by NIH-NAI-07560 and Council for Tobacco Research 1813. D. Barnes is the recipient of Research Career Development Award, NIH-NCI-01226.

[33] Erythroid Differentiation Bioassays for Activin

By RALPH H. SCHWALL and CORA LAI

Introduction

Activin belongs to a family of proteins that affect cellular differentiation in a variety of systems.[1] The members of this family, which includes inhibin and transforming growth factor β, are all cysteine-rich, disulfide-linked dimers.[2] In addition, the amino acid sequence of each member is extremely highly conserved across species, suggesting that each has a fundamental role in development. Of the members of this family, activin is most closely related to inhibin, and in many systems the

[1] S.-Y. Ying, *Endocrine Rev.* **9**, 267 (1988).
[2] J. Massague, *Cell (Cambridge, Mass.)* **49**, 437 (1987).

two proteins have opposite biological activities, for example, activin stimulates, whereas inhibin suppresses, the secretion of follicle-stimulating hormone (FSH).[1] Interestingly, activin and inhibin are also structurally related. Inhibin is a 32K dimer of α and β subunits, and activin is a 24K dimer of inhibin β subunits. Two distinct β subunits have been identified and are referred to as β_A and β_B. These can combine with α to generate inhibin A and inhibin B, or with each other to form activin A, activin AB, and activin B. Both forms of inhibin suppress FSH secretion with similar potencies, and, likewise, the different activins also have comparable potencies.

Although activin was originally discovered because of its ability to stimulate the secretion of FSH in cultured pituitary cells,[1] there are now several pieces of evidence showing that activin also stimulates erythroid differentiation. The first such evidence was the quite unexpected finding that erythroid differentiation factor (EDF), an activity that is produced by THP-1 cells and stimulates the accumulation of hemoglobin in mouse erythroleukemia cells, has the same N-terminal sequence as activin A.[3] Subsequent studies revealed that activin A from ovarian follicular fluid would stimulate erythroid differentiation in the K562 human erythroleukemia cell line.[4] In primary bone marrow cultures, activin A has been found to have no effect alone but to enhance erythropoietin-induced colony formation.[4,5] Lastly, β_A mRNA is expressed in significant amounts in bone marrow,[6] suggesting that activin A may be produced locally in this site.

We have recently been able to produce activin A by recombinant expression[7] and have begun to study its role in erythropoiesis. As part of this effort, we have adapted several assays for erythroid differentiation to assess activin bioactivity. A major advantage provided by these types of assays is that they employ continuous cell lines. Activin bioactivity can also be assessed in cultured pituitary cells; however, since there are currently no stable cell lines that secrete FSH, each pituitary bioassay requires that pituitaries be collected and collagenase-dispersed. The details of the erythroid differentiation assays are described below.

[3] Y. Eto, T. Tsuji, M. Takezawa, S. Takano, Y. Yokogawa, and H. Shibai, *Biochem. Biophys. Res. Commun.* **142**, 1095 (1987).
[4] J. Yu, L. Shao, V. Lemas, A. L. Yu, J. Vaughan, J. Rivier, and W. Vale, *Nature (London)* **330**, 765 (1987).
[5] H. E. Broxmeyer, L. Lu, S. Cooper, R. H. Schwall, A. J. Mason, and K. Nikolics, *Proc. Natl. Acad. Sci. U.S.A.* **85**, 9052 (1988).
[6] H. Meunier, C. Rivier, R. M. Evans, and W. Vale, *Proc. Natl. Acad. Sci. U.S.A.* **85**, 247 (1988).
[7] R. H. Schwall, K. Nikolics, E. Szonyi, C. Gorman, and A. J. Mason, *Mol. Endocrinol.* **2**, 1237 (1988).

General Considerations

We have used primarily the K562 human erythroleukemia cell line, which was derived from the pleural effusion of a patient with chronic myelogenous leukemia in terminal blast crisis.[8] Cells from this lineage behave like multipotential stem cells, capable of being induced along erythroid, myeloid, and megakaryocyte lineages.[9] That activin induces erythroid differentiation in this line was first demonstrated by Yu et al.[4] The cells are easy to handle because they grow in suspension. There are, however, several important points that should be kept in mind. First, erythroid differentiation in K562 cells can be induced by at least 19 agents, all of which are small molecules and most of which are cytotoxic.[10] Thus, one must be careful to include appropriate controls to ensure that the erythroid-differentiating activity of an activin preparation is not due to a small molecular weight component of the buffer. Second, marked differences have been observed between K562 cells obtained from different laboratories,[11] suggesting that the cells can undergo drift after repeated passaging. Cells that we use were obtained from ATCC (Rockville, MD), from which a bank was prepared. After being thawed, cells are used for no longer than 3 months. Third, factors in fetal calf serum can induce erythroid differentiation,[12] so batches of serum must be screened to avoid this potential problem.

A variety of cellular markers can be assayed as an index of erythroid differentiation in K562 cells.[13] The end point most easily measured is hemoglobin, and the easiest method for measuring hemoglobin is to take advantage of its psuedoperoxidase activity. Thus, in the presence of hydrogen peroxide and an appropriate chromogenic substrate, such as benzidine, hemoglobin will catalyze the generation of a colored product. The reaction can be carried out in whole cells and the cells subsequently scored for the presence or absence of hemoglobin, or the reaction can be

---

[8] C. B. Lozzio and B. B. Lozzio, *Blood* **45**, 321 (1975).

[9] J. A. Sutherland, A. R. Turner, P. Mannoni, L. E. McGann, and J.-M. Ture, *J. Biol. Response Modif.* **5**, 250 (1986).

[10] P. T. Rowley, B. M. Ohlsson-Wilhelm, B. A. Farley, and L. LaBella, *Exp. Hematol.* **9**, 32 (1981).

[11] I. W. Dimery, D. D. Ross, J. R. Testa, S. K. Gupta, R. L. Felsted, A. Pollak, and N. R. Bachur, *Exp. Hematol.* **11**, 601 (1983).

[12] S. Sakata, Y. Enoki, S. Tomita, H. Kohzuki, and A. Nakatani, *Exp. Hematol.* **13**, 745 (1985).

[13] J. L. Villeval, P. G. Pelicci, A. Tabillo, M. Titeux, A. Henri, F. Houesche, P. Thomopoulos, W. Vainchenker, M. Garbaz, H. Rochant, J. Breton-Gorius, P. A. W. Edwards, and U. Testa, *Exp. Cell Res.* **146**, 428 (1983).

performed using cellular lysates and the amount of hemoglobin quantitated in a spectrophotometer. Both methods are described below.

Cell Culture

K562 cells are cultured in RPMI 1640 containing 10% heat-inactivated fetal calf serum along with penicillin and streptomycin sulfate. Cells are passaged twice each week. To induce erythroid differentiation, growth medium containing activin (0–100 ng/ml) is added to a 24-well plate, and then each well is seeded with 25,000 cells. The plates are incubated at 37° in a water-saturated environment containing 5% CO_2. Significant differentiation can be observed after 4 days. As a positive control, cells are induced in the presence of 25 μM hemin,[10] which is prepared as a 25 mM stock solution in 1 M NH_4OH and stored in aliquots at $-20°$. Cells should be washed with phosphate-buffered saline (PBS) prior to analysis, especially when cultured with hemin because hemin also has psuedoperoxidase activity.

Identification of Hemoglobin-Containing Cells

One method to assess the extent of erythroid differentiation following induction by activin is to count the proportion of cells that stain positively with benzidine.[10] The reagent consists of 2 mg/ml benzidine dihydrochloride in 0.5% acetic acid to which is added, just before use, hydrogen peroxide (5 μl of a 30% stock solution for each milliliter of reagent). Equal volumes of reagent and cell suspension are mixed in a microtiter plate. After 5 min at room temperature, cells are scored as being either benzidine-positive (blue) or benzidine-negative (yellow). Benzidine is a potent carcinogen and should be handled with caution. As an alternative, the less dangerous compounds 3,3'-dimethoxybenzidine (also known as o-dianisidine) and 3,3',5,5'-tetramethylbenzidine can be used. Staining with tetramethylbenzidine yields the same colors as benzidine. With dimethoxybenzidine, positive cells stain brown to dark yellow.

When stained in suspension, cells decompose within 10–15 min, so they must be scored within a few minutes after initiation of the reaction. As an alternative, cells are placed on microscope slides either by cytocentrifugation or by pipetting a small aliquot of cell suspension onto the center of a slide and then wicking away the buffer with the edge of an absorbent paper towel. In either case, the cells are air-dried and fixed in methanol for 5 min. The slides are then reacted for 1.5 min in a solution of 1% 3,3'-dimethoxybenzidine in methanol, followed by 1.5 min in 1% hydrogen peroxide, 50% methanol.[14] The reaction is stopped by washing for 1 min in deionized water, and the slides are air-dried and mounted. Counterstaining with a Giemsa-like stain (Diff-Quik) may be desirable but may mask the positive cells. This method is more cumbersome than staining in suspension, but the slides can be saved for examination at a later time. However, the alcohol–benzidine method is also less sensitive than the acid–benzidine method, and it is not useful in cells that have been induced for only a short time.

We find that in a sample of uninduced K562 cells, 3–5% are benzidine-positive. This value increases to 30–40% after induction with activin for 4 days. The reason that not all cells become benzidine-positive is unclear, but this type of incomplete response is characteristic of the K562 cell line.[10] It is not due to an accumulation of nonresponsive cells because similar results were obtained in 22 different subclones of the parental line (R. Schwall, unpublished observation).

Measurement of Hemoglobin Accumulation

The methods described above allow the detection of hemoglobin-containing cells but provide no information on the amounts of hemoglobin that are expressed. In initial studies we measured hemoglobin in cell lysates using o-phenylenediamine as substrate.[15] Although hemoglobin accumulation can be measured this way, something in the lysates of K562 cells precipitates during the reaction, leading to opacity in the microtiter wells. We have subsequently adapted a procedure described by Clarke et al.[16] in which such precipitation does not occur.

From each well of the 24-well plate in which differentiation is induced, 0.7 ml of well-mixed cell suspension is transferred to a 1.5-ml microcentrifuge tube and washed 2 times with PBS. After the final wash, as much buffer as possible is aspirated using a 200-μl yellow micropipette tip placed over the end of a $5\frac{1}{2}$ inch Pasteur pipette. The pellet is resuspended in 50 μl distilled water, then subjected to one freeze–thaw cycle. Debris is removed by centrifugation, and 40 μl of the supernatant is transferred to a 96-well microtiter plate. While the samples are thawing and centrifuging, 100 mg of 3,3′,5,5′-tetramethylbenzidine is dissolved in 10 ml of 90% acetic acid. A 1% solution of hydrogen peroxide is freshly prepared by diluting the stock 30% solution with distilled water. After the lysates are added to

---

[14] H. Beug, S. Palmieri, C. Freudenstein, H. Zentgraf, and T. Graf, *Cell* (*Cambridge, Mass.*) **28**, 907 (1982).

[15] J. A. Schmidt, J. Marshall, M. J. Hayman, P. Ponka, and H. Beug, *Cell* (*Cambridge, Mass.*) **46**, 41 (1986).

[16] B. J. Clarke, A. M. Brickenden, R. A. Ives, and D. H. K. Chui, *Blood* **60**, 346 (1982).

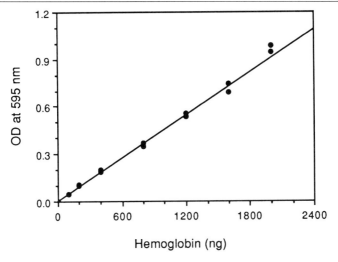

FIG. 1. Standard curve for hemoglobin assay.

the microtiter plate, equal volumes of tetramethylbenzidine and peroxide are mixed, and 160 µl is rapidly added to each well, using a multichannel pipettor. The reaction is allowed to proceed at room temperature for 20 min in the dark. Optical density is then measured at 595 nm in a microtiter plate reader. It should be noted that there appears to be an error in the report by Clarke et al.,[16] who recommend measuring absorbance at 515 nm. This may be the absorbance maximum for benzidine, but we have found that when tetramethylbenzidine is used as substrate, the product has a broad absorbance peak at 590–610 nm.

Standards are prepared using purified hemoglobin diluted in distilled water. A representative standard curve is shown in Fig. 1. The microtiter plate reader we use (SLT Labinstruments, Hillsborough, NC, Model EAR 400 AT) has a built in data-reduction function so that data from the samples can be collected directly in terms of hemoglobin content. Because the standard curve is linear, it is possible to correct for dilutions incurred during preparation of the samples by applying appropriate corrections to the values entered for the hemoglobin standards. In this way, output from the plate reader is expressed directly as micrograms hemoglobin per culture well.

A separate aliquot of the cell suspension is counted in a Coulter counter, and the hemoglobin data can then be corrected for cell number. In our hands, activin typically causes a 5- to 10-fold increase in hemoglobin accumulation, while simultaneously causing a small suppression of cell growth (Table I).

TABLE I
INDUCTION OF HEMOGLOBIN BY ACTIVIN A AND HEMIN[a]

Inducer	Concentration	Hemoglobin (μg/well)	Number of cells (millions/well)	Hemoglobin (μg/million cells)
None	—	0.14	0.63	0.22
Activin A	0.1 ng/ml	0.13	0.59	0.22
	1	0.27	0.63	0.43
	10	0.70	0.53	1.32
	100	0.94	0.48	1.96
Hemin	25 μM	3.04	0.50	6.08

[a] Data are the mean of triplicates which varied by less than 5%.

The most consistent results are obtained if the cells are washed and lysed in 1.5-ml microcentrifuge tubes. However, 4-ml conical centrifuge tubes can also be used. We recommend polystyrene over polypropylene because it is easier to see the pellet while aspirating the final wash. When using the larger tubes, the volume of water used for lysis is increased to 100 μl. The amount of sample transferred to the microtiter plates is also increased to 100 μl, and the amount of reagent is decreased to 100 μl. The reason for these changes is that, in the larger tubes, small amounts of wash buffer remain on the sides of the tube even though the pellet appears dry. When the lysate is then centrifuged, this buffer is forced to the bottom of the tube, causing a small dilution of the lysate. Although it is difficult to quantitate the extent of this dilution, it can be kept consistent so replicates have low sample-to-sample variability.

Another variation on the above method is the use of 0.5-ml conical centrifuge tubes (Sarstedt, Princeton, NJ, No. 73.1055) that can be fitted into a 96-well array in styrofoam racks (Sarstedt, No. 95.1046). The racks fit neatly into centrifuge adaptors designed for 96-well plates. Using this system, larger numbers of samples can be analyzed because all of the procedures can be carried out using multichannel pipettors. The final wash, however, must still be aspirated well by well in order to minimize variability between samples.

[34] Labeling Inhibin and Identifying Inhibin Binding to Cell Surface Receptors

By TERESA K. WOODRUFF, JANE BATTAGLIA, JAMES BORREE, GLENN C. RICE, and JENNIE P. MATHER

Introduction

Few hormones have provided more intense excitement and utter despair to the scientific community than inhibin and activin. In a 1985 review, Channing wondered, "Is there really a biochemical entity with inhibin activity, or is it possible the assay systems are responding to some beguiling artifacts?"[1] An inhibin activity was originally postulated to exist in 1923. With identification hampered by bioassay and purification problems, it was not until 1977 that deJong and Schwartz independently identified inhibin in follicular fluid. A history of the purification of inhibin and activin has been recently reviewed by Vale *et al.*[2]

Inhibin and activin are hormones characterized by their antagonistic action in a plethora of systems. The activities identified to date include regulation of follicle-stimulating hormone (FSH) secretion, erythroid differentiation, oxytocin secretion, corticosterone production, growth hormone biosynthesis, follicular growth, and germ cell differentiation (Table I[3-15]) The problems which investigators have faced in working with inhibin

[1] C. Channing, W. Gordon, W. Liu, and D. Ward, *Proc. Soc. Exp. Biol. Med.* **178**, 339 (1985).
[2] W. Vale, C. Rivier, A. Hsueh, C. Campen, and H. Meunier, *Recent Prog. Horm. Res.* **44**, 1 (1988).
[3] T. Woodruff and K. Mayo, *Annu. Rev. Physiol.* **52**, 807 (1990).
[4] H. Bronxmeyer, L. Lu, S. Cooper, R. Schwall, A. Mason, and K. Nikolics, *Proc. Natl. Acad. Sci. U.S.A.* **85**, 9052 (1988).
[5] P. Sawchenko, P. M. Plotsky, S. W. Pfeiffer, E. T. Cunningham, Jr., J. Vaughan, J. Rivier, and W. Vale, *Nature (London)* **334**, 615 (1988).
[6] N. Billestrup, C. Gonzalez-Manchon, and W. Vale, *Mol. Endocrinol.* **4**, 356 (1990).
[7] J. Mather, K. Attie, T. Woodruff, G. Rice, and D. Phillips, *Endocrinology* in press (1991).
[8] J. D'Agostino, T. K. Woodruff, K. E. Mayo, and N. B. Schwartz, *Endocrinology* **124**, 310 (1989).
[9] M. Hedger, *et al., Mol. Cell. Endocrinol.* **61**, 133 (1989).
[10] D. Robertson and D.deKretser, *Mol. Cell. Endocrinol.* **62**, 307 (1989).
[11] P. LaPolt, *et al., Mol. Endocrinol.* **3**, 1666 (1989).
[12] A. Hsueh, K. Dahl, J. Vaughan, E. Tucker, and J. Rivier, *Proc. Natl. Acad. Sci. U.S.A.* **84**, 5082 (1987).
[13] F. Petraglia, P. Sawchenko, A. Lim, J. Rivier, and W. Vale, *Science* **237**, 187 (1987).
[14] T. Mine, I. Kojima, and E. Ogata, *Endocrinology* **125**, 586 (1989).
[15] Y. Totsuka, M. Tabuchi, I. Kojima, H. Shibai, and E. Ogata, *Biochem. Biophys. Res. Commun.* **156**, 335 (1988).

TABLE I
REPORTED BIOLOGICAL ACTIVITIES FOR INHIBIN
AND ACTIVIN[a]

Activity	Ref.
FSH secretion	2, 3
Hematopoiesis	4
Oxytocin secretion	5
Growth hormone biosynthesis	6
Germ cell proliferation	7
Folliculogenesis	8
Thymocyte and 3T3 growth	9
Oocyte meiotic inhibitor	10
Inhibin mRNA production	11
Steroidogenesis	12
Placental hormone regulation	13
Glucose production	14
Insulin secretion	15

[a] Inhibins: α-β_A or α-β_B; activins: β_A-β_A, β_A-β_B, or β_B-β_B; free monomers: α, β_A, or β_B.

and activin are numerous. The inhibin family includes five hormones based on the dimerization of the three subunits: α (18 kDa), β_A (14 kDa), and β_B (14 kDa). The shared properties of the various hormones severely limit the ability of current radioimmunoassays to discriminate between specific species of inhibins, activins, free monomers, and precursor forms. Moreover, the hormones are usually secreted as a mixture of the dimer, precursor forms, or true aggregates. In addition, fluids obtained from different species or different sources within the same species contain different molecular moieties. Lastly, the inhibin subunits are evolutionarily conserved and share sequence similarity with other growth and differentiation factors including transforming growth factor β (TGF-β), Müllerian inhibiting substance (MIS), and decapentaplegic (DPP).

The difficulties encountered in purifying and radiolabeling inhibin and activin have slowed efforts to identify their receptors. It is possible that more than one receptor exists, each with different affinities for the various subunits and dimers. As of this writing, several laboratories have succeeded in describing properties of an activin A receptor,[15-17] but no report exists concerning inhibin binding.

The purpose of this chapter is to describe several approaches to the

[16] C. Campen and W. Vale, *Biochem. Biophys. Res. Commun.* **157**, 844 (1988).
[17] H. Sugino, T. Nakamura, Y. Hasegawa, K. Miyamoto, M. Igarashi, Y. Eto, H. Shibai, and K. Titani, *J. Biol. Chem.* **263**, 15249 (1988).

labeling of inhibin and activin and the measuring of specific receptor binding. For the sake of brevity and clarity, only data relating to inhibin are described. Activin labeling is similar, and the binding characteristics have been described elsewhere.[15-18] With the hormones having multiple actions on different cell types, we must pose the question of whether multiple receptors exist which utilize different mechanisms of action. The methodologies described below will prove to be invaluable in understanding the complex manner in which the inhibin family of hormones exerts such a wide range of effects on nearly every major organ system (Table I).

Metabolic Labeling of Human Recombinant Inhibin

Because of the high cysteine content (18 per dimer) inhibin can be metabolically labeled with [^{35}S]cysteine. The inhibin α and β_A subunits are stably transfected into Chinese hamster ovary (CHO) cells using the calcium phosphate method. Two cell lines are established: one which produces inhibin A (α-β_A) and activin A (β_A-β_A) and one which produces activin A exclusively.[19]

After 2 to 3 days of cell growth the medium is replaced with 1 ml Dulbecco's modified Eagle's medium (DMEM)/Ham's F12 medium (1:1 v/v) minus cysteine and methionine. Then 100 μCi [^{35}S]cysteine and 100 μCi [^{35}S]methionine (Amersham, Arlington Heights, IL) are added. Cells are labeled for 4 hr at 37° in 5% CO_2. At the end of the labeling period, free [^{35}S]cysteine and [^{35}S]methionine are removed by size-exclusion chromatography or dialysis (final volume of material, 250 μl). Because proteins other than inhibin and inhibin aggregates are also labeled, it is difficult to determine the specific activity of the mature form of inhibin without further purification.

Binding of Inhibin to TM3 Cells

As a preliminary indication of binding, a Leydig cell-derived cell line, TM3, is used to bind the labeled conditioned medium. TM3 cells are grown in DMEM/F12 (1:1) medium supplemented with 10 μg/ml insulin, 10 μg/ml human transferrin (hTF), and 10 ng/ml epidermal growth factor (EGF).[20] Binding is determined using three densities of cells. Eight wells

[18] S. Kondo, M. Hashimoto, Y. Etoh, M. Murata, H. Shibai, and M. Muramatsu, *Biochem. Biophys. Res. Commun.* **161**, 1267 (1989).
[19] R. Schwall, C. Schmelzer, E. Matsuyama, and A. Mason, *Endocrinology* **125**, 1420 (1989).
[20] J. Mather, *Biol. Reprod.* **25**, 243 (1980).

(24-well plate) are treated with labeled conditioned medium in DMEM/F12 medium as described below. The material which is bound is then stripped from its receptor using ammonium acetate (pH 4.0) and analyzed by polyacrylamide gel electrophoresis (PAGE).

Binding of Labeled Hormone to TM3 Cells

1. Wash cells 3 times with serum-free DMEM/F12 (1 : 1).
2. Strip surface-bound agents by treating cells with 1 ml ammonium acetate (pH 4.0) for 10 min; repeat this twice.
3. Wash cells with serum-free medium until the medium color returns to normal (pH 7.4) (i.e., yellow to pink).
4. Add labeled conditioned medium [1×10^5 counts/min (cpm), ~10 µl of labeled medium in 250 µl fresh DMEM/F12].
5. Incubate cells plus labeled medium for various times (range 15 min to 6 hr) at 37°.
6. Wash cells 3 times with serum-free medium.
7. Scrape cells with rubber policeman and collect in scintillation vial.
8. Count bound ligand–receptor complexes in 10 ml scintillant (Betamax, Amersham).

Acid Release of Bound Ligand

1. Follow binding procedure as described above through Step 6.
2. Add 1 ml ammonium acetate (pH 4.0) to the cell monolayer and incubate for 10 min at 37°.
3. Transfer the supernatant to Eppendorf microcentrifuge tubes.
4. Centrifuge at 3500 rpm for 15 min.
5. Freeze the supernatant in an alcohol/dry ice bath for 10 min.
6. Lyophilize the material and analyze the products on a 12.5% Phastgel (Pharmacia, Piscataway, NJ).
7. Dry gel and expose to film.

The results from the binding and binding-release experiments are shown in Fig. 1A,B. Approximately 3% of the input material is bound by the TM3 cell inhibin receptors. Upon acid release, a doublet of 18 kDa (α) and 14 kDa (β_A), is resolved by PAGE, suggesting that inhibin is specifically bound to these cells. Note that only externally receptor-bound material is released by the acid treatment.[21]

[21] J. Mather, J. Saez, and F. Haour, *Endocrinology* **110**, 933 (1982).

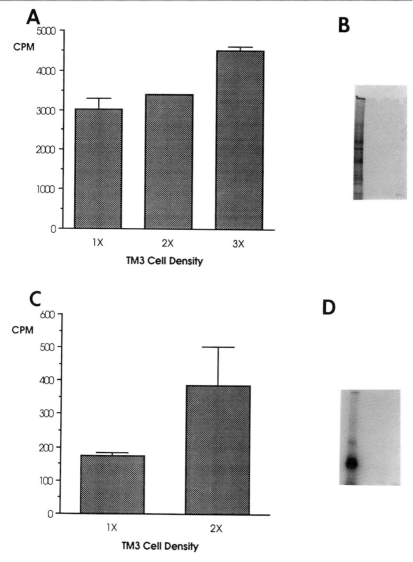

FIG. 1. Binding and acid release of metabolically labeled or iodinated inhibin. (A) Binding of ^{35}S-labeled inhibin to three densities of TM3 cells (n = 2 per condition; 1× cell number equals 1700 cells). (B) Sodium dodecyl sulfate–PAGE analysis of material bound to TM3 cells. (C) Binding of iodinated inhibin to two densities of TM3 cells. (D) PAGE analysis of bound material following acid release. Lane designations are as for (B) except that ^{125}I labeling was used.

Purification of Labeled Inhibin Using Antiinhibin/Activin Antibody and Affi-prep 10 Column Chromatography

The degree of purity of the metabolically labeled ligand will depend on the expression level of the recombinant cell line used. Because the metabolically labeled material is impure, a simple immunoaffinity column purification of the labeled material can be performed to obtain more highly purified ligand (Fig. 2). A polyclonal antibody raised against the N-terminal 32 amino acids of the α chain of the inhibin subunit is coupled via free alkyl or aryl amino groups to Affi-prep 10 (Bio-Rad, Richmond, CA). The medium is then run through a column containing the prepared matrix.

Purification of Labeled Inhibin

1. Slurry a vial (50 ml) of Affi-prep 10 (Bio-Rad) to obtain a uniform suspension.
2. Transfer 5 ml to a Büchner funnel.
3. Wash the matrix with 50 volumes of 10 mM sodium acetate, pH 4.5 (keep moist and wash rapidly, i.e., within 15 min).
4. Transfer the washed gel to the reaction vessel.
5. Dialyze the inhibin antibody against 20 mM HEPES and 150 mM NaCl (pH 7.0).
6. Add the ligand solution to the gel (at least 0.5 ml ligand per 1 ml gel).
5. Rock the ligand plus gel for 1 hr at room temperature.
6. Spin down and decant unbound antibody solution.
7. Block the remaining active esters with 0.1 ml/ml gel of 1 M ethanolamine hydrochloride (pH 8), for 1 hr, at ambient temperature.
8. Wash the gel with 2 volumes of 0.5 M NaCl.
9. Transfer the gel to a column and wash with phosphate-buffered saline (PBS, pH 7.2) until the effluent gives a stable OD_{280} reading. (Approximately 92% of the Affi-prep 10 column capacity is bound using this procedure.)
10. Equilibrate the column with sample buffer: 20 mM HEPES, 150 mM NaCl, pH 7.0.
11. Equilibrate the conditioned medium with sample buffer.
12. Load the conditioned medium onto the antibody column.
13. Wash the column with sample buffer.
14. Elute the bound inhibin with 100 mM acetic acid, pH 3.0, then 100 mM acetic acid, pH 2.0, and finally with 0.02 N HCl, pH 3.0. Inhibin and its precursor forms elute in the final step.

After this single-step purification, an approximate specific activity of

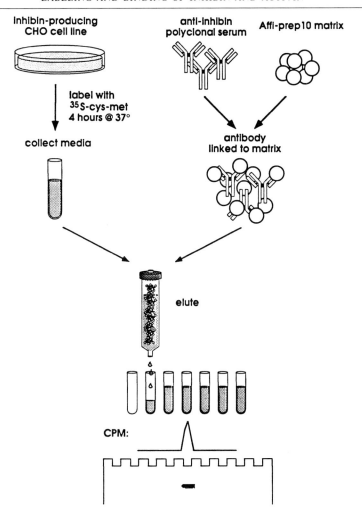

FIG. 2. Schematic diagram of the purification technique for isolating inhibin from conditioned media.

150 cpm/fmol is obtained, representing the activity of inhibin and its three precursor forms. Binding studies using the purified inhibin were performed, and the results were similar to those shown in Fig. 1A,B. Only the mature 32-kDa inhibin is thought to bind to surface receptors. However, we cannot exclude the possibility that the precursor forms are also involved in binding.

Iodination of Recombinant Inhibin

Specific binding and displacement of an ^{125}I-labeled protein is a typical method of assaying binding. However, iodinated inhibin does not bind well to cells. This could be because the iodine blocks specific tyrosines or neighboring sites involved in binding, or it may be due to oxidative damage to the amino acid side chains. We have found that one way to iodinate inhibin which results in limited binding is to use Iodogen (Pierce Laboratories, Rockford, IL). This procedure tends to be less oxidative than other procedures (i.e., chloramine-T). Using this method, the specific activity of the radiolabeled ligand is 1.96×10^5 Ci/mol.

Iodination of Inhibin

1. Prepare Iodogen vials according to the manufacturer's specifications.
2. Add 10 μg of inhibin directly to rinsed microfuge tubes (100 μl).
3. Add 0.5 μCi of Na^{125}I (Amersham).
4. Vortex the sample and place on ice for 15 min.
5. Pipette the iodinated inhibin into a microfuge tube.
6. Rinse the Iodogen tube 1 time with double-distilled water (50 μl).
7. Add 1 ml cold acetone to iodinated inhibin.
8. Precipitate 1 hr on ice.
9. Centrifuge at 12,000 rpm at 4° for 30 min, rinse the precipitate with 500 μl cold acetone, repeat, resuspend in 100 μl double-distilled water.

The binding assay is carried out as described above. Results of binding and acid release of the bound material are shown in Fig. 1C,D. The products (18 kDa and 14 kDa) are identical to those obtained with the ^{35}S-labeled material; however, less than 1% of the input probe is bound to the cells.

Fluorescein Isothiocyanate Conjugation of Inhibin and Analysis of Binding Using Fluorescence-Activated Cell Sorting

The final method we use for detecting inhibin receptors is fluorescence-activated cell sorting (FACS) analysis of fluorescein isothiocyanate (FITC)-conjugated ligands (Fig. 3). FACS analysis of receptor binding provides several advantages. Since measurements are made at a single cell level, populational heterogeneities in binding can be assessed and simultaneously correlated to other end points by multiparameter analysis (i.e., DNA content, lipid content, cell-surface markers). Further, viable cells can be rapidly sorted on the basis of receptor number.[22]

---

[22] R. Chatelier, R. Shcroft, C. Lloyd, E. Nice, R. Whitehead, W. Sawyer, and A. Burgess, *EMBO J.* **5**, 1181 (1986).

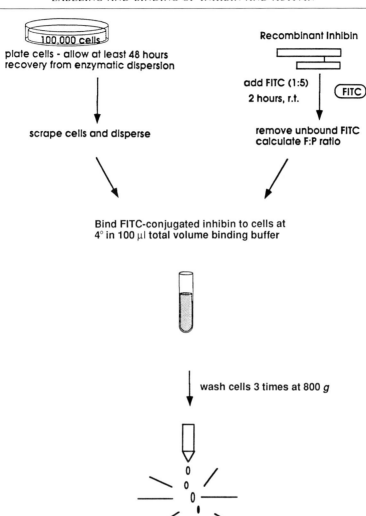

FIG. 3. Flow diagram for FITC labeling of inhibin, inhibin binding to cells, and FACS analysis.

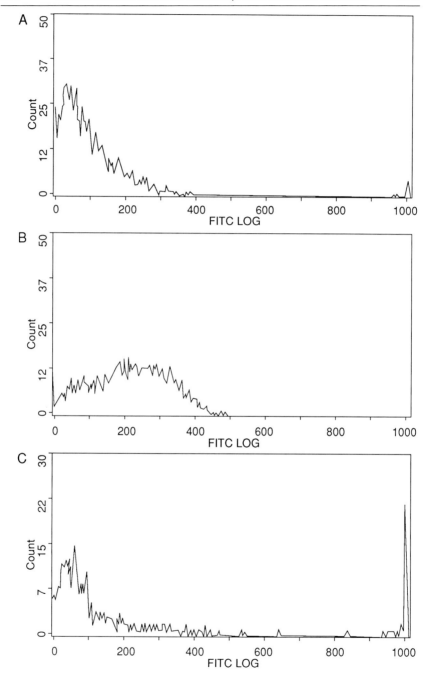

FITC Conjugation of Inhibin

1. Adjust 10 μg of inhibin to pH 8.0 with 1 M sodium carbonate, pH 9.3.
2. Add FITC (Sigma, St. Louis, MO) [1:5 (w/w) protein to FITC] slowly to the solution and incubate while rotating at room temperature for 1 hr (total volume, 100 μl).
3. Separate the FITC-conjugated inhibin from free FITC using size-exclusion chromatography. Measure the final protein concentration and fluorescence with a spectrophotometer (280 nm) and fluorometer (excitation 488 nm, emission 525 nm).
4. Incubate 1×10^5 TM3 cells (in 100 μl) at 4° for 30 min with 100 ng FITC-conjugated inhibin in binding medium [DMEM/F12 containing 0.1% bovine serum albumin (BSA)] in a total volume of 100 μl.
5. Wash the cells plus bound ligand twice with PBS–BSA.
6. Measure the cellular fluorescence using a Coulter Elite flow cytometer. Use an argon ion laser with an excitation wavelength of 488 nm and measure emission using 525 nm (±25) band pass filter.
7. Calculate the number of receptors per cell from a linear conversion of a log fluorescence calibration standard curve using beads of known equivalent soluble fluorescent molecules. Other considerations include the average free-to-bound ratio of the conjugated material and subtraction of nonspecific binding in the presence of 1000-fold excess unconjugated ligand.

The average number of FITC-bound molecules can be directly determined using the following calculation: [(number FITC molecules/cell) − (background number FITC molecules/cell)]([F/p] ratio of conjugated inhibin or activin)$^{-1}$. An example of a binding analysis using TM3 cells is shown in Fig. 4. Figure 4A shows the autofluorescence of the cells, Fig. 4B shows binding of 100 ng FITC-conjugated inhibin, and Fig. 4C shows the competition of FITC-conjugated inhibin with a 1000× excess unlabeled inhibin.

Comparative Analysis of Labeling and Binding Techniques Described

We have described three methods for labeling inhibin: metabolic labeling, iodination, and fluorescein conjugation. The first two methods resulted in poor binding of the ligand to its surface receptor. The labeling of inhibin

---

FIG. 4. FACS analysis of TM3 cells bound with FITC-conjugated inhibin. (A) Autofluorescence of TM3 cells. (B) FITC-conjugated inhibin bound to TM3 cells. (C) Competition of FITC-conjugated inhibin with a 1000-fold excess of unconjugated inhibin.

with FITC resulted in good binding and has proved to be useful to analyze the binding of labeled inhibin to cells with a fluorescence-activated cell sorter. This method should prove useful in studying the characteristics of the inhibin receptors and ultimately provide a route to identify and isolate the receptor(s).

[35] Bioassay, Purification, Cloning, and Expression of Müllerian Inhibiting Substance

By DAVID T. MACLAUGHLIN, JAMES EPSTEIN, and PATRICIA K. DONAHOE

Introduction

Müllerian Inhibiting Substance (MIS) is a glycoprotein homodimer of disulfide-linked 70K–74K subunits with a total molecular weight of approximately 140,000.[1-6] This protein was originally described as a product of the Sertoli cells of fetal testis,[7,8] where it is responsible for the active regression in the male embryo of the Müllerian duct, the anlagen of the uterus, cervix, upper third of the vagina, Fallopian tubes, and coelomic epithelial lining of the ovary. Failure to express full activity of MIS at the critical time in embryogenesis, just following differentiation of the gonad into a testis, results in retained Müllerian structures after birth. There is no known fetal ovarian substance causing Wolffian duct regression in female embryos; rather, the lack of sufficient testosterone in females is responsible for the passive involution of the precursor to the male reproductive tract. MIS action is inhibited by estrogens and epidermal

---

[1] D. A. Swann, P. K. Donahoe, Y. Ito, Y. Morikawa, and W. H. Hendren, *Dev. Biol.* **69**, 73 (1979).
[2] N. Josso, J. Y. Picard, and D. Tran, *Recent Prog. Hormone Res.* **33**, 117 (1977).
[3] G. P. Budzik, D. A. Swann, A. Hayashi, and P. K. Donahoe, *Cell (Cambridge, Mass.)* **21**, 909 (1980).
[4] G. P. Budzik, S. M. Powell, S. Kamagata, and P. K. Donahoe, *Cell (Cambridge, Mass.)* **34**, 307 (1983).
[5] P. K. Donahoe, G. P. Budzik, S. Kamagata, P. Hudson, and M. Mudgett-Hunter, *Hybridoma* **3**, 201 (1984).
[6] J. Y. Picard and N. Josso, *Mol. Cell. Endocrinol.* **34**, 23 (1984).
[7] N. Josso, *Endocrinology* **93**, 829 (1973).
[8] M. G. Blanchard and N. Josso, *Pediat. Res.* **8**, 968 (1974).

growth factor.[9,10] Androgens and progestins, on the other hand, augment MIS activity in the fetal urogenital ridge.[11]

Until recently, MIS was believed to be produced only in testes and essentially only during fetal and early neonatal life. New biological immunohistochemical, immunoassay, and molecular biological studies[12-16] revealing the expression of MIS in the granulosa cells of the ovary, however, indicate that the production of MIS is not male specific as originally believed. MIS is not produced by the ovary until after birth, with a variable onset according to the species under study. MIS production is highest in human males in the perinatal period (50–70 ng/ml), then falls after 10 years of age to basal levels (2–5 ng/ml) that are maintained into adulthood. Preadolescent females begin to produce measurable serum MIS at basal levels similar to those of adult males. MIS falls below detectable levels with menopause.[17] The discovery of MIS as an ovarian product prompted a search for extra Müllerian functions for this hormone. MIS inhibits oocyte meiosis in the 25- to 27-day-old rat,[18,19] and most recently it has been shown to inhibit fetal and neonatal rat lung development as measured by decreased disaturated phosphatidylcholine production, *in vivo* and *in vitro*.[20]

Studies on the molecular mechanism of MIS action have led to the hypothesis that MIS is a specific inhibitor of epidermal growth factor receptor autophosphorylation.[9,21,22] Furthermore, until the discovery of

[9] J. M. Hutson, M. E. Fallat, S. Kamagata, P. K. Donahoe, and G. P. Budzik, *Science* **223**, 586 (1984).
[10] J. M. Hutson, H. Ikawa, and P. K. Donahoe, *J. Pediatr. Surg.* **17**, 953 (1982).
[11] H. Ikawa, J. M. Hutson, G. P. Budzik, D. T. MacLaughlin, and P. K. Donahoe, *J. Pediatr. Surg.* **17**, 453 (1982).
[12] M. Takahashi, M. Hayashi. T. F. Manganaro, and P. K. Donahoe, *Biol. Reprod.* **35**, 447 (1986).
[13] S. Ueno, M. Takahashi, T. F. Manganaro, R. C. Ragin, and P. K. Donahoe, *Endocrinology* **124**, 1000 (1989).
[14] S. Ueno, T. Kuroda, D. T. MacLaughlin, R. C. Ragin, T. F. Manganaro, and P. K. Donahoe, *Endocrinology* **125**, 1060 (1989).
[15] B. Vigier, J. Y. Picard, D. Tran, L. Legai, and N. Josso, *Endocrinology* **114**, 1315 (1984).
[16] R. Voutilainen and W. L. Miller, *J. Mol. Endocrinol.* **1**, 604 (1987).
[17] P. L. Hudson, I. Douglas, P. K. Donahoe, R. L. Cate, J. Epstein, R. Pepinsky, and D. T. MacLaughlin *J. Clin. Endocrinol. Metab.* **70**, 16 (1990).
[18] M. Takahashi, S. S. Koide, and P. K. Donahoe, *Mol. Cell. Endocrinol.* **47**, 225 (1986).
[19] S. Ueno, T. F. Manganaro, and P. K. Donahoe, *Endocrinology* **123**, 1652 (1988).
[20] E. A. Catlin, T. F. Manganaro, and P. K. Donahoe, *Am. J. Obstet. Gynecol.* **159**, 1299 (1988).
[21] J. P. Coughlin, P. K. Donahoe, G. P. Budzik, and D. T. MacLaughlin, *Mol. Cell. Endocrinol.* **49**, 75 (1987).
[22] F. Cigarroa, J. P. Coughlin, P. K. Donahoe, M. R. White, N. Uitvlugt, and D. T. MacLaughlin, *Growth Factors* **1** (2), 179 (1989).

pp63, a hepatic protein that specifically inhibits insulin receptor autophosphorylation,[23] MIS was the only known naturally occurring inhibitor of growth factor autophosphorylation. The sequence homology of the MIS gene and protein with several other known growth factors/inhibitors has incorporated MIS into a large supergene family which includes the transforming growth factors β, activin, inhibin, Vgl, the decapentaplegic complex of *Drosophila*, and the erythropoetic and bone morphogenesis factors.[24] Because this chapter is concerned largely with methodological considerations of MIS research, the reader is referred to several review articles that discuss MIS biology in greater detail.[25-30]

Bioassay for Müllerian Inhibiting Substance

The assay that originally defined the role of MIS in the normal developing male reproductive tract is the 14-day fetal rat urogenital ridge *in vitro* bioassay first described by Picon[31] and modified subsequently by Donahoe *et al.*[32] This system is the standard for all MIS analyses. The regression of the Müllerian duct *in vitro* is only achieved by MIS incubation; no other naturally occurring substance tested mimics this activity.

The assay consists of incubating test substances with single 14-day fetal rat female urogenital ridges containing a Wolffian and Müllerian duct, dissected from female fetuses (Holtzman Laboratories, Madison, WI). Male ridges are not used to avoid the chance of retaining small amounts

[23] P. Auberger, L. Falquerho, J. O. Contreves, G. Pages, G. LeCam, B. Rossi, and A. LeCam, *Cell (Cambridge, Mass.)* **58,** 631 (1989).
[24] R. L. Cate, R. J. Mattaliano, C. Hession, R. Tizard, N. M. Farber, A. Cheung, E. G. Ninfa, A. Z. Frey, D. J. Gash, E. P. Chow, R. A. Fisher, J. M. Bertonis, G. Torres, B. P. Wallner, K. L. Ramachandran, R. C. Ragin, T. F. Manganaro, D. T. MacLaughlin, and P. K. Donahoe, *Cell (Cambridge, Mass.)* **45,** 685 (1986).
[25] P. K. Donahoe, J. M. Hutson, M. E. Fallat, S. Kamagata, and G. P. Budzik, *Annu. Rev. Physiol.* **46,** 53 (1984).
[26] G. P. Budzik, P. K. Donahoe, and H. M. Hutson, in "Developmental Mechanisms: Normal and Abnormal" (J. W. Lash, ed.), p.207. Alan R. Liss, New York, 1985.
[27] R. L. Cate, E. G. Ninfa, D. J. Pratt, D. T. MacLaughlin, and P. K. Donahoe, *Cold Spring Harbor Symp. Quant. Biol.* **51,** 641 (1986).
[28] P. K. Donahoe, R. Cate, D. T. MacLaughlin, J. Epstein, A. F. Fuller, M. Takahashi, J. P. Coughlin, E. G. Ninfa, and L. A. Taylor, *Recent Prog. Hormone Res.* **43,** 431 (1987).
[29] P. K. Donahoe, M. Takahashi, S. Ueno, and T. F. Manganaro, in "Meiosis Inhibition: Molecular Control of Meiosis" (F. Hazeltine and N. First, eds.), p. 153. Alan R. Liss, New York, 1988.
[30] P. K. Donahoe, G. P. Budzik, R. L. Trelstad, B. R. Schwartz, M. E. Fallat, and J. M. Hutson, in "The Role of Extracellular Matrix in Development" (R. L. Trelstad, ed.), p. 573. Alan R. Liss, New York, 1984.
[31] R. Picon, *Action Arch. Anat. Micros. Morphol. Exp.* **58,** 1 (1969).
[32] P. K. Donahoe, Y. Ito, S. Marfatia, and W. H. Hendren, *J. Surg. Res.* **23,** 141 (1977).

of contaminating testes. The ridges are plated on 2% (w/v) agar-coated stainless steel grids inverted to reduce the required medium to 0.7 ml in an organ culture dish (Falcon #3037). Incubation medium, to which MIS or other test substances is added, consists of CMRL medium, 10% female fetal calf serum (to avoid endogenous bovine MIS) plus 1 nM testosterone (to permit contrast of Müllerian duct to the Wolffian duct), 100 U penicillin, and 100 μg streptomycin. All additions are made at a 1:1 dilution or less in the final volume of 0.7 ml, and incubations are carried out for 72 hr in a highly humidified atmosphere of 95% air and 5% CO_2 at 37°.

Samples for routine screening are fixed after the 3-day incubation in phosphate-buffered saline, pH 7.4, containing 15% formalin, dehydrated with graded ethanol washes (95, 100, 100%, 2 hr each), treated with the xylene substitute HISTOCLEAR (National Diagnostics, Manville, NJ), and embedded in paraffin. Serial 8-μm cross sections from the cranial end of the ducts are then stained with hematoxylin and eosin. Multiple sections of each ridge are examined, and the degree of regression is graded from 0 (no regression) to 5 (complete regression). In all cases positive controls with 14-day rat fetal testis and negative controls with maternal rat muscle are examined.

The grading system, as derived from recent *in vitro* observations using human recombinant MIS and slightly modified from the original description,[32] is briefly described below. Grade 0: The Müllerian duct appears as it would in age-matched female embryos of 16 to 17 days. Grade 1: Clearing and/or condensation of the indifferent mesenchymal cells occurs around the basement membrane of the Müllerian ducts. Grade 2: The ducts become irregular in shape and smaller in size; basement membrane is usually lost. Grade 3: The epithelial cells of the ducts become pycnotic, there is no development of the mesenchyme, basement membrane may persist, and the ducts continue to reduce in size. Grade 4: The ducts have been partially replaced by epithelial cells and connective tissue and the duct lumen is lost. Grade 5: Complete regression, with the Müllerian duct being undetectable.

It should be noted that if the incubation is allowed to persist to 6 days in the absence of MIS, the Müllerian duct enlarges to fill the entire ridge, replacing the Wolffian structure completely. MIS must be present for 24 to 36 hr to obtain complete regression of the Müllerian duct as observed at 3 days of incubation. Twelve-hour incubations do no affect the Müllerian duct, whereas 12–24 hr of exposure results in partial regression at 72 hr.[32] Dose–response studies using purified intact human recombinant MIS indicate that maximal activity requires 1–2 μg/ml of the protein. The bioactivity of MIS in this assay has been shown to be inhibited by vanadate at 50 μM, by zinc at concentrations exceeding 0.4 mM, by epidermal growth factor (EGF) at 1 μg/ml, and by a variety of nucleotides including

ATP, GTP, and AMP at millimolar concentrations. EDTA (0.5 mM) or fluoride (1.0 mM) were the only agents tested that exhibited MIS-like activity when added alone to the incubation medium.[9,25,33] These observations led us to speculate that the mechanism of action of MIS might be the result of a regulation of intracellular phosphorylation events, particularly those stimulated by EGF.[9,21,22,25]

Organ culture of the fetal Müllerian ducts can be readily adapted to perform a variety of studies. Autoradiographic, immunohistochemical, and ultrastructural techniques have been used in our laboratory to characterize morphologic changes accompanying duct regression as well as changes in extracellular matrix components using antibodies to fibronectin, collagen type IV, laminin, and heparin sulfate (for review, see Ref. 30). More recently, Josso's laboratory employed fetal or neonatal ovary as the target tissue in organ culture[32] and measured MIS inhibition of aromatase activity and stimulation of seminiferous tubular development over a 3-day period.[34]

Purification

The purification of biologically active Müllerian Inhibiting Substance (MIS) was first achieved using neonatal bovine testis as the starting material. Human recombinant MIS, on the other hand, is purified from culture medium harvested from Chinese hamster ovary (CHO) cells transfected with a human genomic clone (see below). Regardless of the source of MIS, it can be purified by one of two major techniques employed currently in our laboratory. The end products vary in their absolute degree of purity, however. The two approaches, described below, are used since when taken together they provide useful insight into the manner in which MIS is processed to full biological activity.

Serial Chromatography

MIS is purified from the majority of other proteins by precipitation followed by a series of anion and dye-affinity column separations. It is also possible to include a cation-exchange step or a carbohydrate-affinity step since MIS is 13–18% carbohydrate by weight. The original protocol of Budzik *et al.* to purify bovine MIS employed all of the above steps.[3,26] Currently, human recombinant MIS is purified from conditioned CHO cell medium (see below) as follows: 4 liters of medium containing 5% female

---

[33] G. P. Budzik, J. M. Hutson, H. Ikawa, and P. K. Donahoe, *Endocrinology* **110**, 1521 (1982).

[34] B. Vigier, M. G. Forest, B. Eychenne, J. Bezard, O. Garrigori, P. Robel, and N. Josso, *Proc. Natl. Acad. Sci. U.S.A.* **86**, 3684 (1988).

fetal calf serum (i.e., free of bovine MIS) is concentrated 20-fold on a 30-kDa ultrafilter (Amicon, Danvers, MA) and dialyzed against 50 mM NaCl, 10 mM sodium phosphate, 1 mM EDTA, 0.03% sodium azide, pH 8, prior to loading onto a 2.5 × 50 cm DEAE-BioGel A column equilibrated with the same buffer. MIS is eluted in the unbound fraction.

After dialysis to equilibrate the MIS with a buffer containing 10 mM HEPES, 0.01% Nonidet P-40 (NP-40), 0.15 M NaCl, 5 mM 2-mercaptoethanol, pH 8, the sample is loaded onto a 5-ml Matrex Green gel A (Amicon) column. Stepwise elution from this column is achieved with 2 column volumes of loading buffer (Green fraction 1) then with 2 column volumes of this buffer containing 10 mM AMP (Green fraction 2), followed by an elution with 5–10 ml of loading buffer plus 0.35 M NaCl (Green fraction 3). The Green fraction 3, referred to as DG3, is dialyzed in the cold against 20 mM HEPES, 0.001% NP-40, pH 7.4, before use. MIS bioactivity and immunoreactivity as measured in a specific ELISA for MIS[17] are now recovered only in DG3.

The preparations usually contain 15–20% MIS, representing about a 70-fold purification. The starting material contains approximately 1.75 mg/ml total protein, of which MIS is 5 μg/ml. Complete purity would be obtained by 350-fold purification. Using this scheme about 5% of the total MIS is recovered. Although recovery is low using this method, the MIS obtained is active, and activity can be absorbed by polyclonal MIS antibody in all systems responding to this hormone, including Müllerian duct regression,[32] inhibition of oocyte meiosis,[18,19] and inhibition of EGF receptor autophosphorylation in A431 cells[21,22] and fetal lung membranes (E. A. Catlin, N. Uitvlugt, D. M. Powell, P. K. Donahoe, and D. T. MacLaughlin, unpublished observations).

Immunoaffinity Purification

An alternate strategy to the method discussed above is immunoaffinity chromatography as a single-step purification. The antibody employed by our laboratory for this column is a monoclonal termed 6E11 that was raised against a pure preparation of MIS that was itself made homogeneous by an immunoaffinity column step using a different monoclonal (M.10).[35] A description of the use of the monoclonal 6E11 in an ELISA for human MIS and the test of its specificity has been published.[17] Fifty milligrams of the MIS monoclonal 6E11 is partially purified by protein A-Sepharose (Bio-Rad, Richmond, CA) chromatography prior to coupling to column of Affi-Gel 10 (agarose resin, Bio-Rad) according to the instructions provided by the manufacturer. The affinity column is equilibrated with 20 mM

---

[35] J. W. Wallen, R. L. Cate, D. M. Kiefer, M. W. Riemen, D. Martinez, R. M. Hoffman, P. K. Donahoe, D. D. Von Hoff, B. Pepinsky, and A. Oliff, *Cancer Res.* **49**, 2005 (1989).

HEPES, pH 7.4, and MIS-containing CHO conditioned medium, filtered through Whatman #4 filter paper to remove debris and concentrated 20-fold on a Minitan (Millipore, Bedford, MA), system with a 30-kDa exclusion ultrafilter, is loaded at a rate of 1 column volume (5 ml) per hour at 4°. The column is then washed with 100 ml of equilibration buffer and the MIS eluted in 0.5-ml aliquots with 2 ml of either 2 M ammonium or sodium thiocyanate or with 20 mM HEPES, pH 3.0, containing 1 M acetic acid. The resulting MIS fractions are then desalted over a Sephadex G-25 column equilibrated with 20 mM HEPES, 0.15 M NaCl, 10% glucose, pH 7.5, in the former case, or pH-neutralized using sodium or ammonium hydroxide in the latter case. Ammonium hydroxide is used when samples are to be lyophilized, owing to the volatility of the ammonium acetate produced during neutralization. The immunoaffinity purification of MIS results in a 315- to 343-fold purification (i.e., to 90–98% purity), with recoveries ranging from 2.5 to 15% when total protein and ELISA measurements[17] are compared.

We have recently shown that immunoaffinity-purified MIS undergoes heat-activated proteolytic processing, resulting in fragments of 57, 34, 28, and 12.5 kDa on reducing polyacrylamide gel electrophoresis (R. C. Ragin, D. T. MacLaughlin, and P. K. Donahoe, unpublished observation). MIS heated to 37° for up to 8 days contains little remaining intact 70-kDa monomer but retains full biological activity in the organ culture assay. Preliminary sequence analyses show that the 57- and 34-kDA fragments begin at the amino terminus of the MIS monomer and the smallest fragment begins at a consensus monobasic cleavage site at residue 427. Sequence results on the 28-kDa fragment are pending. These results indicate that MIS must be cleaved in at least two positions for these fragments of the hormone to be produced. At this time it is not certain whether MIS itself possesses this cleavage activity or whether a protease copurifies with the hormone. Experiments are underway to address this issue; however, amino-terminal analysis of freshly purified hormone detects only the MIS primary sequence.

It is of interest that, unlike the MIS purified by ion-exchange and dye-affinity chromatography, the fresh preparations are marginally active in single cell and membrane assays but retain activity *in vivo* and in the urogenital ridge bioassays. This result may be explained in part by our recent finding that MIS is processed by proteolytic cleavages. Furthermore, MIS can be processed in a manner similar to TGF-β by exogenous plasmin.[36] Unlike TGF-β, however, MIS appears to require both the

---

[36] R. B. Pepinsky, L. K. Sinclair, E. P. Chow, R. J. Mattaliano, T. F. Manganaro, P. K. Donahoe, and R. L. Cate, *J. Biol. Chem.* **263**, 18961 (1988).

amino- (57 kDa) and carboxy-terminal (12.5 kDa) fragments produced in this manner for full biological activity. The relatively harsh immunoaffinity column elution protocols may denature the approximately 10% of total MIS that is cleaved prior to secretion from the CHO cells, thus rendering it inactive in test systems unable to process the remainder of the MIS by proteolytic cleavage. The absence of mercaptoethanol as a reducing agent and NP-40 as a detergent may also permit irreversible aggregation, thus impeding postpurification processing noted at elevated temperatures.

Cloning and Expression of Human Müllerian Inhibiting Substance Gene

The human MIS gene was isolated and sequenced in collaboration with Cate and colleagues in 1986.[27] Sequences from tryptic peptides of bovine MIS were used to design oligonucleotide probes which were then used to clone the bovine MIS cDNA; the deduced amino acid sequence showed the tryptic peptides to be all in frame. Thereafter a probe from this bovine cDNA preparation was used to screen a human genomic library, to isolate and clone the human genomic sequence, and to deduce the sequence of the human MIS protein.

Bovine MIS purified according to the ion-exchange, carbohydrate- and dye-affinity chromatography protocol[3,26] is further purified by gel electrophoresis run under disulfide bond-reducing conditions, and the protein is electroeluted according to the method of Hunkapiller *et al.*[37] Peptide fragments generated by TPCK-tryptic digestion of each of the carboxymethylated monomers are separated by reversed-phase high-pressure liquid chromatography over a C_{18} (or C_4 in the case of succinylated fragments, see below) column. In all, 14 conserved fragments from each of the subunits were sequenced. In order to obtain the data it is necessary to use three different trypsin digestion protocols. The first involves treatment of each MIS monomer in absolute ethanol at $-20°$ for 16 hr followed by recovery by centrifugation at 6000 g for 30 min and dissolution in 6 M guanidine hydrochloride, 0.1 M Bicine, and 0.1 mM EDTA (pH 8.2), then dialysis against 0.1 M NaHCO$_3$ and 0.1 mM CaCl$_2$. Digestion is accomplished with 6% (w/w) N-tosyl-L-phenylalanine chloromethyl ketone (TPCK)-trypsin according to the protocol of Pepinsky *et al.*[38] Alternatively, pooled and lyophilized 70- and 74-kDa MIS monomers are either treated as above but succinylated prior to digestion according to the

---

[37] M. G. V. Hunkapiller and L. E. Hood, this series, Vol. 91, p. 486.
[38] R. B. Pepinsky, L. K. Sinclair, J. L. Browning, R. J. Mattaliano, J. E. Smart, E. P. Chow, T. Falbel, A. Riboloni, J. Garwin, and B. P. Wallner, *J. Biol. Chem.* **261**, 4239 (1986).

method of Glazer et al.[39] or washed in ethanol as above and dissolved in 2 M urea, 0.1 M NH$_4$NCO$_3$, and 0.1 mM CaCl$_2$ before digestion. Any single protocol is productive of only a few peptide fragments. The sequence data are obtained using 100–500 pmol of isolated peptide fragments by automated Edman degradation as described.[38]

Oligonucleotide cDNA probes are synthesized on an Applied Biosystems 380 A DNA synthesizer, purified by gel electrophoresis, and then end-labeled with [γ-^{32}P]ATP using polynucleotide kinase.[40] Most of the pooled probes produced are 17 to 20 nucleotides long with an initial total degeneracy between 256 and 512, based on variations in the base inserted in the third position of each anticodon.

The oligonucleotide probes are used to screen by Northern analysis according to the method of Chirgwin et al.[41] the polyadenylated RNA isolated by oligo(dT)-cellulose chromatography (Pharmacia PL type 7) of the total mRNA extracted from neonatal bovine testis. Northern analyses are carried out on 10 μg of polyadenylated mRNA run on formaldehyde-agarose gels and transferred, after soaking for 20 min in 25 mM sodium phosphate, to nylon filters. RNA is cross-linked to the filters, hybridized at 45°, and washed 4 times in screening buffer and once in 3.2 M tetramethylammonium chloride in 1% sodium dodecyl sulfate hybridized at 45°.[42] After washing twice in the screening buffer, the filters are rewashed with 75 mM NaCl, 7.5 mM sodium citrate in 1% sodium dodecyl sulfate at 65°.

None of the preparations with this degree of degeneracy is specific since each cross-reacts with mRNA from tissue known not to produce MIS as well as to the positive control neonatal testis material. The degeneracy of these probes, therefore, is reduced as follows. One probe, the 17-mer of 256-fold degeneracy, is synthesized in four subpools of 64-fold degeneracy by splitting it at the proline codon, which has a degeneracy of four (CCN, where N is A, G, C, or T). Therefore, one subpool of 64-fold degeneracy is synthesized with an A at the ambiguous position. The other subpools are synthesized with either a G, T, or C at this position. The subpool that detected MIS mRNA is identified by hybridization to Northern blots containing mRNA from 2-week-old bovine testes. One subpool hybridizes to a 2000-nucleotide transcript in the neonatal sample that is not present in RNA of the older tissues. This subpool is split further

[39] A. N. Glazer, R. J. DeLange, and D. S. Sigman, in "Laboratory Techniques in Biochemistry and Molecular Biology" (T. S. Work and E. Work, eds.), Vol. 4, p. 78. Elsevier, New York, 1976.
[40] A. M. Maxam and W. Gilbert, this series, Vol. 65, p. 499.
[41] J. M. Chirgwin, A. E. Przybla, R. J. MacDonald, and W. J. Rutter, *Biochemistry* **18**, 5294 (1976).
[42] G. M. Church and W. Gilbert, *Proc. Natl. Acad. Sci. U.S.A.* **81**, 1991 (1984).

into four subpools and the Northern analysis repeated. This allows the degeneracy to be reduced to 16. The degeneracy of a 20-mer is reduced from 512 to 32 in a similar fashion. The nucleotide transcript to which these probes hybridized is 2 kilobases (kb) in length.

While these specific probes are being produced, cDNA libraries of mRNA extracted from neonatal bovine testes are constructed. Polyadenylated mRNA is prepared as described above, and, using the technique of Maniatis et al.,[43] the first single-stranded DNA is synthesized using avian myeloblastosis virus (AMV) reverse transcriptase. DNA polymerase I-large fragment is used with minor modifications to synthesize the second strand.[44] After blunt-ending the double-stranded DNA with T4 polymerase, linkers are attached and the complexes purified on a BioGel A-50 m column. The cDNA eluting in the void volume is ligated into the EcoRI restriction enzyme site of λgt10, incorporated into phage, plated, and amplified on Escherichia coli BNN102.[45] Lysis and transfer of phage DNA onto nitrocellulose filters using the technique of Benton and Davis[46] provide a substrate for hybridization with the ^{32}P-labeled oligonucleotide probes. The buffer used for this step, carried out at 45° is 50 mM Tris-HCl, pH 7.5, 1 M NaCl, 0.2% poly(vinylpyrrolidone), 0.2% Ficoll 400, 0.2% bovine serum albumin, 0.1% sodium pyrophosphate, and 1% sodium dodecyl sulfate (SDS) with 10% dextran sulfate and 100 μg/ml mRNA. After washing 3 times with this buffer at 45°, the filters are rewashed with 3.2 M tetramethylammonium chloride and 1% SDS at 50°.

Clones yielding positive signals on the initial screening are plaque purified and the DNA isolated and analyzed on Southern blots using the technique of Maxam and Gilbert.[40] The 2000-base pair (bp) insert is subcloned (PS21) so that a complete sequence can be obtained. The results are validated by isolating and sequencing a genomic clone of the bovine MIS gene from a bovine cosmid library.[24] The bovine MIS cDNA clone pS21 was used as a hybridization probe to screen a human cosmid library consisting of DNA isolated from a human lymphoblastoid cell line GM 1416,48XXXX (Human Genetic Mutant Cell Repository, Camden, NJ) according to the method of Maniatis et al.[43] The clone isolated (chMIS33) was then analyzed by restriction mapping and fragments generated by using PstI and PvuII cleavage subcloned in order to sequence the inserts in the parental genomic clone. The subclones are identified using specific

---

[43] T. Maniatis, E. F. Fritsch, and J. Sambrook, "Molecular Cloning: A Laboratory Manual." Cold Spring Harbor Laboratory, Cold Spring Harbor, New York, 1982.
[44] H. Okayama and P. Berg, Mol. Cell. Biol. **2**, 161 (1982).
[45] T. Huynh, R. A. Young, and R. W. Davis, in "DNA Cloning, A Practical Approach" (D. M. Glover, ed.), IRL, Oxford, 1985.
[46] W. D. Benton and R. W. Davis, Science **196**, 180 (1977).

bovine probes and their inserts sequenced. These steps allow for the determination of the complete sequence of the human gene.

The MIS gene is GC-rich, making certain stretches of the sequence difficult to interpret owing to the abnormal mobility on electrophoresis. The tendency to form GC base pairs during separation is eliminated using the cytosine modification method of Ambartsumayan and Mazo.[47] At this stage the sequence of the bovine and human MIS genes is established and the amino sequences for the polypeptide portion of the MIS monomers deduced.

The production of human MIS by animal cells is accomplished by inserting a 4.5-kb *Afl*II fragment from the original cosmid clone, chmis33, into the expression vectors pBG311 and pBG312. The former uses the late SV40 early promoter and the latter the adenovirus 2-major late promoter to regulate expression. Stable transfection into Chinese hamster ovary (CHO) cells lacking the dihydrofolate reductase (DHFR) gene is accomplished by incubating the cells with 20 μg of DNA per 5×10^5 cells in the presence of calcium phosphate.[48] The human genomic MIS plasmid pBG311.hmis and the pAdD26 plasmid[49] are linearized by *Pvu*I treatment and mixed in molar ratios of 10:1 before precipitating with the calcium phosphate. After the transfection the cells are grown in a ribonucleoside- and deoxyribonucleoside-free medium (minimal essential medium, GIBCO, Grand Island, NY, 410-2000) with 10% fetal bovine serum. Colonies are screened for MIS mRNA by S1 analysis.[50] The production of bioassayable MIS is measured using precipitated conditioned medium from the appropriate clones after 2 days of culture.

Stable clones are selected in medium containing gradually increasing concentrations of methotrexate to stimulate gene amplification. At present the best producing cell line (B9) is grown in 30 nM methotrexate to drive MIS gene expression. Rather than continue to grow the B9 cell line in a static monolayer, either on plastic flasks or cell factories, alternate methods of growing the cells while maximizing MIS production were sought. The one selected as optimal is described below.[51]

The cell line B9 is grown on stainless steel mesh (grade 304L) in 4-liter bioreactors (Techne Inc., Princeton, NJ). Each vessel contains 500 g of meshwork with a 35,000 cm^2 surface area, or roughly 6 times the area of a single 10-stage cell factory and 40 times that of a roller bottle, and

---

[47] N. S. Ambartsumayan and A. M. Mazo, *FEBS Lett.* **114**, 265 (1980).
[48] R. J. Kaufman and P. A. Sharpe, *Mol. Cell. Biol.* **2**, 1304 (1982).
[49] S. Subramani, R. Mulligan, and P. Berg, *Mol. Cell. Biol.* **1**, 854 (1981).
[50] J. Brosius, R. L. Cate, and A. P. Perlmutter, *J. Biol. Chem.* **257**, 9205 (1982).
[51] J. Epstein, E. J. DesJardins, P. L. Hudson, and P. K. Donahoe, *In Vitro Cell. Dev. Biol.* **25**(2), 213 (1989).

produces approximately 3–5 μg MIS/ml of medium after 3–4 days of incubation. The medium in this system is α-modified Eagle's medium lacking ribosides and deoxyribosides but supplemented with 1 g of glucose and glutamine per liter in 5% female fetal serum (bovine MIS-free). As mentioned above, all media contain 30 nM methotrexate. The stainless steel meshwork (American Copper Sponge Inc., Woonsocket, RI) is rinsed in double-distilled water and sterilized in water by steam autoclaving at least 3 times. The meshwork is then rinsed in 95% ethanol and handled in a sterile fashion thereafter. Cells to be plated onto the meshwork are harvested from the plastic monolayer culture flasks by trypsinization and suspended in a small volume of medium with serum. The cells are then washed onto the stainless steel surfaces with a pipette and allowed to stand in an incubator, at 37°, 5% CO_2 in air, with a humidified atmosphere for 45–60 min. The stainless steel meshwork is then immersed into the 4-liter bioreactor. After 3–4 days of incubation, with stirring, the conditioned medium is removed for MIS purification and fresh medium added. Such bioreactor systems have produced MIS for over 3 months from a single innoculation with approximately 5×10^6 B9 cells, reaching confluence at a level of $2-3 \times 10^5$ cells/cm^2 of meshwork.

Thus, after purification of bovine MIS, cloning the bovine and human genes, and transfecting and amplifying a genomic construct, human recombinant MIS can be produced and purified for structure–function and mechanism of action analyses and for testing in a variety of biological systems. We have since shown that cleavage is essential for activity.[36] Conditions are being studied to prevent aggregation of this highly hydrophobic molecule and to permit requisite activation steps that will be favorable for clinical application of this fetal inhibitor/biological modifier as a chemotherapeutic agent.

Section VI

Other Growth Factors and Growth Inhibitors

[36] Purification and Characterization of Recombinant Melanoma Growth Stimulating Activity

By H. Greg Thomas, Jin Hee Han, Eddy Balentien, Rik Derynck, Rodolfo Bordoni, and Ann Richmond

Introduction

Melanoma growth stimulatory activity (MGSA) was originally purified from extracts of serum-free culture medium of Hs294T human malignant melanoma cells and a variety of melanoma tumors. MGSA exerts an autocrine-like growth effect for melanoma cells.[1] Only recently was a method described that provided adequate amounts of purified protein for complete characterization of MGSA secreted by the Hs294T melanoma cells.[2] This factor is structurally related to various polypeptides in the β-thromboglobulin superfamily which are 8–10 kDa basic heparin-binding proteins.[2] Subsequent isolation and characterization of cDNAs revealed a predicted length of 73 amino acids proteolytically derived from a 107 residue precursor. In addition, the nucleotide sequence of the MGSA gene possesses similarity with the human "*gro*" gene and the PDGF-inducible gene *KC*.[3-5] Northern hybridization demonstrates that MGSA/*gro* gene expression is not restricted to melanoma cells but is found in a variety of normal and transformed cells of different origin.[3]

The recently optimized procedure for the purification of MGSA from Hs294T cell culture medium required multiple processes which were laborious and time consuming. Although adequate amounts of protein were obtained for complete characterization and initial biological studies, the total yield of homogeneous material was low for the time and expense required. Therefore, a mammalian MGSA expression system has been developed in order to produce larger quantities of MGSA. The expression system utilizes a plasmid vector construct which places the MGSA gene under the control of the cytomegalovirus promoter in Chinese hamster ovarian (CHO) cells.

---

[1] A. Richmond and H. G. Thomas, *J. Cell. Physiol.* **129**, 375 (1986).
[2] H. G. Thomas and A. Richmond, *Mol. Cell. Endocrinol.* **57**, 69 (1988).
[3] A. Richmond, E. Balentien, H. G. Thomas, G. Flaggs, D. E. Barton, J. Spiess, R. Bordoni, U. Francke, and R. Derynck, *EMBO J.* **7**, 2025 (1988).
[4] A. Anisowicz, L. Bardwell, and R. Sager, *Proc. Natl. Acad. Sci. U.S.A.* **84**, 7188 (1987).
[5] P. Oquendo, J. Alberta, D. Wen, J. L. Graycar, R. Derynck, and C. D. Stiles, *J. Biol. Chem.* **264**, 4133 (1989).

We compare here the purification of MGSA to homogeneity from the Hs294T cells and from stable CHO transfectants which secrete high levels of human MGSA. The proteins obtained from these sources are compared as to size, biological activity, N-terminal amino acids, and recognition by antibodies raised against proteins derived from both cell sources, demonstrating the identical properties and activities of the proteins.

Sources of Growth Factors

Tissue culture materials and media are supplied as previously described.[1] Hs294T cells are obtained from the ATCC (Rockville, MD), and a subline is selected which grows well on serum-free Ham's F10 medium without growth factor supplements. Transfected Chinese hamster ovary cells are selected in complete medium lacking glycine, hypoxanthine, and thymidine and supplemented with 10% dialyzed fetal bovine serum as described.[6]

Assay Methods for Characterizing Melanoma Growth Stimulatory Activity

[^{3}H]Thymidine Incorporation. MGSA bioassays are performed by measuring stimulation of [^{3}H]thymidine incorporation in serum-depleted low density cultures of Hs294T human malignant melanoma cells according to a modification of the procedures of Iio and Sirbasku[7] as previously described.[1] Cells (8000) are plated into 28 × 61 mm glass scintillation vials (Wheaton, Millville, NJ, #22528) in 2 ml of Ham's F10 medium supplemented with 10% fetal bovine serum. Twenty-four hours later, cells are washed with 2 ml of phosphate-buffered saline (PBS: 8 g NaCl, 1.15 g Na_2HPO_4, 0.2 g KCl, 0.2 g KH_2PO_4, 0.1 g $MgCl_2 \cdot 6H_2O$, 0.1 g $CaCl_2$) and placed on serum-free F10 medium containing HEPES (30 mM) and ovalbumin (10 μg/ml). After a 24-hr incubation in serum-free medium the medium is aspirated, and aliquots of growth factor are added in a binding buffer composed of 2 ml of serum-free F10 with HEPES (30 mM) and ovalbumin (10 μg/ml). Eight hours later [^{3}H]thymidine (5 μCi) is added to each vial, and the incubation is continued for an additional 16 hr. The reaction is stopped by the addition of methanol/ethanol (3 : 1). Unincorporated [^{3}H]thymidine is removed by repeated washing with methanol, then 10 ml of Scintiverse II is added to each vial, and the radioactivity incorpo-

---

[6] R. Derynck, E. Balentien, J. H. Han, H. G. Thomas, D. Wen, A. K. Samantha, C. O. Zachariae, P. R. Griffin, R. Brachmann, W. L. Wong, K. Matsushima, and A. Richmond, *Biochemistry* **29**, 10225 (1990).

[7] M. Iio and D. Sirbasku, *Cold Spring Harbor Conf. Cell Proliferation* **9**, 751 (1982).

rated into DNA is counted in a Beckman liquid scintillation counter. Results are compared to controls that receive only binding buffer.

Cell Number Assay. MGSA bioactivity can also be assessed in a cell number assay.[1] Hs294T cells (8000) are seeded into glass scintillation vials as described for the [³H]thymidine assay. After 72 hr the medium is aspirated, and the cells are washed twice with PBS and placed on serum-free F10 containing HEPES (30 mM) and ovalbumin (10 µg/ml). Forty-eight hours later the medium is aspirated, and dilutions of MGSA in a binding buffer composed of serum-free F10 with HEPES (30 mM) and ovalbumin (10 µg/ml) are added. On the third and sixth day after growth factor additions, cells are released by brief treatment with trypsin, and the cell number is determined from an aliquot of suspended cells counted on a hemocytometer.

Denaturing Electrophoresis. Sodium dodecyl sulfate (SDS) gradient gels (linear, 12–18% polyacrylamide) are prepared as described by Laemmli[8] and poured into slabs 0.75 mm thick. Samples of MGSA from various steps of the purification are brought to less than 10 µl or dryness in a Speed Vac (Savant, Hicksville, NY) or a lyophilyzer. Twice-concentrated SDS sample buffer (~20 µl) is added, and the samples are electrophoresed at 35 mA constant current. After electrophoresis the gels are fixed and stained with Coomassie blue R-250 and/or silver (Bio-Rad, Richmond, CA, silver stain kit).

Purification of Natural Melanoma Growth Stimulatory Activity

Step 1: Extraction from Hs294T Conditioned Medium. Natural MGSA was initially purified from serum-free culture medium conditioned by confluent cultures of the Hs294T human melanoma cell line as previously described.[1] The serum-free conditioned medium is collected from P-150 culture flasks using sterile technique, and the protease inhibitor Trasylol (aprotinin) (Sigma, St. Louis, MO) is added (117 kallikrein inhibitory units per liter). The medium is centrifuged at 15,000 g to remove floating cells or cellular debris, then the supernatant is ultracentrifuged (100,000 g) for 30 min to remove subcellular organelles. The resulting supernatant is lyophilized. When 20 liters of conditioned medium has been processed in this manner, the lyophilized powder is extracted with 1 N acetic acid at room temperature, dialyzed exhaustively against 0.17 N acetic acid at 4°, lyophilized, and stored at −80°.

Step 2: BioGel P-30 Chromatography. Acetic acid extracts of approximately 20 liters of lyophilized Hs294T conditioned medium are dissolved

---

[8] U. K. Laemmli, *Nature (London)* **227,** 680 (1972).

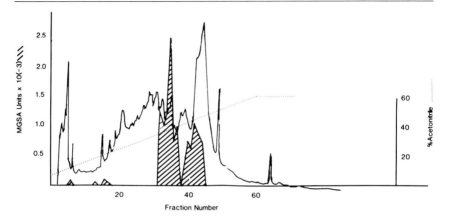

FIG. 1. Reversed-phase HPLC of MGSA on a μBondapak C_{18} column. Fractions eluting from the BioGel P-30 column between the 6K and 14K standards were combined, lyophilized, and dissolved in 2 ml of 6:94:0.05 acetonitrile/water/trifluoroacetic acid. The sample was applied to a μBondapak C_{18} column equilibrated with the same buffer. The gradient consisted of 6–60% acetonitrile for 60 min at 25° with a flow rate of 1.5 ml/min. Sixty fractions were collected. Aliquots of the fractions were used for MGSA bioassays.

in 30 ml of 1 N acetic acid and clarified by ultracentrifugation (100,000 g for 30 min). The resulting supernatant, containing approximately 200 mg of protein, is applied to a 2.5 × 90 cm BioGel P-30 column previously equilibrated with 1 N acetic acid and calibrated with 25K, 14K, and 6K polypeptides. The protein is eluted with 1 N acetic acid at a flow rate of approximately 9 ml/hr at room temperature. Absorbance is monitored at 280 nm, and 80-drop fractions are collected. Greater than 95% of the protein initially loaded onto the column elutes prior to the 25K marker. MGSA bioactivity elutes broadly and resides in two major molecular weight regions (14,000–25,000 and 5000–10,000).[9] The apparent high molecular weight (14,000–25,000) MGSA fractions are combined, and the apparent low molecular weight (5000–10,000) MGSA fractions are combined, with the pools being labeled A and B, respectively.

Step 3: Initial Reversed-Phase Chromatography. Pools A and B from the BioGel P-30 column are lyophilized separately and dissolved in 1–2 ml of acetonitrile/water/trifluoroacetic acid (6:94:0.05). Each sample is injected into a μBondapak C_{18} cartridge column (8 × 100 mm) in a Z module and eluted with a 60-min linear gradient of acetonitrile/water/trifluoroacetic acid (6:94:0.05 to 60:40:0.05 (v/v), pH 2.2) at a flow rate of 1.5 ml/min (Fig. 1). Absorbance is monitored at 206 nm, and 1.5-ml fractions are collected.

[9] A. Richmond, H. G. Thomas, and R. G. B. Roy, this series, Vol. 146, p. 112.

Step 4: Heparin-Sepharose Chromatography. Heparin-Sepharose chromatography has recently been utilized as an alternative to preparative SDS–PAGE for the subsequent purification of MGSA from the initial reversed-phase HPLC fractionation.[2] Bioactive fractions eluting between 31 and 38% acetonitrile are pooled, lyophilized, redissolved in 10 mM Tris, pH 7.5, and applied to a 10 × 50 mm column of heparin-Sepharose CL-6B equilibrated in the same buffer. Material that does not bind to heparin-Sepharose is removed by continuous flow of 10 mM Tris, pH 7.5. A 200-ml linear gradient of 10 mM Tris, pH 7.5, to 10 mM Tris, 1 M sodium chloride, pH 7.5, is followed by elution with 100 ml of 10 mM Tris, 2 M sodium chloride, pH 7.5, and finally elution with 100 ml of 6 M guanidine hydrochloride. Absorbance is monitored at 280 nm, at a sensitivity of 0.1 absorbance units full scale (AUFS). The flow rate is 0.5 ml/min, and 8-ml fractions are collected. Fractions are pooled based on the absorbance at 280 nm, and lyophilized aliquots are subjected to bioassay. The material eluting from 0.1 to 0.3 M sodium chloride provides the more significant stimulation in the cell number assay (~200% control) and is utilized for subsequent reversed-phase HPLC purification.

Step 5: Additional Reversed-Phase Chromatography. Bioactive material from the heparin-Sepharose column is subjected to reversed-phase HPLC fractionation on a Vydac C_{18} column using a linear gradient of 6:94:0.05 to 60:40:0.05 (v/v) acetonitrile/water/trifluoroacetic acid. Absorbance is monitored at 214 nm, and 2-ml fractions are collected. Aliquots of the 2-ml fractions are analyzed by SDS–PAGE and/or bioassay prior to further chromatography. The fractions are then rechromatographed on the Vydac column using a mobile phase containing trifluoroacetic acid or heptafluorobutyric acid. The linear gradients employed consist of 25:75:0.05 to 35:65:0.05 acetonitrile/water/trifluoroacetic acid or 25:75:0.05 to 45:55:0.05 acetonitrile/water/heptafluorobutyric acid. Complete resolution of the individual moieties requires an additional two or three HPLC fractionations using the above gradient containing trifluoroacetic acid but only one additional run when using heptafluorobutyric acid. These modifications significantly increase the yield of homogeneous MGSA from less than 1 µg using preparative SDS–PAGE to 5–10 µg when using these modifications.

Purification of Recombinant Melanoma Growth Stimulating Activity

The purification of recombinant MGSA (rMGSA) from the CHO clone 7 cells utilizes many of the procedures previously mentioned.

Step 1: Extraction from CHO Clone 7 Conditioned Medium. Five liters of serum-free conditioned medium from the CHO clone 7 cells transfected with the expression plasmid pCMV-M23[6] is concentrated by lyophilization, prior to dialysis against 0.17 N acetic acid.

Step 2: BioGel P-30 Chromatography. The concentrate from Step 1 is then subjected to BioGel P-30 chromatography as previously described.[1]

Step 3: Heparin-Sepharose Chromatography. Fractions eluting from the BioGel P-30 column between the ribonuclease A and insulin markers (13.7K and 6K, respectively) are pooled, lyophilized, and subjected to heparin-Sepharose affinity chromatography as previously described.[2]

Step 4: Reversed-Phase Chromatography. Fractions eluting from the heparin-Sepharose column with 0.5 M NaCl are pooled, lyophilized, and dissolved in a solution containing acetonitrile, water, and trifluoroacetic acid (6 : 94 : 0.05) (v/v). Subsequent reversed-phase HPLC is performed on a Vydac Hi-Pore C_{18} column using a 60-min linear gradient from 6 to 60% acetonitrile.[2]

This abbreviated procedure provides homogeneous material eluting at 35% acetonitrile. Purification is probably facilitated by the increased proportion of rMGSA to total protein secreted by the transfected CHO-7 cells. The yield of the purified rMGSA from the initial 5 liters of medium is approximately 60 μg (240 μg/20 liters), based on the relative quantitation of the integrated HPLC peak in comparison with an insulin standard. This quantity of material is much greater than previous recoveries of material from the Hs294T conditioned medium, which also requires a much more labor-intensive purification procedure.

SDS–PAGE (Fig. 2) of the rMGSA suggests a molecular weight of about 8000. This is in apparent contrast to previously reported 13,000[1] and 16,000 values for MGSA from the Hs294T cells. This can be explained, however, by the use of different molecular weight standards. The previously ascribed weight of 13,000 is probably incorrect based on the recent use of molecular weight standards better suited for the determination of molecular weights less than 14K. The nature and relationship of the previously reported 16K moiety to the lower molecular weight MGSA is currently under investigation.

Characterization of Recombinant Melanoma Growth Stimulatory Activity

Amino-Terminal Sequencing. The purified recombinant MGSA is electrophoresed on a 15% SDS–polyacrylamide minigel. The gel is removed from the apparatus, soaked in blotting buffer (24 mM Tris base, 192 mM glycine, 15% methanol), and the protein is electrophoretically transferred

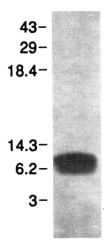

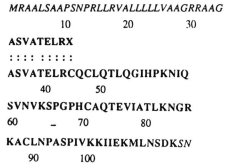

FIG. 2. Purified recombinant MGSA. MGSA eluting at 35% acetonitrile from the Vydac C_{18} column was analyzed by SDS–polyacrylamide gel electrophoresis (12–18% linear polyacrylamide gradient) and stained with Coomassie blue. The positions of the molecular weight markers are shown at left.

for 2 hr at 170 V onto a sheet of poly(vinyl difluoride) (PVDF) (Immobilon, Millipore, Bedford, MA) presoaked in 100% methanol for 1 min and in blotting buffer for 15 min.

The transblotted protein is stained with 0.1% Coomassie blue R-250 in 50% methanol, 10% acetic acid for 1–2 min, destained with 50% methanol, 10% acetic acid (v/v), and air-dried.[10] A 1-mm PVDF strip with the stained protein band is packed into the center of a sequencing column and sandwiched between 1 cm of controlled pore glass and C_{18} reversed-phase packing. The protein can be sequenced by Edman degradation on a liquid-

[10] P. Matsudaira, *J. Biol. Chem.* **262**, 10035 (1987).

phase sequencer. Identical N-terminal sequences are found for MGSA from both Hs294T cells[3] and the transfected CHO-7 cells.

Immunoprecipitation. Additional structural similarity is seen by the precipitation of natural and recombinant MGSA using monoclonal and polyclonal antibodies to MGSA.[11,12] The FB2AH7 monoclonal antibody was developed against Hs294T-derived MGSA and has been used in several applications, namely, immunohistochemistry and immunocytochemistry,[13] enzyme-linked immunoabsorbent assay,[9] and immunoaffinity chromatography.[14] ^{125}I-Labeled rMGSA[15] [100,000 counts/min (cpm)] is incubated overnight with 0.5 ml FB2AH7 antibody in an Eppendorf tube at 4°. Goat anti-mouse IgM (0.5 mg/0.5 ml) is added and the incubation proceeds at 25° for 2 hr. The precipitate is pelleted by microcentrifugation for 15 min. The pellets are washed 3 times with PBS, 0.1% Triton X-100. The pellet is resuspended in 20 μl of 2× Laemmli electrophoresis sample buffer and electrophoresed on a 7.5–15% linear polyacrylamide gel. The gel is removed from the apparatus and stained overnight with Coomassie blue R-250. The next day the gel is destained, dried, and subjected to autoradiography using Kodak XAR-5 film (Rochester, NY).

Results (Fig. 3) indicate the FB2AH7 monoclonal antibody recognizes the ^{125}I-labeled rMGSA moiety and can be used for its detection and/or precipitation in future experiments. The higher molecular weight materials observed in the ^{125}I-labeled rMGSA (Fig. 3, Lane 1) provide an example of the tendency of MGSA to aggregate at neutral pH.[13] This aggregation is noted more frequently with iodinated MGSA. At the present time it is not known if this aggregation actually occurs more frequently in the iodinated preparations, or whether it is simply observed more often because of the greater sensitivity of the autoradiographic techniques.

In analogous experiments, a polyclonal antiserum[12] raised in rabbits against an MGSA-specific oligopeptide can be utilized to immunoprecipitate MGSA from the serum-free conditioned medium of metabolically ^{35}S-labeled CHO clone 7 cells and Hs294T cells. Cells are grown to about 70% confluency in 6-well plates. The monolayers are washed twice with PBS and once with minimal medium and then incubated overnight at 37° in 1.5 ml of serum-free medium lacking cysteine and methionine, or phosphate or sulfate (GIBCO Laboratories, Grand Island, NY). The medium is then supplemented with 100 μCi/ml of L-[^{35}S]methionine. After overnight incu-

[11] D. H. Lawson, H. G. Thomas, R. G. B. Roy, D. J. Gordon, R. K. Chawla, D. W. Nixon, and A. Richmond, *J. Cell. Biochem.* **34**, 169 (1987).
[12] D. Wen, A. Rowland, and R. Derynck, *EMBO J.* **8**, 1761 (1989).
[13] A. Richmond and H. G. Thomas, *J. Cell. Biochem.* **36**, 185 (1988).
[14] H. G. Thomas and A. Richmond, *Arch. Biochem. Biophys.* **260**, 719 (1988).
[15] J. H. Han, H. G. Thomas, and A. Richmond, submitted for publication.

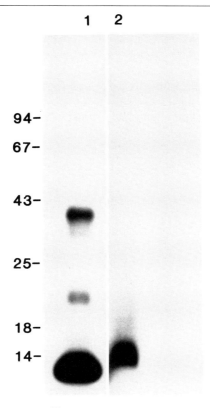

FIG. 3. Immunoprecipitation of ^{125}I-labeled MGSA. Radiolabeled MGSA was immunoprecipitated with the FB2AH7 monoclonal antibody and electrophoresed on a 7.5–15% linear polyacrylamide gel, stained overnight with Coomassie blue R-250, dried, and subjected to autoradiographic analysis. Lane 1 contains ^{125}I-labeled MGSA; lane 2 contains ^{125}I-labeled MGSA immunoprecipitated by the FB2AH7 antibody.

bation the conditioned medium is collected and clarified by centrifugation, and the protease inhibitor phenylmethylsulfonyl fluoride (PMSF) is added to the medium to a 1 mM final concentration.

Immunoprecipitations are carried out as follows. The samples are pretreated with 10 µl of normal rabbit serum and 20 µl of a 1:1 suspension of protein A-Sepharose at 4° for 1 hr. The protein A-Sepharose is removed by centrifugation, and the pretreated sample is reacted overnight at 4° with the rabbit antiserum at a 1:100 dilution in a 500-µl reaction mixture, to which 50 µl protein A-Sepharose is added. The protein A-Sepharose beads are then pelleted and washed 3 times in 0.1% Triton X-100, 0.02% SDS, 150 mM NaCl, 5 mM EDTA, and 10 units/ml of aprotinin. The washed beads are heated with 100 µl of SDS gel electrophoresis loading buffer at 100° for 5 min. The beads are removed by centrifugation, and the supernatant is loaded on a 16.5% tricine–SDS polyacrylamide gel.[16]

O-Glycanase Digestion. Immunoprecipitated MGSA protein is eluted from protein A-Sepharose beads by boiling the samples for 5 min in 60 µl of 0.15% SDS. The protein sample in 50 µl, containing 1 mM calcium acetate, 10 mM D-galactonolactone, 20 mM Tris–maleate, pH 6.0, is digested with neuraminidase at a final concentration of 1 µl/ml for 1 hr at 37°. After digestion, SDS electrophoresis sample buffer is added, and the samples are boiled for 5 min prior to electrophoresis.

Results indicate that similar molecular weight moieties are precipitated from the conditioned medium from both CHO-7 and Hs294T cells. In addition, neither subsequent neuraminidase nor neuraminidase plus O-glycanase treatment of the immunoprecipitated proteins results in a detectable SDS–PAGE mobility shift, indicating a similar lack of O-linked carbohydrates in both proteins. Since the predicted amino acid sequence does not contain a possible N-glycosylation site,[3] we have not investigated the presence of N-linked carbohydrates.

Bioassay. The bioactivity of MGSA derived from CHO-7 cells is assessed by measuring the stimulation of [^{3}H]thymidine incorporation into DNA of serum-depleted low density cultures of Hs294T human malignant melanoma cells and by determining the increase in Hs294T cell number. The incorporation of [^{3}H]thymidine into the melanoma cells is performed as described,[1] except the assay is done in 24-well plates instead of glass vials (Wheaton). The cell number determination is performed as described.[1] MGSA is tested at concentrations from 0.06 to 60 ng/ml. The stimulation of [^{3}H]thymidine incorporation into DNA measured after 16 hr demonstrates a 2- to 3-fold increase in response to nanogram per milliliter levels of MGSA. Determination of Hs294T cell numbers after a 3- or 6-day exposure to such levels of MGSA also demonstrates a 2- to 3-fold increase. These values closely correspond to values attained with Hs294T-derived MGSA as previously reported.[1,2]

Conclusions

Recombinant MGSA/*gro* was purified from the conditioned medium of CHO-7 cells transfected with an MGSA expression vector. The large quantity of MGSA/*gro* produced by these cells, in comparison to that provided from natural sources, facilitated the purification. The modified purification procedure described here resulted in a large increase in recov-

---

[16] H. Schagger and G. Von Jagow, *Anal. Biochem.* **166,** 368 (1987).

ery of bioactive MGSA. Amino-terminal sequencing, SDS–PAGE, immunoprecipitation, and bioassay demonstrate the identical physical, structural, and biological properties of natural and recombinant MGSA. The results demonstrate the advantages and applicability of molecular biological techniques to problems associated with the purification, characterization, and biological investigation of growth factors, for which there are no abundant natural sources.

Acknowledgments

This work was supported by National Cancer Institute Grant Ca34590, Veterans Administration Merit Awards to A.R. and H.G.T., and VA Career Development Awards to R.B. and A.R. We are indebted to Martine Gould, Gary Reece, and Mike Kelleher for excellent technical assistance.

[37] Purification, Cloning, and Expression of Platelet-Derived Endothelial Cell Growth Factor

By Carl-Henrik Heldin, Ulf Hellman, Fuyuki Ishikawa, and Kohei Miyazono

Introduction

Human platelets are a rich source of growth regulatory proteins, the most well characterized being platelet-derived growth factor (PDGF), platelet-derived endothelial cell growth factor (PD-ECGF), and transforming growth factor $\beta 1$ (TGF-$\beta 1$). PDGF stimulates the proliferation mainly of connective tissue cells, and TGF-$\beta 1$ inhibits the growth of most cell types. PD-ECGF is a newly discovered 45-kDa factor with distinct functional properties; it stimulates the growth of endothelial cells of large vessels and capillaries and has angiogenic activity *in vivo* (for a review, see Ref. 1). PD-ECGF occurs in human platelets as well as in placenta. It does not bind to heparin-Sepharose in contrast to endothelial cell mitogens of the fibroblast growth factor (FGF) family. This chapter describes the purification, sequencing, cDNA cloning, and expression of PD-ECGF.

[1] K. Miyazono and C.-H. Heldin, *in* "Peptide Growth Factors and Their Receptors, Handbook of Experimental Pharmacology" (M. B. Sporn and A. B. Roberts, eds.), Vol. 95, p. 125. Springer-Verlag, Berlin and New York, 1990.

Assay for Platelet-Derived Endothelial Cell Growth Factor Activity

The ability to stimulate [^{3}H]thymidine incorporation into porcine aortic endothelial cells is used as an assay for PD-ECGF.[2,3] Endothelial cells are collected from porcine aorta using collagenase digestion.[4] The cells are maintained in 25-cm^2 culture flasks in Ham's F12 medium containing 10% fetal bovine serum and antibiotics and subcultured by 0.25% trypsin treatment at a split ratio of 1:20.

For the assay, cells are seeded sparsely ($\sim 1 \times 10^4$ cells/well) with 0.5 ml of Ham's F12 medium containing 0.5% fetal bovine serum in 24-well tissue culture plates (Costar, Cambridge, MA). After 24 hr of incubation, test samples are added. After an additional 18 hr, cells are pulsed with [^{3}H]thymidine (0.2 μCi/well, 6.7 Ci/mM, 1 mCi = 37 MBq; New England Nuclear, Boston, MA) for 4 hr. Then the cells are fixed with ice-cold 5% trichloroacetic acid for 20 min, washed extensively with water, and solubilized with 200 μl of 1 M NaOH. After 20 min of incubation at room temperature, an equal volume of 1 M HCl is added. The samples are then mixed with 10 ml of Aquazol-2 (New England Nuclear) and counted in a liquid scintillation counter.

Purification of Platelet-Derived Endothelial Cell Growth Factor

A purification procedure resulting in pure PD-ECGF with high yield was established[5] using as starting material a side fraction from the purification of PDGF from human platelets[6] (Fig. 1), namely, a flow-through fraction from CM-Sephadex chromatography. All steps in the purification procedure are performed at 4°, unless otherwise stated.

About 15 liters of the CM-Sephadex flow-through fraction, derived from 1500 units (600 liters) of human blood,[6] is processed at a time. Dry QAE-Sephadex A-50 gel (Pharmacia, Uppsala, Sweden) is added to the starting material (0.7 g/liter). After shaking overnight, the gel is poured into a column (5 × 60 cm; Pharmacia) and washed with 75 mM NaCl, 10 mM phosphate buffer, pH 7.4 ($\sim$4 liters), and PD-ECGF is eluted with 250 mM NaCl, 10 mM phosphate buffer, pH 7.4 ($\sim$2.5 liters).

[2] K. Miyazono, T. Okabe, S. Ishibashi, A. Urabe, and F. Takaku, *Exp. Cell. Res.* **159**, 487 (1985).

[3] K. Miyazono, T. Okabe, A. Urabe, F. Takaku, and C.-H. Heldin, *J. Biol. Chem.* **262**, 4098 (1987).

[4] F. M. Booyse, B. J. Sedlak, and M. E. Rafelson, *Thromb. Diath. Haemorrh.* **34**, 825 (1975).

[5] K. Miyazono and C.-H. Heldin, *Biochemistry* **28**, 1704 (1989).

[6] C.-H. Heldin, A. Johnsson, B. Ek, S. Wennergren, L. Rönnstrand, A. Hammacher, B. Faulders, Å. Wasteson, and B. Westermark, this series, Vol. 147, p. 3.

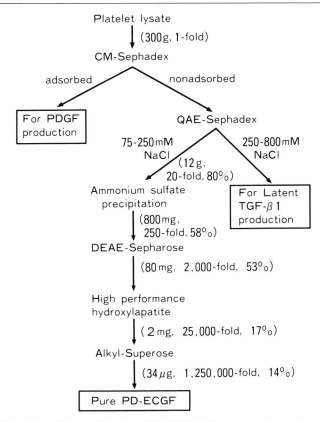

FIG. 1. Flow chart of the purification of PD-ECGF. The amount of protein, purification (-fold), and yield of activity after each step are shown within parentheses. (From Ref. 1.)

The eluate from the QAE-Sephadex step is then subjected to ammonium sulfate precipitation. Solid ammonium sulfate (247 g/liter) is added, and, after 2 hr of equilibration on a magnetic stirrer, the sample is centrifuged at 2075 g for 15 min. The precipitate is resuspended in 50 mM NaCl, 10 mM HEPES, pH 7.0, (~200 ml), treated with 5 mM dithiothreitol at room temperature for 2 hr, and then dialyzed extensively against the same buffer at 4°.

The dialyzate is then subjected to chromatography on a 40-ml DEAE-Sepharose column. The material is pumped through the column at 120 ml/hr, the column washed with 100 ml of 50 mM NaCl, 10 mM HEPES, pH 7.0, and the column then eluted with a linear gradient of NaCl from 50 to 200 mM in 10 mM HEPES, pH 7.0 (400 ml). Fractions of 10 ml are collected and tested for growth-promoting activity on porcine aortic endothelial cells.

The active fractions from the DEAE-Sepharose chromatography (~80 ml) are pooled, filtered through a 0.22-μm filter (Millipore, Bedford, MA), and subjected to chromatography on a hydroxylapatite column. A high-performance hydroxylapatite column (7.8 × 100 mm; Bio-Rad, Richmond, CA), equipped with a guard column (4.0 × 50 mm; Bio-Rad) and attached to a Fast Protein Liquid Chromatography (FPLC) apparatus, is used. The column is equilibrated with 1 mM phosphate buffer, pH 6.8, 50 mM NaCl, 10 μM CaCl$_2$, and the material is loaded at room temperature at a flow rate of 0.5 ml/min. After washing (10 ml), the adsorbed proteins are eluted with a gradient of 1–100 mM phosphate buffer, pH 6.8, in 50 mM NaCl, 10 μM CaCl$_2$ (20 ml); 1-ml fractions are collected and tested for growth-promoting activity on porcine aortic endothelial cells. The column is then washed with 1 M phosphate buffer, pH 6.8, containing 10 μM CaCl$_2$.

The last purification step uses hydrophobic interaction chromatography using an alkyl-Superose column (HR 5/5; Pharmacia) attached to an FPLC apparatus. The active fractions from the hydroxylapatite column are pooled (5 ml), and an equal volume of 2.8 M ammonium sulfate (HPLC-grade; Bio-Rad) in 100 mM phosphate buffer, pH 6.8, is added. The material is applied to the alkyl-Superose column preequilibrated with 1.4 M ammonium sulfate, 100 mM phosphate buffer, pH 6.8, and operated at room temperature at a flow rate of 0.5 ml/min. The sample is eluted with a gradient from 1.4 to 0 M ammonium sulfate in 100 mM phosphate buffer, pH 6.8, as follows; 1.40–1.12 M, 0.5 ml; 1.12–0.56 M, 20 ml; 0.56–0 M, 0.5 ml. Fractions of 0.5 ml are collected and assayed for growth-promoting activity.

Analysis by sodium dodecyl sulfate (SDS)-gel electrophoresis and silver staining of a pool of the active fractions from the last step reveals that the material represents pure PD-ECGF.[5] Sometimes the PD-ECGF preparations contain a second band of slightly lower M_r; this component was found to represent a degraded version of PD-ECGF.[5] The overall increase in specific activity is about 1,200,000-fold at a recovery of 14%.

Protein Sequencing of Platelet-Derived Endothelial Cell Growth Factor

In order to obtain information about the amino acid sequence of PD-ECGF, tryptic fragments of the molecule are prepared.[7] About 40 μg of pure PD-ECGF is first desalted on a C$_4$ narrow-bore reversed-phase col-

---

[7] F. Ishikawa, K. Miyazono, U. Hellman, H. Drexler, C. Wernstedt, K. Hagiwara, K. Usuki, F. Takaku, W. Risau, and C.-H. Heldin, *Nature (London)* **338**, 557 (1989).

umn (Brownlee Aquapore BU300; 2.1 × 30 mm), then eluted with a gradient of acetonitrile in 0.1% trifluoroacetic acid. The desalted PD-ECGF is dried in a Speedvac concentrator and redissolved in 200 µl of 6 M guanidine hydrochloride 0.25 M Tris-HCl, pH 8.5, 1 mM EDTA, containing 100 µg dithiothreitol. After flushing with N_2 for 20 sec, the solution is left at room temperature for 3 hr. Free SH groups are then blocked by incubation with 2 µl of 4-vinylpyridine for another 3 hr at room temperature.

After desalting on the C_4 column and evaporation of the volatile solvent, PD-ECGF is digested for 4 hr at 37° with N-tosyl-L-phenylalanine chloromethyl ketone (TPCK)–trypsin [Sigma, St. Louis, MO; substrate/enzyme ratio 50 (w/w)] in 200 µl of 0.1 M ammonium bicarbonate, containing 2 M urea. The tryptic fragments are immediately loaded onto the same C_4 reversed-phase HPLC column and eluted with a linear gradient of acetonitrile in 0.1% trifluoroacetic acid (Fig. 2). The column temperature is kept at 35°, the flow rate is 100 µl/min, and the effluents are monitored at 220 nm using a diode-array detector (Waters, Milford, MA, Model 990). The resulting pyridylethylcysteine-containing peptides have a characteristic absorption at 254 nm and can therefore easily be detected and distinguished from other peptides. Fractions are collected manually in polypropylene tubes. Nonhomogeneous peptides are rerun on the HPLC narrow-bore reversed-phase system under slightly different conditions.

Peptides are also isolated from PD-ECGF fragmented with staphylococcal V8 protease (Boehringer, Mannheim, Germany) or CNBr (Eastman Kodak, Rochester, NY), using similar separation methods.

The amino acid sequences of the peptides are determined by use of an automated gas-phase sequencer (Applied Biosystems, Foster City, CA, protein sequencer, Model 470A, with an on-line PTH-analyzer, Model 120A). Sequences comprising a total of 382 amino acids are obtained (Fig. 3).

cDNA Cloning of Platelet-Derived Endothelial Cell Growth Factor

Immunoblotting using a specific PD-ECGF antiserum revealed that human placenta is a rich source of PD-ECGF.[7] Total RNA is therefore prepared from a human placenta of 22-weeks gestation by the method of Han *et al.*[8] A cDNA library is then prepared from the poly(A)$^+$ RNA following the method described by Watson and Jackson,[9] with some modi-

---

[8] J. H. Han, C. Stratowa, and W. Rutter, *Biochemistry* **26**, 1617 (1987).
[9] C. J. Watson and J. F. Jackson, in "cDNA Cloning" (D. M. Glover, ed.), Vol. 1, p. 79. IRL, Oxford, 1985.

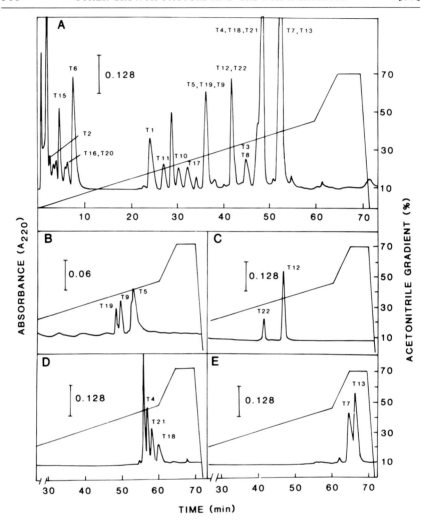

FIG. 2. Separation of tryptic fragments of PD-ECGF on reversed-phase HPLC using a gradient of acetonitrile in 0.1% trifluoroacetic acid (A). Nonhomogeneous material from the chromatography described in (A) was rerun on a Brownlee C_{18} column (2.1 × 30 mm), using an acetonitrile gradient in 0.1% trifluoroacetic acid (B, D, E) or on a C_4 column eluted with a gradient of acetonitrile in 0.15 M NaCl (C).

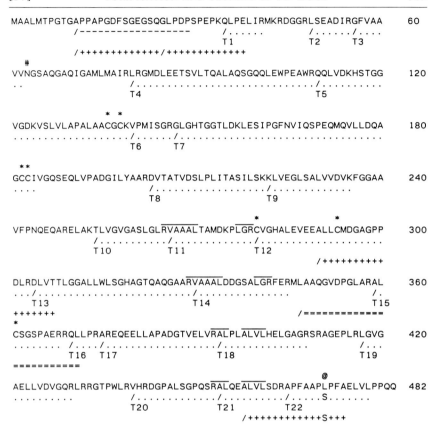

FIG. 3. Amino acid sequence of PD-ECGF. Sequences derived from tryptic digests (/...; designations taken from Fig. 2), N-terminal sequencing (/---), staphylococcal V8 digests (/+++), and a CNBr fragmentation (/===) of PD-ECGF are aligned to the sequence deduced from a cDNA clone. Cysteine residues (*), a potential glycosylation site (#), and a possible site of polymorphism (@) are indicated. Internal repeats are overlined. (Modified from Ref. 7.)

fications. Following the first-strand synthesis, oligo(dG) is added to its 3' end by terminal transferase; second-strand synthesis is primed by addition of an external primer of oligo(dC) as well as internal RNA primers made by RNase H.

Five regions are chosen from the obtained protein sequence (corresponding to amino acids 14–29, 147–176, 266–279, 280–303, and 347–373; Fig. 3), and unique oligonucleotides corresponding to these sequences are deduced according to the method of Lathe.[10] After being synthesized by

[10] R. Lathe, *J. Mol. Biol.* **183**, 1 (1985).

an Applied Biosystems DNA synthesizer 381A, the oligonucleotide probes are labeled by end tailing or primer extension.

About 300,000 independent clones from the placenta λgt10 cDNA library are screened with the probes separately, following the method described by Ullrich et al.[11] Three clones found to be positive with two of the probes are studied further. The cDNA inserts are found to be colinear by restriction mapping. Clone PL8 has the longest insert of 1.8 kilobases (kb); its insert is subcloned into M13mp18 and Bluescript (Stratagene, La Jolla, CA). Deletion mutants are made by the sequential deletion method of Yanish-Peron et al.[12] Both strands of the cDNA are sequenced using the dideoxy-termination method[13]; sequences that are difficult to determine by this procedure are sequenced by the Maxam-Gilbert method.[14]

The obtained deduced sequence (Fig. 3) shows no sequence similarity to other known proteins. The open reading frame codes for a 482 amino acid protein. PD-ECGF seems to undergo a limited processing at the amino terminal since sequencing of the intact protein yields a sequence that starts at amino acid number 11. Notably, PD-ECGF lacks a hydrophobic leader sequence.

Expression of Platelet-Derived Endothelial Cell Growth Factor

NIH 3T3 cells are chosen for expression of PD-ECGF since these cells do not produce PD-ECGF endogenously, nor do they respond to this growth factor. The λPL8 insert is subcloned into the pLJ expression vector to give pLPL8J. This vector has the Moloney murine leukemia virus long-terminal repeat as a promoter and a neomycin resistance gene as a selection marker.[13] pLPL8J, or a control vector, is introduced into NIH 3T3 cells by the calcium phosphate coprecipitation method, and cell lines are selected by neomycin. Transfection with pLPL8J gives a cell line that synthesizes PD-ECGF; it contains in its cell lysate a growth-promoting activity for endothelial cells and a 45-kDa component recognized by PD-ECGF antibodies.[7] Consistent with the lack of signal sequence (Fig. 3), PD-ECGF is not found in the conditioned medium. Control cells, transfected with a vector without insert, are negative.

With this expression system, the authenticity of the cDNA clone was verified. However, the inefficient secretion of PD-ECGF makes this expression system impractical; for production of large quantities of recombinant PD-ECGF, more efficient expression systems are required.

[11] A. Ullrich, C. H. Berman, T. J. Dull, A. Gray, and J. M. Lee, *EMBO J.* **3**, 361 (1984).
[12] C. Yanish-Perron, J. Veira, and J. Messing, *Gene* **33**, 103 (1985).
[13] F. Sanger, S. Nicklen, and A. R. Coulson, *Proc. Natl. Acad. Sci. U.S.A.* **74**, 5463 (1977).
[14] A. M. Maxam and W. Gilbert, this series, Vol. 65, p. 499.

Conclusions

PD-ECGF is a novel endothelial cell mitogen which has angiogenic properties *in vivo*. It is without amino acid sequence similarity with other known proteins. The availability of pure PD-ECGF, a specific antiserum, and cDNA clones will now make it possible to investigate the mechanism of action and *in vivo* function of this factor.

Acknowledgments

We thank Ingegärd Schiller for valuable help in preparation of the manuscript.

[38] Purification and Cloning of Vascular Endothelial Growth Factor Secreted by Pituitary Folliculostellate Cells

By NAPOLEONE FERRARA, DAVID W. LEUNG, GEORGE CACHIANES, JANE WINER, and WILLIAM J. HENZEL

Introduction

Angiogenesis, the growth of blood vessels, is a fundamental biological process in multicellular organisms. It plays a major role in critical physiological events like organogenesis, wound healing, tissue and organ regeneration, and cyclical growth and differentiation of corpus luteum and endometrium.[1,2] Uncontrolled angiogenesis is now recognized as a major pathogenic component of tumor growth and metastasis, rheumatoid arthritis, retinopathies, psoriasis, and retrolental fibroplasia.[1,2] The elucidation of the factors that regulate angiogenesis therefore represents a very important and challenging task of cell and developmental biology and has wide implications in a variety of fields of clinical medicine.

A number of angiogenic factors have been identified, including epidermal growth factor, transforming growth factors α and β, angiogenin, tumor necrosis factor, acidic and basic fibroblast growth factors (aFGF, bFGF), and platelet-derived endothelial cell growth factor (PD-ECGF).[3–5] How-

[1] J. Folkman and M. Klagsbrun, *Science* **235**, 442 (1987).
[2] S. W. Wahl, H. Wong, and N. McCartney-Francis, *J. Cell. Biochem.* **40**, 193 (1989).
[3] A. S. Goustin, E. B. Leof, G. D. Shipley, and H. L. Moses, *Cancer Res.* **46**, 1015 (1986).
[4] D. Gospodarowicz, N. Ferrara, L. Schweigerer, and G. Neufeld, *Endocrine Rev.* **9**, 95 (1987).
[5] F. Ishikawa, K. Miyazono, U. Hellman, H. Drexler, C. Wernstedt, K. Hagiwara, K. Usuki, F. Takaku, W. Risau, and C.-H. Heldin, *Nature (London)* **338**, 557 (1989).

ever, with the exception of FGF and PD-ECGF, these agents have little or no direct mitogenic effect on vascular endothelial cells,[4,5] and their action is likely to be mediated by other angiogenic factors released by inflammatory cells.[1-3] Interestingly, the genes for FGF[6,7] and PD-ECGF[5] do not code for a signal peptide, required for the extracellular transport according to classic secretory pathways.[8] Accordingly, these growth factors are stored intracellularly and are not significantly released in the medium.[5,9-11]

We observed that the medium conditioned by bovine pituitary follicular or folliculostellate cells (FC) is mitogenic for adrenal cortex-derived capillary endothelial cells. These vascular endothelial cells are stimulated to proliferate only by FGF among the known and characterized growth factors.[4] This led us to hypothesize that an endothelial cell growth factor distinct from FGF and possibly any other known growth factor was secreted by FC.[12] In this chapter, we describe methods for the purification and molecular cloning of this novel vascular endothelial growth factor (VEGF).[12,13] VEGF is both a mitogen for cultured vascular endothelial cells and an angiogenic inducer *in vivo*. VEGF is a secreted molecule and demonstrates structural homology with the A and B chains of platelet-derived growth factor. Unlike aFGF and bFGF, VEGF has a narrow target cell specificity, apparently restricted to vascular endothelial cells.

Culture of Pituitary Folliculostellate Cells

Primary cultures of FC are established from bovine adenohypophysial pars distalis or pars tuberalis are previously described[14,15] with minor

---

[6] J. A. Abraham, J. L. Wang, A. Tumolo, A. Mergia, J. Friedman, D. Gospodarowicz, and J. Fiddes, *EMBO J.* **5**, 2523 (1986).

[7] M. Jaye, R. Howk, W. Burgess, G. A. Ricca, I. M. Chiu, M. W. Ravera, S. J. O'Brien, W. E. Modi, T. Maciag, and W. N. Drohan, *Science* **233**, 541 (1986).

[8] D. Perlman and H. O. Avorson, *J. Mol. Biol.* **167**, 309 (1983).

[9] D. Moscatelli, M. Presta, J. Joseph-Silverstein, and D. B. Rifkin, *J. Cell. Physiol.* **129**, 273 (1986).

[10] G. Neufeld, N. Ferrara, L. Schweigerer, and D. Gospodarowicz, *Endocrinology* **121**, 597 (1987).

[11] L. Schweigerer, N. Ferrara, T. Haaparanta, G. Neufeld, and D. Gospodarowicz, *Exp. Eye Res.* **46**(1), 71 (1988).

[12] N. Ferrara and W. J. Henzel, *Biochem. Biophys. Res. Commun.* **161**, 851 (1989).

[13] D. W. Leung, G. Cachianes, W.-J. Kuang, D. V. Goeddel, and N. Ferrara, *Science* **246**, 1306 (1989).

[14] N. Ferrara, P. Goldsmith, D. Fujii, and R. Weiner, this series, Vol. 124, p. 245.

[15] N. Ferrara, D. K. Fujii, P. C. Goldsmith, J. H. Widdicombe, and R. I. Weiner, *Am. J. Physiol.* **252**, E304 (1987).

modification. In brief, four to six pituitaries or 10 to 12 stalk-median eminences are removed from freshly slaughtered steers and placed in sterile vials containing ice-cold calcium- and magnesium-free Hanks' balanced salt solution (HBSS-CMF) supplemented with 25 mM HEPES and antibiotics. The tissues are placed in Petri dishes and washed several times with fresh HBSS-CMF. The anterior lobes are dissected and freed of capsular tissue, while the remainder of the glands is discarded. The anterior pituitary tissue is then minced with scissors and razor blades until the fragments are smaller than 1 mm^3. A similar procedure is followed with stalk-median eminences. The anterior pituitary or stalk-median eminence fragments are transferred into 50-ml conical tubes and washed several times with HBSS-CMF. The fragments are incubated with 40 ml of HBSS-CMF containing 0.5% collagenase (Worthington, Freehold, NJ, type II, 173–240 U/mg) for 50 min at 37°. The fragments are then mechanically dispersed, centrifuged, and resuspended with 25 ml of low glucose Dulbecco's modified Eagle's medium (DMEM) supplemented with 10% fetal bovine serum (FBS), 2 mM glutamine, and antibiotics (growth medium). The suspension is filtered through a nylon screen (450 μm mesh) and then diluted to a density of approximately 1×10^5 cells/ml. Cells are plated into 150-mm tissue culture dishes (Falcon Plastics, Oxnard, CA) and incubated at 37° in a humidified incubator set at 10% CO_2. After 2 or 3 days, when colonies of proliferating cells are clearly established, the cultures are repeatedly washed with a stream of phosphate-buffered saline (PBS) before adding growth medium. This removes unattached fragments and most of the secretory cells. Media are changed every 4 or 5 days thereafter.

Confluent primary cultures, which consist of homogeneous dome-forming cell monolayers,[14–17] are dissociated by exposure to a PBS containing 0.05% trypsin and 0.3% EDTA and passaged into large-scale tissue culture plates (Nunc, Naperville, IL, 24.5 × 24.5 cm) at a split ratio of 1:5, in the presence of growth medium. We demonstrated that the pars tuberalis is the source of dome-forming cells in stalk-median eminence cultures.[14,15] Shortly after confluency, plates are washed 4 times with PBS in order to remove serum components and then incubated in a serum-free medium consisting of low glucose DMEM supplemented with transferrin (10 μg/ml), insulin (5 μg/ml), selenium (10^{-8} M), glutamine (2 mM), and antibiotics. Cultures are maintained in such serum-free conditions for up to

---

[16] N. Ferrara, L. Schweigerer, G. Neufeld, R. Mitchell, and D. Gospodarowicz, *Proc. Natl. Acad. Sci. U.S.A.* **84**, 5773 (1987).

[17] N. Ferrara and D. Gospodarowicz, *Biochem. Biophys. Res. Commun.* **157**, 376 (1988).

6 weeks without significant loss of bioactivity in the medium or appreciable deterioration of the morphology of the monolayer. After 6–8 weeks, cells demonstrate clear signs of senescence (vacuolization, enlargement, polyploidy, etc.). Media (125–150 ml/plate) are collected and replaced every 3 or 4 days. The collected media are centrifuged (1000 g, 15 min at 4°) and stored at −70° or immediately concentrated.

Bioassays

Mitogenic assays on vascular endothelial cells are performed in order to monitor the purification of the growth factor. Bovine adrenal cortex-derived capillary endothelial (ACCE) cells are used in most of the experiments. Primary cultures of ACCE cells can be established according to procedures previously described.[18,19] Stock plates of ACCE cells are maintained in 10-cm tissue culture dishes in the presence of low glucose DMEM supplemented with 10% calf serum, 2 mM glutamine, and antibiotics (growth medium) and weekly passaged at a split ratio of 1 : 50. Basic FGF is added to the cultures at a final concentration of 1 ng/ml, until they become confluent. ACCE cells can be passaged 10–12 times before showing signs of senescence. For mitogenic assays, stock cultures are trypsinized, resuspended in growth media, and seeded at a density of 0.8×10^4 cells/well in 12-well plates (Costar, Cambridge, MA). The plating volume is 2 ml. Fractions to be tested are added to duplicate or triplicate wells in 10-μl aliquots, shortly after plating. After 5 or 6 days, cells are trypsinized and counted in a Coulter Counter (Coulter Electronics, Hialeah, FL). The growth-promoting activity induces a maximal increase in cell number 3- to 7-fold above controls.

Alternatively, human umbilical vein endothelial (HUVE) cells can be used to monitor the purification of the growth factor. For mitogenic assays, HUVE cells are seeded at a density of 1×10^4/well in 12-well plates in the presence of M199 supplemented with 20% FBS, 2 mM glutamine, and antibiotics and counted after 5–6 days. HUVE cells can be isolated as previously described[20] or obtained through commercial sources (Clonetics, San Diego, CA). The major disadvantage of using HUVE cells is their limited life span in culture (2–3 passages).

---

[18] J. Folkman, C. C. Haudenschild, and B. R. Zetter, *Proc. Natl. Acad. Sci. U.S.A.* **6**, 5217, (1979).

[19] D. Gospodarowicz, S. Massoglia, J. Cheng, and D. K. Fujii, *J. Cell. Physiol.* **127**, 121 (1986).

[20] E. A. Jaffe, R. L. Nachman, C. Becker, and C. R. Minick, *J. Clin. Invest.* **51**, 46a (1972).

Purification

Concentration of Conditioned Medium

Four- to six-liter batches of serum-free FC-conditioned medium are subjected to ammonium sulfate precipitation. Ammonium sulfate (500 g/liter) is added in the cold room under constant stirring until the salt is completely in solution. After 8–12 hr, the material is centrifuged (20,000 g, 45 min, 4°). The supernatant is discarded, and the pellet is resuspended with 10 mM Tris-Cl, pH 7.2, 50 mM NaCl (20 ml per liter of initial volume) and dialyzed at 4° against the same buffer for 8–12 hr. The dialyzed material is centrifuged (10,000 g, 15 min at 4°) in order to remove insoluble residue and stored at $-70°$. Alternatively, the medium is concentrated approximately 50-fold by ultrafiltration through Amicon (Danvers, MA) YM10 membranes using a 2.5-liter stir cell unit, with similar results. The purification steps and corresponding yield in bioactivity are outlined in Table I.[12]

Heparin-Sepharose Affinity Chromatography

The concentrated conditioned medium is applied to a 10-ml heparin-Sepharose (H-S) column[21–23] (Pharmacia, Piscataway, NJ) preequilibrated with 10 mM Tris-Cl, pH 7.2, 50 mM NaCl. The absorbance is monitored at 280 nm (Single Path Monitor, Pharmacia). The column is washed with the same buffer until the absorbance becomes negligible and then eluted stepwise with 10 mM Tris-Cl, pH 7.2, containing 0.15, 0.9, and 3 M NaCl. The flow rate is 1.5 ml/min. One-minute fractions are collected. Aliquots are diluted with 0.2% gelatin in PBS and tested for mitogenic activity on vascular endothelial cells.

Eighty to ninety percent of the mitogenic activity of the input is retained by the column. No activity is eluted in the presence of 0.15 M NaCl. In early experiments, we eluted the biological activity with 0.6 M NaCl, but this results in a very broad peak of bioactivity. Elution of the column with 0.9 M NaCl results in a single tight peak of bioactivity immediately following a major protein peak. The most bioactive fractions induce a half-maximal effect on capillary endothelial cell growth at a concentration of 25–50 ng protein/ml and a maximal effect at 150–300 ng protein/ml (Table I). At higher concentrations, we frequently observe an inhibitory effect.

---

[21] Y. Shing, J. Folkman, R. Sullivan, C. Butterfield, J. Curray, and M. Klagsbrun, *Science* **223**, 1296 (1984).
[22] M. Klagsbrun, R. Sullivan, S. Smith, R. Rybka, and Y. Shing, this series, Vol. 147, p. 95.
[23] R. R. Lobb and J. W. Fett, *Biochemistry* **23**, 6295 (1985).

TABLE I
PURIFICATION OF VASCULAR ENDOTHELIAL GROWTH FACTOR[a]

Step	Protein (μg)	Maximal stimulation (ng/ml)	Purification (-fold)	Yield (%)
CM	190,000	2500	1	100
AS	175,000	2500	1	92
H-S	13,000	250	10	68
RP1[b]	25	5	500	6
RP2[c]	4	1	2500	4

[a] From 6 liters of FC-conditioned medium. CM, Conditioned medium; AS, ammonium sulfate precipitate; H-S, heparin-Sepharose; RP1, first reversed-phase HPLC step (acetonitrile gradient); RP2, second reversed-phase HPLC step (2-propanol gradient). Protein concentration determined by Bio-Rad protein assay kit unless noted otherwise.

[b] Protein concentration determined by comparing the relative intensities of bands with standards in silver-stained SDS–polyacrylamide gels.

[c] Protein concentration determined by protein sequencing.

Single-concentration assays therefore can give misleading measures of the bioactivity, and we routinely test each fraction at four different final dilutions (1 : 40,000, 1 : 10,000, 1 : 4000, and 1 : 1000). No mitogenic activity is eluted in the presence of 3 M NaCl. The H-S step results in a 10-fold purification of the growth factor (Table I). The mitogenic activity of the 0.9 M NaCl fractions is very stable at $-70°$. Fractions stored for several months do not show any appreciable loss of bioactivity.

High-Performance Liquid Chromatography

The most bioactive H-S fractions are further purified by reversed-phase HPLC.[12] Fractions are diluted at least 4-fold with water containing 0.1% trifluoroacetic acid (TFA) and applied by multiple injections through a 5-ml loop to a semipreparative Vydac C_4 column (The Separations Group, Hesperia, CA; 10 × 250 mm) preequilibrated in 0.1% TFA/20% acetonitrile. After washing with 10 ml of equilibration buffer, the column is eluted with a linear gradient of acetonitrile (20–45% in 115 min). The flow rate is 2 ml/min. The absorbance is monitored at 210 nm (Model 116 UV detector, Gilson Medical Electronics, Middleton, WI). Fractions of 2 ml are collected. Aliquots of each fraction are diluted with 0.2% gelatin in PBS and tested on endothelial cells at the final dilutions of 1 : 10,000 and 1 : 2000.

The bioactivity elutes as a single peak in the presence of approximately 29% acetonitrile.

The material at this stage is not yet pure, and a silver-stained[24] sodium dodecyl sulfate (SDS)–polyacrylamide gel[25] reveals the presence of several bands. To further purify the growth factor, the most bioactive fractions are diluted 2-fold with 0.1% TFA in water and applied to an analytical Vydac C_4 column (4.6 × 250 mm) preequilibrated with 0.1% TFA/20% 2-propanol. After washing with 3 ml of equilibration buffer, the column is eluted with a linear gradient of 2-propanol (20–45% in 113 min) at a flow rate of 0.6 ml/min. Five hundred microliter fractions are collected. The bioactivity elutes as a single peak in the presence of approximately 29% 2-propanol.

A silver-stained SDS–polyacrylamide gel reveals the presence of a single major band with an apparent molecular mass of ~45 kDa under nonreducing conditions and 22–23 kDa under reducing conditions, indicating that the growth factor is a dimer composed of two subunits of identical molecular weight.[12] The intensity of staining of this band is highly correlated to the mitogenic activity across the bioactivity profile. At this stage, the protein is usually pure as assessed by the presence of a single band on a silver-stained SDS–polyacrylamide gel and a single NH_2-terminal amino acid sequence by microsequencing. If contaminants are still present, as judged by a silver-stained SDS–polyacrylamide gel, the most bioactive fractions are diluted with 0.1% TFA in water and reapplied to the C_4 column. The column is eluted with the same gradient of 2-propanol as the previous step. We found that the IGF-1 binding protein (MW 31,000 NH_2-terminal amino acid sequence EVLFREXPPLTPEEL) elutes in the 2-propanol gradient 2–3 fractions after the peak of VEGF.

This simple procedure consisting of three or four chromatographic steps should yield VEGF pure to homogeneity. However, we found that if FC undergo extensive lysis in culture or are markedly senescent, the increased presence of protein contaminants in the medium requires the introduction of an intermediate purification step after the H-S. A cation-exchange step appears particularly suitable, considering the basic nature of VEGF. We found that a Poly CAT HPLC column (Poly LC, Columbia, MD; 9.4 × 200 mm) provides good resolution and high recovery in bioactivity. The column is equilibrated with 25 mM sodium phosphate, pH 6.5. The 0.9 M NaCl H-S fractions are reconstituted in 25 mM sodium phosphate, pH 6.5, by PD 10 buffer exchange columns, according to the instructions of the vendor (Pharmacia), and applied to the poly CAT

[24] J. H. Morrissey, *Anal. Biochem.* **117**, 307 (1981).
[25] U. K. Laemmli, *Nature (London)* **227**, 680 (1970).

column. This is eluted with a gradient of NaCl (0 to 0.5 M in 30 min and then 0.5 to 1 M in 10 min) at a flow rate of 2 ml/min. The growth-promoting activity elutes as a single peak in the presence of approximately 0.42 M NaCl. The bioactive material is then diluted 4-fold with 0.1% TFA in water and applied to the C_4 reversed-phase column as described above.

Protein Microsequencing

For protein sequencing, approximately 20 pmol of purified protein is directly applied to a gas-phase protein sequenator (Model 470A, Applied Biosystems, Foster City, CA). Edman degradation cycles are carried out, and identification of amino acid derivatives is made by an on-line HPLC column.[26] A single NH_2-terminal amino acid sequence was identified unambiguously. The following residues were identified:

```
1                10                20                30
A P M A E G G Q K P H E V V K F M D V Y Q R S F C R P I E T L V D
```

Comparison of such a sequence with available databanks did not reveal significant similarities with any previously sequenced protein.

Cloning of Vascular Endothelial Growth Factor

Isolation of cDNA Clones Encoding Bovine VEGF

Clones encoding bovine VEGF were isolated from a cDNA library prepared from FC. The methods to prepare and screen cDNA libraries have been extensively described in these volumes and elsewhere. Therefore, here we only outline the general strategy employed.

Total cellular RNA is prepared from confluent FC cultures. Cells (2 × 10^8) are dissociated by exposure to trypsin, washed once with PBS, and then subjected to the guanidine thiocyanate/LiCl RNA extraction procedure as previously described.[27,28] Total RNA (2 mg) is then subjected to oligo(dT) cellulose chromatography[29,30] in order to prepare poly(A)$^+$ RNA. Oligo(dT) and random-primed cDNAs are prepared with a cDNA synthesis kit obtained from Amersham (Arlington Heights, IL). The resulting dou-

---

[26] W. J. Henzel, H. Rodriguez, and C. Watanabe, *J. Chromatogr.* **404**, 41 (1987).
[27] G. Cathala, J.-F. Savouret, B. Mendez, B. L. West, M. Karin, J. A. Martial, and J. D. Baxter, *DNA* **2**, 329 (1983).
[28] R. J. MacDonald, G. H. Swift, A. E. Przybyla, and J. M. Chirgwin, this series, Vol. 152, p. 219.
[29] A. Jacobson, this series, Vol. 152, p. 254.
[30] H. Aviv and P. Leder, *J. Mol. Biol.* **134**, 743 (1972).

ble-stranded DNAs are ligated to hemikinased EcoRI adaptors.[31] The cDNAs with EcoRI cohesive ends are purified by spin dialysis through Sephacryl S500HR before being cloned into λgt10 vectors[32] for the generation of cDNA libraries. The libraries are screened with the probe

5'-CCTATGGCTGAAGGCGGCCAGAAGCCTCACGAAGTGGTGAAGTTCATGGACGTGTATCA

This is a 59-base synthetic probe based on the NH_2-terminal amino acid sequence of VEGF (positions 2–21). Hybridization of the probe, labeled at its 5' end with ^{32}P, is done under nonstringent conditions[33] without any dextran sulfate, and the filters are washed in 0.15 M NaCl, 15 mM sodium citrate, and 0.1% SDS at 50°.

Twenty hybridizing clones were identified[34] in a λgt10 library of 1.5 × 10^6 clones. Two were sequenced for their coding and much of the noncoding region. The DNA sequence analysis was performed by the dideoxy chain termination method with cDNA fragments subcloned into plasmid vectors.[35] The complete cDNA and translated protein sequence of one of these clones (bVEGF.6) is shown in Fig. 1. This cDNA clone contains an open reading frame of 190 amino acids. The NH_2-terminal amino acid sequence determined from the purified native FC-derived VEGF by microsequencing is preceded by 26 amino acids containing a hydrophobic core of 16 amino acids flanked by polar or charged residues, indicative of a secretory signal sequence.[8] This indicates that VEGF is a secreted protein, unlike aFGF, bFGF, or PD-ECGF.[5–7] The mature bovine VEGF monomer is therefore expected to have 164 amino acids with a calculated molecular weight of 19,162. This is smaller than the 23,000 value determined by SDS–PAGE analysis.[12] Such a discrepancy may be accounted for by the presence of a potential glycosylation site at Asn-74, suggesting that VEGF is a glycoprotein.

Further examination of the predicted amino acid sequence of VEGF reveals the presence of clusters of basic amino acids at the position around residues 110, 123, and 163. These could be responsible for the binding of the molecule to heparin, by analogy with what has been proposed for FGF.[6,7] When Northern (RNA) blotting experiments were performed with poly(A)$^+$ RNA isolated from FC, a single hybridization band having the size of 3.7 kilobases (kb) was detected.[13]

A search of several databanks indicates significant similarity between the complete amino acid sequence of VEGF, the sequences of the A and

---

[31] R. Wu, T. Wu, and A. Ray, this series, Vol. 152, p. 343.
[32] J. Jendrisak, R. A. Young, and J. D. Engel, this series, Vol. 152, p. 359.
[33] W. I. Wood, this series, Vol. 152, p. 443.
[34] G. M. Wahl and S. L. Berger, this series, Vol. 152, p. 415.
[35] W. M. Barnes, this series, Vol. 152, p. 538.

```
   1 CAGCGCTGAC GGACAGACAG ACAGACACCG CCCCCTGCCC CAGCGCCCAC CTCCTCCCCG CAGCGCCCAC CAGCGCCCAC CCCAGCGGCA
 101 GGAGCCGGAG CCCGCCCCG GAGGGGGGT GGGCGTCCG GCCGGTGCGC AGCTTGCCT GAAACTTTC TGGGCGTGTC TCGTTCCGGA
 201 GGAGCCGTGG TCCGTGCCGG GCTGCCGAG CCGGATGGCA TGCTCCTCG GCCGGAGG AGCCGCAGTC GCCAGCTGTA GAGAGAAG
 301 AAGAGAAGGA AGAGGAGAAG GGCCGCCGGT GGGCCGAGG CTCTCGGAAG CCGGGCTCAT GACGGGTGA TGCACAGACA GTGCTCCAGC
 401 CGCGGCGCCG CCCAGGCCC TGGCCCGGGC CTCGCTCCG AGAGGAGAG GAGCCCGCCT GGGCGCCAG GAGAGCGGGC CGCCCAGCGG
 501 AGAGGGAGCG CGAGCCGCGC CGGCCCCGGC CAGGCCTCCG AAACC       ATG AAC TTT CTG TCT GTA CAT TGG AGC CTT GCC TTG
 -26                                                            M   N   F   L   L   S   W   V   H   W   S   L   A   L
                                                                     NcoI
 588 CTG CTC TAC CTT CAC CAT GCC AAG TGG TCC CAG GCT GCA CCC ATG GGA GAA GGG CAG AAA CCC GAA GTG GTG AAG
 -12   L   L   Y   L   H   H   A   K   W   S   Q   A   A   P   M   G   E   G   Q   K   P   E   V   V   K
 669 TTC ATG GAT GTC TAC CAG CGC AGC TTC TGC CGT CCC ATC GAG ACC CTG GTG GAC ATC TTC CAG GAG TAC CCA GAT GAG ATT
  16   F   M   D   V   Y   Q   R   S   F   C   R   P   I   E   T   L   V   D   I   F   Q   E   Y   P   D   E   I
 750 GAG TTC ATT TTC AAG CCG TCC TGT GTG CCC CTG ATG CGG GGC TGC TGT AAT GAC GAA AGT CTG GAG TGT GTG CCC
  43   E   F   I   F   K   P   S   C   V   P   L   M   R   G   C   C   N   D   E   S   L   E   C   V   P
 831 ACT GAG GAG TTC AAC ATC ACC ATG CAG ATT ATG CGG ATC AAA CCT CAC CAA AGC CAC ATA GGA GAG ATG AGC TTC CTA
  70   T   E   E   F   N   I   T   M   Q   I   M   R   I   K   P   H   Q   S   H   I   G   E   M   S   F   L
 912 CAG CAT AAC AAA TGT GAA TGC AGA CCA AAG GAT AAA GAT TGT CCC TGT GGG CCT TGC TCA GAG AGA
  97   Q   H   N   K   C   E   C   R   P   K   K   D   K   A   R   Q   E   N   P   C   G   P   C   S   E   R   R
 993 AAG CAT TTG TTT GTA CAA GAT CCG CAG ACG TGT AAA TGT TCC TGC AAA AAC ACA GAC TCG CGT TGC AAG GCG CAG CTT
 124   K   H   L   F   V   Q   D   P   Q   T   C   K   C   S   C   K   N   T   D   S   R   C   K   A   R   Q   L
1074 GAG TTA AAC GAA CGT ACT TGC AGA TGT GAC AAG CCG AGG CGG TGA GC CGGGCTGGAG GAAGGAGCCT CCCTCAGGGT TTCGGAACC
 151   E   L   N   E   R   T   C   R   C   D   K   P   R   R   O
1161 AGAGGTCTCA CCAGGAAAGA CTGACACAGA GCCGCCGCCA CCACCACCA CCACCATGA CAGACAATC CTGAATCCAG
1261 AAACCTGACA TGAAGGAAGA GGAGGCTGTG CGCAGAGCAC TTTGGGTCCG GAGCGTGAGG CTCCGGCAGA AGCATTCATG GGGGGTGAC CAGACACGGT
1361 TCCTCTTGGA ATTGGATTGC CATTTTATTT CTCTTGCTCT TAAATCACCG AGCCCGGAAG ATTAGAGAGT TTTATTTCTG GGATTCCTGT AGACACACCC
1461 ACCCACATAC ATACATACAT TTATATATAT ATATATATAA TATATATAAA AATAATAA TATATTTTAT ATATATAAAA AAA
```

FIG. 1. Nucleotide sequence and deduced amino acid sequence of a bovine VEGF cDNA clone. The protein sequence is numbered starting with 1 at the mature NH_2-terminal alanine. The amino acid sequence derived from microsequencing is underlined. ATG and stop codons found in the 5' noncoding region and the poly(A) signal AAATAAA are also underlined. The putative glycosylation site is boxed.

B chains of platelet-derived growth factor (PDGF), and the sequence of the product of the *sis* oncogene.[36–38] All eight cysteine residues found in the A and B chains of PDGF are conserved in VEGF. However, VEGF contains eight additional cysteine residues within its COOH-terminal region. Whereas PDGF is active on a wide variety of cell types of mesenchymal origin and inactive on endothelial cells,[36–38] VEGF appears to be a highly specialized molecule selective for vascular endothelial cells. This suggests that the structural divergence between PDGF and VEGF was accompanied by a marked functional divergence as well. Also, the region spanning the last four cysteines in VEGF is homologous to a cysteine-rich domain found in the mouse plasma cell membrane glycoprotein mPC-1.[39] Although the significance of these homologies is unclear, it is noteworthy that all the proteins are dimers and that intact disulfide bridges are required for their tridimensional organization.

Isolation of cDNA Clones Encoding Human VEGF

Using the bovine cDNA clone as a probe, several human cDNA libraries were screened. Complementary DNA clones encoding human VEGF were identified in a library prepared from mRNA isolated from differentiated HL60 promyelocytic leukemia cells.[13] For induction of differentiation, HL60 cells (0.8×10^6) are incubated for 4 hr in the presence of phorbol myristate acetate (50 ng/ml), lipopolysaccharide (100 μg/ml), indomethacin (1 mM), and cycloheximide (100 μg/ml). HL60 cells are known to undergo terminal differentiation into monocytes–macrophages in the presence of phorbol esters.[40] One of the clones identified encodes a protein which is 95% identical to bovine VEGF. Human VEGF is expected to have an additional amino acid (165) owing to the insertion of a Gly in position 6. We also identified two less abundant clones coding for a shorter and a longer molecular species of VEGF. These display a deletion of 44 amino acids between positions 116 and 159 and an insertion of 24 amino acids at position 116, respectively. In both cases Asn-115 is replaced by a Lys. This suggests the existence of at least three types of homodimers

---

[36] A. Johansson, C.-H. Heldin, A. Wasteson, B. Westermark, T. F. Deuel, J. S. Huang, P. H. Seeburg, E. Gray, A. Ullrich, G. T. Scrace, P. Stroobant, and M. D. Waterfield, *EMBO J.* **3**, 921 (1984).

[37] H. A. Weich, W. Sebald, H. U. Schairer, and J. Hoppe, *FEBS Lett.* **198**, 344 (1986).

[38] C. Betsholtz, A. Johnsson, C.-H. Heldin, B. Westermark, P. Lind, M. S. Urdea, R. Eddy, T. B. Shows, K. Philpott, A. L. Mellor, T. J. Knott, and J. Scott, *Nature (London)* **320**, 695 (1986).

[39] I. R. Van Driel and J. W. Goding, *J. Biol. Chem.* **262**, 4882 (1987).

[40] E. Huberman and M. F. Callahan, *Proc. Natl. Acad. Sci. U.S.A.* **76**, 1293 (1979).

of human VEGF. Heterodimers might also exist. Alternative splicing of mRNA is a likely explanation for the molecular heterogeneity.

Transient Expression of VEGF in Mammalian Cells

For transient expression of VEGF, human kidney 293 cells were used. Subconfluent 293 cells, cultured in the presence of DMEM/F12 (1:1) supplemented with 10% FBS and 2 mM glutamine, are transfected by the calcium phosphate coprecipitation method[41] with an expression vector containing the bovine or human VEGF cDNA insert. The transcription of the VEGF cDNA is directed by a cytomegalovirus promoter.[13] After overnight incubation, the media are removed and replaced with fresh growth media. After 48 hr, the media are collected and tested for mitogenic activity on adrenal cortex-derived capillary endothelial cells. The media conditioned by 293 cells transfected with the vector pbVEGF.6 and pbVEGF.21, containing, respectively, the bovine and human VEGF cDNA, promote the proliferation of capillary endothelial cells. In contrast, the media conditioned by untransfected 293 cells or cells transfected with vector alone had no mitogenic activity on this cell type.[13] Bovine and human recombinant VEGFs were purified and sequenced. They behaved similarly to native FC-derived VEGF in terms of bioactivity, binding to heparin, and retention in reversed-phase columns.

Characterization of Vascular Endothelial Growth Factor

VEGF is a dimeric glycoprotein, completely inactivated following exposure to reducing agents like dithiothreitol or β-mercaptoethanol. The bioactivity of VEGF was unaffected by heating at 65° for 5 min. Also, the bioactivity was relatively acid stable. Exposure to 0.1% TFA (pH 2) for 2 hr decreased the activity by 25–30%. In order to determine the isoelectric point of VEGF, H-S purified material was subjected to chromatofocusing on a Mono P HR5/5 column (Pharmacia). The column was equilibrated with 25 mM diethanolamine, pH 9.5 and eluted with Polybuffer 96, pH 6, (flow rate 0.8 ml/min) according to the instructions of the vendor. The mitogenic activity was eluted at an apparent isoelectric point of 8–8.5. The peak of bioactivity is relatively broad, suggesting the existence of charge heterogeneity.

As previously described, at least three different molecular species of VEGF can exist by alternative splicing of mRNA. The available evidence indicates that the intermediate molecular species (164 amino acids in bo-

[41] C. Gorman, *in* "DNA Cloning" (D. Glover, ed.), Vol. 2, p. 143. IRL, Oxford, 1985.

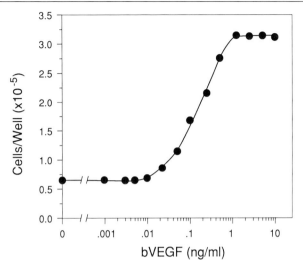

FIG. 2. Dose-responsive prolieration of capillary endothelial cells derived from adrenal cortex in the presence of bovine FC-derived VEGF (bVEGF). bVEGF was added to cells in 5 μl/ml aliquots. Duplicate wells were dissociated by exposure to trypsin and counted in a Coulter counter after 5 days. Variation from the mean did not exceed 10%.

vine and 165 in human VEGF) represents the predominant form of VEGF secreted by both normal and transformed cells. Immunoblot analysis of partially purified preparations of native growth factor using an antibody directed against the amino-terminal portion of VEGF reveals a major band with a molecular mass of ~23 kDa and a minor band with an apparent molecular mass of ~17 kDa. The major band has the same molecular mass as the 164 or 165 amino acid species of recombinant VEGF. The minor band has similar molecular mass as the 121 amino acid of VEGF. However, a 17-kDa band is observed also in lanes containing media conditioned by transfected cells expressing the 165 amino acids species, where it presumably represents a product of proteolysis and/or deglycosylation. Therefore, such lower molecular weight species in native preparations could represent a mixture of the short alternatively spliced form and of a degradation product of the 164 or 165 amino acid precursor.

Purified VEGF stimulated the proliferation of vascular endothelial cells isolated from both small and large vessels.[12] These included bovine adrenal cortex-derived capillary endothelial cells (Fig. 2), cerebral cortex capillary endothelial cells, fetal and adult aortic endothelial cells, and human umbilical vein endothelial cells. Half-maximal stimulation of vascular endothelial cell growth was obtained at 100–150 pg/ml of VEGF (2–3 pM) and a maximal effect at 1 ng/ml (22 pM). VEGF failed, however, to stimulate the proliferation of BHK-21 fibroblasts, adrenal cortex cells, corneal endothelial cells, vascular smooth muscle cells, and human sarcoma cells. This indicates that, unlike aFGF or bFGF,[4] the target cell specificity of VEGF is restricted to vascular endothelial cells.

Purified VEGF was also able to induce a marked angiogenic response when applied to 8-day-old chick chorioallantoic membranes.[13] This *in vivo* effect demonstrates that VEGF is able to trigger the entire chain of events which leads to new blood vessel formation. This requires enzymatic degradation of the basement membrane of a local venule,[42] proliferation of vascular endothelial cells,[18] and their migration toward the angiogenic stimulus.[43]

Conclusions and Perspectives

Unexpectedly, after this work was completed, it became known that vascular permeability factor (VPF), a protein previously identified on the basis of its ability to induce vascular leakage and protein extravasation,[44] rather than as an endothelial cell growth factor, has the same amino acid sequence as the 189 amino acid species of VEGF.[45] It is unclear at this point whether the two shorter forms also induce vascular leakage or whether the 189 amino acid species is an endothelial cell mitogen.

The availability of cDNA clones should also make it possible to address the question of the *in vivo* biological function of VEGF and of its expression in adult and developing tissues in different physiological or pathological circumstances, such as tumor growth and metastasis. Preliminary evidence, using techniques such as Northern (RNA) blotting analysis and *in situ* hybridization, already indicate that the VEGF mRNA is expressed by a wide variety of normal and transformed cultured cells as well as by tissues. The presence of clones encoding VEGF in a library derived from differentiated HL60 leukemia cells suggests that the growth factor is produced by monocytes and macrophages. These cells play a central role in normal and pathological angiogenesis and in wound healing and therefore it is tempting to speculate that *in vivo* such functions are, at least in part, mediated by VEGF. Also, very recent studies[46] provide evidence for the involvement of the growth factor in a major physiological process, the

---

[42] J. L. Gross, D. Moscatelli, and D. B. Rifkin, *Proc. Natl. Acad. Sci. U.S.A.* **80,** 623 (1983).
[43] B. Zetter, *Nature (London)* **285,** 41 (1979).
[44] D. R. Senger, S. J. Galli, A. M. Dvorak, C. A. Perruzzi, V. S. Harvey, and H. F. Dvorak, *Science* **219,** 938 (1983).
[45] P. J. Keck, S. D. Hauser, G. Krivi, K. Sanzo, T. Warren, J. Feder, and D. T. Connolly, *Science* **246,** 1309 (1989).
[46] H. S. Phillips, J. Hains, D. W. Leung, and N. Ferrara, *Endocrinology* **127,** 965 (1990).

cyclical proliferation of capillary vessels which takes place in the ovarian corpus luteum.

Acknowledgments

We thank M. Cronin, H. Niall, and H. Goeddel for their support.

[39] Preparation and Bioassay of Connective Tissue Activating Peptide III and Its Isoforms

By C. W. Castor, E. M. Smith, M. C. Bignall, P. A. Hossler, and T. H. Sisson

Background and Nomenclature

Connective tissue activating peptide III (CTAP-III) is a platelet α-granule-derived growth factor which stimulates synthesis of DNA, hyaluronic acid (HA), sulfated glycosaminoglycan (GAG) chains, proteoglycan monomer, and proteoglycan core protein in human synovial fibroblast cultures. It also stimulates glucose transport, formation of prostaglandin E_2, HA synthetase activity, and plasminogen activator activity.[1-4] Present evidence indicates that CTAP-III is derived from PBP, a biologically inactive precursor protein[5]; the anabolic activities documented for CTAP-III are lost on proteolytic cleavage of the amino-terminal tetrapeptide to form CTAP-III(des 1–4), also named β-TG.[3,6] In this summary, the prefix des indicates a molecule characterized by the removal of one or more functional units, here, amino acids.

Analytical isoelectric focusing coupled with Western blotting showed that CTAP-III prepared rapidly from single donors has multiple isoelectric

[1] C. W. Castor, J. C. Ritchie, M. E. Scott, and S. L. Whitney, *Arthritis Rheum.* **20**, 859 (1977).
[2] C. W. Castor, J. C. Ritchie, C. H. Williams, M. E. Scott, S. L. Whitney, S. L. Myers, T. B. Sloan, and B. Anderson, *Arthritis Rheum.* **22**, 260 (1979).
[3] C. W. Castor, J. W. Miller, and D. A. Walz, *Proc. Natl. Acad. Sci. U.S.A.* **80**, 765 (1983).
[4] C. W. Castor, in "Arthritis and Allied Conditions" (D. J. McCarty, ed.), 11th Ed. p. 242. Lea & Febiger, Philadelphia, Pennsylvania, 1988.
[5] J. C. Holt, M. E. Harris, A. M. Holt, E. Lange, A. Henschen, and S. Niewiarowski, *Biochemistry* **25**, 1988 (1986).
[6] G. S. Begg, D. S. Pepper, C. N. Chesterman, and F. J. Morgan, *Biochemistry* **17**, 1739 (1978).

point variants (isoforms), a finding thought more likely related to posttranslational modification than to preparative artifact or genetic polymorphism. Evidence for three new isoforms of CTAP-III from human platelets has been presented; two are amino-terminal cleavage products, CTAP-III(des 1–13) and CTAP-III(des 1–15).[7] CTAP-III(des 1–13) has a pI of 8.6 and is a relatively stable proteolytic cleavage product that retains the capacity to stimulate [^{14}C]GAG synthesis in human synovial cell cultures. CTAP-III(des 1–15) appears to be an elastase or chymotrypsin cleavage product and identical to NAP-2, an entity thought to have neutrophil activating properties.[7,8]

The major known isoforms of CTAP-III are shown schematically in Fig. 1. Methods important for isolation and measurement of the parent protein and its isoforms as well as for measurement of certain biological activities are presented.

Methods of Analysis

Bioassay of Growth Factors

Cell Culture Methods. Normal human fibroblastic cells are developed from explants obtained at amputation, arthrotomy (synovium and cartilage), and reduction mammoplasty (dermis) as described earlier.[9,10] Tissue is cut into 1–2 mm^2 fragments and transferred to a T-9 or T-15 flask and anchored by a layer of polypropylene net cut to fit the flask. Delnet GS3736 (Applied Extrusion Technologies, Inc., Middletown, DE) replaces perforated cellophane which is no longer available. It is lightweight, nontoxic, transparent, easily manipulated, and can be safely autoclaved when covered with a thin layer of distilled water. Cells attach and grow on both sides of the polypropylene, serving to double the surface area available for explant outgrowth. Cell dispersal is performed by incubating monolayer cultures in 0.025% trypsin (porcine pancreatic, without calcium or magnesium, Sigma Chemical Co., St. Louis, MO) in Hanks' balanced salt solution for 15 min at 37°. Cells may be grown as monlayer cultures in air in T-75 flasks in CMRL 1066 medium (GIBCO, Grand Island, NY) supplemented with 5% human serum and 5% fetal calf serum (FCS), 10% neonatal calf serum, 0.025% sodium bicarbonate, 2 mM L-glutamine, 20 mM N-2-

---

[7] C. W. Castor, D. A. Walz, C. G. Ragsdale, P. A. Hossler, E. M. Smith, M. C. Bignall, B. P. Aaron, and K. Mountjoy, *Biochem. Biophys. Res. Commun.* **163**, 1071 (1989).
[8] A. Walz and M. Baggiolini, *Biochem. Biophys. Res. Commun.* **159**, 969 (1989).
[9] C. W. Castor, *J. Lab. Clin. Med.* **77**, 65 (1971).
[10] C. W. Castor and R. B. Lewis, *Scand. J. Rheum.* **5** (Suppl. 12), 41 (1975).

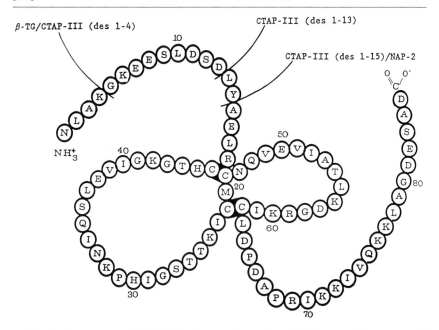

FIG. 1. The structure of CTAP-III is illustrated using the single letter code for amino acid residues. Smaller isoforms are generated by cleavage at the indicated sites. It may be important that this structure has a high degree of homology with more recently described factors including the MGSA, MDNCF (NAP-1/IL-8), 9E3, CHEF GRO, HUMAN GRO, and γ-IP-10 proteins as well as PF-4.

hydroxyethylpiperazine-N'-2-ethanesulfonic acid (HEPES) buffer, penicillin (100 U/ml), streptomycin 100 μg/ml, and gentamicin (5 μg/ml).

[^{14}C]Glucosamine Incorporation Assay. The following procedure provides a measure of [^{14}C]glucosamine incorporation into [^{14}C]GAG, primarily [^{14}C]hyaluronic acid ([^{14}C]HA) in the case of synovial target cells.[11] Cells are plated in 96-well polystyrene microtiter culture plates (Costar, No. 3596, Data Packaging, 205 Broadway, Cambridge, MA; growth surface area/well 0.32 cm^2) at a density of 1×10^4 cells/well. We draw attention to an assay plate "edge effect" and suggest a remedy for this problem. In the [^{14}C]glucosamine incorporation bioassay, significantly higher isotope incorporation ($p < .001$) was consistently observed in media from the 36 wells which make up the outermost rim of the plate in comparison to the interior wells. Initially, it was thought that concentration of well contents due to evaporative loss was the cause of this phenomenon.

[11] C. W. Castor, M. C. Bignall, P. A. Hossler, and D. J. Roberts, *In Vitro* **17**, 777 (1981).

Sealing the edges of the plate did not eliminate the effect; however, it was noted that during incubation in a 37° humidifying chamber, a condensate formed on the lid of the plate over all wells *except* the outermost 36. We speculated that these wells had more direct contact with ambient air and that uneven heating and cooling between outer and inner wells might occur. Since very small temperature changes have been shown to affect synovial cell HA synthesis, this might account for the excess [^{14}C]HA detected in the rim wells in the glucosamine incorporation assay.[12] These wells are now excluded from the assay and instead are filled with 200 µl sterile Milli-Q water to act as an insulating zone between the well cultures and external temperature variations.

The culture medium for the [^{14}C]glucosamine incorporation assay consists of 100 µl semisynthetic medium (L-15, Sigma) containing 1% FCS, 2 mM L-glutamine, penicillin (100 U/ml), gentamicin (5 µg/ml), streptomycin (100 µg/ml), 0.025% sodium carbonate, and 20 mM HEPES buffer, pH 7.6. After 20 to 24 hr of incubation at 37° in a humidified chamber to allow attachment and spreading of cells, the medium is supplemented with another 100 µl of L-15 medium containing 0.5 µCi/ml of uniformly labeled D-[^{14}C]glucosamine hydrochloride (New England Nuclear, Boston, MA; specific activity 250–300 mCi/nmol). Quantities from 5 to 20 µl of test substances or appropriate control vehicles may be delivered with reasonable accuracy with a Pipetman (Gilson Medical Electronics, Middleton, WI; Rainin Instrument Co., Woburn, MA) using beveled tips which improve the delivery of the sample to the well. Microcultures are then incubated for an additional 40 hr at 37° in a humidified chamber, after which 100 µl of medium from each well is spotted on Whatman 3 MM chromatography paper and dried at 32°–37°. To precipitate HA, the spotted 3 MM paper is immersed in 0.1% cetylpyridinium chloride (CPC) (Sigma) containing 1 mM Na$_2$SO$_4$ for 15 min followed by eight 5 min washes in 0.1 N NaOH at room temperature. The washes, which remove unincorporated [^{14}C]glucosamine hydrochloride, are discarded. After the final wash, the paper is dried at 37° and cut into squares. Washed sample squares of ^{14}C assay media are counted in Ecolume scintillation cocktail (ICN Radiochemicals, Irvine, CA). Ecolume, which demonstrates counting efficiencies equivalent to the previously used toluene-based POPOP/PPO system, is biodegradable, nontoxic, nonflammable, and is also easier to dispose of and store.

[^{3}H]Thymidine Incorporation into Fibroblast DNA. Cells are plated, 10^4 cells/microtiter well in 100 µl ESM (Minimal Essential Medium Eagle, Sigma), 1% FCS, with penicillin (100 U/ml), gentamicin (5 µg/ml), strepto-

---

[12] C. W. Castor and M. Yaron, *Arch. Phys. Med. Rehab.* **57**, 5 (1976).

mycin (100 μg/ml), L-glutamine (2 mM), and HEPES (20 mM). The microcultures are incubated in a humidified chamber at 35°–37° for a 20-hr period to allow attachment and spreading of cells.[1] Test samples or vehicles are added (5–15 μl/well), and incubation is continued for 24 hr, at which time tritiated methylthymidine (New England Nuclear, specific activity 50–70 Ci/mmol) is added and the incubation continued for a final 24 hr. Medium is aspirated and discarded, and the cell sheets are washed twice, each separately, with phosphate-buffered saline (PBS, pH 7.0), 5% trichloroacetic acid, and absolute methanol. Cells are then lysed by incubation with 50 μl of 0.3 N NaOH for 1.5–2 hr at 37°. On completion of lysis, the contents of each well are pipeted directly into a 15 × 45 mm glass counting vial containing 4 ml of Ecolume cocktail. The vial is capped and shaken gently to mix before scintillation counting. This method improves ^{3}H counting efficiency on the order of 50–60% compared to counting on glass fiber paper immersed in either Ecolume or toluene–POPOP/PPO cocktail.

Preparation of CTAP-III. CTAP-III and its isoforms are isolated from pellets of human platelets by extraction into acid/ethanol and precipitation with cold acetone.[1-3] In practice, 50 g of pelleted platelets is added to 500 ml of acid ethanol (5 ml of 1.25 N HCl/95 ml ethanol); this is stirred slowly at 4° for 16 hr, then centrifuged (15,000 g, 10 min), and the supernatant fluid is added to 1500 ml of cold (4°) acetone. After 1 hr at 4° the supernatant fluid is removed by aspiration and centrifugation; the precipitated protein is dissolved in 30 ml of 0.5 M acetic acid and dialyzed overnight against 0.2 M acetate buffer, pH 4.0, with protease inhibitors (1 mM Benzamidine and 5 mM EACA). Seventy milliliters of the dialyzed preparation are applied to a Sephacryl S-200 HR column (5 × 72 cm) and eluted with the same buffer and inhibitors in 15-ml fractions at a flow rate of 4.0 ml/min. Fractions containing CTAP-III are identified by immunodiffusion with anti-CTAP-III antisera and pooled.

The isolation process is continued by heparin affinity or immunoaffinity chromatography, or both, in sequence. Partially purified CTAP-III is applied to a heparin affinity column in PBS, pH 7.0; CTAP-III is eluted with 0.3 M sodium chloride and dialyzed against PBS. When CTAP-III is applied to an immunoaffinity column, the column is washed with PBS and CTAP-III is eluted with 0.1 M acetic acid (pH 2.8). To reduce growth factor binding to plastic and glass, either polyethylene glycol or highly purified human albumin is added before lyophilization in amounts sufficient to make the concentration 0.01% in final preparations.

Preparation of Heparin Affinity Columns. Preswollen Bio-Rad (Richmond, CA) Affi-Gel 15 is washed on a scintered glass filter (medium pore size) with 20 column volumes of cold (4°) 10 mM sodium acetate, pH 4.5, using negative pressure to assist flow. After washing with 2 column volumes of 0.1 M HEPES, pH 8.0, the gel is transferred to a centrifuge tube and resuspended in 1 column volume of crude, unbleached heparin (Sigma, St. Louis, MO H5515) at 6 mg/ml dissolved in 0.1 M HEPES, pH 8.0. The tube is tightly sealed and incubated 4 hr at 4° with rocking agitation. Following incubation, the gel is transferred to a scintered glass filter and washed 5 times with 1 column volume aliquots of 0.1 M HEPES, pH 8.0. The incubation solution (heparin ligand) along with the first two washes may be analyzed for unbound heparin to permit estimation of the amount of heparin bound to the gel.

The washed gel is resuspended in 1 column volume of 1.0 M ethanolamine hydrochloride, pH 8.0, transferred to a Corex tube, and incubated at 4° for 2 hr with rocking agitation to block unoccupied binding sites on the gel matrix. After incubation, the gel is again transferred to a scintered glass filter and washed with 10 column volumes of 0.1 M HEPES, pH 8.0, followed by alternating 10 column volume aliquots of 0.1 M sodium acetate, 0.5 M NaCl, pH 4.0, and 0.1 M Tris, 0.5 M NaCl, pH 8.0. Alternating washes with these two solutions are repeated 2 more times. The gel is resuspended in PBS, poured in a glass column of appropriate dimensions, and packed and washed before use. This procedure yields a heparin affinity column with approximately 5.0 mg of heparin bound/ml of gel. Such a column is capable of binding 0.25–0.30 mg CTAP-III/ml gel.

Immunologic Methods

Immunoaffinity Chromatography to Isolate CTAP-III. Recent studies have led to the development of more efficient, higher capacity columns with which to isolate CTAP-III[13] To construct such a column, recombinant protein A-agarose (Repligen, Cambridge, MA), selected for its high IgG binding capacity, is poured with PBS into a 1.5 × 20 cm Econo-column (Bio-Rad, Richmond, CA). Affinity-purified antibody (anti-CTAP-III) in 0.15 M NaCl is then passed over the gel. Unbound antibody is collected and reapplied to the column repeatedly until no further IgG will complex with the protein A matrix as indicated by UV absorbance at 280 nm. Enough gel is used when pouring the column so that the IgG preparation saturated only 50% of the protein A binding sites.

Bound antibody is next cross-linked to the protein A matrix by collecting the gel on a sintered glass filter and washing it with 10 column volumes of 0.2 M triethanolamine in 0.1 M borate buffer, pH 8.2. The gel is then transferred from the sintered glass filter to a solution of 0.2 M triethanolamine in 0.1 M borate buffer containing 50 mM dimethyl pimelimidate

---

[13] T. H. Sisson and C. W. Castor, *J. Immunol. Methods* **127**, 215 (1990).

prepared immediately before this step. The pH of the cross-linking solution is readjusted to 8.2 with concentrated NaOH. The gel cross-linking mixture is continuously agitated for 45 min. The gel–antibody complex is then reacted a second time with freshly made 50 mM cross-linker for another 45 min, at which time the reaction is stopped by first washing the gel on a sintered glass filter with 20 column volumes of 50 mM ethanolamine, pH 8.2, and subsequently reacting the gel with 20 column volumes of the same 50 mM ethanolamine buffer for 5 min. Finally, after terminating the cross-linking reaction, the gel-antibody complex is poured backed into a 1.5 × 20 cm Econo-column with PBS containing 0.02% sodium azide. The procedure for utilizing such an immunoaffinity column has been covered in detail earlier.[14]

Enzyme-linked Immunosorbent Assay for CTAP-III. The CTAP-III antigen is found in high concentrations (10,000–35,000 ng/ml) in serum from freshly clotted blood and at 30–50 ng/ml in the plasma of healthy individuals. Earlier studies showed elevated plasma values for CTAP-III/β-TG antigen in patients with active rheumatoid arthritis, systemic lupus erythematosus, and other types of vasculitis.[15,16] The previously used radioimmunoassay (RIA) is now replaced by the ELISA described below. It should be remembered that the polyvalent antisera used for both RIA and ELISA procedures cannot distinguish between native CTAP-III and any of the isoforms so far described.

Reagents

PBS buffer: 8 g NaCl, 0.2 g KH_2PO_4, 2.17 g $Na_2HPO_4 \cdot 7H_2O$, and 0.2 g KCl in 1 liter Milli-Q water, pH 7.4

PBS–Tween (PBST): PBS buffer plus 0.05% (v/v) Tween 20

Coating buffer: 1.59 g Na_2CO_3, 2.93 g $NaHCO_3$ in 1 liter Milli-Q water, pH 9.6

Phosphate–citrate buffer: 243 ml of 0.1 M citric acid, 257 ml of 0.2 M Na_2HPO_4, and 500 ml Milli-Q water, pH 5.0

Thrombotect solution (platelet-stabilizing medium): Prepare an aqueous solution with 2.5% EDTA, 0.025% 2-chloroadenosine, and 7.0% procaine hydrochloride

Antigen buffer: PBS–Tween plus 0.5% bovine serum albumin (BSA), 10 mM EDTA, pH 7.4 (if background color development is excessive, it may be helpful to prepare this buffer frequently)

Bovine serum albumin, globulin-free, Sigma A7638

---

[14] C. W. Castor and A. R. Cabral, this series, Vol. 163, p. 731.
[15] S. L. Myers, P. A. Hossler, and C. W. Castor, *J. Rheumatol.* **7**, 814 (1980).
[16] D. K. MacCarter, P. A. Hossler, and C. W. Castor, *Clin. Chim. Acta* **115**, 125 (1981).

Biotinylated IgG (anti-CTAP-III antibody)
Vectastain ABC Kit (Peroxidase Standard PK-4000, Vector Laboratories, Burlingame, CA)
H_2O_2, Sigma H-1009
ABTS (2,2′-azinobis-3-ethylbenzthiazolinesulfonic acid, Sigma A-1888)

Blood collection and processing. Blood samples are collected by venipuncture using a 21-guage butterfly needle attached to a 3-way stopcock. Two 3-ml syringes are used; the initial sample is discarded, the second is drawn into a syringe containing 300 μl of the thrombotect solution. Blood and thrombotect solution are mixed by gentle inversion and placed in ice. Plasma is separated from whole blood by centrifugation at 1500 g for 30 min at 4° and stored until assay at −20°.

Biotinylation of IgG. Affinity-purified anti-CTAP-III IgG is prepared by passing rabbit serum containing the antibody over an antigen column constructed as reported.[14] Two to five milligrams of IgG per milliliter is dialyzed versus 0.2 M sodium bicarbonate buffer, pH 8.8, plus 0.15 M NaCl. N-Hydroxysuccinimidobiotin (BNHS, Sigma H-1759) is dissolved in N,N-dimethylformamide (DMF, Sigma D8654) with the aid of a vortex at a concentration of 1 mg BNHS/250 μl DMF. For every 1 mg of IgG, add 20 μl of BNHS in DMF; the reaction mixture is rotated or shaken for 15 min at room temperature. The reaction is stopped by adding 10 μl of 1 M NH_4Cl per milligram of IgG (binds unreactive BNHS). Biotin-labeled antibodies are dialyzed against PBS to remove NH_4Cl and unbound BNHS and stored at 4° with 0.1% azide.

Assay Procedure for Double-Antibody Sandwich ELISA

Coating. Affinity-purified anti-CTAP-III antibody is diluted 1 : 1000 in coating buffer, and 100 μl is pipetted into each microtiter plate well. The antibody dilution is determined by testing a series of antibody dilutions against a known antigen concentration. The dilution chosen is the one showing the best color development with the least background color. One column of wells is set aside as "blank control," and only coating buffer is placed in these wells. These wells are used as a blank for the plate reader before reading the assay well. The plate is covered with plastic wrap (Sysco, Houston, TX), placed in a humidified chamber, and incubated overnight at 37°.

Wash. Wash the plate 3 times with coating buffer.

Back coat wells. To minimize nonspecific adsorption of antigen to the wells, they are incubated with 1% BSA (globulin-free): 200 μl of BSA in coating buffer is placed in each well, and the plate is covered and incubated at 37° in humidified chamber for 1 hr.

Wash. The plate is then washed 3 times with PBST.

Antigen incubation. A standard curve of known CTAP-III concentrations is added to the plate in triplicate, 50 μl/well. Dilutions of antigen are carried out in antigen buffer at 0, 2.5, 5, 10, 25, and 50 ng/ml. A group of 10–20 antigen-coated wells with only antigen buffer added to them are designated as "BKG" or "0" wells. The average of their optical densities (OD) is subtracted from standard and unknowns to obtain a final OD value.

Plasma samples to be assayed are diluted 1:25 and incubated with a glutaraldehyde-cross-linked preimmune rabbit IgG immunoadsorbent overnight. Fifty microliters of each diluted, adsorbed sample is pipetted into duplicate wells. The standards and unknowns are incubated for 1 hr at 37° in a humidified chamber with the plate covered.

Wash. The plate is then washed 3 times with PBST.

Biotinylated IgG incubation. The biotinylated IgG preparation is diluted in antigen buffer, 1:500; 50 μl is pipetted into each well, and the plate is covered, placed in humidified chamber, and incubated for 1 hr at 37°.

Wash. The plate is then washed 3 times with PBST.

Avidin–biotin–horseradish peroxidase incubation. The detection system is introduced by incubating with avidin–biotin–peroxidase using the Vector ABC standard kit (Vector Laboratories). Five microliters of A (avidin) and 5 μl of B (biotin–HRP) are added to 1.0 ml PBS (no Tween) and mixed (this is stable for 72 hr at 4°). The solution is diluted 1:15, 50 μl is pipetted into each well, and the plate is covered, placed in a humidified chamber, and incubated for 1 hr at 37°.

Wash. Next, the plate is washed 5 times with PBST.

Enzyme–substrate incubation. Enzyme substrate solution is prepared by adding 4 mg ABTS to 20 ml of phosphate–citrate buffer, pH 5.0. One then quickly adds 4 μl of 30% H_2O_2, mixes the solution, and adds 50 μl/well. Color is developed by incubating for 1–2 hr at room temperature in the dark.

Measurement of Optical Density. Optical density is measured at 414 nm with a Titertek Multiskan (Flow Laboratories, McLean, VA) and corrected for BKG by subtracting the "0" well average from each value. A standard curve is constructed on semilog paper and used to determine antigen concentrations in unknown samples.

Analytical Electrophoretic Procedures

SDS–Polyacrylamide gel electrophoresis. To assess the homogeneity of CTAP-III preparations and provide information concerning molecular mass, we employ a sodium dodecyl sulfate-polyacrylamide gel electrophoresis (SDS–PAGE) system which separates proteins over the range of 2500 to 90,000 Da.[17] Eight percent polyacrylamide gels containing 8 M urea are cast as slabs (140 × 160 × 1.0 mm) and run vertically in a Protean Bio-Rad cell using a Pharmacia (Piscataway, NJ) ECPS 3000/150 DC power supply. Cooling water supplied to the cell core is maintained at 10° with a Lauda RM3 refrigerated circulating unit. Between 1 and 5 μg of protein per lane is typically run and identified with a silver stain which detects as little as 100 ng of individual proteins.[18]

Analytical Isoelectric Focusing. Analytical isoelectrical focusing (IEF) is carried out on 0.4 mm thick polyacrylamide gels, 125 × 125 mm; the gel composition is 6.25%T, 3%C bisacrylamide, 5% glycerol, and 2% Biolyte ampholytes, pH 3–10. The gel is formed between a glass plate (125 × 125 mm) and the Teflon face of the Bio-Rad Mini IEF casting tray. A 125 × 250 mm strip of permeable polyethylene high-density homopolymer mesh, Delnet RB070-60H (provided by Applied Extrusion Technologies, Inc.), is washed in Tween 20 diluted 1 : 250; this is dried and placed between the glass and Teflon surfaces. Ten milliliters of gel is prepared and degassed by water vacuum for 5 min. Catalysts are then added as follows: 15 μl of 10% ammonium persulfate, 50 μl of 0.1% riboflavin 5′-phosphate, and 3 μl TEMED. The acrylamide solution fills the gel sandwich by capillary action when delivered from a pipette placed at one side of the glass plate. Polymerization occurs in 1 hr when the gel is placed 10 cm from a white fluorescent light.

Gels are run on a Pharmacia flatbed FBE 3000, cooled to 4° by a Lauda RM 3 refrigerated circulating water bath and focused by a Pharmacia ECPS 3000/150 power supply. Samples of 5 to 20 μl are added to a silicone rubber strip containing 11 slots, 7 × 1 mm, 3 mm apart; strips are placed near the midpoint of the gel. One millimeter thick electrode wicks placed 10 cm apart are wet with Serva Anode Fluid 3 and Cathode Fluid 10. Focusing is complete in 3 hr using maximum settings of 2100 V, 6 mA, and 8 W. The initial voltage should not exceed 200 V to avoid band distortion from heating effects. Bio-Rad IEG standards (Cat. No. 161-0310) are used to estimate the pI of protein samples.

Staining and Blotting Techniques. Analytical IEF gels may be stained directly using either Coomassie Brilliant Blue R-250 (C. I. No. 42660) or silver stain.[19] Coomassie blue-stained gels are fixed in 20% trichloroacetic acid for 30 min, then transferred to 0.1% Coomassie Brilliant Blue R-250 dissolved in 45% methanol and 7% acetic acid for 30 min. After destaining

---

[17] B. L. Anderson, R. W. Berry, and A. Telser, *Anal. Biochem.* **132,** 365 (1983).
[18] J. H. Morrissey, *Anal. Biochem.* **117,** 307 (1981).
[19] J. Heukeshoven and R. Dernick, *Electrophoresis* **6,** 103 (1985).

in 45% methanol and 7% acetic acid for 15 min, the gel is reswollen and further destained in 5% methanol and 7% acetic acid for another 15 min. After a final wash in deionized water for 15 to 30 min, the gel is air-dried while taped to a Teflon surface. Bio-Rad IEF standards are diluted 1 : 20 when staining with Coomassie blue and 1 : 100 for the silver stain.

Proteins are transferred to 0.2 μm pore nitrocellulose or Immobilon-P using a semidry blotter (Polyblot, American Bionetics, Inc Hayward, CA) using the buffer system described by Kyhse-Anderson.[20] Transfer conditions were optimized for small proteins (5000 to 25,000 Da). The analytical IEF gel transfer time is 0.3 hr for nitrocellulose and 0.4 hr for Immobilon-P at currents of 0.63 mA/cm² gel area. Proteins transferred from SDS–PAGE gels are run at currents of 2.5 mA/cm² gel area for 0.4 hr for nitrocellulose and 0.5 hr for Immobilon-P. Transferred proteins are immunostained on both membranes as previously described.[14] Proteins adsorbed onto Immobilon-P are Coomassie blue-stained and may be sequenced using an Applied Biosystems (Foster City, CA) 470A/120A gas-phase protein sequinator.[21]

Concluding Comments

It is unlikely that CTAP-III is unique among growth factors with respect to its propensity to undergo cleavage into smaller, biologically active forms. Our recent experience indicates that progress in defining and studying the significance of such forms is facilitated by attention to certain central issues. Key aspects include accomplishing bulk isolation of agonist with a high level of purity, satisfactory electrophoretic separation, blotting, and microsequencing, and access to an adequate spectrum of bioassays. Rapid isolation of multimilligram quantities of CTAP-III is aided by pressure-assisted, high-resolution gel-permeation chromatography. Heparin and immunoaffinity columns used for final purification steps are constructed to provide maximum capacity. Clearly, the ability to microsequence from Coomassie blue-stained Immobilon blots is critical. Analysis of the spectrum of biological activities requires not only access to an array of relevant bioassays, but also adequate amounts of the cleavage products separate from the parent molecule. While we have encountered the cleavage product alone, the more common situation is to find them together, which renders definitive biological activity studies ambiguous. In at least some cases, it may be possible to completely convert the parent compound to the cleavage form enzymatically; porcine elastase,

[20] J. Kyhse-Anderson, *J. Biochem. Biophys. Methods* **10**, 203 (1984).
[21] P. Matsudaira, *J. Biol. Chem.* **262**, 10035 (1987).

for instance, quantitatively converted CTAP-III to CTAP-III(des 1–15)/NAP-2.

Acknowledgments

The authors are indebted to Virginia Castor for preparation of the figure and to Mary Helen Gilbert for preparation of the manuscript. The author is supported by U.S. Public Health Service Grant AR10728, the Michigan Chapter of the Arthritis Foundation, and the University of Michigan Multipurpose Arthritis Center USPHS Grant AR20557.

[40] Purification of Growth Factors from Cartilage

By Yukio Kato, Kazuhisa Nakashima, Katsuhiko Sato, Weiqun Yan, Masahiro Iwamoto, and Fujio Suzuki

Introduction

Cartilage contains factors that stimulate the synthesis of large chondroitin sulfate proteoglycans by chondrocytes in culture.[1–8] These cartilage-derived factors (CDFs) consist of somatomedin-like[2–4] and non-somatomedin growth factors.[4,6,9] They exert various biological effects in cultured chondrocytes, such as stimulation of DNA synthesis in the presence of fibroblast (FGF) and epidermal (EGF) growth factors,[10–12] stimulation of colony formation in soft agar in the presence of 10% serum, and stimula-

[1] Y. Kato, Y. Nomura, Y. Daikuhara, N. Nasu, M. Tsuji, A. Asada, and F. Suzuki, *Exp. Cell Res.* **130**, 73 (1980).
[2] Y. Kato, Y. Nomura, M. Tsuji, R. Watanabe, and F. Suzuki, *Biochem. Int.* **1**, 319 (1980).
[3] Y. Kato, Y. Nomura, M. Tsuji, H. Ohmae, M. Kinoshita, S. Hamamoto, and F. Suzuki, *Exp. Cell Res.* **132**, 339 (1981).
[4] Y. Kato, Y. Nomura, M. Tsuji, M. Kinoshita, H. Ohmae, and F. Suzuki, *Proc. Natl. Acad. Sci. U.S.A.* **78**, 6831 (1981).
[5] Y. Kato, Y. Nomura, M. Tsuji, H. Ohmae, T. Nakazawa, and F. Suzuki, *J. Biochem. (Tokyo)* **90**, 1377 (1981).
[6] F. Suzuki, Y. Hiraki, and Y. Kato, this series, Vol. 146, p. 313.
[7] Y. Hiraki, Y. Yutani, M. Takigawa, Y. Kato, and F. Suzuki, *Biochim. Biophys. Acta* **845**, 445 (1985).
[8] Y. Eilam, A. Beit-Or, and Z. Nevo, *Biochem. Biophys. Res. Commun.* **132**, 770 (1985).
[9] V. Shen, L. Rifas, G. Kohler, and W. A. Peck, *Endocrinology* **116**, 920 (1985).
[10] Y. Kato, R. Watanabe, Y. Hiraki, F. Suzuki, E. Canalis, L. G. Raisz, K. Nishikawa, and K. Adachi, *Biochim. Biophys. Acta* **716**, 232 (1982).
[11] Y. Kato, Y. Hiraki, H. Inoue, M. Kinoshita, Y. Yutani, and F. Suzuki, *Eur. J. Biochem.* **129**, 685 (1983).
[12] Y. Hiraki, Y. Kato, H. Inoue, and F. Suzuki, *Eur. J. Biochem.* **158**, 333 (1986).

tion of proteoglycan synthesis. In this chapter, we describe methods for isolation of non-somatomedin-like growth factors from fetal bovine cartilage.

Assays

Proteoglycan Synthesis. Chondrocytes are isolated from resting cartilage from the ribs of 4-week-old New Zealand White rabbits as described previously.[13] Resting cartilage cells are used because the basal level of proteoglycan synthesis is lower than that of growth-plate chondrocytes. Freshly isolated chondrocytes are seeded at 1×10^4 cells/6-mm microwell in 0.1 ml of Eagle's minimum essential medium (MEM) with 10% fetal bovine serum and 60 µg/ml of kanamycin. When the cells become confluent, they are promptly preincubated for 24 hr in 0.1 ml of a 1:1 (v/v) mixture of Dulbecco's modified Eagle's medium and Ham's F12 medium with 0.3% fetal bovine serum, 32 U/ml of penicillin, and 40 µg/ml of streptomycin (DF medium). They are then transferred for 20 hr to 0.1 ml of DF medium in the presence or absence of CDF. Three hours after the addition of CDF, 10 µl of DF medium containing 0.5 µCi of [^{35}S]sulfate is added. Proteoglycan synthesis is assayed by measuring incorporation of [^{35}S]sulfate into material precipitated with cetylpyridinium chloride after treatment of the medium and cell layers with pronase E.[1,6]

Previous studies have shown that after cultures become confluent, they form multicell layers of well-differentiated, spherical chondrocytes surrounded by an abundant matrix, and that in these cultures added CDF does not increase proteoglycan synthesis beyond the high basal level, probably because of accumulation of endogenous CDF in the matrix.[7] Embryonic chicken[8] and rat chondrocytes[9] can also be used for CDF assay.

Colony Formation by Chondrocytes in Soft Agar. Chick embryo chondrocytes are obtained from 17-day embryonic sternal cartilage. Five to six sterna are separated from soft tissues with forceps and incubated at 37° for 30 min in 10 ml of phosphate-buffered saline (PBS) (calcium- and magnesium-free) containing 2.5 mg/ml of trypsin.[14] Then the sterna are washed 3 times with PBS to remove fibroblastic cells and cut into small pieces (0.5–1.0 mm³) with a scalpel. The tissue fragments are transferred to 10 ml of PBS containing 0.3–1.0 mg/ml (depending on the batch) of crude collagenase (Sigma, St. Louis, MO, type IA) and 2.5 mg/ml of trypsin at 37°. After 1.5 hr, the tissue fragments and cell aggregates are

[13] Y. Shimomura, T. Yoneda, and F. Suzuki, *Calcif. Tissue Res.* **19**, 179 (1975).
[14] Y. Kato, M. Iwamoto, and T. Koike, *J. Cell. Physiol.* **133**, 491 (1987).

dispersed by pipetting in a 10-ml plastic pipette, and the resulting cell suspension is filtered through a nylon sieve (pore size 120 μm). The filtrate is centrifuged at 1500 rpm for 5 min in a Hitachi clinical centrifuge, and the pellet is washed 3 times with DF medium supplemented with 10% fetal bovine serum. The cells are suspended in DF medium supplemented with 10% fetal bovine serum, 0.3% (w/v) tryptose phosphate broth, 32 U/ml of penicillin, and 40 μg/ml of streptomycin (medium A).

A basal layer of 0.25 ml of 0.72% Bacto-agar (Difco, Detroit, MI) in medium A is introduced into a 16-mm petri dish. Chondrocytes (2.5×10^3) are suspended in 0.25 ml of 0.41% Bacto-agar in medium A and used as an overlayer. CDF is added 2 days after cell seeding and then every fourth day. Cultures are maintained in 5% CO_2/95% air at 37°, and after 14 days colonies are counted. A cell colony is defined as a cluster of cells of more than 0.1 mm in diameter.

Radioreceptor Assays of Insulin-Like Growth Factor I (IGF-I) and FGF. Chondrocytes are obtained from resting cartilage of the ribs of 4-week-old rabbits.[13] Cells are seeded at 3×10^4 cells/16-mm microwell in 0.5 ml of MEM supplemented with 10% fetal bovine serum and 60 μg/ml kanamycin. When the chondrocytes become confluent, they are washed twice with MEM supplemented with 20 mM HEPES, pH 7.5, and 0.2% gelatin (buffer A) and then transferred to 0.25 ml of buffer A with various concentrations of cartilage extract or IGF-I in the presence of ^{125}I-labeled IGF-I (2000 Ci/mmol, Amersham, Arlington Heights, IL, 20,000 dpm in 20 μl of water containing 0.03% CHAPS, 0.1% bovine serum albumin, and 0.9% NaCl). The cells are incubated for 2 hr at 4° and then washed 5 times with cold PBS. They are solubilized in 0.5 ml of 0.1 M NaOH/0.2% Triton X-100, and the radioactivity is determined.

In experiments with ^{125}I-labeled basic FGF, chondrocytes are incubated in 0.25 ml of buffer A with various concentrations of cartilage extract or basic FGF in the presence of ^{125}I-labeled basic FGF (1000 Ci/mmol, Amersham, 20,000 dpm in 20 μl of water containing 0.1% CHAPS, 0.1% gelatin, and 0.9% NaCl). The cells are incubated for 90 min at 4° and then washed twice with PBS. They are then incubated at 4° in 0.5 ml of buffer containing 10 mM HEPES, pH 8.0, 0.1% bovine serum albumin, and 0.8 M NaCl. After 5 min, the buffer is removed by aspiration. This procedure is repeated twice. The cells are then solubilized in 0.5 ml of 0.1 M NaOH/ 0.2% Triton X-100, and the radioactivity is determined.

Cartilage

Fetal bovine cartilage obtained from local slaughterhouses within 10 hr of sacrifice was used in previous studies. But it became difficult to

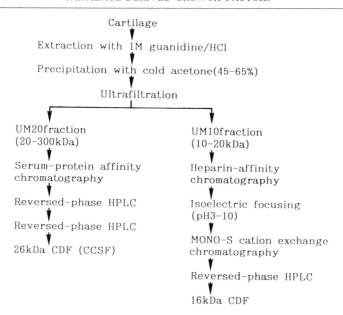

FIG. 1. Procedures for the purification of cartilage-derived factors.

obtain fresh fetal bovine cartilage, so here we report a procedure using fetal bovine cartilage (articular, limb, and rib cartilage) stored for 1 to 3 months at $-20°$. The specific activity of CDF from the frozen cartilage is 3 to 5 times less than that of CDF from fresh cartilage.

Purification of 16-kDa Cartilage-Derived Factor

Cartilage Extract. Cartilage extract is prepared by homogenizing fetal bovine cartilage in 10 volumes of 1 M guanidine hydrochloride/0.1 M 6-amino-n-caproic acid/20 mM 2-(N-morpholino)ethanesulfonic acid (MES), pH 6.0, at $4°$ in a Polytron.[1,6] The homogenate is stirred at $4°$ for 48 hr and then centrifuged at 10,000 g for 20 min. The supernatant is fractionated with cold acetone (45–65%), as described previously[1,6] (Fig. 1). The acetone-fractionated cartilage extract (4 g) is dissolved in 3000 ml of 4 M guanidine hydrochloride/0.1 M 6-amino-n-caproic acid/20 mM Tris-HCl, pH 8.0/1 M NaCl. The solution is filtered through an Amicon (Danvers, MA) XM300 filter. The filtrate is then filtered through an Amicon UM20 filter (20,000 M_r cutoff), and the resulting filtrate is filtered through a UM10 filter (10,000 M_r cutoff). The fractions concentrated with UM10 and UM20 filters are dialyzed for 3–4 days at $4°$ against distilled water and lyophilized. The lyophilized material is suspended in distilled water, and insoluble material is removed by centrifugation. Approximately 0.4 and 1 g (dry weight) of the UM10 and UM20 fractions are obtained from 40 g of the acetone fraction of CDF (from 10 kg cartilage).

Heparin Affinity Chromatography. A column of heparin-5PW (7.5 × 75 mm, Tosoh Co., Tokyo, Japan) connected to a fast protein liquid chromatography (FPLC) system (Pharmacia, Piscataway, NJ) is equilibrated with 25 mM sodium phosphate buffer, pH 7.4, containing 0.15 M NaCl and 0.03% CHAPS. The UM10 fraction (9 mg soluble protein/170 mg dry weight) is applied to the column. After washing with 10 column volumes of equilibration buffer, CDF is eluted with a 0.15–3 M NaCl gradient (30 ml) at a flow rate of 1 ml/min. Fractions of 2 ml are collected. Portions (1–3 μl) of each fraction are used to assay CDF. CDF activity is recovered in the fractions eluted between 0.5 and 1.2 M NaCl. The active fractions are pooled, dialyzed against distilled water, and lyophilized. The recovery of CDF activity from the column of Heparin-5PW is about 60%. About 2.64 mg of protein is obtained.

Isoelectric Focusing. CDF (2.64 mg) partially purified by heparin affinity chromatography is dissolved in 0.2 ml of water. The solution is applied to a thin layer of Sephadex-IEF gel with 2% Pharmalyte, pH 3–10 (Pharmacia). Isoelectrofocusing is carried out in a flat bed apparatus (FBE 3000, Pharmacia) at 2000 V and 15 W at 1°. After 5 hr, 1.5-cm fractions are collected, a portion of each fraction is transferred to 1 ml of water and the pH measured. The rest of each fraction is scraped into a plastic column, and the gels are eluted with 2–4 ml of 0.5 M acetic acid containing 0.1% CHAPS. A portion (0.1 ml) of the eluate is mixed with 0.1 ml of 0.5% bovine serum albumin and dialyzed for 3 days against sterilized saline. Aliquots (1–5 μl) of the dialyzed samples are used for CDF assay. CDF activity is found between pH 6.7 and 8.3 and between pH 8.7 and 9.9. The pH 6.7–8.3 fraction (1.8 mg protein) is dialyzed for 1.5 hr against distilled water at 4°. The pH is adjusted at 7.4 with 0.2 M Na$_2$HPO$_4$, and NaCl is added to a concentration of 0.15 M.

Mono S Cation-Exchange Chromatography. A Mono S HR5/5 column (Pharmacia) is equilibrated with 25 mM sodium phosphate buffer, pH 7.4, containing 0.15 M NaCl and 0.03% CHAPS. CDF (1.8 mg protein) partially purified by isoelectric focusing is applied to the column. After washing with 10 volumes of equilibration buffer, CDF is eluted with a 0.15–1 M NaCl gradient (60 ml) at a flow rate of 1.0 ml/min. Fractions of 2 ml are collected. Portions (1–3 μl) of each fraction are added to cultures grown in 96-well plates to assay of their activity to stimulate proteoglycan synthesis. CDF is eluted in the fractions between 0.2 and 0.5 M NaCl. The recovery of CDF activity from the column of Mono S is about 50%.

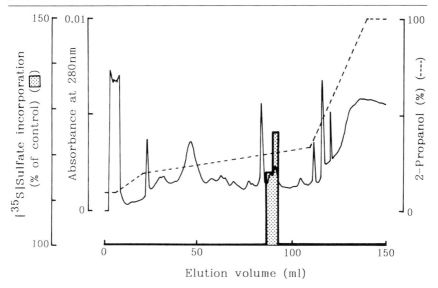

FIG. 2. The active fraction eluted from the Mono S column was mixed with 0.1 volume of 1% trifluoroacetic acid and 0.1 volume of 2-propanol and loaded onto a 3.9 × 150 mm C_{18} reversed-phase column (Waters, μBondasphere). The column was then washed until unbound materials were removed (at a flow rate of 0.5 ml/min with 10% 2-propanol). The column was developed with two linear gradients of 2-propanol (10–18% and 18–33%) in 0.1% trifluoroacetic acid in water. Fractions of 1 ml were collected. CDF activity was eluted with 30.5% 2-propanol.

Reversed-Phase HPLC. Mono S-bound CDF (544 μg) is dissolved in 18 ml of 0.1% (v/v) trifluoroacetic acid in water and applied to a column of μBondasphere C_{18} (5 μm, 100 Å, Waters Ltd., Rochester, MI, 3.9 × 150 mm) which is equilibrated with 0.1% trifluoroacetic acid and 10% 2-propanol in water. A 2-propanol gradient composed of 0.1% trifluoroacetic acid in water as starting buffer and 0.1% trifluoroacetic acid in 100% 2-propanol as limit buffer is used. All solutions are degassed before use. The column is operated at a flow rate of 0.5 ml/min, and the effluent is collected in 1-ml fractions. A portion (100 μl) of each fraction is mixed with 0.2% bovine serum albumin, lyophilized, and dissolved in 100 μl of PBS. Portions (1–5 μl) of the solution are used to assay CDF. The peak of activity is eluted with 30.5% 2-propanol (Fig. 2). The protein in the active fraction is iodinated with ^{125}I by the chloramine-T method and analyzed by SDS–PAGE. One band corresponding to a molecular weight of 16,000 is seen in the autoradiogram in the presence and absence of 2-mercaptoethanol.

Purification of Factors That Stimulate Chondrocyte Colony Formation in Soft Agar

Chondrocytes are unique connective tissue cells that can be maintained in soft agar. Chondrocytes in soft agar retain their differentiated state for more than 3 weeks,[11] although cells maintained on plastic dishes lose their phenotypic characteristics after several generations. Recently, we found that a cartilage extract stimulates the growth of chick embryo chondrocytes in soft agar. The chondrocyte colony-stimulating factor (CCSF) can be partially purified by serum–protein affinity chromatography and reversed-phase HPLC.

Serum Protein Affinity Chromatography. Rabbit serum is fractionated with ammonium sulfate (20–33%), and the precipitated proteins are bound to CNBr-activated Sepharose CL-4B according to the manufacturer's recommended procedure (2 mg protein/ml gel). A column (5 × 11 cm) of serum protein-bound Sepharose CL-4B is equilibrated with PBS. The UM20 fraction (200 mg protein) is applied to the column and washed with 500 ml of PBS. CCSF is eluted from the column with 100–200 ml of 4 M guanidine hydrochloride/Tris-HCl, pH 8.0. The fraction bound to serum protein is dialyzed against distilled water for 2 days at 4° and lyophilized.

Reversed-Phase HPLC. CCSF (50 mg) partially purified by serum protein affinity chromatography is dissolved in 0.5 ml of 0.1% trifluoroacetic acid in water and applied to a column of μBondasphere C$_{18}$ (5 μm, 100 Å, Waters, 3.9 × 150 mm) which is equilibrated with 0.1% trifluoroacetic acid in water. A 2-propanol gradient composed of 0.1% trifluoroacetic acid in water as starting buffer and 0.1% trifluoroacetic acid in 40% (v/v) 2-propanol as limit buffer is used. The column is operated at a flow rate of 0.5 ml/min, and the effluent is collected in 1-ml fractions. A portion (100 μl) of each fraction is mixed with 0.2% bovine serum albumin, lyophilized, and dissolved in 100 μl of PBS. Portions (1–5 μl) of the solution are used to assay CCSF. The peak of activities is eluted with 30% 2-propanol, and the active fraction is rechromatographed on the same column. A portion of the active fraction is labeled with ^{125}I and analyzed by SDS–PAGE under reducing conditions. Three bands corresponding to molecular weights of 22,000–26,000, 14,000–16,000, and 6000–10,000 can be shown by autoradiography. The protein eluted from the 22,000–26,000 band has both CCSF and proteoglycan synthesis stimulating activity. Thus, CCSF is a CDF. Figure 3 shows that CCSF, as well as a crude cartilage extract, stimulates colony formation by chick embryo chondrocytes in soft agar in a dose-dependent manner.

Molecular Weight of Various Cartilage-Derived Factors. The UM10 fraction (300 μg) and the serum protein-bound fraction (300 μg) are each

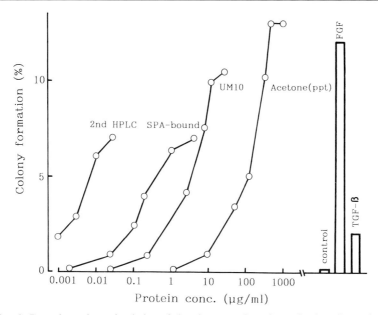

FIG. 3. Dose-dependent stimulation of chondrocyte colony formation in soft agar by the acetone-fractionated cartilage extract, the UM10 fraction, and CCSF purified by serum protein affinity chromatography and reversed-phase HPLC. Basic FGF (0.4 ng/ml) and transforming growth factor β1 (TGF-β1, 3 ng/ml) also stimulated colony formation by chondrocytes in soft agar.

dissolved in 0.1 ml of 10 mM Tris-HCl buffer, pH 8.0, containing 0.1% SDS and 20% glycerol, and incubated for 2 hr at 37°. The samples are then applied on SDS–polyacrylamide (17.8%) gels (running length 5 cm, width 2 cm, thickness 1 mm). After electrophoresis, the gels are sliced (2 mm width). Each slice is cut into three or four pieces, placed in 0.15 ml of PBS in a 1.5-ml plastic centrifuge tube, and allowed to stand for 18 hr at 4°. The gels are washed with 0.1 ml of PBS, and the washing fluid is combined with the eluate. Aliquots (1–2 μl) of the samples are used for CDF assays. For assay of CCSF activity, a portion (0.2 ml) of the samples is mixed with an equal volume of PBS containing 0.2% bovine serum albumin and dialyzed for 2 days against distilled water at 4°, because chondrocytes in soft agar are sensitive to low concentrations of SDS. Aliquots (1–2 μl) of the samples are added to the semiliquid medium of soft agar cultures. In addition, aliquots (5 μl) of the samples are used to test IGF-I activity.

On SDS–PAGE, the UM10 fraction (10–20 kDa CDF) gives three peaks of CDF activity corresponding to molecular weights of 24,000–27,000,

14,000–16,000, and 6000–11,000. The protein with a mobility corresponding to 6–11 kDa also competes with ^{125}I-labeled IGF-I in radioreceptor assays. On SDS–PAGE, the serum protein-bound fraction gives two peaks of CDF activity corresponding to molecular weights of 22,000–26,000 and 14,000–16,000. The 22–26K but not the 14K–16K protein induces colony formation by chondrocytes in soft agar.

Although a crude cartilage extract shows high levels of basic FGF[15] in the radioreceptor assay, no detectable basic FGF is found in the UM20 and UM10 fractions, probably because FGF is inactivated or removed during ultrafiltration. Transforming growth factor β (TGF-β) is reported to be present in cartilage extract[16,17] and to stimulate proteoglycan synthesis by chondrocytes in culture.[18,19] However, TGF-β alone, unlike CCSF, induces only low levels of colony formation by chondrocytes in soft agar at a maximal dose, although it increases the efficiency of colony formation in the presence of basic FGF.[20] Furthermore, the UM20 and UM10 fractions have a very low level of TGF-β on the basis of radioreceptor assays (M. Iwamoto and Y. Kato, unpublished data). Thus the CDFs of 16K and 22–26K seem to be novel growth factors, although their amino acid sequences have not yet been determined.

Acknowledgment

We thank Mrs. Elizabeth Ichihara for assistance in preparation of the manuscript.

[15] R. Lobb, J. Sasse, R. Sullivan, Y. Shing, P. D'Amore, J. Jacobs, and M. Klagsbrun, *J. Biol. Chem.* **261,** 1924 (1986).
[16] L. R. Elingsworth, J. E. Brennan, F. Fok, D. M. Rosen, H. Bentz, K. A. Piez, and S. M. Seyedin, *J. Biol. Chem.* **261,** 12362 (1986).
[17] J. L. Carrington, A. B. Roberts, K. C. Flanders, N. S. Roche, and A. H. Reddi, *J. Cell Biol.* **107,** 1969 (1988).
[18] Y. Hiraki, H. Inoue, R. Hirai, Y. Kato, and F. Suzuki, *Biochim. Biophys. Acta* **969,** 91 (1988).
[19] H. Inoue, Y. Kato, M. Iwamoto, Y. Hiraki, M. Sakuda, and F. Suzuki, *J. Cell. Physiol.* **138,** 329 (1989).
[20] M. Iwamoto, K. Sato, K. Nakashima, H. Fuchihata, F. Suzuki, and Y. Kato, *Biochem. Biophys. Res. Commun.* **159,** 1006 (1989).

[41] Purification, Biological Assay, and Immunoassay of Mammary-Derived Growth Inhibitor

By R. GROSSE, F.-D. BOEHMER, P. LANGEN, A. KURTZ, W. LEHMANN, M. MIETH, and G. WALLUKAT

Introduction

Mammary-derived growth inhibitor (MDGI),[1] a 14.5-kDa polypeptide, has been purified from lactating bovine mammary gland[2-4] and from milk fat globule membranes.[5] Enrichment of inhibitory activity was pursued by a proliferation assay using Ehrlich ascites mammary carcinoma cells (EAC cells). MDGI partially blocks the proliferation of various normal and transformed mammary epithelial cells.[4,6]

The primary structure of MDGI was determined, and it turned out that the inhibitor belongs to a family of structurally related proteins including fatty acid-binding proteins (FABP), cellular retinoid-binding proteins (CRABP and CRBP), adipocyte differentiation associated protein (P422), and myelin P-2.[1,4] There is no relationship between MDGI and other known growth inhibitors. Thus, along with interferons, transforming growth factors β, and tumor necrosis factor, MDGI is one of the few naturally occurring growth inhibitors for mammary epithelium identified so far. Immunologically, MDGI is also related to a fibroblast growth inhibitor, designated as FGR-s.[7] The proteins, homologous to MDGI, basically share two properties in common: they bind hydrophobic ligands such as long-chain fatty acids, retinoids, and eicosanoids, and they are expressed in a differentiation-dependent manner in heart, liver, or intestine.[1] The exact functional meaning of both properties is not clear. For MDGI, likewise,

---

[1] R. Grosse and P. Langen, *Handb. Exp. Pharmacol.* **95,** 249 (1990).
[2] F.-D. Boehmer, W. Lehmann, H.-E. Schmidt, P. Langen, and R. Grosse, *Exp. Cell Res.* **150,** 445 (1984).
[3] F.-D. Boehmer, W. Lehmann, F. Noll, R. Samtleben, P. Langen, and R. Grosse, *Biochim. Biophys. Acta* **846,** 145 (1985).
[4] F.-D. Boehmer, R. Kraft, A. Otto, C. Wernstedt, U. Hellmann, A. Kurtz, T. Mueller, K. Rohde, G. Etzold, W. Lehmann, P. Langen, C.-H. Heldin, and R. Grosse, *J. Biol. Chem.* **262,** 15137 (1987).
[5] R. Brandt, M. Pepperle, A. Otto, R. Kraft, F.-D. Boehmer, and R. Grosse, *Biochemistry* **27,** 1420 (1988).
[6] W. Lehmann, R. Widmaier, and P. Langen, *Biomed. Biochim. Acta* **48,** 143 (1989).
[7] F.-D. Boehmer, Q. Sun, M. Pepperle, T. Mueller, U. Eriksson, J. L. Wang, and R. Grosse, *Biochem. Biophys. Res. Commun.* **148,** 1425 (1987).

both properties were shown to be characteristic.[8] However, lipid binding to MDGI is not necessary for exerting its biological activities in a growth assay or in a β-adrenergic receptor assay with neonatal rat heart myocytes.

Expression of MDGI mRNA and protein is coupled to differentiation in the mammary gland, reaching some maximal level during functional differentiation.[9] The cDNA for MDGI was recently cloned in our group and used for an *in situ* hybridization analysis. The data strongly support a role for MDGI in the normal development of mammary gland.

In this chapter we describe the assays and purification procedures which have been developed to identify and to characterize the biological activities of MDGI. Furthermore, the immunoassay for evaluating MDGI levels and the procedure for cloning a cDNA coding for MDGI are outlined. Finally, we show by use of the *in situ* hybridization technique that MDGI expression is coupled to differentiation in the developing bovine mammary gland.

Growth Inhibitor Assay with EAC Cells

A hyperdiploid line of EAC cells, derived from a mammary carcinoma of mice, is routinely used for monitoring purification (see below) of MDGI. Cells are aspirated from the intraperitoneal cavity of mice 12 to 14 days after transplantation, diluted 5-fold with phosphate-buffered saline (PBS), and centrifuged at 300 g for 5 min. The centrifugation step is repeated once before the cells are resuspended in PBS at a concentration of 2.5 to 3.0 × 10^6 cells/ml. The suspension (0.2 ml) is added to 1.8 ml of medium 199 with Hanks' salts (6.4 mg/ml), supplemented with 4% heat-inactivated calf serum, 2 mM disodium phosphate, 2 mM monopotassium phosphate, 5.2% glucose, 6 mM glutamate, 29 mM sodium bicarbonate, 0.2 mM reduced glutathione, and antibiotics (100 IU penicillin and 50 µg streptomycin per milliliter). Cells are cultured in suspension according to Negelein *et al.*[10] in Warburg vessels in an atmosphere of 1.5% O_2 and 5% CO_2 for 24 hr at 37°. Caution: EAC cells are very sensitive to oxygen and grow *in vitro* only at low oxygen. After 24 hr cells are counted in a hemocytometer; the standard deviations are routinely less than 5%. The increase in cell number during 24 hr is usually 60 to 100%. Maximal inhibition of cell proliferation by purified MDGI is in the range of 35 to 50%. Only cells from the stationary phase of growth *in vivo* respond to MDGI. Cells of that type

[8] F.-D. Boehmer, M. Mieth, G. Reichmann, C. Taube, R. Grosse, and M. D. Hollenberg, *J. Cell. Biochem.* **38**, 199 (1988).
[9] T. Mueller, A. Kurtz, F. Vogel, H. Breter, U. Schneider, U. Engstroem, M. Mieth, F.-D. Boehmer, and R. Grosse, *J. Cell. Physiol.* **138**, 415 (1989).
[10] E. Negelein, I. Leistner, and L. Jaehnchen, *Acta Biol. Med. Ger.* **16**, 372 (1966).

contain 25–35% cells in G_1 phase, 35–45% in G_2, and 15–25% in the S phase of the cell cycle.[6] If cells in the stationary growth phase are preincubated *in vitro* for 4 hr in 4% calf serum and then subjected to the standard 24-hr assay, the cells lose their sensitivity to MDGI. Instead of using serum under the same conditions, 5–10 ng/ml of platelet-derived growth factor (PDGF) can be used to achieve the same effect. Insulin and epidermal growth factor (EGF) also prevent MDGI inhibition if added simultaneously with MDGI under standard conditions.

Purification of Mammary-Derived Growth Factor

Several purification schemes for MDGI have been published.[2-4] Most of the work on biological characterization as well as on the primary and secondary structure has been performed with preparations obtained by the method designated as Method A (see below). The availability of specific antibodies and the knowledge of the close relationship of MDGI to the cardiac fatty acid-binding proteins[4] prompted us to develop a new purification scheme similar to the one described for bovine cardiac FABP by Jagschies *et al.*[11] This method has not been published previously and is designated below as Method B. A comparison of both methods is schematically illustrated in Fig. 1.

Method A

Healthy lactating bovine mammary glands obtained from a slaughterhouse are dissected from most of the fat and connective tissue, cut into 100-g pieces, and frozen with liquid nitrogen. The frozen tissue can be stored at $-20°$ up to 1 month. For purification of MDGI, 0.8–0.9 kg of tissue is partially thawed for 30–60 min at room temperature. All further steps are at 0–4°. The tissue is chopped into 5- to 10-g pieces and homogenized in 1.3 liters of buffer A (10 mM sodium phosphate, pH 7.4, 10 mM NaCl) in a Waring blendor for 2 min. The homogenate is centrifuged for 15 min at 10,000 g (Beckman centrifuge J21, rotor JA20); the supernatant is carefully aspirated with a tubing, avoiding suction of fat and loosely pelleted material. The collected supernatant is then centrifuged for 1 hr at 50,000 g (Beckman rotor 19). The supernatant from this centrifugation step is carefully decanted through four layers of cotton gauze. Under continuous stirring with a magnetic stirrer, solid, finely ground ammonium sulfate is added over 10–15 min to a final concentration of 30% saturation. Stirring is continued for another 15 min, then precipitation is allowed for

[11] G. Jagschies, M. Reers, C. Unterberg, and F. Spener, *Eur. J. Biochem.* **152**, 537 (1985).

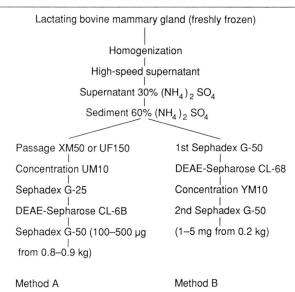

FIG. 1. Schematic illustration of two different purification schemes for mammary-derived growth inhibitor (MDGI).

2 hr. The precipitate is removed by centrifugation for 30 min at 6000 g, and the supernatant undergoes a second precipitation with ammonium sulfate at a final concentration of 60% saturation as described before.

The precipitate is collected by centrifugation for 30 min at 8000 g and dissolved by stirring in 360 ml of buffer A. Insoluble material is removed by centrifugation for 45 min at 75,000 g, and the supernatant of this centrifugation is passed through a 50,000 molecular weight cutoff membrane filter (Amicon, Danvers, MA, XM50 or UF150 filter, VEB Zellstoffwerk Wittenberge, Germany) overnight. The filtrate is then concentrated with an Amicon UM10 or YM10 membrane to about 10 ml and chromatographed over a 1.5 × 30 cm column of Sephadex G-25 (fine), previously equilibrated with 50 mM imidazole/HCl, pH 8.0 (buffer B). Protein eluting in the void volume is collected (~15 ml) and applied to a 2-ml column of DEAE-Sepharose CL-6B, previously equilibrated with buffer B. After application of the sample the column is washed with 20–30 ml of buffer B and then eluted with buffer B containing 50 mM NaCl. The protein-containing fractions (5–8 ml) are collected and chromatographed over a 1.5 × 30 cm column of Sephadex G-25 (fine), previously equilibrated with buffer A. Protein containing-fractions in the void volume are collected, aliquoted in siliconized glass ampoules, and stored frozen at −20°. Material obtained in this way (100–500 μg) is about 95% pure, as judged by polyacrylamide gel electrophoresis, and retains biological activity for 2–3 weeks of storage.

Method B

Method B was developed to circumvent the passage of MDGI through an ultrafiltration membrane and thereby to increase recovery and reproducibility of the preparation. In addition several changes were made to simplify the procedure and to speed up the preparation procedure.

Tissue (0.2 kg) is partly thawed, chopped, and homogenized in a Waring blendor with 600 ml of a buffer C, made, immediately before homogenization, 1 mM in EDTA and 0.1 mM in phenylmethylsulfonyl fluoride (PMSF) by the addition of 1/100 and 1/1000 volume of 0.1 M EDTA, pH 7.4, and 0.1 M PMSF in 2-propanol, respectively, to buffer A. The homogenate is centrifuged at low and high speed as described in Method A and is then subjected to sequential precipitation with ammonium sulfate, also as described above, except that the total time for one precipitation step is only 1 hr. Centrifugation of the precipitates is performed for 15 min at 10,000 g.

The precipitate from the 60% ammonium sulfate step is dissolved in 30 ml of buffer C, then centrifuged for 30 min at 50,000 g, and the supernatant is applied to a 5 × 80 cm column of Sephadex G-50, previously equilibrated with a solution containing 50 mM imidazole/HCl, pH 8.0, 10% glycerol, and 1 mM 2-mercaptoethanol (buffer D). The column is then eluted with buffer D at a flow rate of 80–100 ml/hr; 10-ml fractions are collected. MDGI-containing fractions are identified on the basis of a dot-immunobinding assay with anti-MDGI antiserum. Typically, the material eluting between about 800 and 900 ml contains most of the anti-MDGI-reactive material. These fractions are pooled and applied to a 1.5 × 15 cm column of DEAE-Sepharose CL-6B, equilibrated with buffer D, at a flow rate of 30–40 ml/hr. The column is washed with about 50 ml buffer D and then eluted with 50 mM NaCl in buffer D. The protein-containing fractions are pooled (typically ~30 ml) and concentrated with an Amicon YM10 ultrafiltration membrane to about 5 ml. The concentrate is then applied to a 2.5 × 80 cm column of Sephadex G-50 (fine), equilibrated with 20 mM sodium phosphate, pH 7.4, 140 mM NaCl (PBS). The column is developed with PBS at a flow rate of 30–40 ml/hr. MDGI-containing fractions (typically eluting as a single peak at ~200 ml elution volume, see Fig. 2 for a characteristic chromatogram) are identified on the basis of immunoreactivity, pooled, aliquoted, and stored as described above.

The yield of MDGI using Method B is 1–5 mg for one preparation, that is, 5–25 mg/kg mammary gland tissue. Preparations obtained by Method

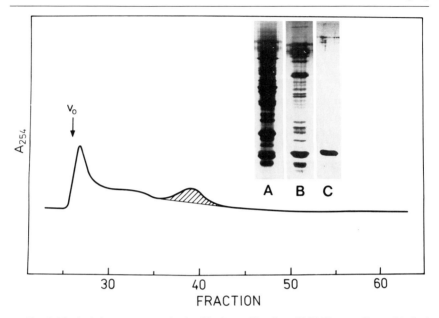

FIG. 2. Typical chromatogram obtained in the purification of MDGI according to Method B. Chromatography of the fractions obtained by DEAE-Sepharose chromatography and subsequent concentration over a 2.5 × 80 cm column of Sephadex G-50 (second Sephadex G-50 chromatography). The indicated fractions contained purified MDGI. In the inset, SDS–polyacrylamide gels are shown. (A) Pooled material after the first Sephadex G-50 step; (B) pooled material after the DEAE-Sepharose step; (C) purified MDGI after the second Sephadex G-50 step. Bands were detected by the silver staining method.

B are indistinguishable from preparations obtained by Method A with respect to inhibition of Ehrlich cell proliferation, inhibition of β-adrenergic supersensitivity in neonatal cardiac myocytes (see below), behavior in sodium dodecyl sulfate (SDS)–polyacrylamide gel electrophoresis, isoelectric focusing, and immunoreactivity.

Immunoassay of Mammary-Derived Growth Inhibitor

A radioimmunoassay for MDGI was developed to estimate MDGI levels in various physiological states of the mammary gland[9] and in mammary epithelial cells.

Radioiodination

MDGI is delipidized by incubating 100–150 μg of MDGI in a total volume of 0.5 ml PBS with 0.5 ml (sedimented gel bed) of Lipidex 1000 (Pachard, Downers Grove, IL) previously equilibrated with PBS for 30 min at 37°. The mixture is centrifuged and the supernatant carefully aspirated. Under these conditions 50–70% of the MDGI protein is recovered in the supernatant. In analytical experiments at least 90% of ^{3}H-labeled palmitic acid prebound to MDGI could be removed by the described Lipidex treatment. Delipidized MDGI is labeled with ^{125}I according to the method of Hunter and Greenwood[12] with the following modifications: 4 μg of delipidized MDGI in about 0.04 ml PBS is mixed with 0.03 ml freshly prepared 250 mM sodium phosphate, pH 7.5, 1/10 volume of 1% SDS and about 1 mCi (37 MBq) sodium [^{125}I]iodide (Institute of Isotopes, Budapest, Hungary, ~2000 Ci/mmol). The reaction is started by the addition of 50 μg chloramine-T dissolved in 10 μl water and quenched after 1 min by addition of 50 μl 2-mercaptoethanol (1 mg/ml). The mixture is then chromatographed over a 5-ml column of Sephadex G-25 (fine), prepared in a pipette tip, equilibrated with 10 mM sodium phosphate, pH 7.4, 10 mM NaCl containing 0.05% gelatin and 0.1% SDS. The radioiodinated MDGI is collected from the void volume (1.5–2.5 ml). Typically, the incorporation of label is 130–160 μCi/μg.

Radioimmunoassay

Ninety-six-well immunoplates (Nunc, Naperville, IL) are coated overnight with protein A by incubating 25 ng protein A/well in 0.1 ml of 50 mM sodium bicarbonate (pH 9.6). After washing with 50 mM Tris, pH 7.5, 200 mM NaCl, 0.2% Tween 20 (TBS/Tween), the nonspecific binding is blocked by incubation with 3% bovine serum albumin (BSA) in TBS/Tween (0.25 ml/well) for 0.5 hr at 25°. The wells are washed 2 times with TBS/Tween. Then incubation with affinity-purified polyclonal rabbit anti-MDGI antibodies (1.0 μg/ml dissolved in 3% BSA in TBS/Tween, 0.25 ml/well) is performed for 4 hr at room temperature. The wells are washed 3 times with TBS/Tween before the samples to be analyzed for immunoreactivity are added for a 2-hr incubation at room temperature. Samples are pretreated either by boiling for 5 min in 1% SDS or by an incubation for 30 min in 0.1% SDS at 37°. The SDS concentration in the assay is reduced below 0.1% SDS by diluting the samples in 3% BSA/Tween. After 2 hr, ^{125}I-labeled MDGI [~25,000 counts/min (cpm) or 0.02 ml/well] is added, and the incubation is continued at 4° overnight. The wells are then extensively washed with TBS/Tween, and the bound radioactivity is dissolved by incubation for 2 hr with 1 N sodium hydroxide. The dissolved radioactive samples are removed and counted in a multicrystal γ counter (Berthold), and data are fitted with an inbuilt routine on the basis of a

---

[12] W. M. Hunter and F. C. Greenwood, *Nature (London)* **194**, 495 (1962).

standard curve run as control on every plate. The range of the assay is 3–300 ng/ml. A maximum of 27.8% of the radioactivity is bound, with a nonspecific binding level of 0.1%.

Inhibition of Growth of Permanent Mammary Carcinoma Cell Lines

The proliferation assay is based on a serum starvation to trigger cells into quiescence followed by a restimulation wtih fresh medium. MDGI is present during the restimulation period for 16–20 hr. Flow cytophotometric measurements with MaCa 2017 mouse mammary carcinoma cells, MATU and T47D human mammary carcinoma cells, or normal human mammary epithelial cells obtained from normal reduction mammoplasty prove the cells to be arrested in G_1/G_0.[6]

In general, the following general protocols are used for the different cell lines. For mammary carcinoma cell lines T47D, MCF7, and MaCa 2017, cells are plated in 24-well plates (10^4–10^5 cells/well) and cultured for 3–5 days in Eagle's minimum essential medium (MEM, 1 ml/well) supplemented with 4% newborn calf serum and 50 μg/ml gentamicin in a humidified atmosphere of 95% air/5% CO_2 at 37° until confluency is reached. The cells are washed with serum-free medium and then receive fresh Eagle's MEM containing 0.2% fetal calf serum. After a 48-hr period, the medium is changed for fresh Eagle's MEM with 10% calf serum for restimulation. Twenty hours later [^{3}H]TdR (2 μCi/ml, 10^{-6} M TdR) is added for 1 hr. The cells are washed twice with PBS, trypsinized, and the suspension dotted on filter paper. The filters are washed once with 10% trichloroacetic acid (TCA). After extensive washing with water they are dried and counted in a dioxane-based scintillation cocktail. Alternatively, fixed cells are solubilized in 1% SDS, 0.1 N NaOH, 2% Na_2CO_3 before aliquots are taken for estimation of [^{3}H]TdR incorporation into DNA.

For MATU cells, cells are grown in Eagle's MEM with 4% newborn calf serum until confluency. Cells are first maintained for 72 hr in a medium with 0.1% calf serum which is changed 3 times and then during an additional starvation period of 48 hr in serum-free Eagle's MEM medium including two medium changes. The cells are then restimulated by adding fresh medium with 4% calf serum. DNA synthesis is measured 24 hr later by adding a [^{3}H]TdR pulse for 1 hr as described before.

For normal human mammary epithelial cells, cells are grown in the standard MCDB 170 medium containing 5 μg/ml insulin, 0.14 μM hydrocortisone, 10 ng/ml EGF, 0.1 mM ethanolamine, 0.1 mM phosphoethanolamine, 5 μg/ml transferrin, 70 μg/ml bovine pituitary extract, and 4% fetal calf serum.[6] Confluent cells are serum starved for 24 hr in MCDB 170 medium without supplements and then restimulated by adding fresh medium plus all supplements for another 24-hr period. Incubation is terminated by a 1-hr [^{3}H]TdR pulse. MDGI is added in a volume of 20 µl in isotonic salt solution with the restimulation medium.

For human and mouse mammary carcinoma cell lines, maximal inhibition ranged from 30 to 50%.[4-6] MCF-7 cells did not respond to MDGI. DNA synthesis in normal human mammary epithelial cells was inhibited up to 70%, depending on the number of passages.[6]

Assay of Mammary-Derived Growth Inhibitor-Induced Modulation of β_2-Adrenergic Receptor Function

The following assay is based on the induction of β-adrenergic supersensitivity in neonatal rat heart myocytes by L-(+)-lactate[13] or arachidonic acid and the inhibition of this effect by MDGI.[14]

Single cells are dissociated from the minced heart ventricles of 1- to 2-day-old Wistar rats with a 0.2% solution of trypsin. Cells are cultured at 37° in an air atmosphere for 3 or 4 days as monolayers on the 17-cm^2 bottom of Mueller flasks (2.4 × 10^6 seeded cells) in 3 ml of Halle SM 20-I medium[15] that is supplemented with 10% of heat-inactivated calf serum and 2 µM fluorodeoxyuridine, the latter to prevent proliferation of nonmuscle cells. The flasks are attached to a platform that is continuously rocked back and forth at a frequency of 2 cycles of ±30° each per minute. The medium is renewed at the end of the first, fourth, and seventh days. On the eighth day the medium is replaced, with washing, by 2 ml of serum-free medium. Three groups of flasks are prepared: (a) control flasks without additions, except the solvents of the reagents indicated below, (b) flasks with L-(+)-lactate in a final concentration of 3 mM (60 µl of a 100 mM solution in the medium described above) or with added arachidonic acid in a final concentration of 10 nM (20 µl of a 1 µM stock solution prepared by dilution in serum-free culture medium of a 0.1 mM pre-stock serum-containing medium, with 0.1% ethanol) for the induction of supersensitivity, and (c) flasks with 0.1–50 ng/ml MDGI in addition to lactate or arachidonic acid.

After another 2 hr of incubation on the rocker platform, the flasks are transferred to the heated stage of an inverted Zeiss (Jena, Germany) microscope, where 10 small circular fields of the cell layer (0.8 mm^2, 10

[13] G. Wallukat and A. Wollenberger, in "Pharmacological Aspects of Heart Disease" (R. E. Beamish, V. Panagin, and N. S. Dhalla, eds.), p. 217. Nijhoff, Boston, Dodrecht, Lancaster, 1987.
[14] G. Wallukat, F.-D. Boehmer, U. Engstroem, P. Langen, M. D. Hollenberg, J. Behlke, W. Kuehn, and R. Grosse, Mol. Cell. Biochem., in press (1989).
[15] P. Chomczynski and N. Sacchi, Anal. Biochem. 162, 156 (1987).

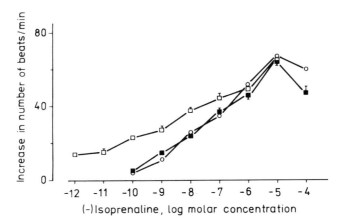

FIG. 3. Dose–response curves for β-adrenergic stimulation in cells without pretreatment (○) or with a 2-hr pretreatment with 10 nM arachidonic acid in the absence (□) or presence of (■) 50 ng/ml MDGI.

mm apart from each other) are inspected through the perforations of a metal template. This device allows the flask to remain at 37° (±0.3°) throughout the experiment. The number of beats of a selected isolated myocardial cell or a group of cells in each of the 10 fields is counted for 15 sec. After measuring the control beating rate, increasing amounts of ($-$)-isoprenaline are added in a volume of 20 μl, starting at a final concentration of 10^{-12} M and continuing in steps of 1 order of magnitude to 10^{-5} M (cumulative dose–response curve). Starting 5 min after each addition of the reagent the beating rate is estimated as described. A typical result of this type of analysis is shown in Fig. 3. The effect of both the inducers of supersensitivity as well as MDGI is most clearly seen at an ($-$)-isoprenaline concentration of 10^{-10}–10^{-9} M, so that dose–response curves for MDGI-dependent modulation are carried out at one of these concentrations.

Isolation of cDNA Encoding Mammary-Derived Growth Inhibitor
Preparation of mRNA

RNA from lactating bovine mammary gland is prepared according to Ref. 16 by homogenization with a tissue mincer (maximum speed, 3 times, 30 sec each) of fresh tissue in a solution composed of 4 M guanidinium thiocyanate, 25 mM sodium citrate, pH 7.0, 0.5% sodium lauroylsarcosi-

[16] H. Aviv and P. Leder, *Proc. Natl. Acad. Sci. U.S.A.* **69**, 1408 (1972).

nate, and 0.1 M 2-mercaptoethanol (10 ml/g fresh tissue). To this suspension 2 M sodium acetate, pH 4.0 (6 ml/10 ml) 6 volumes (60 ml/10 ml) of phenol, and 1.2 volumes (12 ml/10 ml) of chloroform/isoamylalcohol (49 : 1) are added. After intensive mixing the sample is incubated for 15 min at 0°, followed by centrifugation for 20 min at 10,000 g. The aqueous phase is collected and RNA precipitated by adding 2 volumes of ethanol. After storing for 1 hr at $-20°$ the mixture is centrifuged at 10,000 g. The pellet is dissolved in a solution containing 0.5% SDS and 1 M 2-mercaptoethanol and precipitated again with ethanol. A yield of about 1 mg of RNA can be expected from 1 g of tissue. Poly(A) RNA is obtained by two cycles of chromatography on oligo(dT)-cellulose according to a published protocol,[16,17] with a yield of 0.5%. RNA is shown to be intact by using a standard *in vitro* translation assay with 1 μg of poly(A) RNA in a wheat germ system.[18]

Synthesis of Bovine Mammary Gland cDNA Library

The method for cDNA synthesis has been described elsewhere, and here only the general strategy is outlined.[19] Complementary DNA is synthesized according to Gubler and Hoffman.[20] Briefly, single-stranded cDNA is prepared by oligo(dT) priming of a reverse transcriptase reaction. RNase H is used to nick the RNA to prime second-strand synthesis using *Escherichia coli* DNA polymerase and *E. coli* ligase. The double-stranded cDNA is treated with T4 DNA polymerase to obtain flush ends. The cDNA is cloned directly by blunt-end ligation into the *Hinc*II site of the pUC19 plasmid. The cloned cDNA is transformed into *E. coli* DH5α by the method of Hanahan.[19,21]

Screening of cDNA Library

For identification of MDGI clones, a fragment of the bovine mammary gland cDNA is amplified by the polymerase chain reaction[22] using the degenerate oligonucleotide primers 5'-CAAGCTTAAQTTQGAQGAQT-AQATQ-3', designated as A_1, and 5'-GGGAATTQTGNCCPTTCCAQT-

[17] M. Edmonds, M. Vaughn, Jr., and N. Nakazato, *Proc. Natl. Acad. Sci. U.S.A.* **68,** 1336 (1971).
[18] C. W. Anderson, J. W. Straus, and B. S. Dudeck, this series, Vol. 101, p. 635.
[19] H. Okayama, M. Kawaichi, M. Brownstein, F. Lee, T. Yokota, and K. Arai, this series, Vol. 154, p. 3.
[20] K. Gubler and B. J. Hoffman, *Gene* **25,** 263 (1983).
[21] O. H. Hanahan, *J. Mol. Biol.* **166,** 557 (1983).
[22] R. K. Saiki, U. B. Gyllenstein, and H. A. Erlich, "Genome Analysis," p. 141. IRL, Oxford, 1988.

TQTG-3', designated as A_2, corresponding to amino acids at positions 16 to 21 (A_1) and 97 to 101 (A_2) of the described MDGI sequence.[4,23] The oligonucleotides include additional restriction endonuclease sites for HindIII (A_2) and EcoRI (A_2) to facilitate ligation. The amplified fragment of 258 base pairs (bp) is isolated, cleaved with EcoRI and HindIII, and cloned into pUC19 for sequence analysis according to Chen and Seeburg.[24,25] Sequence analysis shows that this fragment codes for the anticipated amino acids 16 to 101 of the reported MDGI amino acid sequence[4] with the following exceptions. At position 41 Leu is exchanged for Thr, a Ser is exchanged for Leu at position 44, and Glu is substituted by His at position 94.

About 80,000 bacterial colonies of the bovine mammary gland cDNA library are probed with the 258-bp fragment. Duplicate nitrocellulose filters are screened under low stringency hybridization conditions to allow the identification of incompletely matched clones which may represent less abundant MDGI sequence variants. Nine cDNA clones are isolated, and all of them show an insert length of approximately 680 bp. Nucleotide sequence analysis performed by using the dideoxy chain termination method[24,25] reveals identical sequences for all clones (Fig. 4). The open reading frame codes for the published sequence for MDGI,[4] although differences were found at six positions. In positions 41 and 44 the cDNA codes for Thr and Glu, respectively, instead of Leu and Ser as found in the amino acid sequence analysis. The other differences at positions 13 (Asp for Ser), 15 (Lys for Glu), 94 (His for Gln), and 128 (Thr for Val) represent ambiguous determinations in the reported protein sequence, where the amino acids predicted from the cDNA sequence were also found but at lower yield. The Met residue at position 1 and the Ala at position 133 of the cDNA deduced sequence were not found by protein sequencing and are possibly removed by posttranslational protein processing.

Dot-Blot Analysis for Sequence Variants

In order to investigate the possible appearance of nonabundant cDNA species or genes coding for different MDGI forms, the approach of selective hybridization[26,27] with a collection of synthetic oligonucleotides is

---

[23] A. Kurtz, F. Vogel, K. Funa, C.-H. Heldin, and R. Grosse, *J. Cell Biol.* **110**, 1779 (1990).
[24] F. Sanger, S. Nicklen, and A. R. Coulson, *Proc. Natl. Acad. Sci. U.S.A.* **74**, 5463 (1977).
[25] E. Y. Chen and P. H. Seeburg, *DNA* **4**, 165 (1985).
[26] B. J. Conner, A. A. Reyes, C. Monu, K. Itakura, R. L. Teplitz, and R. B. Wallace, *Proc. Natl. Acad. Sci. U.S.A.* **80**, 278 (1983).
[27] M. Verlaan-deVries, M. E. Bogaard, H. van der Elst, J. H. van Boom, A. J. van der Eb, and J. L. Bos, *Gene* **50**, 313 (1986).

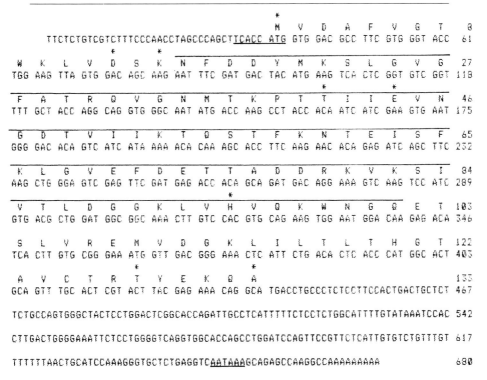

FIG. 4. Nucleotide and deduced amino acid sequences of MDGI cDNA. Residues that are overlined correspond to the amplified fragment used as a screening probe. The nucleotide sequence underlined at the beginning of the open reading frame represents the eukaryotic translation initiation consensus sequence. The polyadenylation signal in the 3' nontranslated region is also underlined. Amino acids that are not identical with those of the reported MDGI sequence are labeled by asterisks.

chosen. The fixed positions of the differences between cDNA-deduced and protein-derived MDGI sequences makes it possible to use this approach to screen directly for sequence variants. The assay is based on the fact that a fully matched DNA hybrid has a higher thermal stability than a hybrid with a mismatched base pair. By use of the primers 5'-CAAGCTTGTNG-AQGCNTTQGTN-3' and 5'-GGGAATTCKATNACNGTPTCQCC-3' (designated as B_1 and B_2 and coding for residues 2 to 6 and 46 to 51, respectively), an MDGI fragment spanning amino acids 2 to 51 was amplified from both cDNA and genomic DNA by the polymerase chain reaction. Under the selected hybridization conditions only the oligomers DK (GTGACAGCAAGAAT) and TE (ACCACAATCATCGAAGTG), which are complementary to the described cDNA sequence, hybridize with the amplified bovine mammary DNA. The oligomers derived from the protein sequence did not hybridize with either fragment.

Localization of Mammary-Derived Growth Inhibitor Gene Transcription in Mammary Glands by *in Situ* Hybridization

In situ hybridization techniques are used when localization of gene expression is of importance, and especially when regulated gene expression is studied in heterogeneous cell populations, as outlined in this series.[28] The complete MDGI cDNA is subcloned in pGEM3Z (Promega, Madison, WI). After linearizing the plasmid pGEM/MDGI with *Hin*dIII, the insert is transcribed *in vitro* with T7 polymerase, yielding a ^{35}S-labeled antisense RNA probe with a specific activity of 10^9 cpm/μg.

Bovine mammary tissue at different stages of development is collected and stored frozen at $-70°$ until sectioning. Slides coated with γ-aminopropyltriethoxysilane (Sigma, St. Louis, MO) are used to lift 6-μm cryostat sections, which are then fixed in freshly prepared 4% paraformaldehyde for 1 min and dehydrated in 70% ethanol. Dehydrated sections may be stored for months at 4° in 70% ethanol. Hybridization is performed under the conditions described[28] with the following modifications. The prehybridization treatment consists of rehydration of the sections in 2× SSC, acetylation in 0.25% acetic anhydride, 0.1 M triethanolamine, pH 8, followed by immersing the slides for 30 min in 0.1 M Tris-HCl, pH 7.0, 0.1 M glycine. After rinsing the slides in 2× SSC and an incubation in 2× SSC, 50% formamide at 55° for 5 min, the sections are finally hybridized in a solution composed of 2× SSC, 50% formamide, 10% dextran, 10 mM dithiothreitol (DTT), 1 μg/ml *E. coli* tRNA, 1 μg/ml sheared herring sperm DNA, 2 μg/ml BSA, and 1 × 10^6 cpm of ^{35}S-labeled RNA probe at 50° for 3 hr. The slides are then washed in 2× SSC, 50% formamide at 52° for 30 min with 3 buffer changes, rinsed in 2× SSC, and then incubated in RNase solution (100 μg/ml RNase A, 1 μg/ml RNase T1) at 37° for 30 min. After a final washing in 2× SSC, 50% formamide at 52° for 5 min and 2× SSC at room temperature for 10 min, sections are dehydrated in 70, 80, and 90% ethanol. Autoradiography is performed with an NBT-2 nuclear track emulsion (Kodak, Rochester, NY) melted at 43° and diluted 1 : 1 in water. The slides are exposed for 4–5 days in the dark at 4° and are then developed and counterstained with hematoxylin and eosine, dehydrated, mounted, and covered with coverslips.

In order to follow MDGI transcription during development, tissue

---

[28] F. Baldino, Jr., M.-F. Chesselet, and M. E. Lewis, this series, Vol. 168, p. 761.

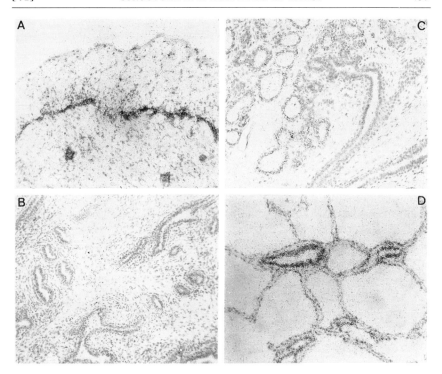

FIG. 5. Distribution of MDGI RNA in embryonic, virgin, pregnant, and lactating mammary tissues. MDGI transcripts were detected by *in situ* hybridization, using cryostat sections hybridized with a ^{35}S-labeled antisense RNA probe. Shown are sections counterstained with hematoxylin. (A) Sections of mid-stage mammary epithelial rudiment; MDGI transcripts are evenly distributed in epithelial cells of the embryonic mammary rudiment. (B) Virgin mammary gland; no MDGI transcripts are detectable. (C) Longitudinal section of midpregnant mammary gland. (D) Cross section of the proximal part of lactating mammary tissue.

sections are taken from embryonic, virgin, pregnant, and lactating mammary glands. Analysis of a 5-month-old female bovine embryo reveals clearly detectable transcripts in the mammary epithelial rudiment (Fig. 5A). In the immature, resting mammary gland of the virgin animal, MDGI mRNA is not detected (Fig. 5B). The bovine mammary gland in the fifth month of pregnancy is undergoing deep morphological and functional changes (Fig. 5C). In Fig. 5C a branching duct is shown which terminates with developing lobuloalveoli. During this stage of development ductal and alveolar epithelial cells in combination with myoepithelial cells are forming the lobuloalveolar gland. In the ductal epithelial cells, MDGI transcription is rather low, whereas epithelial cells of the developing alveoli contain MDGI transcripts. The terminally differentiated lactating mammary gland is characterized by ducts branching into large, active secretory lobuloalveolar structures (Fig. 5D). In contrast to the pregnant stage, both the alveolar and ductal cells of terminally differentiated mammary gland transcribe the MDGI gene, with higher expression in ductal epithelial cells. Other cell types such as fibroblasts, macrophages, myoepithelial cells, and blood cells are not found to transcribe the MDGI gene.

Conclusions and Perspectives

The methods described here can be used to isolate easily sufficient amounts of MDGI for further studies of biological activities (or for raising antibodies) and mechanisms of action. MDGI was detected in the lactating bovine mammary gland by conducting a growth inhibition assay with Ehrlich ascites mammary carcinoma cells. However, the rather low degree of inhibition of different mammary carcinoma cells, the increased responsiveness of normal human mammary epithelial cells, and the developmentally regulated expression of MDGI in the mammary gland indicate a role for MDGI in the normal growth of mammary epithelial cells. It is therefore necessary to develop new biological assays for answering questions about the involvement of MDGI in the physiological regulation of proliferation and differentiation of normal mammary epithelial cells.

[42] Assay and Purification of Naturally Occurring Inhibitor of Angiogenesis

By PETER J. POLVERINI, NOEL P. BOUCK, and FARZAN RASTINEJAD

Assays for Detection of Angiogenic Activity *in Vivo*

The development of a number of *in vivo* bioassays of angiogenesis has greatly aided in identifying and elucidating the mechanism of action of a variety of positive and negative regulators of angiogenesis. These include, among others, the hamster cheek pouch,[1] dorsal air sac,[2] rabbit ear chamber,[3] iris and avascular cornea of rodent eye,[4-7] the chick chorioallantoic

---

[1] G. K. Klintworth, *Am. J. Pathol.* **73**, 691 (1973).
[2] J. Folkman, E. Merler, C. Abernathy, and G. Williams, *J. Exp. Med.* **133**, 275 (1971).
[3] J. C. Sandison, *Am. J. Anat.* **41**, 475 (1928).
[4] M. A. Gimbrone, Jr., S. B. Leapman, R. S. Cotran, and J. Folkman, *J. Natl. Cancer Inst.* **50**, 219 (1973).

membrane (CAM),[8] and dorsal mouse skin assays.[9] All of these methods allow for direct inspection of an area that is either devoid of blood vessels initially or has a vascular pattern clearly distinguishable from newly formed capillaries.

The bioassays used most often involve CAM and cornea, and both have their advantages and disadvantages. The principal advantages of the CAM assay are that it is relatively inexpensive and lends itself to large-scale screening of samples. Also, because the CAM can be manipulated outside the eggshell, it is possible to monitor and photograph evolving responses on a daily basis. The major disadvantage of this assay system is that since the CAM already contains a well-developed vascular network there may be some difficulty, particularly for inexperienced examiners, in distinguishing vasodilatation, which invariably occurs following manipulation of the membrane, from neovascularization. The cornea assay, on the other hand, because the cornea is avascular, avoids the problem of interpretation inherent in the other bioassays. Any vessels penetrating into the corneal stroma can be readily identified as newly formed. With the aid of a stereomicroscope parameters of capillary growth such as sprout and hairpin-loop formation and the development of corneal edema and inflammation can be documented in considerable detail. The main disadvantages of this model system are that, in comparison to the CAM assay, it is relatively expensive and since only one or at most two samples can be evaluated per cornea it is not a practical screening assay. Because our laboratory employs the rat cornea model almost exclusively, it is described in detail.

Preparation of Conditioned Media for Intracorneal Assay

A wide variety of tissues, cells, and cell extracts or media conditioned by cells grown in culture have been examined for angiogenic stimulatory or inhibitory activity in the cornea. When possible, cell extracts or conditioned media are the preferred form of material to be assayed. One can more accurately determine the dosage/potency of a particular test sample and precisely control the quantity and positioning of materials within the corneal stroma, after incorporation into one of several slow-release noninflammatory polymers. Since the inhibitor we are studying is elabo-

---

[5] M. A. Gimbrone, Jr., R. S. Cotran, S. B. Leapman, and J. Folkman, *J. Natl. Cancer Inst.* **52**, 413 (1974).
[6] P. J. Polverini, R. S. Cotran, M. A. Gimbrone, Jr., and E. M. Unanue, *Nature (London)* **269**, 804 (1978).
[7] P. J. Polverini and S. J. Leibovich, *Lab. Invest.* **51**, 635 (1984).
[8] M. T. Vu, C. F. Smith, P. C. Burger, and J. K. Klintworth, *Lab. Invest.* **53**, 499 (1985).
[9] Y. A. Sidkey and R. Auerbach, *J. Exp. Med.* **141**, 1084 (1975).

rated in culture media, we will describe the preparation of samples of conditioned media (CM) for assay.

Two 100-mm dishes of cells are grown in Dulbecco's modified Eagle's medium (DMEM) containing 10% donor calf serum until 80–90% confluent, then rinsed twice with phosphate-buffered saline (PBS) and refed with 10 ml of serum-free DMEM. After an 8-hr incubation at 37° in a CO_2 incubator, the medium is replaced with fresh DMEM, and conditioned medium (CM) is collected after an additional 48- to 72-hr incubation. The CM is collected and centrifuged at 2500 rpm to remove cell debris and thereafter stored and handled in the cold. Bovine serum albumin to 0.1 mg/ml may be added to prevent proteolytic degradation.

Prior to biological assay the CM is concentrated and desalted. This can be done most conveniently by dialysis using Spectra Por 3 tubing (molecular weight cutoff 3500). In our experiments, dialysis against 3 changes of precooled Milli-Q-purified water permits desalting without significant denaturation of the inhibitor or the angiogenic factors. Other dialysis solutions may need to be employed as the nature and stability of the angiogenic mediators from different sources can vary. Alternatively, desalting and concentration (usually 20- to 40-fold for assaying in the cornea) can be carried out using a stirred cell ultrafiltration unit or a Centricon microconcentrator (both from Amicon, Danvers, MA) as described below.

The desalted medium should be rapidly concentrated by freezing at $-80°$ and lyophilized to dryness. Prior to assay, samples are dissolved in PBS; multiple samples may be adjusted to be equivalent either based on cell number at the time of medium collection or based on protein concentration (i.e., 1 ml per 1×10^7 cells or 1 mg/ml protein). For CM from baby hamster kidney (BHK) cells, the final volume is usually approximately 1/20 of the starting volume. The resuspended samples should be clarified by centrifugation for 5 min at 15,000 g, aliquoted, and stored at $-80°$. Mock CM carried through the procedure in parallel but without cells serves as a control for medium preparation. Standardization of culture conditions for the preparation of CM is absolutely essential in order to minimize variations in responses with a particular batch of material being assayed. It is also important to determine the minimum concentration of material needed to induce or inhibit angiogenesis. Attention to these details early on is essential if one is to avoid misinterpreting responses. For testing of our angiogenesis inhibitor we include one of several known mediators of angiogenesis, most often basic fibroblast growth factor (FGF), which we purchase from commercial sources. We also test each new batch to determine the minimum quantity (usually 20–50 ng per cornea) that will elicit an angiogenic response.

Samples to be assayed are then incorporated into one of several slow-release noninflammatory polymers. The two most often used are poly-2-hydroxylethylmethacrylate (Hydron, Lot No. 110, Interferon Sciences, Inc., New Brunswick, NJ) and ethylene-vinyl acetate copolymer (Aldrich Chemical, Milwaukee, WI).[10] Both materials work with equal effectiveness. Our laboratory uses Hydron exclusively. Sterile casting solutions are prepared by dissolving the Hydron powder in absolute ethanol (12%, w/v) at 37° with continuous stirring for 24 hr. An equal volume of Hydron and sample (50%) are combined, and 10 μl of solution is pipetted onto the surface of a sterile 3.2 mm diameter, 1.2 cm long Teflon (Du Pont Corp., Wilmington, DE) rods glued to the surface of a petri dish. After drying for 1–2 hr the approximately 2 mm diameter disks can be stored at 4°.

Formation of Corneal Pocket

The surgical procedure used to form a corneal pocket is essentially identical to that first described by Gimbrone et al.[5] We routinely use imbred F344 rats, although any rat strain is suitable. Male or female rats weighing 150–200 g are anesthetized with sodium pentobarbital (29 mg/kg body weight). The eyes are gently proptosed and secured in place by clamping the upper eyelid with a nontraumatic hemostat. Using a No. 11 Bard Parker blade, a 1.5 mm incision is made approximately 1 mm from center of the cornea into the stroma but not through it. Depending on experience this procedure can be done with or without the use of a dissecting scope. A curved iris spatula (No. 10093-13, Fine Science Tools, Inc., Belmont, CA), approximately 1.5 mm in width and 5 mm in length, is then inserted under the lip of the incision and gently blunt-dissected through the stroma toward the outer canthus of the eye. Slight finger pressure against the globe of the eye helps steady it during dissection. The spatula is premarked so that the shaft does not penetrate laterally into the stroma more than 2.5 mm. Once the corneal pocket is made the spatula is removed, and the distance between the limbus and base of the pocket is measured to make sure it is no closer than 1 mm.

We generally keep the pocket base between 1 and 1.5 mm from the limbus. Extending the pocket depth any closer than this often results in a false-positive response. Also, if the depth of incision is too close to the inner surface of the cornea, nonspecific inflammation invariably occurs. The first 24–48 hr after implantation are critical. If nonspecific inflammation is to occur it will manifest during this time. In such cases corneal clouding and the presence of a yellowish exudate signal inflammation. As

---

[10] R. Langer and J. Folkman, *Nature (London)* **263**, 797 (1976).

long as asepsis is maintained and trauma during surgery is minimized, nonspecific inflammation will rarely be a problem. Even in the most carefully executed procedure some transient corneal edema will occur. However, this usually resolves within 24 hr.

Just before implanting, pellets are rehydrated with a drop of sterile lactated Ringers solution. Pellets are positioned down to the base of the pocket, which then seals spontaneously. No more than half of the pocket should be occupied with implant material. If more than this is occupied, the resultant transient edema will cause spontaneous expulsion of the implant. Corneas are examined daily or on alternate days with the aid of a stereomicroscope to monitor responses.

Scoring of Vascular Responses

The corneal bioassay is best regarded as a qualitative assay, although a number of quantitative measures have been devised for use with this model system.[8,11] Responses are usually scored on the day animals are sacrificed, 5–7 days after implantation. Positive responses are recorded when sustained ingrowth of capillary loops or sprouts is detected. Negative scores are assigned to responses either where no growth is detected or when an occasional sprout or hairpin loop is detected without evidence of sustained growth. Occasionally (<10% of the time) we encounter responses that are neither unequivocally positive nor negative with samples which are normally positive or inhibitory. In these instances we grade responses as +/−. Several factors account for this, for example, slight differences in the position of implants within the cornea and variations in the quantity and quality of material being tested. No matter how experienced one is with this technique a certain degree of variability can be expected. Obviously this must be kept at a minimum. Figures 1 and 2 are examples of positive and negative corneal responses 7 days after implanting Hydron pellets containing 50 ng of basic FGF (Fig. 1) and basic FGF with purified angiogenesis inhibitor (Fig. 2).

Preparation of Colloidal Carbon-Perfused Whole Mounts

Permanent records of vascular responses are made following perfusion with India ink. Any commercially available source of waterproof India ink is acceptable. Perfusion is accomplished with a simple pressure vessel capable of maintaining a pressure of 120 mm Hg. We routinely perfuse via the abdominal aorta, introducing 100–200 ml warm (37°) lactated Ringers solution per 150-g rat. Once the snout of the animal has completely blanched, approximately 20–25 ml of ink is injected until the head and

---

[11] M. Sholley, G. P. Ferguson, H. R. Seibel, J. L. Montour, and J. D. Wilson, *Lab. Invest.* **51**, 624 (1984).

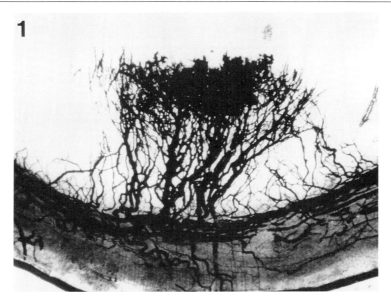

FIG. 1. Positive angiogenic response induced by 50 ng of basic FGF after 7 days. Note the brushlike network of capillary sprouts. Magnification: ×30.

thoracic organs have completely blackened. Eyes are then carefully enucleated and placed in half-strength Karnovsky's fixative or neutral buffered formalin for 24 hr. The next day corneas are dissected free from the surrounding globe and underlying iris, bisected, and loosely mounted between two glass slides where they are gently flattened. Corneas are photographed with a dissecting microscope equipped with a camera. If so desired one can quantitate the response. A simple method described by Sholley et al.[11] is as follows. Vessel length is measured directly using 4 × 5 transilluminated photographic negatives at a magnification of 10×. Three radially oriented measurements are taken using a vernier caliper; two of these measurements include vessels present at the periphery of the radius, and the third includes the largest vessels along the center of the radius. The three measurements are averaged to provide a single length of measure for each response. Differences between groups are compared using a Students t test.

Assays for Angiogenic Activity *in Vitro*

With the development of reproducible methods for the isolation and culture of endothelial cells, it is possible to examine, in a quantitative fashion, a number of endothelial cell functions that mimic one or more of

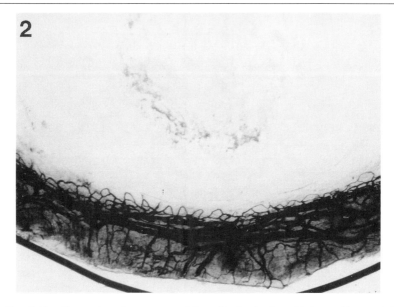

FIG. 2. Negative angiogenic response at 7 days after implanting a Hydron pellet containing 50 ng of basic FGF and 125 ng of gp140 inhibitor. Note the complete absence of capillary ingrowth. Magnification: ×30.

the steps that contribute to capillary vessel formation *in vivo*.[12] Assays that measure endothelial cell proliferation,[13] directional migration (chemotaxis),[14] and endothelial cell proteolytic enzyme activity[15] are now routinely performed. In addition it is possible to induce cultured endothelial cells to organize into hollow, tubelike structures and for single cells to form ringlike structures characteristic of capillaries *in vivo* depending on the matrix components on which endothelial cells are grown and the type of angiogenic mediator to which the cells are exposed.[16–19] Endothelial cells from both large and small vessels can be obtained from a variety of animal and human sources, and many are available from several commer-

[12] J. Folkman and M. Klagsbrun, *Science* **235,** 442 (1987).
[13] Y. Shing, J. Folkman, R. Sullivan, C. Butterfield, J. Murray, and M. Klagsbrun, *Science* **223,** 1296 (1984).
[14] S. U. Orredson, D. R. Knighton, H. Scheuenstuhl, and T. K. Hunt, *J. Surg. Res.* **35,** 249 (1983).
[15] M. Presta, D. Moscatelli, J. Joseph-Silverstein, and D. B. Rifkin, *Mol. Cell Biol.* **6,** 4060 (1986).
[16] R. Montesano, L. Orci, and P. Vassalli, *J. Cell Biol.* **97,** 1648 (1983).
[17] J. A. Madri, S. K. Williams, T. Wyatt, and C. Mezzio, *J. Cell Biol.* **97,** 153 (1983).
[18] D. S. Grant, K.-I. Tashiro, B. Segui-Real, Y. Yamuda, G. R. Martin, and H. K. Kleinman, *Cell* **58,** 933 (1989).
[19] R. Montesano, J.-D. Vassalli, A. Baird, R. Guillemin, and L. Orci, *Proc. Natl. Acad. Sci. U.S.A.* **83,** 7297 (1986).

cial sources (American Type Culture Collection, Rockville, MD; Clonetics Corporation, San Diego, CA). Our laboratory routinely assays for endothelial cell chemotaxis and proliferation using bovine adrenal gland capillary endothelial (BCE) cells, although large vessel endothelium can also be used. Procedures for the isolation and culture of both types of endothelial cells have been described in detail elsewhere.[20] One should be aware that the *in vivo* angiogenic properties of a particular mediator may not exhibit corresponding angiogenic activity *in vitro*.[12,21,22]

Endothelial Cell Chemotaxis

Endothelial cell migration is routinely assayed in our laboratory using a 48-well modified Boyden chamber (Nucleopore Corp., Pleasanton, CA) equipped with Nucleopore membranes (5 μm pore size) that have been soaked overnight in 3% acetic acid, incubated for 2 hr in 0.1 mg/ml gelatin, rinsed in sterile water, dried under sterile air, and stored for up to 1 month. BCE cells usually not older than passage 10 are preferred. We have found that older cultures show greater variability in responsiveness to chemotactic stimuli. Also, we frequently find differences in the responsiveness of endothelial cells to chemotactic stimuli from isolate to isolate, so comparisons must be made within each experiment.

Chambers are set up using 5×10^5 cells suspended in 1.5 ml of DMEM supplemented with 0.1% fetal bovine serum (FBS). The bottom wells are filled with 25 μl of the cell suspension and covered with the gelatin-coated membrane. The chamber is assembled and then inverted and incubated for 2 hr to permit adherence of BCE cells to the membrane surface. This modification of the standard assay is important since the inhibitory activity may interfere with adhesion of BCE to the membrane surface. The chamber is then turned upright, 50 μl of test material is dispensed into the top wells, and the chamber is incubated for an additional 2–3 hr. The membrane-bound cells are carefully washed with buffered saline and stained with Diff-Quick stain (American Scientific Products, McGaw Park, IL), and membranes are mounted with Permount (Fisher Scientific, Fairlawn, NJ) or taped to the slide with the surface to which the cells have migrated up. The number of cells that have migrated per 10 high powered ($\times 100$) fields are counted. By focusing through the membrane it is possible to distinguish the surface of the membrane to which cells have migrated.

[20] J. Folkman, C. C. Haudenschild, and B. R. Zetter, *Proc. Natl. Acad. Sci. U.S.A.* **76**, 5217 (1979).
[21] A. B. Roberts, M. B. Sporn, R. K. Assoian, J. M. Smith, N. S. Roach, L. M. Wakefield, U. I. Heine, L. A. Liotta, V. Falange, J. H. Kehrl, and A. S. Fauci, *Proc. Natl. Acad. Sci. U.S.A.* **83**, 417 (1986).
[22] M. Fràter-Schroder, G. Müller, W. Birchmeier, and P. Böhlen, *Biochem. Biophys. Res. Commun.* **137**, 295 (1986).

Endothelial Cell Proliferation

Endothelial cell proliferation is assayed according to the method of Shing et al.[13] Early passage BCE cells are suspended at a concentration of 2×10^4 cells/well in 24-well dishes in 0.5 ml of DMEM containing 10% FBS. Twenty-four hours later wells are washed and refed with DMEM with 0.1% serum. Forty-eight hours later cells are trypsinized and counted with an automated cell counter (Coulter Electronics, Hialeah, FL).

Purification of Angiogenesis Inhibitor

Purification of Angiogenesis Inhibitor from Baby Hamster Kidney Fibroblast Conditioned Medium

Preparations of inhibitor[23] typically start with 300 ml of serum-free medium conditioned for 48 hr as described previously. Lower amounts of starting material can lead to a significant loss of activity during purification. All steps of purification are performed in a cold room or with the sample on ice.

Following low-speed centrifugation to remove cellular debris, the CM is concentrated using a stirred cell, such as an Amicon Model 8400, fitted with a YM30 ultrafiltration membrane (molecular weight cutoff 30K, Amicon). The volume is reduced in about 2 hr to approximately 10 ml under a nitrogen pressure of 20 psi which takes about 2 hr. Fifty milliliters of Milli-Q water is then added to the cell and the volume again reduced to 10 ml under pressure. The filtration cell is then depressurized and the contents stirred for 15 min to dislodge proteins that may have accumulated on the membrane under pressure.

Salt fractionation should immediately follow ultrafiltration. A saturated solution (100%) of ammonium sulfate at 4° is added dropwise with stirring to bring the final concentration to 30% ammonium sulfate. The sample is stirred for an additional 30 min and the precipitate pelleted at 15,000 g for 30 min. The supernatant is transferred to a new tube, and more salt is added to bring the final ammonium sulfate concentration to 40%. The pellet resulting from centrifugation of this solution, the 30–40% cut, contains the inhibitory activity and may be stored frozen at this stage without loss of activity, providing a convenient stopping point.

For chromatography, a prepackaged anion-exchange column of Mono Q HR5/5 (Pharmacia, Piscataway, NJ) is fitted to a Pharmacia fast protein liquid chromatography (FPLC) system composed of two p-500 pumps, a

---

[23] F. Rastinejad, P. J. Polverini, and N. P. Bouck, *Cell (Cambridge, Mass.)* **56**, 345 (1989).

LCC500 controller, and a UV-M monitor. The column is preequilibrated in 50 mM Tris, pH 7.5 (running buffer). Traces of salt solution are removed from the tube containing the salt cut pellet with a cotton swab before resuspending the pellet in as small a volume of running buffer as possible. This sample is clarified 3 times by microcentrifuge spins with tube changes. The preparation can then be directly applied to the anion-exchange column without any significant effect of the salt on the binding of the protein. The sample is loaded via a 0.5-ml sample loop and the column washed with 2 ml of running buffer. Next, a sodium chloride gradient of 0–2 M in running buffer is run at the maximum flow rate, usually 1 ml/min, with the salt concentration increasing at the rate of 50 mM/ml. When 2-ml fractions are collected, the inhibitor elutes as a sharp protein peak at 0.4 M salt, usually in one fraction.

The ion-exchange fraction is applied to a lectin affinity column without desalting. The column, which binds carbohydrate moieties containing branching mannose structures,[24] is composed of 1 ml of Sepharose 4B conjugated to lentil lectin (Pharmacia) packed in a disposable Bio-Rad (Richmond, CA) column. It is washed with 10 bed volumes of start buffer (50 mM Tris, pH 7.5, 0.4 M sodium chloride). The flow is established either by gravity or by slight air pressure applied with a fitted syringe. Following addition of sample, the column is washed with 10 bed volumes of start buffer. The bound inhibitor is specifically eluted by addition of 5 bed volumes of 10% α-methyl-D-mannopyranoside (Sigma, St. Louis, MO) in start buffer.

The column eluate is concentrated and dialyzed into PBS with a Centricon 30 microconcentrator (Amicon) using centrifugation in a Sorvall SS34 rotor at 5000 rpm. Two-milliliter aliquots of the column eluate are sequentially added to the same microconcentrator and the volume reduced to 0.2 ml. Three 2-ml additions of PBS are used to remove the mannopyranoside. The purified protein is stored at $-80°$, with 20% glycerol if extremely dilute. A summary of the purification procedure is outlined in Fig. 3.

Comments

Using the above procedure, a single 140K band on a reducing sodium dodecyl sulfate (SDS)-polyacrylamide gel is obtained. Although a single lectin affinity step can yield a nearly pure product, it is not recommended as an early step in the purification because of the high cost of the resin-conjugated lectin and the large losses that occur in subsequent steps owing to the small amount of protein in the sample. The YM30 membrane used for

---

[24] M. J. Hayman, J. J. M. Rowe, P. S. Steward, and D. C. Williams, *FEBS Lett.* **29,** 185 (1973).

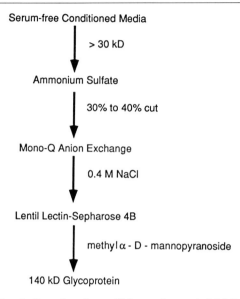

FIG. 3. Procedure for purifying angiogenesis inhibitor.

ultrafiltration is preferred because of its low protein affinity and consistent retention of the activity. The YM100 membrane with a molecular weight cutoff of 100K retains the activity initially and provides a faster flow rate; however, this membrane becomes permeable to the 140K inhibitor after extended application of pressure. It should be noted that pressures exceeding 20 psi can rapidly block the membrane pores and slow down the overall flow rate. The major problem experienced with the anion-exchange chromatography is clogging of the column with repeated use. To prolong the life of the column, the sample and all buffers are passed through a 0.22-μm filter and degassed prior to use.

Samples obtained after each step of purification are assayed for total protein and for activity using the migration assay as outlined above. A dose–response curve is generated by assaying dilutions of inhibitory sample against a standard attractant such as basic FGF at 50 ng/ml. A unit of activity is defined as the amount of sample able to produce 50% inhibition of migration, as read from a graph of log percent inhibition of maximum migration for that day against amount of sample in the assay, and specific activity calculated.

[43] Derivation of Monoclonal Antibody Directed against Fibroblast Growth Regulator

By JOHN L. WANG

Introduction

Treatment of sparse, proliferating cultures of 3T3 cells with medium conditioned by exposure to density-inhibited 3T3 cultures resulted in an inhibition of growth and division in the target cells when compared to a similar treatment with unconditioned medium.[1] This growth inhibitor activity was fractionated by ammonium sulfate precipitation, gel-filtration, and ion-exchange chromatography,[2,3] yielding one preparation designated FGR-s (13K), which stands for fibroblast growth regulator, soluble form. Analysis of the chemical and biological properties of this factor indicated that (1) it consists of a single polypeptide chain (M_r 13,000; pI ~10) when analyzed by polyacrylamide gel electrophoresis in the presence of sodium dodecyl sulfate (SDS) under both reducing and nonreducing conditions; (2) it is an endogenous cell product, synthesized by the 3T3 cells and secreted or shed into the medium; (3) the dose–response curve of growth inhibition shows 50% inhibition at a concentration of approximately 3 ng/ml, corresponding to about 0.23 nM; and (4) it is not cytotoxic and its effects on target cells are reversible.

Using a partially purified preparation of the growth inhibitor as immunogen, we have carried out *in vitro* immunization of rat splenocytes and have obtained a monoclonal antibody, designated 2A4, that specifically bound FGR-s (13K).[4] In this chapter, we describe in detail the derivation of the monoclonal antibody specifically directed against the FGR-s (13K) inhibitory protein.

Isolation of Immunogen Fraction

Cell Culture and Preparation of Conditioned Medium

Swiss 3T3 fibroblasts (American Type Culture Collection, CCL92, Rockville, MD) are cultured in Dulbecco's modified Eagle's medium (DMEM) supplemented with 100 U/ml penicillin, 100 μg/ml streptomycin,

---

[1] P. A. Steck, P. G. Voss, and J. L. Wang, *J. Cell Biol.* **83**, 562 (1979).
[2] P. A. Steck, J. Blenis, P. G. Voss, and J. L. Wang, *J. Cell Biol.* **92**, 523 (1982).
[3] Y.-M. Hsu and J. L. Wang, *J. Cell Biol.* **102**, 362 (1986).
[4] Y.-M. Hsu, J. M. Barry, and J. L. Wang, *Proc. Natl. Acad. Sci. U.S.A.* **81**, 2107 (1984).

and 10% (v/v) calf serum (Microbiological Associates, Walkersville, MD). The cells are incubated in Corning plastic tissue culture flasks (Corning Glass Works, Nos. 25100, 25110, 25120, Corning, NY), maintained at 37° in a humidified atmosphere of 10% CO_2. The 3T3 cells are passaged and used for a maximum of 3 months, after which they are discarded and a fresh sample grown from frozen cultures kept at $-80°$.

Routinely, our stock culture of 3T3 cells is passaged every 2 days, before the cells reach a confluent state (5×10^4 cells/cm^2). The growth medium is first removed, and the cells are washed and then incubated in phosphate-buffered saline (PBS) with 0.25% trypsin (Nutritional Biochemicals, Cleveland, OH) and 4×10^{-4} M Versene at 37° for 10 min. The cells are dislodged from the growth surface, transferred, and centrifuged at 1320 g for 3 min. The trypsin solution is removed, and the cells are resuspended in DMEM–10% calf serum and seeded at the desired density.

We have "conditioned" growth medium by exposing it to a confluent monolayer of cells and then tested its capacity to support growth and cell division in sparse cultures. Serum-free conditioned medium [CM(SF)] is prepared according to the protocol diagrammed in Fig. 1. Source cells, which are used to condition the medium, are seeded at an initial density of 2×10^4 cells/cm^2 and allowed to grow for at least 3 days. When cells reach a confluent monolayer, the medium is removed and "stimulation feeding" is performed by adding fresh DMEM–10% calf serum (0.1 ml/cm^2 of growth surface). The purpose of this feeding is to induce one final round of cell division and assure a confluent monolayer. After 24 hr, the medium is again removed, and the cells are washed twice with serum-free DMEM. Then DMEM (10 ml/150 cm^2 of growth area) is added to the cultures as medium to be conditioned. After 24 hr the medium is transferred and centrifuged at 1470 g for 10 min to remove cellular debris and particulate material. The supernatant is collected and used as CM(SF). In all experiments, serum-free unconditioned medium [UCM(SF)] is prepared in parallel from the same batch of DMEM and incubated under the same conditions, but in the absence of cells (Fig. 1).

Radioactive growth inhibitor preparations are isolated from source cells cultured as described above. During the period of conditioning (Fig. 1), [^{35}S]methionine is added (100 μCi/ml, 1000 Ci/mmol, Amersham Corp., Arlington Heights, IL) to DMEM containing unlabeled methionine at 3 μg/ml (1/10 of the concentration of methionine normally found in DMEM).

Assays of Growth-Inhibitory Activity

Target cells used to test growth-inhibitory activity are routinely seeded in DMEM containing 10% calf serum at an initial density of 5×10^3 cells/

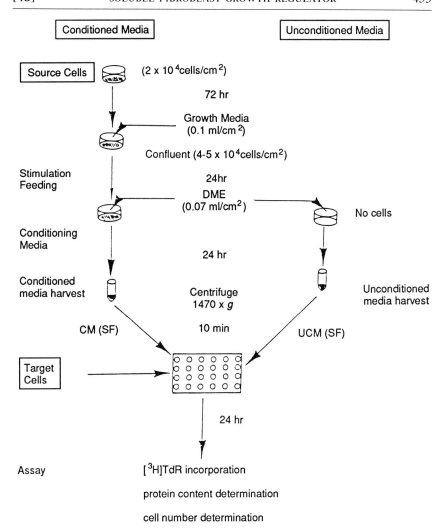

FIG. 1. Protocol used in the preparation of serum-free conditioned medium [CM(SF)] and unconditioned medium [UCM(SF)]. Cell densities are expressed as the number of cells per square centimeter of growth surface in tissue culture dishes. The volumes of medium added at the various steps are expressed in this diagram as the milliliters of medium used per unit area of growth surface.

cm² in a 96-well culture dish (Costar, No. 3596, Cambridge, MA). After overnight incubation, the cells are deprived of serum for 24 hr. Then the medium is removed, and the test fraction is added (75 µl) along with 150 µl DMEM containing 5% (v/v) calf serum. DNA synthesis is assayed 24 hr later with a pulse of [³H]thymidine (1 µCi/culture, 1.9 Ci/mmol, Schwarz-Mann, Orangeburg, NY) for 3 hr at 37°. After the pulse the radioactive medium is removed, and the cells are washed 3 times with cold PBS and once with 10% trichloroacetic acid. The cells are then solubilized with 0.2 ml of 1% (w/v) SDS in 0.1 N NaOH. After incubation at 37° for 10 min, cell lysates are added to 2 ml of scintillation cocktail [1 g dimethyl-1,4-bis(2-5-phenyloxazolyl)benzene, 8 g 2,5-diphenyloxazole; 1333 ml Triton X-100, and 2666 ml of toluene] for scintillation counting.

To assay increases in cell number after treatment with growth inhibitor fractions, target cells are prepared in the same fashion in 6- or 24-well culture dishes (Costar, No. 3406 or 3524). Cultures are then treated with the inhibitory fraction or a control fraction. At various times thereafter, the cells are washed 3 times with PBS and then removed from the growth surface by trypsin treatment. The cells are centrifuged, resuspended in PBS, diluted with trypan blue [0.08% (w/v) in PBS], and counted in a corpuscle counting chamber (Hausser Scientific, Blue Bell, PA).

The viability of cells treated with inhibitor fractions is determined while the cells remain attached to the growth surface. After removal of growth medium, the cells are incubated with trypan blue for 10 min at room temperature. The staining solution is then removed, and the viable cells are counted using an Olympus inverted microscope.

Fractionation of Conditioned Medium Components

Solid ammonium sulfate is added to serum-free conditioned medium to a saturation of 80% at room temperature. All subsequent operations are performed at 4°. The ammonium sulfate-precipitated mixture is centrifuged at 12,400 g for 15 min and the supernatant decanted. The precipitate collected from 800 ml of serum-free conditioned medium is resuspended in 2 ml of DMEM and fractionated on a column (1.4 × 110 cm) of Sephadex G-50 (Pharmacia, Piscataway, NJ) equilibrated with DMEM. Fractions of 1.6 ml are collected. FGR-s represents the material eluting from the Sephadex G-50 column at a position corresponding to a molecular weight of 10,000–15,000. This region (Component C) contains material previously shown to be enriched in specific biological activity.[2] Polyacrylamide gel electrophoretic analysis in the presence of SDS is performed according to the procedure of Laemmli[5]; the acrylamide concentration of the running

[5] U. K. Laemmli, *Nature (London)* **227**, 680 (1970).

gel is 16%. The gels, stained with the silver reagent,[6] show that the principal components of the FGR-s fraction are polypeptides of approximate M_r 10,000 and 13,000. This preparation is used for the immunization of rat spleen cultures for the generation of antibody-secreting cells, as well as for the screening of hybridoma clones.

FGR-s can be further fractionated by ion-exchange chromatography on DEAE-cellulose.[3] To carry out ion-exchange chromatography, two minor modifications of the above procedure are made. First, gel filtration on Sephadex G-50 columns is performed in 5 mM Tris, pH 8.0, instead of DMEM. This allows the direct application of the effluent from the Sephadex column onto the ion-exchange column. Second, a wider range of fractions, centered approximately at FGR-s, is pooled and subjected to ion-exchange chromatography. The rationale for this is that we do not assay the individual fractions from the Sephadex G-50 column for activity or for polypeptide content on polyacrylamide gels. Therefore, the precise position corresponding to the FGR-s fractions is not determined and is compensated for by including material in fractions adjacent to FGR-s. This allows us to save material which would have been used in the assays and to save time, thereby minimizing losses of material owing to adsorption to test tubes.

A column (0.8 × 2 cm) of DEAE–cellulose (Pharmacia) is equilibrated with 5 mM Tris, pH 8.0. The pooled material corresponding to the FGR-s fractions of the Sephadex G-50 column is applied to the ion-exchange column. After washing with starting buffer, a gradient (0 to 0.5 M NaCl in 100 ml of 5 mM Tris, pH 8.0) is used to develop the column. Fractions of 1.7 ml are collected. When the fractions eluting from the ion-exchange column are assayed for growth-inhibitory activity, only the material flowing through the column (Component A) exhibits activity; the remainder of the components fail to show any appreciable activity. The sum of the growth-inhibitory activity in Component A accounts for 80% of the total activity applied to the column. There is a 6-fold enrichment in terms of specific activity in this fractionation step.

Polyacrylamide gel electrophoresis analyses are carried out on the fractions derived from the DEAE–cellulose column. Component A yields a single polypeptide, migrating at a position corresponding to a molecular weight of 13,000. Identical results are obtained irrespective of whether the polyacrylamide gel electrophoresis is carried out under reducing [with 5% (v/v) 2-mercaptoethanol] or nonreducing conditions. Using known amounts of cytochrome c (M_r 12,500) as a standard for the silver staining

---

[6] C. R. Merril, M. L. Dunau, and D. Goldman, *Anal. Biochem.* **110**, 201 (1981).

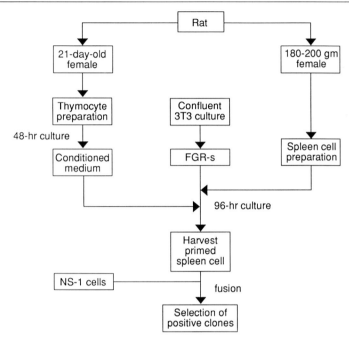

FIG. 2. Procedure used for the generation of monoclonal antibodies directed against components of FGR-s using the *in vitro* immunization method.

technique, we estimate that 1 liter of conditioned medium derived from 3T3 cultures yields approximately 1 μg of the 13,000-dalton polypeptide.

Generation of Hybridoma Cells

In Vitro Immunization with FGR-s

The immunization procedure is adapted from Luben and Mohler[7] and is outlined in Fig. 2. Thymocytes from a 21-day-old female Sprague-Dawley rat (Charles River Co., Wilmington, MA) are isolated and used to condition the medium [5×10^6 cells/ml; 40 ml of DMEM containing 20% (v/v) fetal calf serum (Hazelton Dutchland, Denver, PA), 50 μM 2-mercaptoethanol, 2 mM glutamine] for 48 hr at 37° in a humidified atmosphere of 10% CO_2. The cells are then removed by centrifugation (1320 g for 3 min). The supernatant medium is sterilized by filtration using a 0.2-μm filter. The conditioned medium, which is to be used for the culture of

[7] R. A. Luben and M. A. Mohler, *Mol. Immunol.* **17**, 635 (1980).

rat spleen cells in the presence of the immunogen, FGR-s, can be stored for up to 14 days at 4°.

Spleen cells from a female Sprague-Dawley rat (~200 g) are suspended at a density of 1.0×10^7 cells/ml in DMEM containing 2 mM glutamine and 50 μM 2-mercaptoethanol. The spleen cell suspension (4 ml) is first mixed with 1 ml of FGR-s (~100 ng/ml) and incubated at 37°. After 45 min, 5 ml of thymocyte-conditioned medium (resupplemented with 2 mM glutamine and 50 μM 2-mercaptoethanol), 0.5 ml fetal calf serum, and 0.5 ml horse serum (GIBCO, Grand Island, NY) are added. The mixture is then cultured at 37° for 96 hr. The immune spleen cells are harvested and fused with the mouse myeloma cell line, P3/NS1/1-Ag4-1, as described below.

Comment. The initial incubation of the immunogen fraction, FGR-s, with the splenocytes to be immunized in the absence of serum is an important modification of the general procedure as described by Luben and Mohler.[7] Because the amount of antigen available is low, we sought to maximize the probability of binding of the antigen by specific immunoglobulin receptors on the lymphocytes without interference by the large amounts of protein contained in serum. Thus, thymocyte-conditioned medium (which contains fetal calf serum) and freshly added fetal calf serum and horse serum are introduced only after the immunogen has had a chance to interact with the lymphocytes. Although the initial incubation in the absence of serum may result in a decrease of viability of the lymphocytes, it has been our experience that successful immunization with limiting amounts of antigen (e.g., ~100 ng) could best be achieved by this serum-free exposure to the lymphocytes.

In Vivo Immunization of Rats with 3T3 Cells

We have also carried out immunization of rats with 3T3 cells to generate monoclonal antibodies that bind to whole 3T3 cells but are not specifically directed against the FGR-s fraction. Confluent monolayers of 3T3 cells (5×10^4 cells/cm^2) are scraped with a rubber policeman and washed with 10 ml of PBS. The cells are resuspended in PBS at a density of 1.0×10^7 cells/ml and emulsified with an equal volume of Freund's complete adjuvant (Difco, Detroit, MI). Female Sprague-Dawley rats (~200 g) are immunized intraperitoneally with 1-ml aliquots of the emulsion. The rats are given booster injections of similar doses in Freund's incomplete adjuvant at days 8 and 31. When the serum from an immunized rat is assayed as positive in the primary screening assay (see below), the rat is boosted again and sacrificed 7 days later. Immune spleen cells are prepared and fused with mouse myeloma cells.

Fusion of Immune Spleen Cells with Myeloma Cells

Procedures for the preparation of the fusion partners to generate hybridomas, for the fusion event, and for selection of hybrid cells in hypoxanthine–aminopterin–thymidine medium have been detailed in this series,[8] as well as in a volume on general immunological methods.[9] We follow these procedures closely. The mouse myeloma cell line we use is P3/NS1/1-Ag4-1 (commonly known as NS-1) and is obtained from the Cell Distribution Center of The Salk Institute (La Jolla, CA). It is checked for continued resistance to 8-azaguanine prior to use. NS-1 cells in log-phase growth ($\sim 10^5$ cells/ml) with a viability greater than 95% (trypan blue exclusion assay[9]) are used.

Immune spleen cells (from either *in vitro* immunization with FGR-s or *in vivo* immunization with 3T3 cells) and NS-1 myeloma cells are mixed at a ratio of 10:1 in a 50-ml centrifuge tube and centrifuged at 400 g for 10 min at room temperature to form a tight pellet. Polyethylene glycol 1500 (Gallard Schlessinger, Carle Place, NY) is added as a 50% (w/w) solution as detailed.[9] After fusion and washing, the cell suspension is distributed into individual wells of 96-well culture dishes (Costar, No. 3596) to yield master plates for selection.[9]

After selection, individual cultures are screened for production of antibodies directed against 3T3 cells using the primary screening assay (see below). Positive cultures are expanded by transferring cells to 1-ml cultures in 24-well dishes (Costar, No. 3524). Cells in positive cultures after two rounds of expansion are cloned in soft agar.[9]

Screening for Relevant Antibody-Producing Hybridomas

Primary Screening Assay

Antisera from the immunized rats or supernatants from selected hybridoma cultures are tested by the primary screening assay (Fig. 3). 3T3 cells (3×10^5 cells/well) are seeded in 96-well polyvinyl plates (Costar, No. 2595) and cultured at 37° for 24 hr. All further steps are done at room temperature. The attached 3T3 cells are fixed with 200 μl of 0.25% (w/v) glutaraldehyde in PBS for 10 min. The cells are washed 3 times in PBS, and 200 μl of bovine serum albumin [3% (w/v), BSA] in PBS is added, incubated for 1 hr, and then washed as above. Antisera or supernatants from hybridoma cultures are added (50 μl), incubated for 1 hr, and washed.

[8] G. Galfrè and C. Milstein, this series, Vol. 73, p. 3.
[9] B. B. Mishell and S. M. Shiigi, "Selected Methods in Cellular Immunology." Freeman, San Francisco, California, 1980.

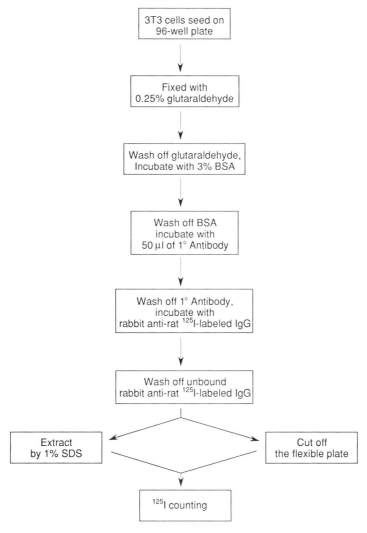

FIG. 3. Outline of the primary screening assay. The primary (1°) antibody is either serum or supernatant from hybridoma cultures.

Finally, rabbit anti-rat ^{125}I-labeled immunoglobulin [5×10^8 counts/min (cpm)/mg; 1×10^5 cpm; 50 μl] is added and incubated for 1 hr. After unbound radioactivity is removed by washing, the polyvinyl plates are cut into individual wells, and the amount of radioactivity in each well is measured using a LKB RIA γ counter. This assay detects, in general, any rat antibody reactive against any component of 3T3 cells.

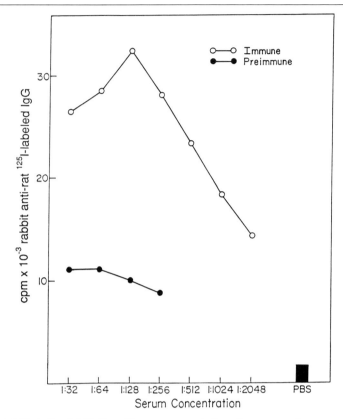

FIG. 4. Effect of serial dilution of immune serum in the primary screening assay. Immune and preimmune sera were serially diluted in PBS as indicated and then tested in the primary screening assay.

Figure 4 shows the results of the primary screening assay using the serum collected from the rat immunized *in vivo* with 3T3 cells whose spleen cells were subsequently used for fusion. Serial dilutions of the immune serum (1 : 128 to 1 : 2048) result in a linear decrese of the binding of ^{125}I-labeled antibody. On the other hand, higher concentrations of immune serum (1 : 64 and 1 : 32 dilutions) also show lower binding. The reason for this decrease at high concentrations is not known. Nonetheless, it is obvious that the primary screening assay can clearly distinguish normal serum (preimmune) from serum containing antibodies directed against 3T3 cells.

Hybridomas which produce antibodies reacting with fixed 3T3 cells are detected by the primary screening assay and are cloned in soft agar.

The clones and their secreted products are numbered according to the sequence of establishment of the clones.[4] Supernatants from the clones are tested using the primary screening assay; the clones exhibit 1.7- to 4.4-fold greater binding activity compared to that seen using the supernatant from the parental myeloma NS-1 cell line.

Screening Assays for Hybridoma Cultures Producing Antibodies Directed against FGR-s

A modification of the primary screening assay has been developed to detect hybridoma clones secreting antibodies directed specifically against a partially purified protein fraction (e.g., FRG-s in place of 3T3 cells in Fig. 3). This assay requires the efficient adsorption of the specific antigen fraction onto solid surfaces to minimize the amount of antigen required during the screening process. Moreover, because of the limited amount of material available, this screening assay is used only on hybridoma clones that show a positive reaction in the primary screening assay, thus indicating that these clones may be "putative positives" in terms of being clones which secrete monoclonal antibodies directed against our antigen of interest, FRG-s.

Unlabeled FGR-s (200 μl of 100 ng/ml solution) is added to individual wells of Immulon-2 micro-ELISA plates (Dynatech, Alexandria, VA). Approximately 8 ng (~40% of amount added) binds to each well. The wells are washed 3 times in PBS, and BSA (3% w/v) in PBS is then added and incubated for 1 hr. The wells are washed again as above. Supernatants (50 μl) from hybridoma cultures or immunoglobulins purified from such supernatants are added and incubated for 1 hr. Finally, rabbit ^{125}I-labeled antibodies directed against rat immunoglobulin (5×10^8 cpm/mg; 1×10^5 cpm; 50 μl) are added and incubated for 1 hr. After unbound radioactivity is removed by washing, the plates are cut into individual wells and the amount of radioactivity in each well is measured in a γ counter.

The supernatants from three hybridoma lines, designated DC4, 3C9, and 2A4, show positive reactions when assayed with FGR-s on Immulon-2 plates. Compared to the supernatant of NS-1 myeloma cultures, each of the hybridoma supernatants exhibit 1.5- to 1.7-fold more binding. This binding is specific. When bovine serum albumin is substituted for FGR-s on the Immulon-2 plates, the binding of the supernatants from DC4, 3C9, 2A4, and NS-1 cells are all the same. The supernatants of clones 2A4 and 3C9 are also positive when assayed with whole 3T3 cells fixed to the plates, exhibiting 1.4- to 1.6-fold more binding than the corresponding supernatant from NS-1 cells.[4]

Quite the opposite results are obtained with hybridoma clones derived from rats immunized *in vivo* with whole 3T3 cells and screened and selected on the basis of binding to whole 3T3 cells. A representative clone, designated 104, is used here for illustrative purposes. Clone 104 shows a strong reaction (2.4-fold) when assayed on whole 3T3 cells but a negligible reaction when assayed on FGR-s. These results suggest that the products of clones 2A4 and 3C9 are monoclonal antibodies directed against some component of FGR-s, which, in turn, is a constituent of whole 3T3 cells.[4]

Properties of Monoclonal Antibody 2A4

Isolation and Characterization of Antibody 2A4

Rabbit antibodies directed against rat immunoglobin are coupled to Affi-Gel 10 (Bio-Rad, Richmond, CA) using the procedure of Ikeda and Steiner.[10] Supernatant from the hybridoma clone 2A4 is passed through the column (1 × 12 cm) 3 times. Material bound nonspecifically is removed by washing with 0.1 M phosphate buffer (pH 8.0). The specifically bound material is eluted with 0.1 M citrate buffer (pH 3.0), dialyzed overnight against PBS with 3 changes of buffer, and subjected to SDS–polyacrylamide gel electrophoresis in the presence of 2-mercaptoethanol (5%). Two polypeptides, corresponding to the heavy (M_r 55,000) and light (M_r 23,000) chains of immunoglobulin G, are observed. All subsequent experiments are performed wtih the purified IgG molecule, designated Antibody 2A4. The isolated Antibody 2A4 is stored at $-20°$.

Similar results are obtained with the antibody secreted by hybridoma clone 104. This rat IgG molecule is also purified, and Antibody 104 is used in all experiments parallel to Antibody 2A4 as a control antibody.

Identification of Target of Antibody 2A4

Antibody 2A4 is coupled to Affi-Gel 10 as described above.[10] The affinity column (0.6 × 4 cm) is used to fractionate FGR-s preparations to determine the molecular identity of the antigenic target of Antibody 2A4. When a [^{35}S]methionine-labeled preparation of FGR-s is fractionated over this Antibody 2A4 column, the unbound flow-through fraction yields one predominant polypeptide (M_r 10,000). The material bound by Antibody 2A4 also shows one polypeptide band (M_r 13,000). These results indicate that Antibody 2A4 is specifically directed against the M_r 13,000 polypeptide in the FGR-s fraction.[4]

[10] Y. Ikeda and M. Steiner, *J. Biol. Chem.* **251**, 6135 (1976).

Effect of Antibody 2A4 on Growth-Inhibitory Activity of FGR-s

To test the possibility that Antibody 2A4 can neutralize the growth-inhibitory activity of FGR-s, the effect of the inhibitor preparation on [^{3}H]thymidine incorporation in target 3T3 cells is assayed in the presence and absence of the purified antibody. The growth-inhibitory effect of FGR-s on 3T3 cells is dependent on the concentration of the ligand added.[2] In the present assay, 60% inhibition is obtained with 20 μl of the FGR-s preparation. When the effect of FGR-s is assayed in the presence of Antibody 2A4 (25 ng/ml), however, the inhibition is completely abrogated. The level of DNA synthesis in this case is the same as that of control cultures, without any FGR-s or Antibody 2A4. Similarly, the addition of 25 ng/ml of Antibody 2A4 reduces the inhibitory effect of 50 μl of FGR-s from 80 to 35%. Thus, when the effect of FGR-s is assayed in the presence of Antibody 2A4, there is always a higher level of [^{3}H]thymidine incorporation (i.e., a reduced level of growth inhibition).

The effects of Antibody 2A4 on reversing the inhibition of FGR-s are specific. Antibody 104, which is not reactive with FGR-s polypeptides, fails to yield the same effect. These observations suggest that the results obtained with Antibody 2A4 are most probably not due to a growth factor contaminating the immunoglobulin fraction. The results on the neutralization of growth-inhibitory activity of FGR-s by Antibody 2A4, coupled with the fact that this antibody specifically recognizes the M_r 13,000 polypeptide, strongly suggest that this polypeptide carries growth-inhibitory activity.[4] This conclusion is further supported by experiments that showed depletion of the growth-inhibitory activity from preparations of FGR-s (13K) on Antibody 2A4 affinity columns.

Depletion of Growth-Inhibitory Activity from FGR-s (13K) on Affinity Column Containing Antibody 2A4

Rabbit antibodies directed against rat immunoglobulin (40 mg) are coupled to Affi-Gel 10 (2 ml of beads).[10] The derivatized gel is used to prepare two different columns: (1) the supernatant (50 ml) of hybridoma clone 2A4 is passed over one column (0.4 × 2.5 cm) 3 times to bind Antibody 2A4 (Antibody 2A4 column); (2) the supernatant of the parent myeloma NS-1 cell line is passed over the other column (NS-1 control column). Material bound nonspecifically is removed by washing with 0.1 M phosphate buffer (pH 8.0). Finally, a preparation of FGR-s (13K) (8 ng/ml, 2 ml) is percolated through either the Antibody 2A4 column or the NS-1 control column. The original FGR-s (13K) preparation, as well as the pooled flow-through fractions, representing material not bound by the respective affinity columns, are assayed for growth-inhibitory activity.

The results show that the growth-inhibitory activity of FGR-s (13K) is depleted in the material passed through the Antibody 2A4 column.[3] In contrast, the NS-1 control column has little effect on the growth-inhibitory activity. These results strongly suggest that FGR-s (13K) is directly responsible for the observed growth-inhibitory activity. The data also argue against any indirect mechanisms, such as stimulation of a growth factor receptor by Antibody 2A4, in the observed neutralization of FGR-s effects in the growth-inhibition assays.[4]

Acknowledgments

This work was supported by Grant GM 27203 from the National Institutes of Health and by Faculty Research Award FRA-221 from the American Cancer Society. I thank Dr. Yen-Ming Hsu for all his help and Mrs. Linda Lang for the preparation of the manuscript.

Section VII

Techniques for Study of Growth Factor Activity

[44] Iodination of Peptide Growth Factors: Platelet-Derived Growth Factor and Fibroblast Growth Factor

By ANGIE RIZZINO and PETER KAZAKOFF

Introduction

Six methods are commonly used to radiolabel growth factors with iodine. Three of the methods employ the chloramine-T reagent or modifications of this reagent, namely, Iodogen (1,3,4,6-tetrachloro-3α,6α-diphenylglycouril) and Iodobeads (a chloramine-T-derivatized nonporous polystyrene bead).[1-4] The other three commonly used methods employ lactoperoxidase (usually in the form of Enzymobeads, Bio-Rad, Richmond, CA), iodine monochloride, or the Bolton–Hunter reagent.[5-7] The amino acids labeled by the six methods and the reactive species involved are indicated in Table I.

Selection of an appropriate iodination method is influenced by several factors. One of the most important factors is the stability of the growth factor. Although the chloramine-T method is often used because it is one of the easiest and least expensive, this method can destroy the biological activity of many proteins. This is an important consideration in the case of fibroblast growth factor (FGF). Consequently, methods employing milder conditions are more appropriate for some growth factors. Whether the growth factor possesses tyrosine residues and whether iodination of tyrosine residues alters the biological activity are important considerations when labeling platelet-derived growth factor (PDGF). Porcine platelet PDGF (pPDGF) is a homodimer of the B chain and, thus, lacks tyrosine residues. PDGF isolated from human platelets (hPDGF) is primarily a heterodimer of the A- and B chains, but can contain 10–30% B:B homodimers.[8] Thus, only about 70–90% of the hPDGF will be labeled when the method employed selectively labels

[1] W. M. Hunter and F. C. Greenwood, *Nature (London)* **194**, 495 (1962).
[2] F. C. Greenwood, W. M. Hunter, and J. S. Glover, *Biochem. J.* **89**, 114 (1963).
[3] P. J. Fraker and J. C. Speck, *Biochem. Biophys. Res. Commun.* **80**, 849 (1978).
[4] M. A. K. Markwell, *Anal. Biochem.* **125**, 427 (1982).
[5] J. J. Marchalonis, *Biochem. J.* **113**, 299 (1969).
[6] A. S. McFarlane, *Nature (London)* **182**, 53 (1958).
[7] A. E. Bolton and W. M. Hunter, *Biochem. J.* **133**, 529 (1973).
[8] A. Hammacher, U. Hellman, A. Johnsson, A. Ostman, K. Gunnarsson, B. Westermark, A. Wasteson, and C.-H. Heldin, *J. Biol. Chem.* **263**, 16493 (1988).

TABLE I
AMINO ACIDS LABELED AND REACTIVE SPECIES GENERATED DURING
METHODS COMMONLY USED TO IODINATE PROTEINS

Iodination method	Amino acid labeled	Reactive species
Chloramine-T	Primarily tyrosine[a]	Conversion of I^- to I^+
Iodobeads	Primarily tyrosine[a]	Conversion of I^- to I^+
Iodogen	Primarily tyrosine[a]	Conversion of I^- to I^+
Lactoperoxidase	Primarily tyrosine[a]	Conversion of I^- to I_2
Iodine monochloride	Primarily tyrosine[a]	Conversion of I^- to I^+
Bolton–Hunter	Primarily lysine[b]	Bolton–Hunter reagent

[a] Under some conditions, other amino acids, in particular histidine, are labeled.
[b] Amino-terminal amino groups may also be labeled.

tyrosine residues. This is likely to affect the determination of the biological activity exhibited by iodinated PDGF (see below). Still another important factor is the amount of protein to be labeled. Some methods, for example, Iodobeads, generally yield better results when large amounts of protein are used. Cost is another important consideration. The Bolton–Hunter method, which is a gentle labeling method, is far more expensive than the other methods commonly used to label growth factors. Lastly, the choice of the iodination reaction vessel can influence the selection of the iodination method. In this regard, the manufacturer of Iodogen recommends coating the reagent onto glass prior to its use. Unfortunately, many growth factors cannot be easily recovered after they bind to glass, which makes it difficult to estimate the amount of iodinated growth factor recovered, which in turn makes it difficult to estimate the radioactive specific activity of the growth factor. Although polypropylene can be coated with Iodogen,[9] variation in the degree of coating can be a problem.

With the possible exception of the lactoperoxidase method, hPDGF has been iodinated by each of the methods described above.[10–13] In the author's laboratory, the chloramine-T and the Bolton–Hunter methods

[9] B. Y. Rubin, S. L. Anderson, S. A. Sullivan, B. D. Williamson, E. A. Carswell, and L. J. Old, *Proc. Natl. Acad. Sci. U.S.A.* **82**, 6637 (1985).
[10] D. F. Bowen-Pope and R. Ross, *J. Biol. Chem.* **257**, 5161 (1982).
[11] J. S. Huang, S. S. Huang, B. Kennedy, and T. F. Deuel, *J. Biol. Chem.* **257**, 8130 (1982).
[12] C.-H. Heldin, B. Ek, and L. Ronnstrand, *J. Biol. Chem.* **258**, 10054 (1983).
[13] D. T. Graves, H. N. Antoniades, S. R. Williams, and A. J. Owen, *Cancer Res.* **44**, 2966 (1984).

are used to label hPDGF and pPDGF, respectively.[14] Basic FGF (bFGF) and acidic FGF (aFGF) have been reported to be iodinated with Chloramine-T, Iodogen, Iodobeads, and lactoperoxidase.[15-18] In the author's laboratory, the lactoperoxidase method is used to label bovine bFGF and bovine aFGF.[19] This chapter describes the use of the chloramine-T method and the Bolton–Hunter method to label PDGF and describes the use of the lactoperoxidase method (Enzymobead) to label aFGF and bFGF.

Iodination of Human Platelet-Derived Growth Factor by the Chloramine-T Method

Materials

Carrier-free hPDGF (R&D Systems Inc., Minneapolis, MN)
Resuspension buffer: 25% acetonitrile, 0.1% trifluoroacetic acid
$Na^{125}I$ (New England Nuclear, Boston, MA): 16.5 Ci/mg, 100 mCi/ml
Sodium phosphate: 0.4 M, pH 7.4
Sodium phosphate: 0.1 M, pH 7.4
Chloramine-T (Eastman Kodak, Rochester, NY)
0.1 M Tris-HCl, pH 8.4
Saturated tyrosine solution: in 0.1 M Tris-HCl, pH 8.4; made fresh for each iodination (100 mg/ml is sufficient)
KI: 60 mM
Urea solution: 1.2 g/ml in 1 M HCl; made fresh for each iodination
Column equilibration buffer: 5 mM acetic acid, 30 mM KI, 0.2% (w/v) bovine serum albumin (BSA; ICN ImmunoBiologicals, Costa Mesa, CA, Cat. No. 81-001)
Sephadex G-25 column: 0.7 × 15 cm glass column (1.4 g Sephadex G-25 preswelled, 4–12 hr, in 13 ml of column equilibration buffer)
Trichloroacetic acid: 15% (w/v) TCA, ice-cold
Concentrated bovine serum albumin: 360 mg/ml BSA (ICN ImmunoBiologicals, Cat. No. 81-001) in water

---

[14] A. Rizzino, P. Kazakoff, E. Ruff, C. Kuszynski, and J. Nebelsick, *Cancer Res.* **48,** 4266 (1988).
[15] M. Vigny, M. P. Ollier-Hartmann, M. Lavigne, N. Fayein, J. C. Jeanny, M. Laurent, and Y. Courtois, *J. Cell. Physiol.* **137,** 321 (1988).
[16] G. Neufeld and D. Gospodarowicz, *J. Biol. Chem.* **260,** 13860 (1985).
[17] S. S. Huang and J. S. Huang, *J. Biol. Chem.* **261,** 9568 (1986).
[18] B. B. Olwin and S. D. Hauschka, *Biochemistry* **25,** 3487 (1986).
[19] A. Rizzino, C. Kuszynski, E. Ruff, and J. Tiesman, *Dev. Biol.* **129,** 61 (1988).

Procedure. The following steps are performed at room temperature.

1. Carrier-free hPDGF is resuspended in resuspension buffer at a concentration of 100 µg/ml. The amount of PDGF to be iodinated is added to a 1.5-ml Eppendorf vial in which the iodination will occur. (The procedure described in this section has been used to successfully label as little as 2 µg of hPDGF and as much as 20 µg of hPDGF.)
2. An amount of 0.4 M sodium phosphate equal to one-third the volume of hPDGF is added to the reaction vial and mixed.
3. Na^{125}I (0.125 mCi ^{125}I per microgram hPDGF) is added to the reaction vial and mixed. (We have observed that the amount of Na^{125}I used may need to be varied with some preparations of PDGF.) This step and subsequent steps should be performed in a hood that is externally vented.
4. Immediately before the reaction is to be initiated, a 1 mg/ml chloramine-T solution in 0.1 M sodium phosphate is prepared.
5. Chloramine-T (1.5 µg per microgram hPDGF) is added to the reaction vial to initiate the reaction. The reaction mixture is incubated for 1.75 min and is mixed 3 times during this period. After 1.75 min, an equal amount of chloramine-T (1.5 µg per microgram hPDGF) is added, and the mixture is allowed to incubate for an additional 1.75 min, again with mixing 3 times during this period.
6. At the end of the second 1.75-min incubation, the reaction is terminated by adding in quick succession 20 µl saturated tyrosine, then 200 µl of 60 mM KI, and finally 200 µl of the urea solution. The reaction mixture is then incubated for 5 min.
7. The reaction mixture is added to the Sephadex G-25 column. The reaction vial is rinsed with column equilibration buffer (volume equal to the reaction mixture), and this rinse is also added to the column. The column is eluted with column equilibration buffer, with fractions of approximately 10 drops being collected.
8. The percentage of TCA-precipitable counts in each fraction is determined. For this step, a small volume from each fraction is added to ice-cold 15% TCA, and the precipitate is collected by filtration or centrifugation. The radiolabeled growth factor peak elutes first and exhibits greater than 85% TCA-precipitable counts. This peak is followed by a second peak of unincorporated radioactive material.
9. The concentrated BSA stock solution is used to increase (by 3 mg/ml) the concentration of BSA in the radiolabeled growth factor peak fractions to an approximate final concentration of 5 mg/ml. The labeled growth factor is aliquoted and stored at $-20°$ (frost-free freezers should not be used to store the labeled growth factors).

Comments. It is best to characterize the radiolabeled growth factor by two methods, one that examines the binding characteristics of the labeled growth factor and another that compares the biological activity of the labeled growth factor to its unlabeled counterpart. The binding of labeled PDGF can be examined by determining the ability of unlabeled PDGF to compete with the binding of the labeled PDGF to membrane receptors. In our laboratory, unlabeled PDGF, at concentrations ranging from 0.5 to 200 ng, is used to compete with the binding of 1 ng ^{125}I-labeled PDGF to NR-6-R cells. The residual binding observed in the presence of the 200-fold excess of unlabeled hPDGF is defined as nonspecific binding.

hPDGF labeled by the chloramine-T procedure described above generally exhibits 10–25% nonspecific binding, but this is dependent on several factors, in particular, cell density. For most of the cell lines examined in our laboratory, nonspecific binding (e.g., to NR-6-R cells) is higher at lower cell densities (e.g., 7000 cells/cm^2). Presumably, this occurs because the labeled growth factor exhibits greater nonspecific binding to the tissue culture plastic than it does to cells. Thus, as cell density increases, the level of nonspecific binding decreases. Another important parameter is the radioisotopic specific activity of the labeled PDGF. In general, the level of nonspecific binding becomes unacceptably high (>30%) when PDGF is labeled at specific activities above 40 μCi/μg. The reason for this is unclear, but alterations in the biological activity of the PDGF are also likely to be observed at the higher specific activities. In regard to the amount of ^{125}I incorporated, it should be noted that Na^{125}I is added at a concentration such that it is limiting. This not only reduces radioisotope exposure, but permits one to manipulate the final radioisotopic specific activity of the growth factor. Thus, by increasing the amount of Na^{125}I used during the reaction, one increases the amount of ^{125}I incorporated.

Preparations of ^{125}I-labeled PDGF that exhibit acceptable levels of nonspecific binding can be tested for biological activity by comparing the ability of labeled and unlabeled PDGF to stimulate [^{3}H]thymidine incorporation by quiescent 3T3 cells. hPDGF labeled by the chloramine-T procedure described above generally exhibits relatively little change in biological activity. In this regard, the potential to observe a change in the biological activity is dependent on the extent to which hPDGF contains B:B homodimers. If hPDGF contains large amounts of B:B homodimers and is labeled by the chloramine-T method, relatively small decreases (~2- to 3-fold) in biological activity are expected, since the PDGF B:B dimers should not be labeled by the chloramine-T method.

Examination of Binding of ^{125}I-Labeled Platelet-Derived Growth Factor

Materials

Cells: For routine testing of ^{125}I-labeled hPDGF, NR-6-R cells are plated overnight in medium supplemented with 10% calf serum.[20] For reasons indicated above, the cell density at the time of the binding assay should be in the range of 60,000 to 100,000 per well (24-well tissue culture plate). (Other cell lines can be used. Most fibroblast cell lines exhibit PDGF receptors.)

Binding buffer: Ham's F12 medium (GIBCO, Grand Island, NY, Nutrient mixture F12) supplemented with 2.5 g/liter BSA (Sigma, St. Louis, MO, A-7888) and 25 mM HEPES (Research Organics Inc., Cleveland, OH), pH 7.4 (adjusted with NaOH). The binding buffer is filter-sterilized and stored at 4°. (All solutions containing HEPES that are used in the procedures described in this chapter employ dilutions of a 1.5 M HEPES stock solution that is adjusted to pH 7.4.)

^{125}I-labeled PDGF: diluted to an appropriate concentration in binding medium just prior to the experiment.

PDGF: diluted to appropriate concentrations and added to binding medium containing ^{125}I-labeled PDGF.

Lysing buffer: 0.1% (v/v) Triton X-100, 10% (v/v) glycerol, 20 mM HEPES (from a 1.5 M stock solution, pH 7.4), 0.01% BSA (Sigma, A-7888). The lysing buffer is stored at 4°.

Procedures

1. Just prior to performing the binding assay, the cells are washed 3 times with 1 ml of cold binding buffer.

2. Cold binding buffer (0.5 ml/well) containing ^{125}I-labeled PDGF (e.g., 1 ng/well) and varying amounts of unlabeled PDGF as competitor (ranging from 0.5 to 200 ng/well) is added to the cells.

3. The cells are incubated for 4 hr at 4° with gentle rocking on a Bellco (Vineland, NJ) rocker platform (Cat. No. 7740-10010) at a setting of 5. (Alternatively, the binding assay can be performed at room temperature for 1 hr.)

4. At the end of the incubation period, the cells are washed 3 times with cold binding buffer (1.5 ml).

5. The last wash is removed, and then 600 μl lysing buffer is added to each well followed by vigorous oscillation for 20 min with a Bellco orbital shaker (Cat. No. 7744-01010; setting of 7.5) to ensure complete lysis.

[20] A. Rizzino and E. Ruff, *In Vitro Cell. Dev. Biol.* **22**, 749 (1986).

6. A portion (550 μl) of the lysate from each well is pipetted into a 6 × 50 mm glass tube. The tubes are placed in snap-cap vials and counted with a γ counter.

Examination of Biological Activity of ^{125}I-Labeled Platelet-Derived Growth Factor

Procedure

1. NR-6-R cells are plated at a density of 3 × 10^4 cells/16-mm tissue culture well and cultured in 1 ml of Dulbecco's modified Eagle's medium (DMEM) containing 10% calf serum.

2. After 4 days, the medium is removed and the cells are refed with fresh medium composed of a 1 : 1 mixture of DMEM/Ham's F12 containing 2% calf plasma (Irvine Scientific, Irvine, CA).

3. After 2 days, the growth factor to be tested is added directly to each well in a volume not exceeding 5% of the total culture medium volume.

4. After 18–22 hr, the medium is removed and replaced with 1 ml of medium (DMEM with 10% calf serum) containing [^{3}H]thymidine (1 μCi/ml). Then the cells are incubated 2 hr at 37° with gentle rocking.

5. The cells are washed twice with 0.8 ml of ice-cold (4°) 5% TCA and then solubilized with 0.8 ml of 0.25 N NaOH. To ensure complete lysis, the cells are subjected to rapid rotation on a Bellco orbital shaker (Cat. No. 7744-01010) at a setting of 7.5 for 20 min. Then the solubilized samples are neutralized with 0.2 ml of 1 N HCl.

6. [^{3}H]Thymidine incorporation is determined by adding 0.75 ml of solubilized materials from each well to 5 ml of scintillation cocktail and counting with a β counter. Unlabeled hPDGF exhibits an ED$_{50}$ of 2–4 ng/ml, and one observes about a 20-fold stimulation of thymidine incorporation at the optimal concentration of hPDGF.

Iodination of Human and Porcine Platelet-Derived Growth Factor with Bolton–Hunter Reagent

Materials

Carrier-free PDGF: hPDGF or pPDGF (R&D Systems, Inc.)
Resuspension buffer: 25% acetonitrile, 0.1% trifluoroacetic acid
Bolton–Hunter reagent: 1 mCi/vial (New England Nuclear, NEX-120H)
Borate buffer: 0.2 M, pH 8.5
Borate buffer: 0.1 M, pH 8.5

Glycine–borate buffer: 0.2 M glycine in 0.1 M borate buffer
Gelatin solution: 10 mg/ml; prepared fresh for each iodination
Column equilibration buffer: 0.2% gelatin in 5 mM acetic acid
Sephadex G-25 column: 0.7 × 15 cm glass column (1.4 g G-25 pre-swelled, 4–12 hr, in 13 ml of column equilibration buffer)
Trichloroacetic acid: 15% ice-cold TCA

Procedure. The following steps are performed at the temperature of an ice bath.

1. Carrier-free PDGF is resuspended in resuspension buffer at 400 μg/ml. This is accomplished by adding 25 μl of resuspension buffer to a vial containing 10 μg PDGF. The vial is allowed to stand at room temperature for 10 min and then is vortexed at a speed sufficient to wash the walls of the vial. Vortexing is repeated twice during the following 10 min. Next, an equal volume of 0.2 M borate buffer is added to the vial. This reduces the PDGF concentration to 200 μg/ml and the borate concentration to 0.1 M. The resuspended PDGF is then transferred to a 1.5-ml polypropylene Eppendorf tube and stored at $-20°$ until needed.

2. The iodination reaction is performed in the vial containing the Bolton–Hunter reagent. Prior to the iodination reaction, the Bolton–Hunter reagent is air-dried in an appropriate hood to remove benzene. After the benzene has evaporated, the vial is placed on ice.

3. The iodination reaction is initiated by adding the thawed borate-buffered growth factor to the vial containing the Bolton–Hunter reagent. The reaction mixture is incubated on ice for 3.5 hr with mixing at least every 30 min. (The extent of iodination can be varied by changing the incubation time. This time period for the procedure described above was selected because it yields ^{125}I-labeled PDGF with a specific activity of 30–40 μCi/μg.)

4. The reaction is terminated by adding 500 μl of the glycine–borate buffer. The reaction vial is incubated for 5 min on ice. Fifty microliters of the gelatin solution is then added to the vial as carrier. (BSA should not be used because it absorbs labeled hydroxyphenyl proprionic acid.)

5. The reaction mixture is added to a Sephadex G-25 column. The reaction vial is rinsed with an equal volume of column equilibration buffer, and this rinse is also added to the column. The column is eluted with column equilibration buffer, and fractions of approximately 10 drops are collected.

6. The percentage of TCA-precipitable counts in each fraction is determined using 15% TCA as indicated above. The labeled growth factor peak elutes first and exhibits over 85% TCA-precipitable counts. The protein peak fractions are combined, aliquoted, and stored at $-20°$.

Comments. The biological activity and the binding capacity of the ^{125}I-labeled PDGF prepared by the Bolton–Hunter method can be determined as described above for hPDGF labeled by the chloramine-T method. In our laboratory, the Bolton–Hunter reagent has been used to label hPDGF and pPDGF. ^{125}I-Labeled PDGF of high quality, in terms of radioisotopic specific activity and low nonspecific binding, has been obtained with both hPDGF and pPDGF. In the case of hPDGF, the procedure described above has yielded ^{125}I-labeled PDGF with specific activities ranging from 33 to 48 μCi/μg and nonspecific binding ranging from 11 to 23%.

In some cases, hPDGF labeled with the Bolton–Hunter reagent has been observed to exhibit a decrease in biological activity. This is particularly evident for preparations exhibiting radioisotopic specific activity above 50 μCi/μg. Interestingly, although the biological activity appears to decrease as radioisotopic specific activity increases, nonspecific binding does not increase. In addition, the competition by unlabeled PDGF is virtually identical for ^{125}I-labeled PDGF preparations of high and low radioisotopic specific activity. The only difference in the binding is the total number of counts that bind to the cells. In general, there appears to be a direct relationship between the number of counts that bind to the cells and the radioisotopic specific activity of ^{125}I-labeled PDGF using the Bolton and Hunter reagent. Thus, it appears that biologically inactivated PDGF does not bind either specifically or nonspecifically.

In regard to regulating the radioisotopic specific activity of the labeled growth factor, time is the critical factor. By increasing the incubation period, one can increase the radioisotopic specific activity of the labeled growth factor.

Iodination of Acidic and Basic Fibroblast Growth Factor by the Lactoperoxidase Method

In our laboratory, bFGF and aFGF are radiolabeled with lactoperoxidase (Enzymobeads) because aFGF and bFGF are easily inactivated and the lactoperoxidase method is one of the mildest labeling methods. The procedure described below is used for bFGF. This procedure, with minor modifications (indicated in the text below), is also used to prepare labeled aFGF.

Materials

Lyophilized carrier-free bFGF or lyophilized carrier-free aFGF (R&D Systems, Inc.), resuspended in sterile water at 500 μg/ml

Buffer A: 4 M NaCl, 10 mM HEPES (from a 1.5 M stock solution of HEPES, pH 7.4)

Enzymobeads resuspended in 0.5 ml water as per the manufacturer's instructions; aliquots (50 μl) are stored for up to 3 months at −20° and are used only once

Na^{125}I (New England Nuclear): 16.5 Ci/mg, 100 mCi/ml D-Glucose solution: 2% in water; filter-sterilized and allowed to racemize to β-D-glucose overnight (the iodination reaction requires β-D-glucose)

Buffer B: 10 mM HEPES (from a 1.5 M stock solution, pH 7.4), 0.6 M NaCl, 0.1% BSA (ICN ImmunoBiologicals, Cat. No. 81-001) (for aFGF, buffer B contains 0.3 M NaCl in place of 0.6 M NaCl)

Sephadex G-25 column: 0.7 × 15 cm glass column (1.4 g Sephadex G-25 preswelled in 14 ml of buffer B)

Trichloroacetic acid: 15% ice-cold TCA

Heparin-Sepharose column: prepared according to Pharmacia's instructions. One-tenth gram of heparin-Sepharose is resuspended in 5 ml of buffer B and allowed to swell for 15–30 min. Next, it is washed on a sintered glass filter with small volumes of buffer B to a total of 20 ml/0.1 g. The heparin-Sepharose column is then poured, and at least 10 ml of buffer B is run through the column.

Washing buffer: 10 mM HEPES (from a 1.5 M stock solution, pH 7.4), 0.9 M NaCl, 0.1% BSA (ICN ImmunoBiologicals, Cat. No. 81-001) (for aFGF, washing buffer contains 0.7 M NaCl instead of 0.9 M NaCl)

Elution buffer: 10 mM HEPES (from a 1.5 M stock solution, pH 7.4), 2.5 M NaCl, 0.1% BSA (ICN ImmunoBiologicals, Cat. No. 81-001) (for aFGF, elution buffer contains 2.0 M NaCl instead of 2.5 M NaCl)

Procedure

1. The reagents are added to the reaction vessel in the following order: 10 μg bFGF, 26 μl buffer A, 50 μl Enzymobeads, 20 μl Na^{125}I (2 mCi), and 25 μl glucose solution (glucose is added last since it initiates the iodination reaction). The reaction mixture is mixed immediately after addition of the glucose. [Note: If more or less bFGF is to be labeled, use proportionally larger or smaller volumes of bFGF, buffer A, and Na^{125}I (0.2 mCi Na^{125}I per microgram bFGF), but the amount of glucose and Enzymobeads added is not modified.]

2. The reaction mixture is incubated for 1 hr at 22°–25° with mixing at 15-min intervals. (Higher or lower temperatures will influence the final level of iodine incorporated, since this is an enzymatic procedure.)

3. The reaction is stopped by loading the reaction mixture onto the Sephadex G-25 column. The reaction vial is rinsed with equilibration buffer

(equal to the volume of the reaction mixture), and the rinse is loaded onto the Sephadex G-25 column. The labeled protein is eluted from the column with buffer B, with 10-drop fractions being collected.

4. The radiolabeled growth factor peak is located by determining the TCA precipitability of each fraction using 15% TCA as described above.

5. The radiolabeled growth factor peak from the Sephadex G-25 column is loaded onto the heparin-Sepharose column.

6. The labeled growth factor is recycled over the column 10 times to ensure that as much as possible has bound. The heparin-Sepharose column is then washed with approximately 20 ml of buffer B (until only low counts elute from the column; determined with a Geiger Mueller Counter). The remaining buffer is run into the column without letting the top of the column go dry.

7. The column is then washed with washing buffer in steps of 200 and 500 μl. (This step decreases the level of nonspecific binding exhibited by the labeled growth factor.)

8. ^{125}I-Labeled bFGF is eluted from the column with elution buffer in steps of 200, 500, 500, 200, 200, and 200 μl. Labeled bFGF usually elutes from the column in the 1000–1600 μl after the first 200-μl step. (If larger amounts of bFGF are labeled, e.g., 15–20 μg, then the elution buffer is added in steps of 200 μl, 1 ml, 1 ml, 500 μl, 200 μl, and 200 μl.)

9. The TCA precipitability of the fractions containing the growth factor is determined using 15% TCA as described above. This information is used to determine the yield and thus the protein concentration.

Comments. Na^{125}I is the limiting component of the lactoperoxidase iodination reaction, and the amount of Na^{125}I used in the above protocol yields aFGF and bFGF with specific activities of 100–150 μCi/μg. One can increase the specific activity of the labeled growth factor by increasing the amount of Na^{125}I added. We have not observed any increase in nonspecific binding when the specific activity of these growth factors was increased up to 250 μCi/μg. In regard to nonspecific binding, we have observed that nonspecific binding is cell line dependent. Nonspecific binding to NRK-49F cells shows little effect of cell density, whereas there is a clear-cut increase as cell density increases for 3T3/A-31, NR-6-R, WI-38, and CCL-205 cells. Finally, as done by others,[21] our binding assay usually includes a high salt wash step. However, the usefulness of this step should be determined for each cell line or primary culture.

---

[21] D. Moscatelli, *J. Cell. Biol.* **107,** 753 (1988).

Examination of Binding of ^{125}I-Labeled Basic and Acidic Fibroblast Growth Factor

Materials

Cells: NR-6-R cells or other cells that are available (most fibroblast cell lines exhibit FGF receptors)

Binding buffer: Ham's F12 medium (GIBCO, Nutrient mixture F-12) containing 2 g/liter IgG (Sigma, G-5009) and 25 mM HEPES (Research Organics, Inc., from a 1.5 M stock solution, pH 7.4) and 1.2 g NaCl in place of 1.2 g of NaHCO$_3$. The pH of the binding buffer is adjusted to 7.4. It is filter-sterilized and stored at 4°. [The removal of NaHCO$_3$ from the medium prevents an increase in pH during the binding reaction. With this modified binding buffer, we have observed significant increases (40–50%) in the binding bFGF but only a small positive effect on the binding of aFGF.]

Salt wash buffer: 10 mM HEPES (from a 1.5 M stock solution, pH 7.4), 2.0 M NaCl

Lysing buffer: 0.2% (v/v) Triton X-100, 10% (v/v) glycerol, 25 mM HEPES (from a 1.5 M stock solution, pH 7.4), 0.1% BSA (Sigma, A-7888)

^{125}I-Labeled FGF: diluted to an appropriate concentration in binding medium just prior to the experiment.

FGF: diluted to appropriate concentrations and added to binding medium containing ^{125}I-labeled FGF.

Procedure

1. The cells are set up in 16-mm wells (20,000 cells/well) the day before the binding assay is performed. In our laboratory, the cells are generally plated in medium containing 1% calf plasma rather than in medium containing 10% serum. This reduces the amount of nonspecific binding.

2. Just prior to performing the binding assay, the cells are washed 3 times with cold binding buffer.

3. The last wash is removed and replaced with cold binding buffer (0.5 ml/well) containing ^{125}I-labeled FGF (e.g., 0.5 ng/well) and varying amounts of unlabeled FGF as competitor. (For routine testing, unlabeled FGF is added at several concentrations ranging from 0.5 to 1000×)

4. The cells are incubated for 4 hr at 4° with gentle rocking on a Bellco rocker platform (Cat. No. 7740-10010) at a setting of 5.

5. At the end of the incubation period, the cells are washed 3 times with cold binding buffer (1.5 ml/well).

6. The last wash is removed, and 1 ml of salt wash buffer is added to each well. In the case of NR-6-R cells, this wash should *not* remain on the

cells for longer than 1 min, since it causes the cells to detach. (The salt wash reduces the binding to low affinity sites and, in the case of NR-6-R cells, is usually used only during ^{125}I-labeled aFGF binding assays. For some cells, this salt wash will also reduce the binding of ^{125}I-labeled bFGF to low affinity sites.)

7. Lysing buffer (600 μl) is added to each well followed by vigorous oscillation for 20 min with a Bellco orbital shaker (Cat. No. 7744-01010; setting of 7.5) to ensure complete lysis.

8. A portion (550 μl) of each lysate is then pipetted into a 6 × 50 mm glass tube. The tubes are placed in snap-cap vials and are counted in a γ counter.

Examination of Biological Activities of ^{125}I-Labeled Basic and Acidic Fibroblast Growth Factor

In our laboratory, the biological activities of ^{125}I-labeled aFGF and bFGF are determined by the procedure described above for PDGF. The ED_{50} of unlabeled bFGF with NR-6-R cells is approximately 50 pg/ml. The ED_{50} of unlabeled aFGF with NR-6-R cells is about 4-fold higher than that observed for bFGF.

Acknowledgments

We thank Bradley Olwin and Paul DiCoreleto for providing protocols used in their laboratories to label aFGF and human PDGF, respectively. Over a period of several years, these protocols were modified and have given rise to the chloramine-T and the lactoperoxidase protocols described in this chapter. We also thank members of our laboratory, in particular Eric Ruff and Charles Kuszynski, for their efforts during the development of the protocols described in this chapter and Heather Rizzino for helpful comments during the preparation of this chapter. Work in our laboratory was supported by grants from the National Institute of Child Health and Human Development (HD19837, HD21568), the Nebraska Department of Health (89-51), the National Cancer Institute (Laboratory Research Center Support Grant CA36727), and the American Cancer Society (Core Grant ACS SIG-16).

[45] Localization of Peptide Growth Factors in the Nucleus

By Bruno Gabriel, Véronique Baldin, Anna Maria Roman,
Isabelle Bosc-Bierne, Jacqueline Noaillac-Depeyre,
Hervé Prats, Justin Teissié, Gérard Bouche,
and François Amalric

Introduction

Work in several laboratories has shown that various peptide hormones and growth factors are associated with nuclei of target cells. This has usually been demonstrated by biochemical or immunocytological detection of the hormone or growth factor in the nucleus. For platelet-derived growth factor (PDGF) and related molecules,[1,2] the signal for nuclear targeting has also been identified.

The accumulated data have been largely ignored because they do not fit with the current concept that all endocytosed polypeptides are destroyed by lysosomes, while the signal transduction is carried out by second messengers. Nuclear localization of peptide hormones and growth factors requires the existence of an alternative transport pathway, resulting in delivery of the polypeptide to the nucleus rather than to lysosomes. In addition, nuclear localization suggests that these polypeptides may exert some of their biological effects directly at the nuclear level.[3,4] In this chapter, we describe the procedures used to demonstrate the localization of basic fibroblast growth factor (bFGF) in the nucleus and the nucleolus of proliferative primary cultures of adult bovine aortic arch endothelial (ABAE) cells; bFGF is mitogenic for a wide variety of mesoderm- and neuroectoderm-derived cells.[5] The methodologies consist of the use of radioiodinated growth factors followed by cellular fractionation and utilization of immunocytochemistry and indirect immunofluorescence techniques. Because of the low level of fluorescence, a video-enhanced fluorescence microscope coupled to a digitized image processor[6] is used to quantify fluorescein-conjugated anti-FGF antibodies at the single cell level and to compare fluorescence intensities in the various domains of the cell

---

[1] B. A. Lee, D. W. Maher, M. Hannink, and D. Donoghue, *Mol. Cell. Biol.* **7**, 3527 (1987).
[2] D. W. Maher, B. A. Lee, and D. Donoghue, *Mol. Cell. Biol.* **9**, 2251 (1989).
[3] S. J. Burwen and A. L. Jones, *Trends Biochem. Sci.* **12**, 159 (1987).
[4] G. Bouche, N. Gas, H. Prats, V. Baldin, J. P. Tauber, J. Teissié, and F. Amalric, *Proc. Natl. Acad. Sci. U.S.A.* **84**, 6770 (1987).
[5] D. Gospodarowicz, G. Neufeld, and L. Schweigerer, *Cell Differ.* **19**, 1 (1986).
[6] S. Inoué, "Video Microscopy." Plenum, New York, 1986.

or between different cells. Changes in both time and space can be readily determined by mathematical image processing.

Cell System and Measurement of DNA Synthesis

Adult bovine aortic endothelial (ABAE) cells are established from the aortic arch according to Gospodarowicz et al.[7] Quiescent sparse endothelial cells are obtained as follows: endothelial cells are seeded at low density [10^5 cells per 10 cm diameter plastic petri dishes (Nunc, Roskilde, Denmark)] in Dulbecco's modified Eagle's medium (DMEM), supplemented with 10% calf serum (Intermed, Roskilde, Denmark) and 1 ng/ml bFGF, and routinely cultured at 37° (10% CO_2 atmosphere). After 72 hr, the cells are washed twice with serum-free DMEM supplemented with transferrin (10 μg/ml) (Sigma, St. Louis, MO), and cultures are continued in the same medium for 48 hr. At this point, these primary cell cultures are arrested in the G_1 phase of the cell cycle. The $G_1 \rightarrow$ mitosis transition is obtained by stimulation of quiescent cells with serum-free DMEM containing only bFGF (5 ng/ml). Synchronously growing ABAE cells provide a good system to study the influence of different bFGF preparations on cell growth. bFGF purified from bovine pituitaries[8] or human recombinant bFGF is used for the experiments. These two bFGF preparations correspond to the 146 amino acid form.

For measurement of DNA synthesis, cells undergoing the $G_1 \rightarrow$ mitosis transition are pulse-labeled for 15 min with [*methyl*-^{3}H]thymidine (Amersham; 10 μCi/ml; 47.5 Ci/mmol) at different times after bFGF stimulation. The rate of DNA synthesis is measured by cell counting and determination of the [^{3}H]thymidine incorporated into trichloroacetic acid-insoluble material.[9]

Indirect Immunofluorescence

Preparation of Antibodies

Female rabbits (1.5 kg) are immunized with 100 μg of human recombinant bFGF in 1 ml of sterile phosphate-buffered saline (PBS; 0.15 *M* NaCl, 10 m*M* sodium phosphate buffer, pH 7.5) diluted 1 : 1 in Freund's complete adjuvant by dorsal subcutaneous injection at multiple sites.[10] The rabbits

---
[7] D. G. Gospodarowicz, J. Moran, D. Braun, and C. R. Birdwell, *Proc. Natl. Acad. Sci. U.S.A.* **73**, 4120 (1976).
[8] D. Gospodarowicz, this series, Vol. 147, p. 106.
[9] F. Amalric, M. Nicoloso, and J. P. Zalta, *FEBS Lett.* **22**, 62 (1972).
[10] J. L. Vaitukaitis, this series, Vol. 73, p. 46.

are boosted at 2- and 4-week intervals by intradermal injection of 50 μg of bFGF diluted in Freund's incomplete adjuvant. Two weeks after the last booster injection, rabbits are bled every 2 weeks. The serum is stored frozen at $-70°$. Preimmune serum is obtained before immunization of the rabbits.

Antibody Screening

An enzyme-linked immunoabsorbent assay (ELISA) has been developed for detection of anti-human bFGF antibodies. A 96-well polystyrene microtiter plate (Nunc) is coated with 50 μl of purified human bFGF (0.5 μg/ml of PBS) and left for 16 hr at 4°. Wells are washed 3 times with PBS containing 0.1% Tween 20 (PBS/Tween) and filled with PBS/Tween (2 hr at 20°) to saturate the remaining protein binding sites. Plates are then incubated for 2 hr at 37° with 100 μl of serial dilutions of antiserum in PBS/Tween. Wells are washed with PBS/Tween (3 times) and incubated for an additional 2 hr at 37° with 100 μl alkaline phosphatase-conjugated goat anti-rabbit IgG (Promega Biotec, Madison, WI). Following this incubation, wells are washed as above, and 100 μl of 1 mg/ml p-nitrophenyl phosphate (Sigma) in 1 M Tris, pH 8.8, is added. Color development is measured after 30 min at 405 nm using a Titertek Multiscan ELISA plate photometer (Flow Lab., Puteaux, France).

Characterization of Affinity-Purified Anti-Human bFGF IgG

The IgG fraction of the serum is obtained by chromatography on a protein A-Sepharose column (Pharmacia, Uppsala, Sweden). The bound immunoglobulins are eluted with 0.1 M glycine-HCl, pH 2.5, buffer, and the fractions are neutralized immediately with 1 M Tris and dialyzed overnight against PBS. Affinity-purified anti-human bFGF IgG is prepared by applying purified IgG to a column of human recombinant bFGF conjugated to agarose beads (Affi-Gel 10; Bio-Rad Laboratories, Richmond, CA)[11] or to a column of human recombinant bFGF conjugated to AH-Sepharose (Pharmacia) according to the manufacturer's instructions.[4] In both cases, affinity-purified antibodies are then eluted with 0.1 M glycine-HCl, pH 2.5, neutralized with 1 M K_2HPO_4, and extensively dialyzed against PBS. The eluates from the affinity columns are collected and subjected to two additional rounds of chromatography on the bFGF affinity column to prepare anti-human bFGF depleted IgG for use in control studies.

[11] J. Joseph-Silverstein, S. A. Consigli, K. M. Lyser, and C. Verpault, *J. Cell. Biol.* **108**, 2459 (1989).

The evaluation of affinity-purified anti-human bFGF IgG is carried out by dot-blot analysis. For dot-blot analysis, purified human bFGF (from 10 to 20 ng) or other unrelated purified proteins are absorbed on a nitrocellulose membrane (0.1 μm, Schleicher and Schuell, Dassel, Germany) using a dot-blot apparatus (Bio-Rad). The filters are washed in blocking buffer [3% bovine serum albumin (BSA fraction V, pH 7, IBF, Paris, France), 0.15 M NaCl, 10 mM Tris-HCl, pH 7.5] for 2 hr at 20° to saturate additional protein binding sites. These filters are then incubated at 4° overnight with affinity-purified anti-human bFGF IgG or preimmune rabbit IgG in blocking buffer. Subsequently, after extensive washing with blocking buffer, the antibody-treated sheets are incubated for 1 hr at 37° with ^{125}I-labeled protein A (Amersham) to detect antigen–antibody complexes, then extensively washed in blocking buffer and put on film for autoradiography. ELISA and dot-blot analyses are two complementary techniques that enable a rapid determination of the sensitivity and the specificity of the antibodies. The antibodies must not recognize antigens other than bFGF in the total extract of the ABAE cells.

Subcellular Localization of bFGF by Immunofluorescence Microscopy

Cells that are grown on glass coverslips are fixed by one of the two following methods before immunodetection.

Fixation without Permeabilization. Cell monolayers are fixed with 3% paraformaldehyde–PBS for 15 min at 4°, washed with PBS–0.5% BSA (2 times, 5 min each), and then excess paraformaldehyde is removed by treatment with 50 mM NH$_4$Cl–PBS at 4° for 20 min.

Fixation Followed by Methanol Permeabilization. The cell monolyaers are incubated with absolute methanol at 20° for 2 min, then washed twice with PBS and twice with PBS–0.5% BSA at 4°.

Immunodetection

Coverslips treated as above are then incubated at 37° for 1 hr with the affinity-purified anti-human bFGF IgG diluted to 60 μg/ml in PBS–0.5% BSA. In all cases, nonimmune rabbit IgG is included at the same dilution as controls. After 3 washes with PBS–0.5% BSA, cells are further incubated for 1 hr at 37° with fluorescein-conjugated goat anti-rabbit IgG (Nordic Immunological Laboratories, Lausanne, Switzerland) diluted 1:80 according to the manufacturer's instructions. Finally, the coverslips are extensively washed with PBS–0.5% BSA (2 times, 5 min each) and with PBS alone (2 times, 5 min), mounted on glass slides, and examined in a Leitz Ortholux II microscope equipped for epifluorescence with a 100-W mercury lamp. Micrographs are obtained after an exposure of 2–4 min (Kodak Timax 400 ASA, Rochester, NY) (Fig. 1). In parallel, the nucleolar compartment is characterized using IgG raised against nucleolin,[4] a nucleolar-specific antigen. To rule out the possibility that reorganization of the nucleus and nucleolus after bFGF stimulation is responsible for nonspecific trapping of antibody within these organelles, controls are performed with an unrelated antibody (e.g., antitubulin) using an identical technique.[4]

Fluorescence Localization and Quantification

Digitized Video Microscopy

Figure 2 shows a diagram of the imaging system used in our laboratory. Cells are observed under a Leitz fluorescence inverted microscope. The light source for fluorescence observations is an HBO 100W2 (Osram, Munich, Germany), and wavelengths are selected by a Leitz H3 filter block. Two video monitoring setups are connected to the microscope, a charge coupled device (C.C.D.) camera (Panasonic, Gsaka, Japan) associated with a color monitor (Sony, Fellbach, Germany) for direct or phase-contrast observations (high light level) and a light-intensifying camera (Lhesa, Pontoise, France) associated with a black-and-white monitor (RCA, Lancaster, PA) for fluorescence observations (low light level). The two video setups are connected by means of a selector to a digitizer (Info'Rop, Toulouse, France) driven by a computer (CPU 68010, Motorola, Tempe, AZ). In this way, the video signal is converted to a matrix with a 8-bit gray scale, namely, 256 different light levels. The size of this matrix is selected to be 256×256 to shorten computer calculation times. The software library (Trimago, Ifremer, Paris, France) contains the major routines for digital image processing. The following peripherals are connected to the computer: a hard disk (85 Mo) to store the matrix, a color monitor (Sony) on which the resulting images and associated graphs are displayed, a color printer (Canon, Le Blanc-Mesnil, France) for printing color graphs, and another printer (Epson, Levallois-Perret, France) to print the data (subroutines used or numerical pixel values).

Digital Image Processing

The first step of the analysis is the selection of cells and their digitalization. Images which quantify the dark level and illumination heterogeneity are digitalized.

1. Glass coverslips of fixed cells are mounted on glass slides and observed by fluorescence under oil immersion with a magnification of 63 (Leitz Wetzlar objective)

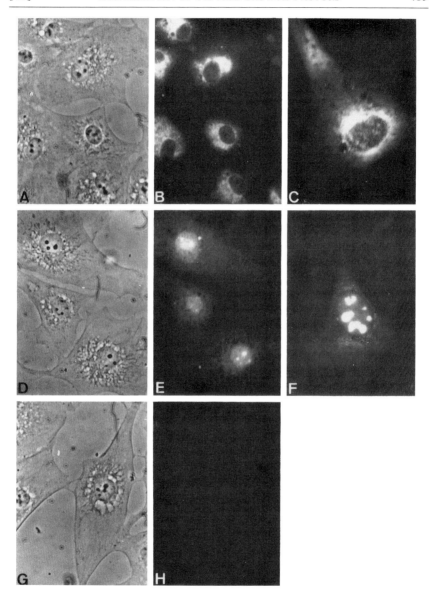

FIG. 1. Indirect immunofluorescence staining of exponentially growing cells. (B, C, E, F) Staining with affinity-purified anti-bFGF IgG. (H) Control with preimmune rabbit IgG. (A, D, G) Corresponding phase-contrast photos. Magnifications: B, E, ×220; C, F, ×660. (B, C) Fixation without permeabilization. Exponentially growing ABAE cells in the presence of bFGF (5 ng/ml culture medium) show low diffuse cytoplasmic staining with intensive

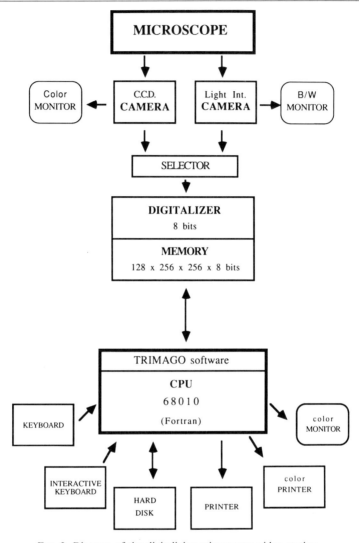

FIG. 2. Diagram of the digitalizing microscopy video station.

fluorescence located around the nucleus. Nuclei and nucleoli appear barely stained. (E, F) Fixation followed by methanol permeabilization. Under these conditions, most of the cytoplasmic label is lost during permeabilization, which allows the detection of bFGF in the nucleus and in the nucleolus.

2. The gain of the light-intensifying camera is fixed with a manual adjustment. Thus, the light levels are directly comparable between each digitalized image after mathematical correction. The gain is set so as to avoid saturation of the camera and to keep the signal arising from unlabeled cell intrinsic fluorescence at the background level of detection.

3. Isolated cells are selected by microscope observation and then digitized.

4. For each selected cell, the same image is digitalized 12 times with a frequency of 1 image/sec.

5. All cell images are obtained by averaging the 12 uncorrected images in order to improve the signal-to-noise ratio (Wiener's law).

6. Several images of both labeled and unlabeled cells are stored.

7. In the same way, the image corresponding to the blank level of the camera (Ib), that is, fluorescence observation with no object between the light source and the camera, is digitalized and stored.

8. A reference illumination map (Im) is obtained by fluorescence observation of a slide with a fluorescein-conjugate dye solution and is digitalized by the same technique.

Mathematical Image Processing

The aim of the mathematical treatment[12] is to compare fluorescence levels between each cell compartment or between different cells. Mathematical corrections are carried out in order to reduce or eliminate the contribution of noise to the light level. Just after the digitalization step, the encoded light level value (If) for a point M of coordinates x and y, is given by the following equation:

$$If(x,y) = KC(x,y)Im(x,y) + Ib(x,y)$$

where K is a constant which depends on the apparatus and adjustments (manual gain value), $C(x,y)$ the concentration of fluorescent dye at the point M, $Im(x,y)$ the incident light level (illumination heterogeneity), and $Ib(x,y)$ the blank light level (dark level). Thus, after correction for illumination and background noise, a direct relationship between light levels and dye concentration may be obtained.

1. A mathematical enlargement with linear extrapolation or with duplication and suppression may be used if cells cover less than 50% of the image area. In this way, the computer transforms an image area of 128 × 128 to an image with a size of 256 × 256. The mathematical method effectively suppresses areas containing little information, such as regions where no cells are present.

---

[12] K. R. Castleman, "Digital Image Processing." Prentice-Hall, Englewood Cliffs, New Jersey, 1979.

2. If a cell image is transformed by a mathematical enlargement, it is important to process the background image (Ib) and the illumination map (Im) in the same way. For each cell analysis, three equally enlarged images are produced which represent the cell (Ii), the background (Ibi), and the illumination map (Imi).

3. The image (Ibi) is subtracted from each image (Ii) and (Imi) to allow correction for the background:

$$Ii1 = Ii - Ibi \quad \text{and} \quad Imi1 = Imi - Ibi$$

4. Each corrected image ($Ii1$) is divided by its associated illumination map ($Imi1$) to correct for heterogeneity in incident light level and camera sensitivity

$$Ii2 = Ii1/Imi1$$

5. Contrast in images may be improved by spreading their pixel values from 0 to 255 using a Look Up Table (L.U.T.). The same L.U.T. must be used for all images.

After such processing, all the pixel values are directly related to the local concentration of dye. Figure 3a–c shows differences between initial (Ii) and processed ($Ii2$) cell images.

Image Analysis

Fluorescence Localization. Several routines are used to analyze the fluorescence in the cells: (1) histograms which represent the distribution of pixel light levels characterizing the change in fluorescence in each cell; (2) horizontal or vertical line scans which depict unidirectional (1D) light level distributions allowing observation of the fluorescence compartmentation in the cell (Fig. 3d shows such a representation indicating relative fluorescence levels of the different cell compartments, e.g., cytoplasm, nucleus, nucleolus); (3) Pseudocolor fluorescence intensity maps which display dye distribution in the cells.

Fluorescence Quantification. Using pseudocolor maps, it is also possible to quantify the fluorescence. The total fluorescence intensity $F(r)$ of an area may be represented by

$$F(r) = \int_x \int_y I(x,y) \, dx \, dy \tag{1}$$

where $I(x,y)$ is the fluorescence level of a point M of coordinates x and y. This equation assumes that digitalized fluorescence images are representative of a planar concentration of dyes, that is, the microscope field depth is null or independent of z. Equation (1) may be written as

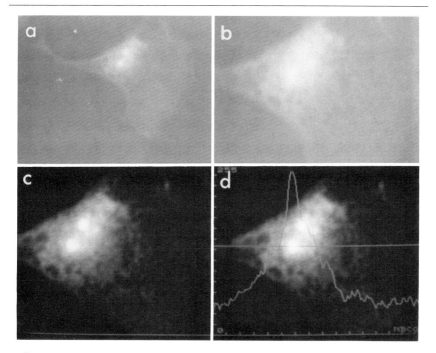

FIG. 3. Image contrast improvement through data processing. (a) Original image; (b) enlargement through pixel expansion of the region of interest; (c) image b processed through background subtraction, shading distorsion correction, and contrast linear enhancement; (d) 1D light level distribution along the indicated horizontal line. Image d indicates a strong accumulation of the fluorescent antibodies in the nucleolus.

$$F = NI_o \qquad (2)$$

where N is the total number of pixels included in the area, and I_o the average value of fluorescence in the area. This fluorescence quantification gives information on the number of dye molecules owing to the linear relationship between fluorescence level and dye concentration in the final corrected cell images.

1. The cells that are observed under fluorescence and digitalized are subsequently observed using phase contrast. An image of each cell is digitalized.

2. Each cell subcompartment surface is quantified by the corresponding number of pixels.

3. From fluorescence-corrected images, the average and the standard error of the pixel values are calculated for each fluorescent area.

4. Quantification is then obtained by multiplying the number of pixels by the average pixel value for the different cell areas.

The main advantage of this video technique is that pixel light levels, and thus the number of dye molecules, may be compared in the various domains of the cell or between different cells.

Ultrastructural Immunocytochemistry

Exponentially growing cells are fixed at 4° in 0.1% glutaraldehyde and 2% formaldehyde in PBS (pH 7.5, for 15 min). The cells are treated with 10 mM sodium borohydride, dehydrated in ethanol, and embedded in Lowicryl K4M, using the low temperature procedure.[13] Ultrathin sections are placed on nickel grids and preincubated for 30 min at 20° with normal goat serum at 10% to block the nonspecific sites of protein absorption. Excess liquid is removed, and the grids are incubated overnight at 4° with affinity-purified anti-bFGF IgG (diluted at 1 : 1000) in PBS (10% BSA). The grids are washed and incubated for 1 hr at 20° with goat anti-rabbit IgG coupled to colloidal gold (GAR 15, Jansen Life Science, Beerse, Belgium) diluted 1 : 20 in 20 mM Tris-HCl buffer, pH 8.2, 0.5% BSA. After several washes with Tris-HCl buffer, the grids are rinsed with double-distilled water and dried. The sections are contrasted with uranyl acetate and viewed in a Jeol JEM 200 CX electron microscope at 80 KV. Control sections from the same resin block are incubated only with PBS and preimmune IgG (Fig. 4).

Biochemical Localization of bFGF in ABAE Cells

Iodinated bFGF is utilized to provide biochemical support for the microscopic observations and to quantify the amount of growth factor accumulated in each cellular compartment.

Iodination of bFGF

Purified bFGF is radioiodinated using the chloramine-T method.[14] Three micrograms of bFGF in 10 μl of 50 mM sodium phosphate buffer, pH 7.2, is added to 1 mCi of Na^{125}I (Amersham) and 10 μl of a freshly prepared solution of chloramine-T (Sigma) (5 mM in 50 mM sodium phosphate buffer, pH 7.2). After 1 min at 20° the reaction is stopped by adding 20 μl of 5 mM sodium bisulfite. Free Na^{125}I is separated from ^{125}I-labeled bFGF by chromatography on Sephadex G-25 equilibrated with 50 mM

---

[13] H. Yeh, G. F. Pierce, and T. M. Deuel, *Proc. Natl. Acad. Sci. U.S.A.* **84**, 2317 (1987).
[14] W. M. Hunter and F. C. Greenwood, *Nature (London)* **194**, 495 (1962).

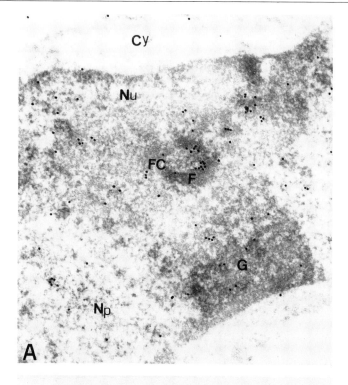

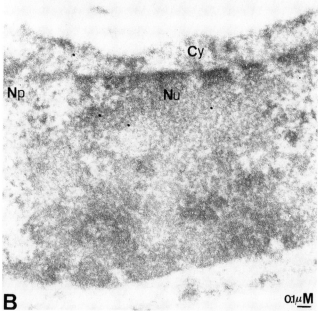

TABLE I
Intracellular Distribution of ^{125}I-Labeled bFGF in Growing and Stimulated Cells

Cells	Nucleus		Cytoplasm	
	Amount (pg)[a]	Molecules[b]	Amount (pg)[a]	Molecules[b]
Growing[c]	55 ± 10	1800	610 ± 53	20,400
Stimulated[c]	70 ± 12	2400	706 ± 45	24,000

[a] Values (picograms per 10^6 cells) calculated from the specific activity of radiolabeled bFGF are the means ± S.E.M. from four experiments.
[b] The number of bFGF molecules per subcellular fraction corresponding to one cell was calculated using Avogadro's number and the specific activity of bFGF.
[c] Growing cells refer to an asynchronous primary culture of sparse cells; stimulated cells, G_1-arrested cells stimulated to growth ($G_1 \to S$) by the addition of bFGF.

sodium phosphate buffer, pH 7.2, containing 0.25% BSA. The analysis of ^{125}I-labeled bFGF by SDS–PAGE (15% acrylamide) followed by autoradiography reveals a single band at 18.4 kDa. The specific activity of ^{125}I-labeled bFGF in this experiment is 50,000 to 150,000 counts/min (cpm)/ng. ^{125}I-Labeled bFGF has the same mitogenic activity as unlabeled bFGF.

Cell Fractionation

Exponentially growing ABAE cells are stimulated for 2 hr with ^{125}I-labeled bFGF, 5 ng/ml of culture medium. At this time, cells are washed 3–5 times with PBS and harvested by trypsinization, with 0.05% trypsin (Intermed), 0.025% EDTA in PBS at 4° over 10 min. The trypsin solution is discarded, and the cells are harvested in 5 ml of DMEM–10% calf serum (Intermed) and then centrifuged for 10 min (400 g at 4°). All the cell fractionation steps are carried out at 4°. Cells (2×10^6) are disrupted for 15–30 sec (Ultra Turrax, Janke and Hunkel, Staufen, Germany) in 10 ml of medium A (0.3 M sucrose, 60 mM KCl, 15 mM NaCl, 1.5 mM spermine, 0.5 mM spermidine, 14 mM 2-mercaptoethanol, 5 mM EGTA, 2 mM

Fig. 4. Ultrastructural localization of bFGF in the nucleolus of exponentially growing ABAE cells by immunogold staining. (A) Immunocytochemical localization of bFGF with affinity-purified anti-bFGF IgG. (B) Control with preimmune rabbit IgG. Nu, Nucleolus; Np, nucleoplasm; Cy, cytoplasm; FC, fibrillar center; F and G are, respectively, fibrillar and granular components. (A) The gold particles show some enrichment around the fibrillar center in dense fibrillar components and are also scattered throughout the granular components. Significant labeling is also observed in the nucleoplasmic network. (B) Low background is observed with preimmune IgG.

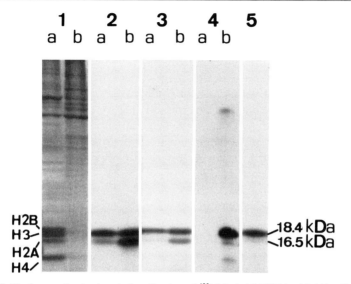

FIG. 5. Nuclear and cytoplasmic localization of ^{125}I-labeled bFGF in ABAE cells. Two hours after the addition of ^{125}I-labeled bFGF (5 ng/ml), ABAE cells were harvested and fractionated into nuclear (a) and cytoplasmic (b) fractions. Total nuclear proteins corresponding to 10^6 cells and 1/10 of the cytoplasmic proteins were then analyzed on a 15% SDS–polyacrylamide slab gel. After protein fixation and Coomassie blue staining (lanes 1), gels were dried and exposed for 24 hr for autoradiography. Lanes 2 show G_1-arrested ABAE cells stimulated by ^{125}I-labeled bFGF; lanes 3 show exponentially growing cells. For lanes 4, a control experiment, ^{125}I-labeled bFGF (5 × 10^4 cpm) was added to disrupted ABAE cells. Cellular fractionation and analysis of proteins were carried out as described in the text. For lanes 5, an aliquot of ^{125}I-labeled bFGF recovered in the culture medium after stimulation of the cells was analyzed as described above. ^{125}I-labeled bFGF is detected in nucleus and cytoplasm both in exponentially growing cells and in synchronized cells undergoing the G_1 → S transition. ^{125}I-Labeled bFGF is detected in essentially full length form (18.4 kDa) in the nucleus, whereas in the cytoplasm two molecular forms of the growth factor (18.4 and 16.5 kDa) are found in an identical molar ratio. As shown in lanes 4, ^{125}I-labeled bFGF is not detected in nuclei, and little degradation of growth factor occurs during cell fractionation when the growth factor is added to disrupted cells.

EDTA, 20 mM HEPES, pH 7.5) containing a cocktail of protease inhibitors [1 mM phenylmethylsulfonyl fluoride (PMSF), 10 μg/ml of leupeptin, 0.02% NPG (Cemulsol SFOS, Persan, France), and Trasylol] and centrifuged at 800 g for 5 min. The supernatant represents the crude cytoplasmic fraction. The nuclear pellet is washed twice in medium A. The purity and integrity of nuclei preparations are monitored by microscopy under phase contrast.

The radioactivity of ^{125}I-labeled growth factor bound to the particular fractions is measured in a γ counter, or using γ-vials and a liquid scintillation counter. The number of bFGF molecules bound is calculated from Avogadro's number and the specific activity of the bFGF (Table I).

Aliquots of crude cytoplasmic fraction and nuclear pellets are precipitated by 15% trichloroacetic acid (30 min, 4°) and centrifuged at 15,000 g for 30 min. The pellets are then dissolved under highly reducing conditions (1 mM Tris-HCl, pH 8, 10 mM EDTA, 2% SDS, 1.5 M 2-mercaptoethanol, 30% glycerol) and analyzed by 15% SDS-PAGE and autoradiography (Fig. 5). In control experiments, the same amount of ^{125}I-labeled bFGF is added to untreated cells in the cell fractionation mixture. When intact nuclei are prepared from control cells, no ^{125}I labeling should be recovered in the nuclear fraction (Fig. 5).

Concluding Comments

A combination of three methods provides an unambiguous technique for nuclear and subnuclear localization of growth factors in cell cultures. Using synchronously growing ABAE cells stimulated with bFGF it is possible to study the cellular uptake of growth factor during the cell cycle and to quantify the accumulation of growth factor in subcellular compartments.

[46] Antiphosphotyrosine Antibodies in Oncogene and Receptor Research

By DAVID F. STERN

Introduction

Protein-tyrosine kinases regulate cell proliferation and differentiation. These kinases include receptors for several peptide growth factors and many of the oncogene products. The original, and most direct, method for detecting tyrosine phosphorylation of proteins was separation of ^{32}P-labeled amino acids from partial protein hydrolyzates on thin-layer plates.[1] More recently, antibodies have been developed that react specifically with phosphotyrosine.[2-4] Use of these antibodies for immunoprecipitation or

[1] J. A. Cooper, B. M. Sefton, and T. Hunter, this series, Vol. 99, p. 387.
[2] A. H. Ross, D. Baltimore, and H. Eisen, *Nature (London)* **294,** 654 (1981).
[3] M. Ohtsuka, S. Ihara, R. Ogawa, T. Watanabe, and Y. Watanabe, *Int. J. Cancer* **34,** 855 (1984).
[4] M. P. Kamps and B. M. Sefton, *Oncogene* **2,** 305 (1988).

immunoblotting of tyrosine-phosphorylated proteins is much more convenient than phosphate-labeling followed by phosphoamino acid analysis. Furthermore, antiphosphotyrosine (anti-Ptyr) antibodies have the unique advantage of being preparative reagents; they can be used to purify tyrosine-phosphorylated proteins. We summarize here the diverse applications of anti-Ptyr antibodies in growth factor receptor and oncogene research. This chapter is intended to complement two related chapters in a separate volume of this series. Kamps discusses methods for preparing anti-Ptyr, methods for immunoblotting and immunoprecipitation, and the specificity of these sera.[5] Lindberg and Pasquale describe the use of anti-Ptyr for immunoscreening of DNA libraries to clone tyrosine kinase genes.[6] A review on this subject was recently published.[7]

Obtaining Antibodies

Monoclonal anti-Ptyr antibodies are available from the American Type Culture Collection (Rockville, MD, as hybridomas) and commercial sources including ICN (Costa Mesa, CA), Boehringer Mannheim (Indianapolis, IN), and Oncogene Science (Manhasset, NY). Monoclonal antibodies vary in the ability to precipitate or react in immunoblots with individual phosphoproteins. Polyclonal sera can be raised either against tyrosine-phosphorylated proteins or synthetic immunogens containing Ptyr or Ptyr analogs.[7] Sera raised against different immunogens recognize a similar spectrum of proteins in immunoblots.[4]

Anti-Ptyr sera are compatible with a range of immunodetection techniques including immunoblotting,[4] ELISA,[3] and immunofluorescence.[8] Moreover, these sera have been employed successfully for immunopurifications, either through immunoprecipitation using *Staphylococcus aureus* or after covalent linking to a solid support for affinity purifications. As with any antisera, the cross-reactivity of anti-Ptyr is always a potential pitfall. One anti-Ptyr monoclonal antibody can react with phosphohistidine and nucleoside monophosphates.[9] Although careful affinity purification can minimize the problem of cross-reactivity, in studying any novel immunoreactive protein it is advisable to directly analyze ^{32}P-labeled phosphoamino acids[1] in order to be certain that Ptyr is present.

[5] M. P. Kamps, this series, Vol. 200, in press.
[6] R. A. Lindberg and E. B. Pasquale, this series, in press.
[7] J. Y. J. Wang, *Anal. Biochem.* **172,** 1 (1988).
[8] J. R. Glenney, Jr., W. S. Chen, C. S. Lazar, G. M. Walton, L. M. Zokas, M. G. Rosenfeld, and G. N. Gill, *Cell (Cambridge, Mass.)* **52,** 675 (1988).
[9] A. R. Frackelton, Jr., A. H. Ross, and H. N. Eisen, *Mol. Cell. Biol.* **3,** 1343 (1983).

Activity of Tyr Kinases

The basal forms of receptor tyrosine kinases contain phosphoserine and phosphothreonine but no Ptyr. Ligand binding activates the intrinsic kinase activity. This results in autophosphorylation, so that the receptor now contains phosphotyrosine. Transphosphorylation of substrate proteins including the presumed effectors of receptor function occurs simultaneously. The ligand-induced Tyr phosphorylations are transient; receptor and substrate Tyr phosphorylation are back to basal levels within 30 min of activation.[10]

We have used anti-Ptyr to monitor epidermal growth factor (EGF)-stimulated phosphorylation of the EGF receptor and the related *neu* protein.[11] EGF-stimulated receptor autophosphorylation and phosphorylation of substrates can be detected in total protein extracts of solubilized fibroblasts by immunoblotting with anti-Ptyr. Cells are lysed in hot electrophoresis sample buffer[4,5,11] to minimize proteolysis and the effects of phosphatases. The sensitivity of this assay is improved by consecutive immunoprecipitation with anti-EGF receptor then immunoblotting with antiphosphotyrosine.[11] Using this technique, receptor stimulation by subsaturating levels of EGF can be detected. After stimulation with saturating levels of EGF, the EGF receptor can be detected in 10^5 rat fibroblasts. It is noteworthy that the fibroblast cell lines (Rat-1, FR3T3) used for these experiments contain only approximately 10,000 EGF receptor molecules per cell,[12] 2 orders of magnitude lower than the receptor number found in the A431 tumor cells typically used for EGF receptor studies.

Activation of EGF Receptor in Rat-1 or FR3T3 Cells

Rat fibroblasts are seeded on 35-mm culture dishes and allowed to grow to confluence. The culture medium is replaced with Dulbecco's modified Eagle's medium (DMEM) containing 0.1% calf serum, and the desired amount of EGF and cells are incubated at 37° for 7–10 min. All remaining steps are performed on ice or in the cold room. Cells are washed twice with cold Dulbecco's phosphate-buffered saline (PBS, Ca^{2+}- and Mg^{2+}-free) and lysed by the addition of 0.5 ml phosphate-buffered RIPA [10 mM sodium phosphate buffer, pH 7–7.4; 150 mM NaCl; 1% (v/v) Triton X-100; 1% (w/v) sodium deoxycholate; 0.1% (w/v) sodium dodecyl sulfate; 1% aprotinin] containing 100 μM sodium orthovanadate (Sigma, St. Louis, MO). The cells are scraped into the buffer and incubated for 20

---

[10] T. Hunter and J. A. Cooper, *Cell (Cambridge, Mass.)* **24**, 741 (1981).
[11] D. F. Stern and M. P. Kamps, *EMBO J.* **7**, 995 (1988).
[12] D. F. Stern, P. A. Heffernan, and R. A. Weinberg, *Mol. Cell. Biol.* **6**, 1729 (1986).

min. The lysates are transferred to microcentrifuge tubes and cleared by centrifugation for 5 min at 4° at top speed in a Brinkmann microcentrifuge, and the supernatants are transferred to new tubes. Antireceptor antibodies are added (the incubation time depends on the antibody; we use 2 hr). Immune complexes are collected by rotation for 30 min with *Staphylococcus aureus* [50 μl, 10% (v/v) in RIPA] and centrifugation for 15 sec in the microcentrifuge. Extensive washing of immunoprecipitates is not necessary. We wash once with RIPA and once with PBS. Samples are stored at −70° before electrophoresis in 7.5% acrylamide–0.173% bisacrylamide Laemmli gels. Detailed procedures for immunoblotting with polyclonal anti-Ptyr are described elsewhere.[4,5] We use rabbit anti-poly(Ptyr,Gly,Ala) conjugated to keyhole limpet hemocyanin and affinity purified as described.[4,5]

Although the receptor kinases do not contain basal Ptyr, members of the *src* family of tyrosine kinases often do. Anti-Ptyr immunoreactivity can in some cases still be used as a measure of activation of these proteins: the basal Tyr-phosphorylated sites of p56^{lck} are only poorly recognized in immunoblots by anti-Ptyr, whereas the site phosphorylated as a consequence of kinase activation is highly immunoreactive.[5]

Anti-Ptyr can be used instead of [^{32}P]ATP in measuring the activity of kinases *in vitro*. Immune complex kinase assays are performed in the presence of nonradioactive ATP (100 μM) as the phosphate donor, and the outcome is visualized by immunoblotting with anti-Ptyr (S. A. Rosenzweig, personal communication, 1990).

Immunoprecipitations with Anti-Ptyr

Anti-Ptyr antibodies can be used to immunoprecipitate Tyr-phosphorylated proteins. This is useful for isolating subpopulations of receptors that may be active because of a modification or association with another protein, for estimating the fraction of a population that contains Ptyr, for identifying Tyr-phosphorylated substrates and determining their biological activities, and for preparative isolation of receptors. Immunoprecipitated proteins can be detected by metabolic labeling, functional assays such as immune complex kinase assays, or immunoblotting with anti-Ptyr or antipeptide antibodies. Polyclonal anti-Ptyr antibodies are compatible with immunoprecipitation buffers such as RIPA which are not completely denaturing. We typically use phosphate-buffered RIPA for immunoprecipitation of metabolically labeled proteins, and TG [1% Triton X-100; 10% (v/v) glycerol in PBS] for immune complex kinase assays.[11,12] Sodium orthovanadate (100 μM, Sigma) is included in lysis and washing buffers to inhibit tyrosine phosphatases.

Bacterial Colony and Plaque Screens

Because anti-Ptyr immunoblotting is a fast and sensitive screen for activity, it can be used to screen for mutations in tyrosine kinases. Kipreos and co-workers[13] mutagenized a bacterial expression vector encoding an active *abl* kinase. Since bacterial colonies producing active protein were recognized by anti-Ptyr, it was possible to use these antibodies to screen for temperature-sensitive kinase activity. The mutants obtained were subsequently expressed in animal cells and retained the temperature-sensitive phenotype. This approach should be useful for screening mutants in any tyrosine kinases that can be expressed as functional proteins in bacteria.

Anti-Ptyr antibodies can be used in immunoscreens to identify genes encoding tyrosine kinases.[6,14] Plaques or colonies expressing tyrosine kinases can be identified with anti-Ptyr, which recognizes autophosphorylated kinases or substrate proteins. Since this assay relies on function (production of Ptyr) rather than nucleic acid homology, it will identify tyrosine kinase genes that are only distantly related to known kinases. This has resulted in identification of new kinase genes, and it may also lead to identification of kinase genes in distantly related organisms.

Growth Factors without Receptors

Purification of most peptide growth factors preceded the identification and purification of the cognate receptors. For example, highly purified platelet-derived growth factor (PDGF) was available long before it was possible to obtain specific antibodies against the PDGF receptor. The use of anti-Ptyr antibodies made it possible to conduct many studies of PDGF receptor function in the absence of anti-PDGF receptor sera.[15] In these experiments anti-Ptyr was used to affinity-purify Tyr-phosphorylated forms of the PDGF receptor. Eventually, anti-Ptyr was used to purify sufficient amounts of the PDGF receptor from PDGF-treated cells to permit microsequencing of the receptor. This was followed by library screening and cloning of the PDGF receptor cDNA.[16,17]

---

[13] D. T. Kipreos, G. J. Lee, and J. Y. J. Wang, *Proc. Natl. Acad. Sci. U.S.A.* **84**, 1345 (1987).
[14] S. Kornbluth, K. E. Paulson, and H. Hanafusa, *Mol. Cell. Biol.* **8**, 5541 (1988).
[15] A. R. Frackelton, Jr., P. M. Tremble, and L. T. Williams, *J. Biol. Chem.* **259**, 7909 (1984).
[16] T. O. Daniel, P. M. Tremble, A. R. Frackelton, Jr., and L. T. Williams, *Proc. Natl. Acad. Sci. U.S.A.* **82**, 2684 (1985).
[17] Y. Yarden, J. A. Escobedo, W.-J. Kuang, T. L. Yang-Feng, T. O. Daniel, P. M. Tremble, E. Y. Chen, M. E. Ando, R. N. Harkins, U. Francke, V. A. Fried, A. Ullrich, and L. T. Williams, *Nature (London)* **323**, 226 (1986).

Receptors without Growth Factors

A number of protooncogenes (including *neu, met, ros, kit*) encode proteins that resemble growth factor receptors but for which no ligand has yet been identified. Since all tyrosine kinase growth factor receptors autophosphorylate after ligand binding, anti-Ptyr antibodies can be used in assays to screen for the ligands. We employ this assay in screens for the *neu* ligand.[11] Cells expressing the putative receptor are incubated with extracts or conditioned medium that may contain the ligand at 37° for 7 min, chilled to 4° by washing with cold phosphate-buffered saline, and then lysed for immunoprecipitation with receptor-specific antibodies. Immunoprecipitations are carried out in the presence of 100 μM sodium orthovanadate exactly as described above for the EGF receptor, and the precipitates are analyzed by immunoblotting with anti-Ptyr. In experiments with the EGF receptor, we have found that this simple assay can detect levels of EGF or transforming growth factor α (TGF-α) present in crude tissue extracts. In principle, the assay could be modified as an ELISA and used in purification of the ligand. One problem that has arisen in the *neu* system comes from our finding that the *neu* protein is itself a substrate for the EGF receptor kinase.[11,12] Since the assay would detect EGF in cells that express both the *neu* protein and the EGF receptor, cell lines that lack EGF receptors are employed in final screens. Other approaches to this problem have been described elsewhere.[18]

Substrate Identification with Anti-Ptyr

Although anti-Ptyr sera provide a useful tool for receptor analysis, the most unique application has been the identification and purification of Tyr-phosphorylated substrates. Prior to the availability of anti-Ptyr, random surveys of Tyr-phosphorylated proteins required metabolic labeling of cells with [^{32}P]P$_i$, separation of labeled proteins, and characterization of the phosphorylated amino acids on individual proteins.[1] Although alkaline washes eliminated much of the background arising from other phosphorylations, many of the remaining proteins still contained phosphoserine and phosphothreonine. Furthermore, the preparative use of these techniques was limited by the capacity of the one- and two-dimensional separation gels and the destructiveness of alkaline hydrolysis. In contrast, surveys of substrates with anti-Ptyr do not require use of large amounts of ^{32}P and are exceedingly sensitive. Comparative studies have shown that different tyrosine kinases cause phosphorylation of some similar and some unique substrates. Since some transformation-defective tyrosine kinases still

[18] Y. Yarden and R. A. Weinberg, *Proc. Natl. Acad. Sci. U.S.A.* **86**, 3179 (1989).

cause protein phosphorylation, it may be possible to correlate phosphorylation of particular species uniquely with transformation.[4] Many substrates for tyrosine kinases have now been identified (including calpactin, GAP, the cdc2 protein, PI-3 kinase, a phospholipase C, the *neu* protein, and several cytoskeletal proteins), so it is often possible to identify the substrates tentatively by their electrophoretic mobilities.

When activities for key substrates can be predicted, anti-Ptyr can be used to purify the substrates. For example, Wahl *et al.* hypothesized that a phospholipase C regulated by tyrosine phosphorylation exists in A431 cells, where EGF regulates phosphatidylinositol turnover.[19] They were able to identify a phospholipase C activity that bound anti-Ptyr after EGF treatment of these cells.

Studies of Human Disease

Anti-Ptyr antibodies may ultimately prove useful in the study of human disease. Tyrosine kinases will contain Ptyr when overexpressed[20] and when activated by ligand or by mutation. All of these mechanisms of kinase activation may turn out to have diagnostic significance. For example, receptor protooncogenes such as *neu* and *erbB* are often amplified in human tumors, and alteration of c-*abl* is diagnostic for chronic myelogenous leukemia. We have found that the *neu/erbB-2* protein can be detected by Ptyr immunoblotting of total lysates of breast tumor cell lines that contain an amplified *neu/erbB-2* gene. The identity of an immunoreactive receptor is suggested by its electrophoretic mobility, and can be confirmed by immunoblotting with the appropriate antireceptor antibody. By comparing the immunoreactivity with antireceptor antibody to the reactivity with anti-Ptyr, a measure of receptor activity (*in vivo* autophosphorylation) is obtained.

Preparation of Samples for Immunoblotting from Tissue

We have used Ptyr immunoblotting in analysis of pathological specimens of human tumors. Tissue samples are quick-frozen in liquid nitrogen and stored in liquid nitrogen or at $-70°$. Approximately 100 mg of tissue is sliced from a specimen that is kept cold on dry ice, dipped in liquid nitrogen, and then crushed in a Bessman tissue pulverizer (Fisher Scientific, Fairlawn, NJ) that has been chilled on dry ice. The powder is quickly poured into a tube containing 0.5 ml of lysis buffer [20 mM sodium phosphate, pH 7.0, 8% (w/v) sodium dodecyl sulfate, 1 mM EDTA, 6 M urea,

[19] M. I. Wahl, T. O. Daniel, and G. Carpenter, *Science* **241**, 968 (1988).
[20] D. F. Stern, M. P. Kamps, and H. Cao, *Mol. Cell. Biol.* **8**, 3969 (1988).

with 1/100 volume aprotinin and 1/100 volume of 10 mM PMSF added fresh] that has been preheated to 100°, vortexed, and incubated at 100° for 5 min. Samples are stored at $-70°$. Prior to analysis by SDS-polyacrylamide gel electrophoresis, samples are combined with 3 volumes of 0.133 M dithiothreitol (DTT)–6.7% (v/v) 2-mercaptoethanol–13% (v/v) glycerol–0.02% (w/v) bromphenol blue, and incubated at 100° for 2 min. The quality and relative concentrations of protein lysates are estimated by staining gels with Coomassie Brilliant Blue.

Phosphotyrosine blotting can also permit identification of new receptors that may be involved in carcinogenesis. Using antiphosphotyrosine, Giordano, Comoglio, and co-workers found that a gastric carcinoma cell line expresses a tyrosine-phosphorylated cell surface protein.[21] This protein has tyrosine kinase activity. Since *in vivo* tyrosine phosphorylation of the protein was inhibited by acid treatment, it appeared that the phosphorylation is stimulated by a factor produced by the same cells. Further analysis showed that this protein is the product of the protooncogene *met*.[22] Finally, since the spectrum of proteins phosphorylated varies with different oncogene products,[4,20] it may also be possible to identify ligand-activated or mutated kinases by their characteristic signature of phosphorylations.

---

[21] S. Giordano, M. F. Di Renzo, R. Ferracini, L. Chiado-Piat, and P. M. Comoglio, *Mol. Cell. Biol.* **8,** 3510 (1988).

[22] S. Giordano, C. Ponzetto, M. F. Di Renzo, C. S. Cooper, and P. M. Comoglio, *Nature (London)* **339,** 155 (1989).

[47] Assays for Bone Resorption and Bone Formation

By G. R. MUNDY, G. D. ROODMAN, L. F. BONEWALD,
R. O. C. OREFFO, and B. F. BOYCE

Assays for Bone Resorption

The organ culture methods for assessing bone resorption have been routine assays for a number of bone biology laboratories for many years.[1-3] Recently, additional techniques have been added. These include long-term marrow culture to study osteoclast formation, culture of isolated osteoclasts, and, most recently, *in vivo* histomorphometric assays.

Fetal Rat Long Bone Assay

The following assay measures the release of previously incorporated ^{45}Ca from fetal rat long bones.[4] It is a widely used but difficult bioassay. Pregnant rats are injected with 200 μCi of ^{45}Ca on the eighteenth day of gestation. The following day the fetuses are removed, and the mineralized shafts of the radii and ulnae are dissected free from the cartilagenous ends and placed in organ culture. After a 24-hr preculture period to allow for the exchange of loosely bound ^{45}Ca with stable calcium in the medium, the bones are cultured for a further 48 hr in the presence of test or control substances. Four of six bones are cultured per group, and bone resorbing activity is expressed as treated to control ratios of ^{45}Ca released from test and control bones. Statistical significance is assessed using Student's *t* test for unpaired samples.

Mouse Calvaria Assay

An alternative to the fetal rat long bone assay is the technique which uses neonatal mouse calvariae previously labeled with ^{45}Ca. The method is described in detail by Gowen *et al.*[5] It is helpful to use the mouse

---

[1] G. R. Mundy, R. A. Luben, L. G. Raisz, J. J. Oppenheim, and D. N. Buell, *N. Engl. J. Med.* **290**, 867 (1974).
[2] G. R. Mundy, L. G. Raisz, R. A. Cooper, G. P. Schechter, and S. E. Salmon, *N. Engl. J. Med.* **291**, 1041 (1974).
[3] K. J. Ibbotson, S. M. D'Souza, K. W. Ng, C. K. Osborne, M. Niall, T. J. Martin, and G. R. Mundy, *Science* **221**, 1292 (1983).
[4] L. G. Raisz, *J. Clin. Invest.* **44**, 103 (1965).
[5] M. Gowen, O. D. Wood, E. J. Ihrie, M. K. B. McGuire, and R. G. G. Russel, *Nature (London)* **306**, 378 (1983).

calvarial system in addition to the fetal rat long bone system because there are slight differences between these two particular assays. For example, platelet-derived growth factor stimulates bone resorption in the mouse calvarial assay but when added to fetal rat long bones produces no effect. Similarly, epidermal growth factor seems to act through a different mechanism in the two assays, by stimulating prostaglandin synthesis in the mouse calvaria and independently of prostaglandin synthesis in the fetal rat long bone system. The most striking difference between the two assays is with transforming growth factor β, which stimulates bone resorption via prostaglandin synthesis in mouse calvariae, but inhibits bone resorption in fetal rat long bones.[6] The disadvantage of the mouse calvarial assay is that it requires larger volumes of medium, which can be a distinct disadvantage if the assay is being used for monitoring purification of a factor.

Marrow Cell Culture System

As part of the assessment of factors which affect osteoclast activity, the generation of osteoclasts in the modified Dexter marrow culture system can be used. This technique is described in detail.

Obtaining Marrow. After informed consent is obtained, human marrow and bone biopsy specimens are prepared from normal donors or patients. Marrow is aspirated from the posterior iliac crest under xylocaine anesthesia. Marrow is collected in 1 ml of preservative-free heparin (1000 μl/ml) and processed as described below.

Culture of Marrow Cells. Human marrow cells are collected and placed in α minimal essential medium (αMEM) containing 5% fetal calf serum and 100 units/ml heparin. Marrow buffy coat is collected by centrifugation, and the cell suspension is layered over Ficoll-Hypaque density gradients. The gradients are centrifuged at 400 g for 30 min at 12° and the mononuclear cells collected from the interface. The mononuclear cells are washed 2 times with αMEM and the viability and nucleated cell number determined. Marrow cells are then cultured in the absence or presence of osteotropic hormones in MEM plus 20% horse serum at 37° in a humidified atmosphere at 4% CO_2 in air. Cultures are fed weekly by removing one-half of the supernatant cells and replacing the medium with an equal volume of fresh medium. No attempt is made to replace the nonadherent cells which have been removed. Cultures are inspected daily for the appearance of multinucleated cells with an inverted phase microscope, and 20 random (200× magnification) microscopic fields are counted on alternate days to determine the kinetics of multinucleated cell formation.

[6] J. P. Pfeilschifter, S. Seyedin, and G. R. Mundy, *J. Clin. Invest.* **82,** 680 (1988).

Morphologic Studies on Cultured Marrow Cells. Human multinucleated marrow cells are examined by both light and electron microscopy for morphologic characterization of osteoclasts. Multinucleated cells are assessed for the presence of numerous nuclei, well-developed ruffled borders in the presence of bone, clear zones, Golgi, pleiomorphic mitochondria, lysosomes, vacuoles, and acid phosphatase activity. The effects of osteotropic factors on these ultrastructural features may also be assessed.

Cultures are stained with Wright's stain to score multinucleated cells with at least three nuclei at the end of the culture period. In this manner it is possible to compare the number of multinucleated cells per culture in cultures treated with osteotropic factors and control cultures.

Autoradiography Experiments on Cultured Marrow Cells. To determine if osteotropic factors affect cell replication, autoradiography experiments may be performed. Bone marrow cells are prepared as described above and grown in the presence of the factor for 6 days on glass coverslips. After 6 days the medium is removed and replaced with fresh medium containing factor and 1 μCi/ml [^{3}H]thymidine (New England Nuclear, Boston, MA). Nonadherent cells are removed with the spent medium, recovered by centrifugation, and replaced with the fresh medium. The cells are incubated with the [^{3}H]thymidine for 48 hr, following which the medium is again replaced with fresh medium containing 1,25-dihydroxyvitamin D_3 at a concentration of 10^{-8} M. The culture is then continued for a further 2 weeks before fixation in 5% glutaraldehyde for 15 min. The coverslips are then mounted on glass slides and processed for autoradiography as described by Roodman *et al.*[7] Results are expressed as the percentage of multinucleated cells containing one or more labeled nuclei and also as the number of labeled nuclei per cell (assessed from the first 50 labeled cells).

Isolated Osteoclasts

Isolation and Culture of Osteoclasts. Osteoclasts may be obtained from newborn rats or fetal baboons according to techniques previously described.[8] Neonatal rats are sacrificed by decapitation and the femurs and tibias explanted, dissected free from surrounding soft tissue, and cut transversely across the epiphyseal margins. The bones of each rat are curetted with a scalpel blade into 1 ml of medium and vigorously agitated with a pipette. the supernatants are added to 12–16 bone slices in Linbro multiwell dishes. The cells are incubated on the bone slices for 15 min at 37°. Bone slices are then removed, washed, and placed in separate wells

[7] G. D. Roodman, J. J. Hutton, and F. J. Bollum, *Biochim. Biophys. Acta* **425,** 478 (1976).
[8] P. M. J. McSheehy and T. J. Chambers, *Endocrinology* **118,** 824 (1986).

containing 100 μl of αMEM and fetal calf serum. After the putative osteoclastotropic factors are added to the wells containing bone slices and osteoclasts, the wells are incubated for 24 hr before quantification of cell numbers and assessment of bone resorption.

Osteoclasts are also isolated from adult chickens maintained on a calcium-free diet, distilled water for 2 weeks. The technique has been described by Zambonin-Zallone et al.[9] The hypocalcemic chickens are sacrificed, the femurs and tibias explanted and curetted, and the osteoclasts removed from the endosteal surfaces. The cellular suspension containing osteoclasts and contaminating marrow cells are plated into microtiter plates (96 or 48 wells) and onto bovine or human bone slices or sperm whale dentine (if resorption of a calcified matrix is to be assessed) or coverslips. The cells may be treated with arabinose β-D-cytosine furanoside to remove all cycling cells. The resulting population is approximately 92–98% osteoclasts. With this technique, the osteoclast yield may range anywhere from 1 to 9 million per chicken. On day 2, the cells are washed and may be stimulated with vitamin A (1 μM) (or other stimulators) for 24 hr, after which time various factors are added, and the cell supernatants or calcified matrices are harvested 24, 48, or 72 hr later.

Osteoclasts may also be isolated from lactating rats. We have developed this technique because of the need for isolating mammalian osteoclasts in reasonable numbers. Previous work from other groups has used neonatal rat osteoclasts, but the yields are relatively small. We have adopted a similar approach for isolating osteoclasts from adult chickens, that is, we have used a calcium-deficient diet which is a stimulus for osteoclastic bone resorption. The pregnant female is maintained on a calcium-free diet beginning approximately 1 week before and maintained until 2 weeks after parturition. The animal is then sacrificed, the femur, tibia, and humerus removed, the bones split, and the marrow and endosteal surface removed by scraping. The cells are immediately added to microtiter plates containing bone slices or to coverslips. After 30 min all nonadherent cells are washed from the plate and factor is added. Six hours later the samples are harvested and processed for detection of bone resorption or stimulation. Our present yield of osteoclasts is over 1000 per rat.

Assessment of Bone Resorption. At the conclusion of the 24-hr culture period, the bone slices are removed and cell numbers assessed by toluidine blue staining (10 ng/ml) after glutaraldehyde fixation. The cells are washed 2 times with phosphate-buffered saline (PBS), then treated with 0.2% aqueous OsO_4; alternatively, the bone slices may be dehydrated through graded alcohols, critical point dried from CO_2, sputter coated with gold,

---

[9] A. Zambonin-Zallone, A. Teti, and M. V. Primavera, *Anat. Embryol.* **165**, 405 (1982).

and examined under a scanning electron microscope. The number of osteoclastic excavations is counted and the area of excavation delineated by a continuous border assessed by tracking the outline of the cavity onto a digitizing table input into an Apple IIE microcomputer. The depth of cavities is calculated by measurement of the change in parallax of a point in the center of each of the first four concavities encountered for a variable, using a tilt change of $10°$ between photographs at a magnification of at least 2000. The approximate volume of bone resorbed can then be derived from values of the total area of bone surface and the average depth of resorption on that slice. In most experiments, sperm whale dentine is used rather than bone slices, since sperm whale dentine has a smoother surface. Bone slices are prepared from specimens of human femoral cortical bone, cleaned of adherent soft tissues, with a cortex cut to $0.3 \times 0.4 \times 0.01$ cm slices with a low speed saw. The slices are cleaned by ultrasonication for 20 min in sterile distilled water, washed in ethanol, and stored dry at room temperature.

Detection of Osteoclast Activation by an Increase in Tartrate-Resistant Acid Phosphatase. Osteoclasts cultured on coverslips are fixed in fresh citrate/acetate solution (20 ml of 0.38 mol/liter citrate in 30 ml acetone) for 30 sec at room temperature. The coverslips are rinsed in distilled water 3 min, air-dried, and incubated in Napthol AS-BI, phosphoric acid, tartaric acid, and Fast Garnet GBC salt for 1 hr at 37° in the dark. After 1 hr the coverslips are rinsed in distilled water, air-dried, and mounted in Euparol.

The acid phosphatase activity of osteoclasts may be quantitatively measured by using fluorescent spectroscopy. Osteoclasts (5×10^4) are placed into 3.5-cm diameter petri dishes and treated as described previously. The cells are harvested by scraping with a rubber policeman in 0.5 ml PBS. The cell lysate is sonicated and aliquots of 100 μl added to 100 μl of 2 mM methylumbelliferyl phosphate (MUP) pH 5.0, in 0.48 M acetate buffer (0.48 M sodium acetate, 0.48 M ethanol, pH 5). Samples are incubated for 30 min at 37° and the reaction stopped by the addition of 1.5 ml stop solution (50 mM glycine, 5 mM EDTA, pH 10.4). Fluroescence is measured at emission 448 nm and excitation at 360 nm. Conditions are standardized, activity expressed in micromoles MUP hydrolyzed per microgram protein, and the protein content measured by the technique of Lowry et al.[10]

Quantitative Bone Histomorphometry

Animals are either infused or injected with potential osteotropic factors, and sections of bone are examined by quantitative bone histomor-

---

[10] O. H. Lowry, N. J. Rosebrough, A. L. Fen, and R. J. Randall, *J. Biol. Chem.* **193,** 265 (1951).

phometry. Sections are examined after being embedded in paraffin. The approaches are the following.

Examination of Calvaria. Following infusions for varying periods by an Alzet osmotic minipump or by local injection over the calvaria, the calvariae are removed and a 4 mm wide strip is cut parallel and anterior to the lambdoid suture through the coronal plane. The following variables are measured on 3 μm thick decalcified sections using a digitizing tablet and the Bioquant image analysis system: (1) total bone area (i.e., bone and marrow between inner and outer periosteal surfaces); (2) the number of osteoclasts within the marrow cavity (expressed per square millimeter of the total bone area); (3) the extent of resorption, either active (with osteoclasts present) or total rough crenated surface, along the interface between bone and bone marrow (expressed as a percentage of the total length of this interface). These histomorphometric measurements are performed on the injection side of the calvariae in a standard length of the segment of bone between the sagittal suture and the muscle insertion at the lateral border of each bone.

The calvarial bone is thin and less compact than tibia cortex (for example), and it has numerous openings for blood vessels which pass into the marrow space. Thus, it is more likely that the injected factors will permeate through to the underlying marrow space where they will come into contact with osteoblast and osteoclast precursors as well as hematopoietic cells. Thus, this protocol may also provide information on the interactions among these cell populations.

Examination of Vertebrae. Trabecular bone in the vertebral bodies may also be examined to assess the systemic effects of infused factors. Following infusions for variable periods by an Alzet osmotic minipump or by local injection over the calvariae, lumbar vertebrae are removed intact, decalcified, and embedded in paraffin. The following variables are measured: (1) trabecular bone volume; (2) extent of active and total resorption surface, and (3) number of osteoclasts per square millimeter of the cancellous space.

Assays for New Bone Formation

Supernatants harvested from tumor cell cultures are added to osteoblast-like cell culture systems to assess collagen synthesis, alkaline phosphatase content, and [^{3}H]thymidine, [^{3}H]uridine, and ^{3}H-labeled amino acid incorporation. Collagen synthesis is assayed by the technique described by Peterkofsky and Diegelmann[11] and described in more detail below. Alkaline phosphatase content is measured according to the stan-

---

[11] B. Peterkofsky and R. R. Diegelmann, *Biochemistry* **10**, 988 (1971).

dard technique of Bessey et al.[12] using p-nitrophenyl phosphate as substrate.

Mitogenesis Assay

The method of Williams[13] is used. Rat osteosarcoma cells (UMR-106) are plated at 2×10^4 cells/well in Dulbecco's modified Eagle's medium (DMEM) plus 10% fetal bovine serum (FBS) in 96-well microtiter plates. Twenty-four hours later cells are washed and preincubated in serum-free Ham's F12 medium for 24 hr. Test substances are added for 22 hr. Each well is pulse-labeled with tritiated thymidine (0.2 μCi) for 2 hr, then washed and incubated for 2 hr with Ham's F12 plus unlabeled thymidine (1 mg/ml). The contents of each well may be removed using cotton-tipped applicators that are then washed in cold 5% trichloroacetic acid (TCA) and 95% ethanol. Incorporated [^{3}H]thymidine is measured by scintillation counting.

Alkaline Phosphatase Activity

The method of Majeska et al.[14] may be used. Rat osteosarcoma (17/2.8) cells are plated at 5×10^4 cells/well in Ham's F12 medium plus 7% FBS in 24-well plates, then washed, and test substances added for an additional 24 hr. Plates are washed twice with PBS. Two hundred microliters water is added, and the plates are freeze–thawed 3 times. The lysate is scraped from each well with a rubber policeman. Cell suspensions at a 1/10 dilution are incubated at 37° for 20 min with phosphatase substrate 104 (Sigma, St. Louis, MO) and the reaction stopped with 2 ml of 0.1 M NaOH. Alkaline phosphatase activity is calculated by measuring the cleaved substrate at an optical density of 410 nm in a Beckman dual-beam spectrophotometer and corrected for protein concentration. This assay can be modified for use in 96-well microtiter plates. With this adaptation, which includes the use of a microplate reader and computer-assisted calculation with statistical analysis of the data, the effects of multiple factors on alkaline phosphatase activity can be determined reliably and accurately using serial dilutions of the factors.

Measurement of Collagen and Protein Synthesis

The method of Peterkofsky and Diegelmann[11] is used to measure the incorporation of [^{3}H]proline into collagenase-digestible protein. Briefly, mouse calvaria or osteoblast-like cell cultures (UMR-106 or 17/2.8) are

[12] O. Bessey, O. Lowry, and M. Brock, *J. Biol. Chem.* **164**, 321 (1946).
[13] M. R. Williams, *Cell. Immunol.* **9**, 435 (1973).
[14] R. J. Majeska, S. B. Rodan, and G. A. Rodan, *Exp. Cell Res.* **111**, 465 (1978).

labeled with [5-³H]proline for 4 hr, and then labeled protein is TCA-precipitated. Total and collagenase-digestible protein counts are determined and expressed per microgram of DNA (diaminobenzoic acid determination, Kissane and Robbins[15]) or per microgram of protein. The calculation of percent collagen synthesis takes into account the fact that collagen contains 5.4 times as much proline as do other cellular proteins.[16]

Human Bone Cell Culture

The technique is that described by Beresford *et al.*[17] Trabecular fragments dissected from human bone obtained at surgery are washed extensively in sterile PBS to remove attached tissue. The fragments are then further dissected into particles 3–5 mm in diameter for seeding as explants into 9-cm tissue culture dishes (Falcon, Oxnard, CA), 0.2–0.6 g of bone per dish. Explants are cultured in 5–10 ml of DMEM containing 10% (v/v) fetal calf serum (FSC), 50 units/ml penicillin, 15 mg/ml streptomycin, and 2 mM glutamine (GIBCO, Grand Island, NY), in an atmosphere of 95% air and 5% CO_2 at 37°. The medium is changed at 24 hr and thereafter at 5-day intervals. Cell outgrowths reproducibly occur within 7 days, and confluence is obtained by 21–28 days with this system. At confluence, bone explants are replated by aseptically transferring the trabecular fragments to petri dishes containing fresh medium and cultured as before. The confluent monolayer is then washed twice with 5 ml of trypsin–EDTA (0.5 and 0.2 g/liter, respectively) (GIBCO) at 37° until the cells become detached (~5 min). The suspended cells are transferred to sterile universal containers (Corning Glass Works, Corning, NY) in which there is an equal volume of medium plus 10% FCS to inhibit tryptic activity. Cells are pelleted by centrifugation for 10 min at 400 g, washed in medium containing FCS, then recentrifuged. The final pellet is resuspended in culture medium, and the cells are dispersed by repeated aspiration through a sterile wide-bore needle (19 gauge), before passaging into 3.5-cm tissue culture wells (6-well plates) at 10^4 cells per well. At first passage, the bone cell cultures respond to 1,25-dihydroxyvitamin D_3 and parathyroid hormone (PTH), express high alkaline phosphatase activity, and synthesize both collagenous and noncollagenous components of the extracellular matrix.[18,19]

[15] J. M. Kissane and E. Robins, *J. Biol. Chem.* **233**, 184 (1958).
[16] B. Peterkofsky, *Arch. Biochem. Biophys.* **152**, 318 (1972).
[17] J. N. Beresford, J. A. Gallagher, J. W. Poser, and R. G. G. Russell, *Metab. Bone Dis. Relat. Res.* **5**, 229 (1984).
[18] J. N. Beresford, J. A. Gallagher, J. W. Poser, and R. G. G. Russell, *Metab. Bone Dis. Relat. Res.* **5**, 229 (1984).
[19] A. Valentin, P. D. Delmas, P. M. Chavassieux, C. Chenu, S. Saez, and P. J. Meunier, *J. Bone Miner. Res.* **1**, 139 (1986).

Organ culture

The organ culture technique for studying factors which influence type-I collagen synthesis has been described in detail previously.[20,21] The frontal and parietal bones are removed from 21-day fetal rats and split along the sagittal suture. The calvariae are first placed in a sterile petri dish containing a chemically defined medium (BGJb) and precultured in an incubator at 37° in an atmosphere of 5% CO_2 in air for a period of 3–4 hr. The purpose of the preculture is to remove loose fragments of bone and connective tissue as well as endogenous hormones that are in contact with the bone. After the preculture each half-calvarium is placed in a 25-ml sterile vial containing 2 ml of chemically defined medium and 4 mg/ml of bovine serum albumin. The vials are then gassed with 5% CO_2 for 1 min, sealed with rubber stoppers, and placed in a continuously shaking water bath (20–40 oscillations/min) at 37° for periods up to 96 hr. Labeled proline is added, usually for the last 2 hr of the culture period. Different substances can be added to the culture medium for variable periods of time and their effects on bone collagenase-digestible and noncollagen protein synthesis determined. In cultures lasting longer than 24 hr, the bones are transferred to fresh culture medium daily.

At the end of the incubation period, the media are acidified and the calvariae are extracted with 5% trichloroacetic acid, acetone, and ether, weighed (dry weight), and then homogenized in 0.5 M acetic acid. Aliquots of bone homogenate are treated with repurified bacterial collagenase and the labeled [^{3}H]proline incorporated into collagenase-digestible and noncollagen protein measured according to the method of Peterkofsky and Diegelmann[11] The enzyme preparation should not contain any proteolytic activity on noncollagen substrates. In some experiments, labeled proline and hydroxyproline are separated by thin-layer chromatography of the collagenase digest and the medium. This provides a check for completeness of collagenase digestion and determination of the degree of hydroxylation of collagen. The medium is also analyzed for labeled collagen content to determine the proportion of collagen synthesized in the bone that is released into the medium. If this occurs, the collagen released is further analyzed by gel chromatography to determine if it consists of large molecules of fragments.[22]

---

[20] J. W. Dietrich, E. M. Canalis, D. M. Maina, and L. G. Raisz, *Pharmacologist* **18**, 234 (1976).

[21] L. G. Raisz, D. M. Maina, S. C. Gworek, J. W. Dietrich, and E. M. Canalis, *Endocrinology* **102**, 731 (1978).

[22] T. L. Chen and L. G. Raisz, *Calcif. Tissue Res.* **17**, 113 (1975).

[48] 3-[(3-Cholamidopropyl)dimethylammonio]-1-propane Sulfonate as Noncytotoxic Stabilizing Agent for Growth Factors

By YUHSI MATUO, NOZOMU NISHI, KUNIO MATSUMOTO, KAORU MIYAZAKI, KEISHI MATSUMOTO, FUJIO SUZUKI, and KATSUZO NISHIKAWA

Introduction

Growth factors are present in tissues, cells, or culture media at an extremely low content (usually less than about 0.5 μg/g or about 0.1 ng/ml). The recovery of such factors during purification may be unexpectedly low. In addition, growth factors can exert biological effects at extremely low concentrations [picomolar to nanomolar (pM–nM) levels]. Growth factors, when used in dilute, highly purified form, are easily lost by irreversible adsorption to surfaces of experimental materials such as containers and chromatographic carriers. The loss should be minimized by using a surfactant that has low cytotoxicity for cultured mammalian cells. We recently found that CHAPS, a zwitterionic detergent, is less cytotoxic than many other mild detergents and can stabilize growth factors.

General Properties

CHAPS is a mild detergent useful for solubilization of nonhistone chromosomal proteins[1] and membrane proteins[2] including receptors.[3] As shown in Table I,[4] CHAPS has useful properties for polypeptide research: (1) it does not alter the charge of proteins since it retains its zwitterionic property over a wide pH range owing to the 3-(alkyldimethylammonio)-1-propane sulfonate, designated sulfobetaine,[5] (2) it is easily removed because of its high critical micelle concentration (CMC), and (3) it does not interfere significantly with protein assays.

---

[1] Y. Matuo, S. Matsui, N. Nishi, F. Wada, and A. A. Sandberg, *Anal. Biochem.* **150**, 337 (1985).
[2] E. M. Bailyes, A. C. Newby, K. Siddle, and J. P. Luzio, *Biochem. J.* **203**, 245 (1982).
[3] K. A. Paganelli, A. S. Stern, and P. L. Kiliani, *J. Immunol.* **138**, 2249 (1987).
[4] P. K. Smith, R. I. Krohn, G. T. Hermanson, A. K. Mallia, F. H. Gartner, M. D. Provenzano, E. K. Fujimoto, N. M. Goeke, B. J. Olson, and D. C. Klenk, *Anal. Biochem.* **150**, 76 (1985).
[5] L. M. Hjelmeland, *Proc. Natl. Acad. Sci. U.S.A.* **77**, 6368 (1980).

TABLE I
GENERAL PROPERTIES OF CHAPS[a]

Formula	$C_{32}H_{58}N_2O_7S$
Name	3-[(3-Cholamidopropyl)dimethylammonio]-1-propane sulfonate
Molecular weight	614.9
CMC[b]	0.46% (7.4 mM)
Charge	Zwitterionic (positive charge of quaternary ammonium ion and negative charge of sulfonate ion)
Solubility	Soluble in water, methanol, and dimethyl sulfoxide
	Insoluble in acetone, acetonitrile, benzene, chloroform, N,N-dimethylformamide, ethanol, hexane, toluene, xylene
Absorbance	1% water solution, $A_{280\text{ nm}} = 0.029$, $A_{260\text{ nm}} = 0.035$
Protein assay	Lowry's method: not disturbed at concentrations less than 0.05%[1]
	Bradford's method: not disturbed at concentrations less than 1%[1]
	Smith's method: not disturbed at concentrations less than 1%[4]

[a] All the properties were examined with CHAPS obtained from Sigma Chemical Co. (St. Louis, MO).
[b] Critical micelle concentration.

Noncytotoxic Properties

The cytotoxic effects of CHAPS on cultured mammalian cells vary for different cell types and also different culture conditions even within the same cell type. As a typical example, the results with normal human keratinocytes cultured in the absence of serum are shown in Table II.[6] The cytotoxicity of CHAPS was examined by morphological observation and assays of cell growth and DNA synthesis. CHAPS at concentrations lower than 0.001% hardly influenced the morphology and the growth of human keratinocytes; CHAPS at a concentration of 0.01% slightly inhibited their growth (Table II). DNA synthesis-stimulating activities of insulin, epidermal growth factor (EGF), and hypothalamic extract using normal human keratinocytes were only slightly affected by the addition of 0.01% CHAPS, but were inhibited by 0.1% CHAPS. The data on maximum noncytotoxic concentrations so far obtained with various kinds of cultured mammalian cells are summarized in Table III.[7,8] CHAPS at 0.001% is not cytotoxic in terms of DNA synthesis and cell growth using a variety of mammalian cell lines. By contrast, the noncytotoxic concentrations of

[6] J. J. Wille, Jr., M. R. Pittelkow, G. D. Shipley, and R. E. Scott, *J. Cell. Physiol.* **121**, 31 (1984).
[7] Y. Matuo, N. Nishi, Y. Muguruma, Y. Yoshitake, Y. Masuda, K. Nishikawa, and F. Wada, *In Vitro Cell. Dev. Biol.* **24**, 477 (1988).
[8] Y. Matuo, N. Nishi, Y. Muguruma, Y. Yoshitake, Y. Masuda, K. Nishikawa, and F. Wada, *Cytotechnology* **1**, 309 (1988).

TABLE II
EFFECT OF CHAPS ON GROWTH OF NORMAL
HUMAN KERATINOCYTES[a]

CHAPS (%)	Cell growth (% of control)
0	100.0
0.0001	97.6
0.001	99.5
0.01	95.0
0.1	26.1

[a] Human keratinocytes were cultured according to the method of Wille et al.[6] with some modifications. Briefly, normal skin was cut into about 4 mm pieces and floated on dispase solution (500 U/ml; Godo Shusei Co., Ltd., Tokyo, Japan) overnight at 4°. The epidermal sheets were incubated in 0.25% trypsin for 10 min at 37°. Epidermal cells were suspended in optimized nutrient medium MCDB 153 (Kyokuto Co., Ltd., Tokyo, Japan) supplemented with EGF (10 ng/ml), insulin (5 μg/ml), hydrocortisone (50 μM), ethanolamine (0.1 mM), phosphoethanolamine (0.1 mM), and bovine hypothalamic extract (150 μg/ml).

Triton X-100 and Nonidet P-40 (NP-40) were estimated to be less than 0.0001% using a DNA synthesis assay. Triton X-100 and NP-40 inhibit 25–30% of the DNA synthesis activity of calf serum with BALB/c 3T3 cells even at 0.0001%.[7] Assuming that samples are diluted 100 times before an assay, maximally 0.1% CHAPS can be used during the preparation and isolation of growth factors. In addition, CHAPS is useful for the solubilization of water-insoluble proteins such as nonhistone chromosomal proteins at 1% at pH 8—9.[1] This suggests that it would also be useful for the study of water-insoluble growth factors or other biologically active polypeptides including growth factor receptors.

Stabilization during Purification

In general, inclusion of CHAPS does not disturb amino acid analysis and sequence analysis of N-terminal amino acids.[7,8] In addition, CHAPS in the sample can be removed by reversed-phase HPLC.[8]

TABLE III
NONCYTOTOXIC PROPERTIES OF CHAPS FOR VARIOUS TYPES OF MAMMALIAN CELLS

Cells	Time[a]	Maximum concentration (%)[b]	Assay method	Ref.
Established cells				
BALB/c 3T3 fibroblasts	20 hr	0.01	DNA synthesis	7, 8
MC3T3-E1	20 hr	0.01	DNA synthesis	8
osteoblasts	14 days	0.002	Cell growth	8
AT-3 prostate	20 hr	0.01	DNA synthesis	8
	7 days	0.002	Cell growth	8
NPK-49F cells	7 days	0.003	Colony formation	—
SC-3 androgen-dependent cells	2 days	0.002	DNA synthesis	—
BRL cells	4 days	0.001	Cell growth	—
Primary cultured cells				
Rat prostate epithelial cells	8 days	0.003	Cell growth	8
Human	7 days	0.01	Cell growth	—
keratinocytes	19 hr	0.001	DNA synthesis	—
Chondrocytes	1 day	0.005	Cell growth	—

[a] Time exposed to CHAPS.
[b] Maximum noncytotoxic concentrations of CHAPS are estimated by the respective dose–response curves except for SC-3 cells, BRL cells, and chondrocytes.

Purification of Androgen-Induced Growth Factors

Shionogi carcinoma (SC 115) derived from a mouse mammary tumor has been maintained *in vivo* for 25 years as an androgen-dependent tumor. One of the cloned cell lines (SC-3) recently established from SC 115 has been observed to be remarkably growth-stimulated by testosterone in serum-free media. Nonomura and Sato *et al.*[9,10] recently found a fibroblast growth factor (FGF)-like factor in serum-free conditioned medium from testosterone-stimulated SC-3 cells. SC-3 cells (5×10^5/100-mm dish) are cultured in Ham's F12/Eagle's minimum essential medium (MEM) (1:1, v/v)–0.1% bovine serum albumin (BSA) supplemented with 10^{-8} M testosterone. After cells reach semiconfluency ($5-7 \times 10^6$/dish), the conditioned medium is collected. The conditioned medium (1.6 liters) is concentrated to 50–100 ml with Pericon fibers. The concentrated sample is applied

[9] N. Nonomura, N. Nakamura, N. Uchida, S. Noguchi, B. Sato, T. Sonoda, and K. Matsumoto, *Cancer Res.* **48**, 4904 (1988).
[10] B. Sato, N. Nakamura, S. Noguchi, N. Uchida, and K. Matsumoto, in "Progress in Endocrinology" (H. Imura *et al.*, eds.), p. 99. Elsevier, Amsterdam, 1988.

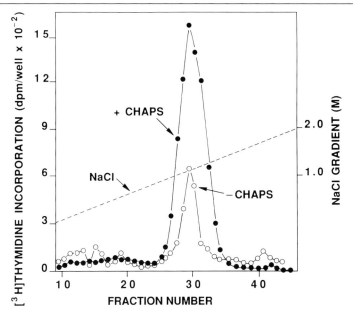

FIG. 1. Purification of an androgen-induced FGF-like growth factor by heparin–Sepharose column chromatography in the presence or absence of CHAPS. Growth factors present in the concentrated serum-free conditioned medium obtained from testosterone (10^{-8} M)-stimulated SC-3 cells were separated into fractions by heparin–Sepharose column in the presence (●) or absence (○) of 0.1% CHAPS as described in the text and Table IV. The growth-stimulatory activity of each fraction on SC-3 cells was estimated by [^{3}H]thymidine incorporation into the cells as follows. SC-3 cells were plated into a 96-well plate (3×10^3 cells/well) containing 0.15 ml MEM added with 2% dextran coated charcoal-treated fetal calf serum (FCS). On the following day (day 0), the medium was changed to 0.15 ml serum-free medium [Ham's F12/MEM (1:1, v/v)–0.1% BSA] containing 2% (3 μl) and 20% (30 μl) of the eluted fraction obtained in the presence and absence of 0.1% CHAPS, respectively. On day 4, the cells were pulsed with [^{3}H]thymidine (0.15 μCi/0.15 ml/well) for 2 hr at 37°, and the radioactivity incorporated into the SC-3 cells was measured. Values are the mean of three determinations.

to a heparin–Sepharose column (0.9 cm i.d. × 2.4 cm) previously equilibrated with 10 mM phosphate buffer containing 0.1 M NaCl (pH 7.4) and 0.1% CHAPS (Fig. 1).

Recovery of the activity was about 60% in the presence of 0.1% CHAPS in five different experiments, whereas it was as low as 4% in the absence of CHAPS. The active fractions obtained in the presence of 0.1% CHAPS (5 ml) are applied to a Sephadex G-100 column in the presence or absence of 0.1% CHAPS (Fig. 2). One major activity peak was eluted at about 38 kDa in the presence of 0.1% CHAPS, whereas no distinct peak was detected in the absence of CHAPS.

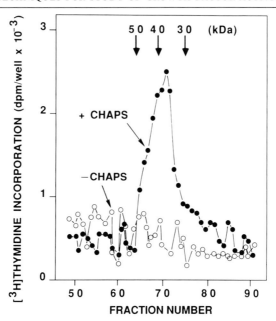

FIG. 2. Sephadex G-100 analysis of the androgen-induced FGF-like growth factor in the presence or absence of CHAPS. Partially purified androgen-induced FGF-like growth factor eluted from the heparin–Sepharose column in the presence of 0.1% CHAPS as shown in Fig. 1 was separated into fractions by Sephadex G-100 column chromatography in the presence (●) or absence (○) of 0.1% CHAPS. The active fractions eluted from heparin–Sepharose column (5 ml) were applied to a Sephadex G-100 column preequilibrated with 10 mM phosphate buffer (pH 7.4) containing 0.6 M NaCl in the presence or absence of 0.1% CHAPS, and elution was performed with the same buffer (see Table IV). The growth-stimulatory activity of each fraction on SC-3 cells was estimated by [^{3}H]thymidine uptake into the cells on day 4 as described in the legend to Fig. 1, using 0.15 ml of the serum-free medium containing 33% (0.5 ml) of the dialyzed fraction eluted from the column. Values are the mean of three determinations.

Purification of Growth Inhibitor from Rabbit Serum

Miyazaki et al.[11,12] previously found that rat and mouse sera contain a growth inhibitor which is much more active on a nonmalignant rat liver cell line, BRL, than on its tumorigenic counterpart RSV-BRL, whereas rabbit serum contains a growth inhibitor which preferentially acts on RSV-

[11] K. Miyazaki, K. Mashima, N. Yamashita, J. Yamashita, and T. Horio, In Vitro Cell. Dev. Biol. **21**, 62 (1985).
[12] K. Mashima, T. Kimura, K. Miyazaki, J. Yamashita, and T. Horio, Biochem. Biophys. Res. Commun. **148**, 1215 (1987).

BRL. Recently, the tumor cell growth inhibitor (TGI) has been purified successfully from rabbit serum with the use of CHAPS (T. Kimura *et al.*, submitted for publication).

For the purification of TGI, rabbit serum is fractionated by ion-exchange chromatographies on a DEAE-Toyopearl column (pH 7.4) and on a CM–Sepharose CL-6B column (pH 6.0), hydrogen-bond chromatography on a Sepharose CL-6B column preequilibrated with 2 M ammonium sulfate, and then affinity chromatography on a concanavalin A-Sepharose column. TGI is enriched about 50-fold by these procedures. The partially purified TGI is then subjected to molecular-sieve HPLC on a TSK gel G3000SW column in the presence of 1% CHAPS. This chromatography effectively separates TGI from a majority of contaminating proteins. By repeating this chromatography twice, TGI is purified to show a homogeneous protein band with a molecular weight of about 56,000 on sodium dodecyl sulfate (SDS)-polyacrylamide gel electrophoresis under reducing or nonreducing conditions.

These HPLC procedures enriched TGI 60-fold. Most of the TGI was recovered in each of the HPLC purifications. However, even in the presence of 0.1% CHAPS, TGI was adsorbed to the column, and very little was recovered, just as in the absence of CHAPS. TGI is labile in 1 M acetic acid (pH 2.3), 6 M urea, or 6 M guanidine hydrochloride but very stable in 1% CHAPS. The purified TGI inhibited the growth of a variety of malignantly transformed cell lines including human carcinoma cell lines at a dose of 20 ng/ml or higher. Several examples of stabilization of growth factors during purification are summarized in Table IV.

Stabilization during Storage and Dilution

Of the growth factors so far discovered, FGF (fibroblast or heparin-binding growth factors) are known as extremely adhesive polypeptides that adhere strongly to surfaces. Acidic FGF (or HBGF-1, 0.4 ml of 0.5 μg/ml) at high purity is not stable when stored in the absence of 0.1% CHAPS.[7] The loss of activity is much higher with basic FGF (HBGF-2) than with acidic FGF. For complete stabilization of purified FGF at low concentrations, the inclusion of 0.1% CHAPS containing 0.5% BSA or 0.2% gelatin is needed for basic FGF, and 0.1% CHAPS containing 0.5% BSA and dithiothreitol (0.5–1 mM) for acidic FGF. We found that 40 ng/ml of basic FGF is stable under these conditions for at least 1 month at $-20°$ or $4°$. Lyophilization of FGF in the absence of CHAPS and BSA should be avoided; 25 μg of basic FGF was completely inactivated by lyophilization, whereas 0.1% CHAPS and 0.5% BSA can stabilize basic

TABLE IV
STABILIZATION OF GROWTH FACTORS WITH CHAPS

Sample[a]	Recovery[b]	Conditions
Crude bFGF	69% (0.1% CHAPS) 15% (−)	TSK G2000SW (7.5 mm i.d. × 30 cm); 10 mM Tris-HCl (pH 7.5)–0.5 M NaCl
aFGF	20% (0.1% CHAPS) 12% (−)	TSK G2000SW (7.5 mm i.d. × 30 cm); 10 mM Tris-HCl (pH 7.5)–0.5 M NaCl
[125]I-Labeled aFGF	76% (0.1% CHAPS)	TSK G200SW (7.5 mm i.d. × 30 cm); 10 mM Tris-HCl (pH 7.5)–0.5 M NaCl
IGF-II (2 μg)	85% (0.1% CHAPS)	TSK G200SW (7.5 mm i.d. × 30 cm); 10 mM Tris-HCl (pH 7.5)–0.5 M NaCl
Crude bFGF	72% (0.1% CHAPS) 30% (−)	Heparin 5PW (7.5 mm i.d. × 7.5 cm); 10 mM Tris-HCl (pH 7.5)–0.5 M NaCl (gradient to 3 M NaCl)
CDF	Improved (0.03% CHAPS)	Mono S

[a] aFGF, Acidic FGF or heparin-binding growth factor-1; bFGF, basic FGF or heparin-binding growth factor-2; FGF, fibroblast growth factor; CDF, cartilage-derived factor; IGF-II, insulin-like growth factor II.

[b] Recoveries in the presence of CHAPS and in the absence of CHAPS (−).

FGF. CHAPS is also useful for stabilization of acidic and basic FGF during iodination using chloramine-T.[13]

To prevent loss of diluted preparations of all growth factors that easily adhere to glass or plastic, we recommend the following (in order of preference): (1) 0.1% CHAPS containing 0.5% BSA or 0.2% gelatin, (2) 0.1% CHAPS alone, and (3) 0.5% BSA alone. One should keep in mind that albumin is not merely an inert protein; albumin itself and/or contaminants may show unexpected biological activities.[14] Parathyroid hormone can be stabilized with 0.1% CHAPS. Thus, these losses are not specific for growth factors, but also occur with all polypeptides. However, as long as we handle polypeptides at high concentrations (greater than several hundred micrograms per milliliter), we can ignore the loss caused by interaction between polypeptides and surfaces of experimental materials.

[13] K. Matsuzaki, Y. Yoshitake, Y. Matuo, H. Sasaki, and K. Nishikawa, *Proc. Natl. Acad. Sci. U.S.A.* **86,** 9911 (1990).

[14] R. Melsert, J. W. Hoogerbrugge, and F. F. G. Rommerts, *Mol. Cell. Endocrinol.* **64,** 35 (1989).

[49] Transgenic Mouse Models for Growth Factor Studies

By NORA SARVETNICK

Introduction

The recently developed technology of transgenic mice has already been employed for assessment of the *in vivo* capabilities of the products of growth factor genes. This represents a very fruitful area of investigation in the growth factor field. Transgenic mouse studies allow the answering of questions regarding the biological actions of growth factors that could not be addressed by other methodologies. As the procedure for the creation of transgenic mice[1] and the methodologies required for molecular biological techniques[2] are available in comprehensive manuals, this chapter focuses on the experimental design required to investigate the activities of specific growth factors *in vivo* in transgenic mice. The parameters for establishing and understanding the resulting phenotypic variations in transgenic progeny are also discussed.

General Guidelines

The phenotype resulting from expression of any growth factor in transgenic mice will vary depending on the *in vivo* cellular source. This is of course at the discretion of the investigator who designs the specific chimeric gene for injection into fertilized zygotes. Several strategies have been employed and are discussed in greater detail below. These include the expression of the growth factor in the cells that normally express it by introducing an unmanipulated gene fragment into the fertilized eggs. The second general strategy is the expression of the growth factor in a large number of cells in the body to attempt to achieve high circulating levels. The third general strategy is to express the growth factor in a very restricted group of cells. Each general method has distinct applications.

Transgenes are constructed within a plasmid vector utilizing standard recombinant DNA techniques. Often, several independent cloning steps are required to combine desired promoter, coding, and terminator sequences. After each step care should be taken in order to verify the

---

[1] B. Hogan, F. Constantini, and E. Lacy, "Manipulating the Mouse Embryo." Cold Spring Harbor Laboratory, Cold Spring Harbor, New York, 1986.
[2] J. Sambrook, E. F. Fritsch, and T. Maniatis, "Molecular Cloning," 2nd Ed. Cold Spring Harbor Laboratory, Cold Spring Harbor, New York, 1989.

isolation of the proper recombinant molecule. In joining promoters to coding sequences, the respective DNA fragments should be combined after the initiation of transcription in the promoter, and before the start of translation in the coding sequences. The presence of possible translation start sites occurring before the desired site must be avoided. Additionally, terminator sequences are often joined to the 3' untranslated region of the gene. Although their necessity has not been proven, they are thought to increase the efficiency of expression. This is generally accomplished utilizing viral sequences, such as the terminator from the SV40 T antigen or the hepatitis B virus. Although regulatory sequences are generally located upstream of coding regions, downstream sequences are occasionally necessary to obtain regulated expression in transgenic mice. In one case downstream sequences have been utilized to obtain adult erythroid expression of β-globin genes in transgenic mice.[3]

The specific growth factor may be encoded in the transgene by either genomic, cDNA, or synthetic oligonucleotide sequences, which are expressed via a choice of promoters as discussed below. The use of oligonucleotide-derived synthetic genes has not yet been reported in transgenic mouse studies, but it may find utility in producing desired variants of certain growth factors. In general, reproducibly higher levels of expression can be achieved with genomic rather than cDNA sequences. This is thought to be due to the presence of an enhancing effect of mRNA splicing on the yield of the gene product[4]; additionally, regulatory regions may also be present in the intervening sequences. In some cases, however, the size of such genomic DNA segments or the unavailability of convenient restriction enzyme sites renders this task very difficult. In this case the use of cDNA fragments is necessary. Additionally, cDNA sequences are more easily manipulated in strategies to produce specific mutations in the gene and its product. If desired, an intron flanked by splicing sites may be engineered between the promoter and the coding sequences or at the 3' end of the gene before the terminator region.

It is sometimes advisable to test the completed construct before embarking on the time- and labor-intensive process of creating and characterizing transgenic mouse lines. The construct can be transfected into appropriate tissue culture cell lines in which the promoter is predicted to be active, and the culture supernatant assayed for the presence of the desired molecule. Depending on the availability of relevant cell lines, the results

---

[3] R. R. Behringer, R. E. Hammer, R. L. Brinster, R. D. Palmiter, and T. M. Townes, *Proc. Natl. Acad. Sci. U.S.A.* **84,** 7056 (1987).

[4] R. L. Brinster, J. M. Allen, R. R. Behringer, R. E. Gelinas, and R. D. Palmiter, *Proc. Natl. Acad. Sci. U.S.A.* **85,** 836 (1988).

of these pilot experiments should be taken to indicate the ability of the transgene to direct expression *in vivo*. Finally, before injecting the transgene into the fertilized zygotes, the transgene should be isolated from the plasmid sequences in order to obtain optimum expression; this should be taken into account when planning the construct to ensure the presence of adequate (preferably unique) restriction enzyme sites at the plasmid junctures.

Strategies for Growth Factor Production

Use of the Native Gene

In this scheme the entire genomic DNA encoding a growth factor, including the upstream regulatory sequences, is used as the transgene. This allows additional copies of the gene to be present in the genome, which may result in higher levels of expression. This method has been utilized mainly to study the fidelity of expression of the promoters, using gene products from a different species so that they can be distinguished from the endogenous product. Owing to the uncertainty in the level of production from introduced transgenes, this method is not recommended to study the effects of overproduction of a growth factor. It does have utility, however, in directing the expression of a gene thought to be missing or defective in strains of mutant mice in order to complement that defect. This has been successful in the infertile hypogonadal mice, which were restored to fertility after the introduction of the gonadotropin-releasing factor gene.[5]

Generalized Overexpression

The gene encoding the growth factor is fused to a promoter which allows the expression of the factor in a wide variety of tissues. An example of this is the mouse mammary tumor virus long terminal repeat (MMTV LTR) which allows the high level of expression in secretory epithelial cells. Other possibilities for promoters inducing widespread expression are actin, H-2, or the SV40 promoter. This strategy was employed to achieve very high levels of expression of the granulocyte–macrophage colony-stimulating factor (GM-CSF) growth factor from the MMTV LTR in transgenic mice, which resulted in a variety of phenotypes.[6] This strat-

---

[5] A. J. Mason, S. L. Pitts, K. Nikolics, E. Szonyi, J. N. Wilcox, P. H. Seeberg, and T. A. Stewart, *Science* **234**, 1372 (1986).

[6] R. A. Lang, D. Metcalf, R. A. Cuthbertson, I. Lyons, E. Stanley, A. Kelso, G. Kannourakis, D. J. Williamson, G. K. Klintworth, T. J. Gonda, and A. Dunn, *Cell (Cambridge, Mass.)* **51,** 675 (1987).

egy of expression can also be used with a promoter that shows cell specificity, such as the immunoglobulin (Ig) promoter, such as been reported for the interleukin-6 (IL-6) growth factor, yielding plasmacytosis in the transgenic mice.[7] Although the Ig promoter is cell specific, it directs expression to B cells that circulate throughout the body and yields high serum levels of the molecule. These methods resulting in diffusely increased levels of a given growth factor allow the maximum opportunity for phenotypic perturbations. It is thus useful when a generalized effect is desired, or when the specificities of a molecule are not well characterized and *in vivo* overexpression may uncover new properties since the overexpressed growth factor influences many different cell types. However the analysis of the resulting animals can be complex since the effects are exerted on all tissues. Additionally, untoward effects of some factors, especially when being expressed at high levels, may lead to lethality prior to breeding age in some instances.

Site-Directed Expression

In this scheme, the gene encoding the growth factor is fused with a regulatory region of a gene which is expressed by a defined cell type *in vivo*. An example of this cell specificity is the insulin promoter, which directs expression to the pancreatic β cells. This promoter was utilized to obtain tissue-specific expression of γ-interferon (IFN-γ) in transgenic mice.[8] These mice suffered from inflammatory insulin-dependent diabetes mellitus, induced by overexpression of this lymphokine growth factor in transgenic mice. The localized expression of IFN-γ in this manner allowed the uncovering of poorly understood properties of the lymphokine. The pathogenic effects are only exerted in the pancreas; no other organ was consistently affected by the pathology, and circulating levels of the lymphokine could not be detected. Nerve growth factor (NGF) has also been expressed from the insulin promoter, yielding a unique pattern of hyperinnervation of the islets of Langerhans.[9] Such targeted expression allows the analysis of the local effects on a specific organ, owing to the short biological half-life of many of the growth factors. Organ- or cell type-specific expression leads to ease of analysis; however, there is less chance of a detectable phenotype in the mice. Additionally, the promoter must be well characterized and chosen carefully to facilitate expression in the cell type desired.

[7] S. Suematsu, T. Matsuda, K. Aozasa, S. Akira, N. Nakano, S. Ohno, J. Yamamura, T. Hiranao, and T. Kishimoto, *Proc. Natl. Acad. Sci. U.S.A.* **86,** 7547 (1989).
[8] N. Sarvetnick, D. Liggitt, S. L. Pitts, S. E. Hansen, and T. A. Stewart, *Cell (Cambridge, Mass.)* **52,** 773 (1988).
[9] R. H. Edwards, W. J. Rutter, and D. Hanahan, *Cell (Cambridge, Mass.)* **58,** 161 (1989).

Parameters for Analysis of Offspring

The methodology for assessment of the effects of any growth factor depends on the nature of the specific questions asked in an individual experiment. This chapter focuses on standard methodology required for most studies. More specialized surgical experiments are not covered.

General Phenotypic Characterization

The most informative studies are derived from the analysis of heritable defects which result from transgenesis. The founder mice acquired for each construct are bred to nontransgenic mice whose progeny are tested for the presence of the transgene. Following the attainment of a number of transgenic mice, they are examined for any notable characteristics. Ideally, phenotypic variation is manifested early but does not prove detrimental to breeding capabilities until after 6 months of age, giving time for the lines to be easily maintained. This is not always the case. In some instances phenotypic variation is notable very early and the animals are either dead or too ill to be bred by the time sexual maturity ensues at 2 months of age. Some forethought is advisable to avoid expressing high levels of potentially toxic molecules. Additionally, certain phenotypic variations are not manifested until late in life, rendering lengthy studies time-consuming and expensive. Within individual lines derived from founder animals there will probably be some variation of the penetrance of the transgene. The analysis of several independent lines of transgenic mice is necessary for most studies, since the individual transgenes can be subject to "chromosomal position" effects. In this case some lines fail to express the integrated gene, whereas others may express it aberrantly, owing to traits that are dependent on the position of integration of an individual transgene. These possibilities are lessened by the analysis of multiple independent lines of transgenic mice. Both male and female mice should be analyzed since phenotypic variation due to gender differences is also possible.

Assessment of Levels of Expression

RNA. The assessment of the levels of transcription can be most easily approached by the analysis of total RNA prepared from the individual tissues of the transgenic mice. Tissue RNA is isolated and subjected to either Northern blot or solution hybridization analysis, using a probe specific for the growth factor transgene introduced. The analyses require the simultaneous control of a nontransgenic littermate for comparative purposes. Additionally, the tissues sampled will depend on the promoter utilized. However, they should include tissues other than those where transcription might normally be expected. The transcriptional analysis should be performed on more than one line of the transgenic mice, as well as on both male and female mice of several age categories.

In some cases the information derived by total tissue RNA analysis is not sufficiently detailed for satisfactory data analysis. For instance, it may be necessary to document expression in the different cell types within an individual organ, the specifics of which may be critical to the interpretation of the work. For this reason *in situ* hybridization of tissue section RNA to transgene-specific probes should be utilized. The information from the tissue distribution analysis obtained above is useful for planning the *in situ* hybridization experiments. Protocols for such experiments are described by Wilcox *et al*.[10] In the case of the insulin–IFN-γ transgenic mice, expression of the transgene limited to the islets of Langerhans was documented by *in situ* hybridization, as islet cell-type specificity cannot be analyzed by Northern analysis of pancreatic RNA.

Protein. As the transgenic mice are usually produced to detect the effects of expression of a growth factor, a quantitation of the amount of the factor in the blood of transgenic mice may be measured and compared to that of nontransgenic littermates. Such an analysis may be performed by radioimmunoassay, enzyme-linked immunoassay, or a biological assay for the effects of the growth factor. These results may provide a good screening method for initially differentiating high expressing lines that could show increased phenotypic effects. Although this analysis is applicable in cases where generalized expression is directed, in experiments in which local expression is induced appreciable blood levels may not be present even though cell-specific production is occurring. In such cases immunohistochemical analysis of tissue sections from the organs in which expression was directed (as well as those in which it was not) can be performed with antisera directed against the growth factor, as described by Bullock and Petrusz.[11] This methodology is not universally applicable since some molecules are difficult to visualize by immunohistochemistry owing to the unavailability of proper antisera or instability of the antigen. We have not been able to detect immunoreactive IFN-γ in the islets of the insulin–IFN-γ transgenic mice, although the RNA was demonstratable by *in situ* hybridization, and the expression of this molecule resulted in a strong phenotypic effect.

---

[10] J. N. Wilcox, C. E. Gee, and J. L. Roberts, this series, Vol. 124, p. 510.
[11] G. R. Bullock and P. Petrusz (eds.), "Techniques in Immunocytochemistry," Vols. 1, 2, and 3. Academic Press, New York, 1981, 1983, 1985.

Pathological Analysis

Some of the studies will give rise to grossly detectable pathology, although many will not. When phenotypic changes are evident, or when analysis of some of the mice in a line are desired, biopsies may be taken or animals sacrificed for histological analysis. Portions of all organs should be placed in fixative, such as 10% buffered formalin, and processed for paraffin embedding, followed by sectioning and staining for microscopic analysis. In general, another portion of the organs should be snap-frozen and saved for other studies, such as those described above. Microscopic examination of tissue sections from all organs should be performed initially and pathologic consultation obtained to identify lesions. Many further histological studies may be performed, depending on the experimental situation and the lesions identified. The evolution of the pathology with age is often very instructive as to the effect of the introduced transgene. Therefore, once the pathology is identified, a time course of its development should be examined. In studying the insulin–IFN-γ transgenic mouse lines, pancreatic pathology was analyzed histologically at 2-week intervals to study the progressive nature of the inflammatory lesions.[12] This work was important in understanding the mechanism of tissue destruction in these mice. It is critical to evaluate multiple animals since individual phenotypic variations exist.

Other Studies

Many additional forms of analysis are available depending on the aims of an experiment. In general, all of the modalities available to physicians can be modified and performed on the mice, including blood chemistries, hormone levels, physiological tests, and radiographic analysis. Such studies, performed in conjunction with the molecular characterization of the transgenic mice, allow the definition of the transgenic mice as new biological systems in which to study the properties and effects of growth factors *in vivo*.

---

[12] N. Sarvetnick, J. Shizurv, D. Liggitt, L. Martin, B. McIntyre, A. Gregory, T. Parslow, and T. Stewart, *Nature* **346,** 844 (1990).

Section VIII

Cross-Index to Prior Volumes

[50] Previously Published Articles from *Methods in Enzymology* Related to Peptide Growth Factors

Methods Related to Specific Growth Factors

Vol. XXXVII [6]. Measurement of Somatomedin by Cartilage *in Vitro*. W. H. Daughaday, L. S. Phillips, and A. C. Herington.
Vol. XXXVII [7]. Erythropoietin: Assay and Study of Its Mode of Action. E. Goldwasser and M. Gross.
Vol. XXXVII [36]. Preparation of Epidermal Growth Factor. S. Cohen and C. R. Savage, Jr.
Vol. 80 [46]. γ-Subunit of Mouse Submaxillary Gland 7 S Nerve Growth Factor: An Endopeptidase of the Serine Family. K. A. Thomas and R. A. Bradshaw.
Vol. 98 [22]. Receptor-Mediated Endocytosis of Epidermal Growth Factor. H. T. Haigler.
Vol. 98 [25]. Isolation of a Phosphomannosyl Receptor from Bovine Liver Membranes. D. C. Mitchell, G. G. Sahagian, J. J. Distler, R. M. Wagner, and G. W. Jourdian.
Vol. 98 [49]. Spontaneous Transfer of Exogenous Epidermal Growth Factor Receptors into Receptor-Negative Mutant Cells. M. Das, J. Feinman, M. Pittenger, H. Michael, and S. Bishayee.
Vol. 99 [41]. Purification of the Receptor for Epidermal Growth Factor from A-431 Cells: Its Function as a Tyrosyl Kinase. S. Cohen.
Vol. 109 [3]. Assaying Binding of Nerve Growth Factor to Cell Surface Receptors. R. D. Vale and E. M. Shooter.
Vol. 109 [8]. Methods for Studying the Platelet-Derived Growth Factor Receptor. D. F. Bowen-Pope and R. Ross.
Vol. 109 [9]. Binding Assays for Epidermal Growth Factor. G. Carpenter.
Vol. 109 [14]. Affinity Cross-Linking of Receptors for Insulin and the Insulin-like Growth Factors I and II. J. Massagué and M. P. Czech.
Vol. 109 [58]. Purification of Human Platelet-Derived Growth Factor. E. W. Raines and R. Ross.
Vol. 109 [59]. Isolation of Rat Somatomedin. W. H. Daughaday and I. K. Mariz.
Vol. 109 [60]. Purification of Human Insulin-like Growth Factors I and II. P. P. Zumstein and R. E. Humbel.
Vol. 109 [61]. Purification of Somatomedin-C/Insulin-like Growth Factor I. M. E. Svoboda and J. J. Van Wyk.
Vol. 146. Peptide Growth Factors, Part A.

Vol. 147. Peptide Growth Factors, Part B.
Vol. 163 [55]. Fibroblast Chemoattractants. A. E. Postlethwaite and A. H. Kang.
Vol. 163 [56]. Growth Factors Involved in Repair Processes: An Overview. D. Barnes.
Vol. 163 [57]. Lymphocyte- and Macrophage-Derived Growth Factors. S. M. Wahl.
Vol. 163 [58]. Connective Tissue Activating Peptides. C. W. Castor and A. R. Cabral.

Cell Culture Methods Related to Growth Factors

Vol. LVIII [5]. Media and Growth Requirements. R. G. Ham and W. L. McKeehan.
Vol. LVIII [6]. The Growth of Cells in Serum-Free Hormone-Supplemented Media. J. Bottenstein, I. Hayashi, S. Hutchings, H. Masui, J. Mather, D. B. McClure, S. Ohasa, A. Rizzino, G. Sato, G. Serrero, R. Wolfe, and R. Wu.
Vol. LVIII [11]. Measurement of Growth and Viability of Cells in Culture. M. K. Patterson, Jr.
Vol. LVIII [19]. Cell Cycle Analysis by Flow Cytometry. J. W. Gray and P. Coffino.
Vol. LVIII [20]. Cell Synchronization. T. Ashihara and R. Baserga.
Vol. LVIII [22]. Autoradiography. G. H. Stein and R. Yanishevsky.
Vol. LVIII [28]. Cell Fusion. R. H. Kennett.
Vol. 79 [46]. The Culture of Human Tumor Cells in Serum-Free Medium. D. Barnes, J. van der Bosch, H. Masui, K. Miyazaki, and G. Sato.
Vol. 103, Section III. Preparation and Maintenance of Biological Materials.
Vol. 109, Section IV. Preparation of Hormonally Responsive Cells and Cell Hybrids.
Vol. 145 [16]. Bone Cell Cultures. J. Sodek and F. A. Berkman.
Vol. 145 [17]. Hormonal Influences on Bone Cells. T. J. Martin, K. W. Ng, N. C. Partridge, and S. A. Livesey.
Vol. 146. Peptide Growth Factors, Part A.
Vol. 147. Peptide Growth Factors, Part B.
Vol. 185 [43]. Optimizing Cell and Culture Environment for Production of Recombinant Proteins. J. P. Mather.

General Methods

Vol. XXXVI [1]. Theory of Protein–Ligand Interaction. D. Rodbard and H. A. Feldman.

Vol. XXXVII [1]. Statistical Analysis of Radioligand Assay Data. D. Rodbard and G. R. Frazier.
Vol. XXXVII [2]. General Considerations for Radioimmunoassay of Peptide Hormones. D. N. Orth.
Vol. XXXVII [16]. Methods for Assessing Immunologic and Biologic Properties of Iodinated Peptide Hormones. J. Roth.
Vol. 92. Monoclonal Antibodies and General Immunoassay Methods.
Vol. 99 [42]. Detection and Quantification of Phosphotyrosine in Proteins. J. A. Cooper, B. M. Sefton, and T. Hunter.
Vol. 99 [43]. Base Hydrolysis and Amino Acid Analysis for Phosphotyrosine in Proteins. T. M. Martensen and R. L. Levine.
Vol. 101, Section III,B. Introduction of Genes into Mammalian Cells.
Vol. 124 [24]. Assay and Purification of Protein Kinase. C. T. Kitano, M. Go, U. Kikkawa, and Y. Nishizuka.
Vol. 124 [25]. Mixed Micelle Assay of Protein Kinase. C. R. M. Bell, Y. Hannun, and C. Loomis.
Vol. 124 [35]. *In Situ* cDNA : mRNA Hybridization: Development of a Technique to Measure mRNA Levels in Individual Cells. J. Wilcox, C. E. Gee, and J. L. Roberts.
Vol. 141 [36]. Hormone- and Tumor Promoter-Induced Activation or Membrane Association of Protein Kinase C in Intact Cells. T. P. Thomas, R. Gopalakrishna, and W. B. Anderson.
Vol. 141 [37]. Protein Kinase C-Mediated Phosphorylation in Intact Cells. L. A. Witters and P. J. Blackshear.
Vol. 146. Peptide Growth Factors, Part A.
Vol. 147. Peptide Growth Factors, Part B.
Vol. 152. Guide to Molecular Cloning Techniques.
Vol. 173 [44]. Measurement of Transport versus Metabolism in Cultured Cells. R. M. Wohlhueter and P. G. W. Plagemann.
Vol. 182. Guide to Protein Purification.
Vol. 185 [39]. Vectors used for Expression in Mammalian Cells. R. J. Kaufman.
Vol. 185 [41]. Methods for Introducing DNA into Mammalian Cells. W. A. Keown, C. R. Campbell, and R. S. Kucherlapati.
Vol. 195, Section II. Guanine Nucleotide-Dependent Regulatory Proteins.
Vol. 197. Phospholipases.

Author Index

Numbers in parentheses are footnote reference numbers and indicate that an author's work is referred to although the name is not cited in the text.

A

Aaron, B. P., 406
Aaronson, S. A., 97, 98, 129, 272, 273(5), 274(5), 275, 276(15), 277
Abraham, J. A., 96, 97, 98, 99(25), 101(25), 102(25), 104(26), 117, 119(6), 392, 399(6)
Abrahmsén, L., 3, 5, 7(5, 6, 7, 9, 19), 8, 9(6), 10(5, 6), 12(1, 6), 18(6)
Adelaide, J., 97, 126, 127
Aden, D. P., 161
Adnane, J., 137
Agarwal, L. A., 334
Agata, M., 130
Agid, Y., 36, 37(14)
Ahmed, H., 147
Akasaka, K., 203
Akira, S., 522
Alberta, J., 373, 378(5)
Alexander, S., 333
Allen, G., 239
Allen, J. M., 520
Aloe, L., 48
Amalric, F., 98, 104(29), 480, 481, 482(4), 484(4)
Amann, E., 106
Amaral, D. G., 36, 37, 47
Ambartsumayan, N. S., 368
Amsbaugh, C., 290, 293(4)
Amsbaugh, S. C., 274, 275(11)
Anderson, B. L., 414
Anderson, B., 405, 409(2)
Anderson, C. W., 435
Anderson, H., 194
Anderson, J. K., 317, 327
Anderson, S. L., 468
Anderson, S., 7, 19
Anderson, W. B., 531
Andersson, C., 10

Ando, M. E., 73, 498
Ando, Y., 260
Angerer, L. M., 181
Angerer, R. C., 181
Anisowicz, A., 373
Antoniades, H. N., 468
Anzano, M. A., 323, 328
Aozasa, K., 522
Aquino, A. M., 96
Arai, K., 129, 435
Arakawa, T., 98, 109(39), 112(39)
Armes, L. G., 102
Armour, R., 317
Armstrong, D. A., 46
Armstrong, D. M., 35, 36(7), 38, 42(38), 43(38), 44(7), 46(8)
Armstrong, R. M., 318
Asada, A., 416, 417(1), 419(1)
Ashihara, T., 530
Assoian, R. K., 304, 316, 317, 322, 323, 327, 328
Auberger, P., 360
Auburger, G. R., 37
Auerbach, R., 441
Auffrey, C., 73
Auger, K. R., 78, 79, 80, 83(1)
Ausubel, F., 228
Auv, H., 434, 435(16)
Aviv, H., 50, 52(24), 398
Avorson, H. O., 392, 399(8)
Axel, R., 62, 63(9), 74
Ayer-Lelievre, C., 38

B

Baastrup, B., 11, 13(36), 14(36), 15(36)
Bachur, N. R., 342
Badley, J. E., 50, 53(26)

Baggiolini, M., 406
Bagley, C. J., 10
Bahl, O. P., 333
Bailyes, E. M., 511
Baird, A., 97, 98, 99(14, 16), 102, 104(14), 138, 139, 140(7, 8, 9), 144(8, 9), 145, 146, 147(8, 9), 148
Baker, A. E., 194, 196
Bala, R. M., 17
Bald, L. N., 312
Baldin, V., 480, 482(4), 484(4)
Baldino, F., Jr., 438
Baldwin, J. H., 317
Balentien, E., 373, 374, 380
Balland, A., 19
Ballard, F. J., 10
Baltimore, D., 117, 119(5), 494
Bancroft, F. C., 58
Bandtlow, C., 49
Banks, A. R., 98, 109(39), 112(39)
Barany, G., 188
Barbour, R., 96
Barde, Y. A., 35
Barde, Y.-A., 48
Bardwell, L., 373
Bargmann, C. I., 273, 274, 276
Barnes, D. W., 337
Barnes, D., 329, 337, 338(2, 3), 530
Barnes, W. M., 399
Baron, M., 203
Barr, P. J., 98
Barr, P., 138, 139(9), 140(9), 144(9), 147(9)
Barrach, H. J., 326
Barrell, B. G., 99, 102(44)
Barrett, J. N., 35
Barry, J. M., 451, 462(4), 464(4)
Barton, D. E., 171, 373, 380(3)
Bartus, R. T., 36, 46(13), 47(13)
Baserga, R., 530
Bashayee, S., 529
Basilico, C., 97, 125
Baskin, D. G., 31
Bassas, L., 26, 29(2)
Batchelor, P. E., 35, 46(8)
Batchelor, P., 36
Bates, B., 97, 129
Bates, G., 77
Battifora, H., 274, 275(13), 290, 291(2)
Baxter, J. D., 21, 50, 398
Beavo, J. A., 139
Bechtel, P. J., 139

Becker, C., 394
Becker-Andre, M., 49
Beckmann, M. P., 75
Begg, G. S., 405
Behlke, J., 433
Behringer, R. R., 520
Beit-Or, A., 416, 417(8)
Bekesi, E., 147
Bell, C. R. M., 531
Bell, G. I., 175, 176, 185(4)
Bell, G. J., 49
Bell, J. R., 20, 226, 304, 322, 327
Bellot, F., 226
Benitz, W. E., 334
Benjamin, T. L., 80
Bennett, C. D., 194
Bennett, C., 97, 99(15)
Bennett, G. N., 105
Benton, W. D., 21, 367
Bentz, H., 317, 318, 424
Beresford, J. N., 509
Berg, P., 122, 178, 367, 368
Berg, T. M., 336
Berger, S. L., 399
Berger, W. H., 192
Berkman, F. A., 530
Berlot, C. H., 69
Berman, C. H., 19, 390
Berman, C., 49
Bernfield, M., 334
Berry, R. W., 414
Bertics, P. J., 140, 233, 236
Bertonis, J. M., 360, 367(24)
Bessey, O., 508
Betsholtz, C., 73, 74(4), 401
Beug, H., 344
Bezard, J., 362
Bhaumick, B., 17
Bignall, M. C., 406, 407
Bignon, J., 304, 308(10)
Bilich, M., 5, 7(5), 10(5)
Billestrup, N., 347
Birchmeier, W., 447
Birdwell, C. R., 481, 482(7)
Birnbaum, D., 97, 126, 127
Birren, B. W., 181
Bishop, G. A., 50, 53(26)
Bishop, J. M., 233
Bjorklund, A., 36, 37, 38, 42(26, 27), 46(39)
Bjornson-Childs, C., 329
Blacher, R., 74

Blaker, S. M., 35, 46(8)
Blam, S. B., 98
Blanchard, M. G., 358
Blenis, J., 451
Blombäck, B., 11, 13(38), 14(38)
Blumberg, D. D., 50
Blumenthal, D. I., 109, 112(65)
Bodary, S. C., 49
Bodmer, S., 303, 317
Boedtker, H., 55, 293
Boehmer, F.-D., 425, 426, 427(2, 3, 4), 430(5), 433, 436(4)
Bogaard, M. E., 436
Böhlen, P., 97, 98, 99(14), 102(14), 104(14), 138, 139, 145, 148, 447
Bohn, W. H., 192
Bolivar, F., 3
Bollum, F. J., 504
Bolton, A. E., 307, 467
Bond, H. M., 7
Bonewald, L. F., 304, 308(9)
Booyse, F. M., 384
Bordoli, R. S., 102
Bordoni, R., 373, 380(3)
Borrello, I., 171
Bos, J. L., 436
Botchan, M., 176
Bothwell, A. L. M., 117, 119(5)
Bothwell, M. A., 63, 65, 69, 71(12)
Bothwell, M. V., 61, 62(3)
Bothwell, M., 36, 46(13), 47(12, 13), 61, 171
Bottenstein, J., 530
Bouche, G., 480, 482(4), 484(4)
Bouck, N. P., 448
Bowen-Pope, D. F., 72, 468, 529
Boyer, H. W., 3
Brachmann, R., 185, 374
Bradley, J. D., 98, 138, 139(9), 140(9), 144(9), 147(9)
Bradley, J. G., 200, 214, 218(1), 219(1), 220(1), 221(2)
Bradshaw, R. A., 48, 49, 56(14), 529
Brandt, R., 425, 430(5), 433(5)
Branum, E. L., 328, 331
Braun, D., 481, 482(7)
Brayton, C. F., 334
Breckenden, A. M., 344, 345(16)
Brennan, J. E., 324, 424
Brent, R., 228
Breter, H., 426
Breton-Gorius, J., 342

Brewer, M. T., 97, 104(28)
Brewer, P. S., 96
Brewitt, B., 31
Bringman, T. S., 187, 188(8), 192
Brinster, R. L., 520
Brobjer, M., 11, 13(36), 14(36), 15(36)
Brock, M., 508
Bronxmeyer, H., 347
Brookes, S., 129, 137
Brosius, J., 106, 368
Brown, A. M. C., 64
Brown, J. P., 200, 202(3)
Brown, K. D., 105
Brown, S. S., 324
Brown, V. I., 289
Browning, J. L., 260, 261, 262, 266(11), 365, 366(39)
Brownstein, M., 435
Broxmeyer, H. E., 341
Bruce, G., 38, 42(38), 43(38)
Brunner, A. M., 335, 336
Brunner, G., 339
Buchok, J. B., 213
Buck, C. R., 61, 62(3), 171
Budkley, A., 96
Budzik, G. P., 358, 359, 360, 362, 363(21)
Buell, D. N., 502
Bullick, G. R., 524
Burdick, D., 11, 12(39), 13(39), 16(39)
Burgess, A. W., 203, 214
Burgess, W. H., 91, 96, 97(4), 102, 104(4), 108(4), 138, 158, 161(1), 265, 266(18), 272(18)
Burgess, W., 392, 399(7)
Burke, R. L., 74
Burlingame, A. L., 98
Burne, C., 260, 262, 266(11), 267, 271
Burnier, J. P., 11, 12(39), 13(39), 16(39)
Burow, S. A., 259
Burstein, Y., 11, 13(37)
Burwen, S. J., 480
Butterfield, C., 91, 395, 446, 448(13)
Butters, T. D., 251, 255(5)
Buxser, S., 69
Bye, J., 21

C

Cabral, A. R., 411, 412(14), 530
Cachianes, G., 392, 399(13), 401(13), 404(13)

Callahan, M. F., 401
Calvert, I., 147
Campbell, C. R., 531
Campbell, I. D., 203
Campen, C., 348, 349(16)
Canalis, E. M., 510
Canalis, E., 318, 416
Cantley, L. C., 78, 79, 80, 83(1)
Cantley, L., 78, 79(3, 4, 10), 82(10), 83(9), 267, 269, 271(24)
Cao, H., 500
Capon, D. J., 99, 101(43)
Caput, D. F., 176
Caput, D., 74, 98, 104(29)
Carlson, G. M., 139
Carlsson-Skwirut, C., 10
Carman, F. R., Jr., 206
Carpenter, C. D., 240, 241(25)
Carpenter, C. L., 79
Carpenter, G., 209, 214, 240, 256, 258(11), 263, 500, 529
Carr, S. A., 102
Carrillo, P., 317, 323
Carrington, J. L., 328, 424
Carswell, E. A., 468
Carter, P., 11, 12(39), 13(39, 40), 16
Castleman, K. R., 487
Castor, C. W., 405, 406, 407, 408, 409(1, 2, 3), 410, 411, 412(14), 530
Cate, R. L., 261, 360, 363, 364, 365(27), 367(24), 368, 369(36)
Cathala, G., 21, 50, 398
Catlin, E. A., 359, 363(20)
Cecchini, M. A., 119, 120(8), 121(8)
Centrella, M., 318
Cesar, L. B., 96
Cesarini, G., 9
Chahwala, S. B., 267, 269, 271(24)
Chalon, P., 98, 104(29)
Chambaz, E. M., 171
Chamberlain, S. H., 98
Chambers, T. J., 504
Chambon, P., 243
Chamness, G. C., 290, 291(3), 294(3), 297(3)
Chandler, C. E., 62
Channing, C., 347
Chao, M. V., 61, 62(5), 71(5), 72(5)
Chao, M., 61
Chasin, L., 62, 63(9)
Chatelier, R., 354

Chavassieux, P. M., 509
Chawla, R. K., 380
Chegini, N., 206
Cheietz, S., 317, 327
Chen, C., 231
Chen, E. Y., 20, 73, 187, 226, 303, 304, 317, 322, 327, 336(7), 436, 498
Chen, E., 19, 20, 22, 25, 226, 228(4), 273
Chen, K. S., 36, 46
Chen, T. L., 510
Chen, W. C., 233
Chen, W. S., 240, 241(25), 495
Cheng, J., 110, 115(67), 394
Chenu, C., 509
Chernausek, S. D., 17
Chesselet, M.-F., 438
Chesterman, C. N., 405
Cheung, A., 360, 367(24)
Chevray, P. Y., 274, 276(9)
Chiado-Piat, L., 501
Childs, C. B., 316
Childs, J., 98
Chinkers, M., 194
Chirgwin, J. M., 50, 292, 366, 398
Chiu, I. M., 392, 399(7)
Chiu, M. L., 96
Chomczynski, P., 433
Chow, E. P., 260, 261, 262, 264(6), 266(6, 11), 267, 269, 271, 360, 364, 365, 367(24), 366(39), 369(36)
Chu, C., 74
Chui, D. H. K., 344, 345(16)
Church, G. M., 366
Cigarrao, F., 359, 362(22), 363(22)
Cinti, D. L., 251
Claesson, G., 11, 13(38), 14(38)
Claesson-Welsh, L., 73, 74(4), 75, 76
Clark, G. M., 275, 290, 291(3), 294(3), 297(3)
Clarke, B. J., 344, 345(16)
Clarke, N. G., 82
Claude, P., 36, 47(12)
Clauser, E., 20
Clegg, D. O., 49
Cline, M. J., 274, 275(13), 290, 291(2)
Clowes, A., 334
Cochet, C., 171, 233
Cockley, K. D., 192
Coffey, R. J., 303, 327, 328, 329(20), 336(7)
Coffey, R. J., Jr., 96, 97(7)
Coffino, P., 530

Coghlan, J. P., 175
Cohen, S., 194, 200, 209, 214, 233, 240, 251, 256, 258(11), 259, 261, 263, 264(8), 266(12), 529
Collins, C., 19, 20, 22, 25, 226, 228(4)
Comoglio, P. M., 501
Conner, B. J., 436
Connolly, D. T., 323, 404
Consigli, S. A., 482
Constantini, F., 519
Contreves, 360
Cook, C. L., 206
Cooke, R. M., 203
Cooper, C. S., 501
Cooper, J. A., 141, 142(13), 200, 233, 237, 336, 494, 496, 499(1), 531
Cooper, R. A., 502
Cooper, S., 341
Cooper, T. G., 108
Copeland, T. D., 147
Corese, R., 197
Corp, E., 31
Cortese, R., 7, 9
Cotran, R. S., 441
Coughlin, J. P., 359, 362(21, 22), 363(21, 22)
Coughlin, S. R., 79
Coulier, F., 97, 127
Coulson, A. R., 73, 99, 102(44), 390, 436
Coulson, J. D., 334
Courtois, Y., 147, 469
Cousens, L. C., 138, 139(9), 140(9), 144(9), 147(9)
Cousens, L. S., 98, 138, 171
Coussens, L., 20, 21, 226, 273
Cowing, D., 272, 278
Cowley, G., 192
Cox, K. H., 181
Cozzari, C., 48
Crabb, J. W., 102, 158, 159(2), 162(2)
Crane, R. K., 251
Crawford, R. J., 175
Crea, R., 3
Crego, B., 214
Creutz, C. E., 260
Crimmins, D. L., 238
Crkenjakov, R., 55
Crochet, J., 328
Cronin, M. T., 234
Cross, S. H., 126
Crumpton, M. J., 260

Cuatrecasas, P., 17, 18
Cuello, A. C., 36
Culpepper, A., 126
Cunningham, E. T., Jr., 347
Curatola, M., 97
Curray, J., 395
Cuthbertson, R. A., 521
Czech, M. P., 17, 529

D

D'Agostino, J., 347
D'Amore, P., 97, 418
D'Souza, S. M., 502
Dahl, H. H., 74
Daikuhara, Y., 416, 417(1), 419(1)
Daniel, T. O., 73, 498, 500
Danielpour, D., 309, 314(20), 315(20)
Darak, K., 201
Darling, T. L. J., 48
Darlington, B. G., 129
Dart, L. L., 309, 314(20), 315(20)
Das, M., 529
Dasch, J. R., 311, 318
Dasch, J., 323
Daughaday, W. H., 529
Davidson, D. A., 31
Davidson, J. M., 96
Davidson, S., 91, 102
Davis, J. G., 289
Davis, K. L., 36
Davis, R. W., 21, 367
Davis, R., 21
Dawid, I., 243
Dawson, R. M. C., 82
de Crombrugge, B., 327
de Crombrugghe, B., 186, 303, 314(2), 317, 323(11)
de la Sale, H., 19
De Lapeyriere, O., 126, 127
de Lapeyriere, O., 97
De Larco, J. E., 328, 331
de Martin, R., 303, 317
De Pablo, F., 26, 29(2)
De St. Groth, S. F., 189
De, B. K., 263, 266(12)
Decker, S. J., 272, 278
Dedman, J. R., 260
Defeo-Jones, D., 196, 197

Deisenhoer, J., 7
deKretser, D., 347
DeLange, R. J., 365, 366(37)
DeLarco, J. E., 206, 208, 263
DeLarco, J., 214
DeLean, A., 313
DeLeon, D., 181
Delli Bovi, P., 97, 125
Delmas, P. D., 509
DeLustro, B. M., 192
DeLustro, B., 202
Delwart, E., 99, 101(43)
Demetriou, A. A., 96
Denhardt, D. T., 21, 73
Denoroy, L., 97, 99(14, 16), 102(14), 104(14), 139, 145, 148
Dente, L., 9
Dernick, R., 414
Derynck, R., 185, 187, 188(8), 192, 200, 303, 304, 317, 322, 327, 336(7), 373, 374, 378(5), 380
DesJardins, E. J., 368
Deteresa, R. M., 38, 42(38), 43(38)
Deuel, T. F., 401, 468
Deuel, T. M., 490
Devey, K., 95
Di Fiore, P. P., 275, 276(15), 277
Di Marco, 276
Di Renzo, M. F., 501
Diamond, D., 293
Dickson, C., 97, 129, 137
Diegelmann, R. R., 507, 508(11), 510(11)
Dietrich, J. W., 510
Dignam, J. D., 248
Dillard, P. J., 303, 327
Dimery, I. W., 342
Dina, D., 74
Dinele, K., 317
DiSalvo, J., 97, 99(15)
DiSorbo, D., 149, 161
Distler, J. J., 529
Dobashi, K., 289
Donahoe, P. K., 358, 359, 360, 361(32), 362, 363, 364, 365(27), 367(24), 368, 369(36)
Donoghue, D., 480
Doolittle, R. E., 214
Doolittle, W. F., 107
Dorsa, D. M., 31
Douglas, I., 359, 363(17), 364(17)
Douillard, J. Y., 151

Downes, C. P., 78, 79, 82(10)
Downward, J., 21, 233, 236
Drapeau, G. R., 236
Drebin, J. A., 272, 278, 281, 282, 283(6), 285, 286(4), 287(7)
Drexler, H., 386, 387(7), 389(7), 391, 392(5), 399(5)
Dreyer, W. J., 237
Drohan, W. N., 392, 399(7)
Drohan, W., 171
Drucker, B., 78, 79(3)
Drust, D. S., 260
Dudeck, B. S., 435
Dull, T. J., 19, 20, 21, 49, 175, 225, 226, 390
Dunau, M. L., 455
Dunn, A., 521
Dunnett, S. B., 36, 42(26)
Dusemund, B., 326
Dvorak, A. M., 404
Dvorak, H. F., 404
Dynan, W. S., 242, 243

E

Earp, H. S., 17, 185, 235
Eaton, D. H., 304, 322, 327
Eaton, D. L., 99, 101(43), 231
Eaton, D., 231
Ebendahl, T., 38
Ebendal, T., 37, 49
Ebert, M. H., 221
Ebina, Y., 20
Eddy, R., 401
Edelman, G. M., 281
Edery, M., 20
Edgar, D., 35
Edwards, P. A. W., 342
Edwards, R. H., 522
Edwards, S. B., 27, 28(7)
Eilam, Y., 416, 417(8)
Eisen, H. N., 83, 495
Eisen, H., 494
Eisenberg, S. P., 98
Ek, B., 73, 74(4), 75, 334, 384, 468
Elder, J. H., 333
Elingsworth, L. R., 424
Ellingsworth, L. R., 318, 323, 324
Ellingsworth, L., 311
Ellis, L., 20

Elmblad, A., 8, 9, 10(29)
Enfors, S.-O., 5, 7(6), 9(6), 10(6), 12(6), 18(6)
Engel, J. D., 399
Engstroem, U., 426, 433
Engström, A., 15
Engvall, E., 312, 313(28)
Enoki, Y., 342
Eppstein, D. A., 203
Epstein, J., 368
Erikson, R. L., 78
Eriksson, A., 73, 74(4), 76
Eriksson, U., 425
Erlich, H. A., 435
Ernst, T., 337, 338(2)
Esch, F., 97, 98, 99(14, 16), 102, 104(14), 138, 139, 145, 148
Escobedo, J. A., 73, 79, 498
Estrada, C., 153
Estratiadis, A., 21
Etherington, D. J., 304, 308(8)
Eto, Y., 341, 348, 349(17)
Etoh, Y., 349
Etzold, G., 425, 427(4), 433(4), 436(4)
Evans, R. M., 240, 241(25), 341
Ewing, W. R., 95
Eychenne, B., 362
Ezban, M., 74

F

Fabricant, R. N., 263
Fahnestock, M., 48, 49
Fairclough, P., 251
Falbel, T., 365, 366(39)
Falcone, D. J., 334
Fallat, M. E., 359, 360, 362(9, 25)
Faloona, F. A., 59
Falquerho, L., 360
Fanger, B. O., 304
Fanning, P., 98, 117, 119(2), 123(2), 124(2, 3)
Farber, N. M., 360, 367(24)
Farley, B. A., 342, 343(10), 344(10)
Farris, J., 138, 139(9), 140(9), 144(9), 147(9)
Faulders, B., 75, 334, 384
Fava, R. A., 261, 263(7)
Favalaro, J., 74
Fayein, N., 469
Feder, J., 323, 404

Feige, J. J., 138, 139(9), 140(9), 144(9), 145, 147(9)
Feige, J.-J., 138, 139(8), 140(8), 144(8), 145, 146(18), 147(8)
Feinman, J., 529
Feldman, H. A., 530
Felsted, R. L., 342
Fen, A. L., 506
Fendly, B. M., 289, 312
Feng, D. F., 214
Fernandez-Pol, J. A., 256
Ferracini, R., 501
Ferrara, N., 96, 97(1), 104(1), 108(1), 138, 391, 392, 393, 395(12), 396(12), 397(12), 399(12, 13), 401(13), 403(12), 404
Fett, J. W., 395
Ficher, W., 38
Fiddes, J. C., 96, 97, 98, 99(25), 101(25), 102(25), 104(26), 109, 112(65), 117, 119(6)
Fiddes, J., 392, 399(6)
Figari, I. S., 312
Figlewicz, S. P., 31
Finch, P. W., 97, 129
Findeli, A., 19
Finkelstein, R., 272, 278
Fiol, C. J., 239, 241
Fire, A., 248
Fischer, W., 36, 46(39)
Fisher, B., 290
Fisher, C., 192
Fisher, D., 49, 56(14)
Fisher, E. R., 290
Fisher, R. A., 360, 367(24)
Flaggs, G., 373, 380(3)
Flanders, K. C., 309, 314(20), 315(20), 318, 328, 331, 334(31), 336(31), 424
Flier, J. S., 18
Florkiewicz, R. Z., 98, 104(35)
Flügge, U. I., 18
Foeller, C., 261
Fok, F., 424
Folkman, J., 91, 93, 95, 96, 97(2), 108(2), 117, 119, 123(7), 148, 391, 392(1), 394, 395, 404(18), 440, 443, 446, 447, 448(13)
Fong, N. M., 176
Fong, N., 175
Fonnum, F., 46
Fontana, A., 303, 317
Fordis, C. M., 178

Forest, M. G., 362
Forsberg, G., 11, 13(36), 14(36), 15, 16(44)
Foster, T., 5
Foulkes, J. G., 303, 327
Fox, C. F., 259
Fox, G. M., 98, 109(39), 112(39)
Frackelton, A. R., 83
Frackelton, A. R., Jr., 495, 498
Fraker, P. J., 467
Francis, G. L., 10
Francke, U., 19, 20, 73, 171, 226, 228(4), 273, 373, 380(3), 498
Franklin, G., 201
Fraser, B. A., 102
Fràter-Schroder, M., 447
Frazier, G. R., 531
Frazier, W. A., 48
Frei, K., 317
Freidman, P. L., 49
Frelinger, J. A., 50, 53(26)
Frencke, U., 22, 25
Freshney, R. I., 180
Freudenstein, C., 344
Frey, A. Z., 260, 360, 367(24)
Frey, A., 262, 266(11)
Fried, V. A., 73, 138, 171, 498
Friedman, G. B., 80
Friedman, J., 97, 98(25), 99(25), 101(25), 102(25), 104(26), 117, 119(6), 392, 399(6)
Friedman, P. L., 38
Friedman, R., 119, 123(7)
Fritsch, E. F., 57, 59(32), 73, 107, 121, 228, 230(6), 367, 519
Fritze, L., 334
Froesch, E. R., 17
Frolik, C. A., 304, 323, 328, 331
Frost, E., 178
Fryburg, K., 138, 139(9), 140(9), 144(9), 147(9)
Fryklund, L., 9, 10(29), 14, 15(42)
Fryling, C., 214
Fu, J., 95
Fuchihata, H., 424
Fujii, D. K., 110, 115(67), 392, 393(15), 394
Fujii, D., 392, 393(14)
Fujimoto, E. K., 255, 511
Fujita-Yamaguchi, Y., 18, 19, 20, 22, 25, 226, 228(4)
Fujiwara, H., 129, 130
Fukui, T., 80

Fukui, Y., 78
Funa, K., 436
Fuquay, J. I., 337
Furusawa, M., 125, 129, 130

G

Gage, F. H., 35, 36, 37, 38, 42, 43(38), 46, 47
Gage, G. H., 35, 36(7), 38(7), 44(7)
Galfrè, G., 151, 458
Gallagher, J. A., 509
Galli, J., 175
Galli, S. J., 404
Galluppi, G. R., 317
Galyean, R., 194
Garbaz, M., 342
Garcea, R. L., 78, 79(4)
Gardiner, G. R., 130
Garofalo, L., 36
Garramone, A. J., 175
Garrigori, O., 362
Garsky, V. M., 196
Gartner, F. H., 255, 511
Garwin, J., 365, 366(39)
Garycar, J., 317
Gas, N., 480, 482(4), 484(4)
Gash, D. J., 360, 367(24)
Gash, D. M., 36, 46(13), 47(13)
Gatenbeck, S., 5, 6(18), 7(8), 9(18), 10(8)
Gates, R., 259
Gaudray, P., 137
Gaugitsch, H., 303
Gee, C. E., 524, 531
Gelinas, R. E., 520
Gelmann, E. P., 290
Gentry, L. E., 335, 336
Gerhart, J. C., 199
Ghosh, S., 97
Ghosh-Dastidar, P., 259
Gibbs, L., 36, 47(12)
Gifford, A., 123
Gilbert, W., 366, 367(40), 390
Gill, G. N., 140, 233, 234, 236, 237, 239(17), 240, 241, 495
Gimbrone, M. A., Jr., 440, 441
Gimenez-Gallego, G., 97, 99(15)
Giordano, S., 501
Girbau, M., 26, 29(2)
Gitschier, J., 99, 101(43)

Givol, D., 171, 226
Glazer, A. N., 365, 366(37)
Glenney, J. R., Jr., 495
Glenney, J., 260
Glover, J. S., 467
Go, M., 531
Goding, J. W., 66, 312, 401
Godolphin, W., 274, 275(12), 289, 290, 291(1), 294(1), 297(1)
Godpodarowicz, D., 394
Godsave, S. F., 129
Goeddel, D. V., 187, 192, 304, 317, 322, 327, 392, 399(13), 401(13), 404(13)
Goeke, N. M., 255, 511
Goetz, I. E., 153
Goff, S. P., 123
Goldfarb, M., 97, 126, 127, 129
Goldine, I. D., 20
Goldman, D., 221, 455
Goldsmith, P. C., 392, 393(15)
Goldsmith, P., 392, 393(14)
Goldwasser, E., 529
Gonda, T. J., 521
Gonzalez-Carvajal, M., 49
Gonzalez-Carvajarl, M., 38
Gonzalez-Manchon, C., 347
Gopalakrishna, R., 531
Gordon, D. J., 380
Gordon, W., 347
Gorga, F. R., 80
Gorka, J., 238
Gorman, C., 231, 247, 248(15), 341, 402
Gospodarowicz, D. G., 481, 482(7)
Gospodarowicz, D., 91, 96, 97, 98, 99(14, 16, 25), 101(25), 102(14, 25), 104(1, 14, 26), 108, 110, 115(67), 117, 119(6), 129, 138, 139, 145, 148, 149, 151, 153(4), 391, 392, 393, 399(6), 404(4), 469, 480, 481
Götz, F., 7
Gotz, R., 49
Gould, K. L., 139
Goustin, A. S., 328, 391, 392(3)
Gowen, M., 502
Graf, L., 20
Graf, T., 272, 344
Graham, F. L., 121
Graham, M. K., 31
Grant, D. S., 446
Graves, D. J., 139
Graves, D. T., 468

Gravstock, P. M., 10
Gray, A., 19, 20, 21, 25, 49, 175, 226, 228(4), 231, 273, 390
Gray, E., 401
Gray, J. W., 530
Gray, M. E., 263
Graycar, J. L., 303, 327, 336(7), 373, 378(5)
Greco-O'brine, B., 260, 264(6), 266(6), 267(6)
Green, S. H., 69, 70(17)
Greene, L. A., 48, 69, 70(17)
Greene, M. I., 272, 278, 281, 282(6), 283(6), 285, 286(4), 289
Greenwald, F. C., 431
Greenwood, F. C., 304, 467, 490
Gregory, H., 203, 214
Griffin, P. R., 374
Grob, P. M., 69
Gross, E., 10, 13(35), 14(35)
Gross, J. L., 404
Gross, M., 529
Grosse, R., 425, 426, 427(2, 3, 4), 430(5), 433, 436
Grunberger, G., 17
Grundstroem, T., 243
Grunfeld, C., 20, 226
Gubler, K., 435
Guillemin, R., 97, 99(14, 16), 102, 104(14), 138, 139, 145, 148
Gundersen, R. W., 35
Gunnersson, K., 467
Gupta, S. K., 342
Guss, B., 5, 6(18), 7, 9(18)
Gusterson, B., 192
Gutterman, J. U., 192
Guyton, J. R., 334
Gworek, S. C., 510
Gyllenstein, U. B., 435

H

Haashi, K., 60
Haber, M. T., 80
Haendler, B., 303
Hagg, T., 42, 47
Hagiwara, K., 386, 387(7), 389(7), 391, 392(5), 399(5)
Haigler, H. T., 263, 265, 266(18), 272(15, 18), 529

Hains, J., 404
Halaban, R., 97
Hall, K., 10, 14, 15
Hallewell, R. A., 98
Ham, R. G., 338, 530
Hamamoto, S., 416
Hamer, D. H., 176
Hamilton, K. K., 259
Hammacher, A., 75, 334, 384, 467
Hammarberg, B., 9
Hammer, R. E., 520
Hamuro, J., 289
Han, J. H., 374, 380, 387
Han, K. H., 204
Hanafusa, H., 78, 498
Hanahan, D., 522
Hanahan, O. H., 435
Hanauske, A. R., 213
Hanks, S. K., 317
Hannink, M., 480
Hannun, Y., 531
Hansen, J., 74
Hansen, P., 327
Hansen, S. E., 522
Hansen, T. P., 303
Haour, F., 350
Harakas, N. K., 323
Hardison, R. C., 21
Hare, D. L., 119, 120(8), 121(8)
Harkins, R. N., 73, 498
Haroutunian, V., 36
Harris, A. L., 79
Harris, C. C., 304, 308(13), 328, 329(21), 331, 336(21)
Harris, M. E., 405
Hartikka, J., 35, 36, 37(5)
Hartmanis, M., 10, 11, 13(36), 14(36), 15, 16(44)
Hartog, K., 74
Harvey, V. S., 404
Hascall, V. C., 326
Hasegawa, Y., 348, 349(17)
Hash, J. H., 214
Hashimoto, M., 349
Hass, P. E., 231
Hatch, J., 109, 112(65)
Hattori, R., 259
Hattori, Y., 138
Haudenschild, C. C., 394, 404(18)
Haudenschild, C., 93

Hauschka, S. D., 469
Hauser, S. D., 404
Hawkins, P. T., 79
Hayashi, A., 358, 362(3)
Hayashi, I., 530
Hayashi, M., 359
Hayflick, J. S., 21
Hayflick, J., 231
Hayman, M. J., 344, 449
Hazan, R., 290
Heath, J. K., 129
Heath, W. F., 203
Hebda, P. A., 96
Hedén, L.-O., 9, 10(29)
Hedger, M., 347
Heffernan, P. A., 496, 497(12), 499(12)
Hefti, F., 35, 36, 37(5), 38(16), 44(16), 47(5)
Heilmeyer, M. G., Jr., 240
Heine, U. I., 318
Heisermann, G. J., 237, 239(17), 241
Heldin, C. H., 332, 333, 334(32, 38), 336
Heldin, C.-H., 72, 73, 74(4), 75, 76, 334, 383, 384, 385(1, 5), 386, 387(7), 389(7), 391, 392(5), 399(5), 401, 425, 427(4), 433(4), 436, 467, 468
Hellman, U., 5, 7(19), 332, 334(32), 386, 387(7), 389(7), 391, 392(5), 399(5), 467
Hellmann, U., 425, 427(4), 433(4), 436(4)
Hellweg, S., 37
Helmrich, A., 337, 338(3)
Hempstead, B. L., 61, 62(5), 71(5), 72(5)
Henderson, L. E., 147
Hendren, W. H., 358, 360, 361(32), 362(32)
Hendrickson, A., 27, 28(7)
Hendrickson, J. E., 96, 97(7)
Henri, A., 342
Henrichson, C., 8
Henschen, A., 236, 405
Henzel, W. J., 138, 392, 395(12), 396(12), 397(12), 398, 399(12), 403(12)
Henzel, W., 19, 20, 22, 25, 226, 228(4)
Herington, A. C., 529
Herkenham, M., 26, 28(6)
Herman, B., 185
Hermanson, G. R., 255
Hermanson, G. T., 511
Herrera, R., 20, 226
Herschman, H. R., 181
Hersh, L. B., 38, 42(38), 43(38)
Herzenberg, L. A., 171

Herzenberg, L., 61, 62(2), 311
Hessel, B., 11, 13(38), 14(38)
Hession, C., 260, 261, 262, 266(11), 360, 367(24)
Heukeshoven, J., 414
Heumann, R., 37, 49
Hewick, R. M., 97, 237
Heynecker, H. L., 3
Higgins, G. A., 46
Hill, F., 97, 99(14, 16), 102(14), 104(14), 139, 145, 148
Hill, J. M., 26, 33(1), 34(3), 35(1, 3)
Hill, K. E., 96
Hirai, H., 125
Hirai, R., 424
Hiraki, Y., 416, 417(6, 7), 419(6), 424
Hiranao, T., 522
Hirohashi, S., 137
Hirose, T., 3
Hirota, T., 137
Hjelmeland, L. M., 511
Hjerrild, K. A., 97, 98(25), 99(25), 101(25), 102(25), 117, 119(6)
Hofer, E., 303, 317
Hofer-Warbinek, R., 303, 317
Hoffman, B. J., 435
Hoffman, R. M., 363
Hoffman, T., 151
Hoffmann-Posorske, E., 240
Hofmann, C. A., 175, 185(4)
Hogan, B., 519
Hogg, D., 11, 13(38), 14(38)
Hogue-Angeletti, R. A., 48
Holets, V. R., 38, 49
Hollenberg, M. D., 17, 426, 433
Holley, R. W., 317, 331
Hollingshead, P., 99, 101(43), 231
Holmgren, A., 5, 7(5, 8), 8, 9(12), 10
Holmgren, E., 5, 7(5), 9, 10(5, 29)
Holt, A. M., 405
Holt, J. A., 274, 275(12), 289, 290, 291(1), 294(1), 297(1)
Holt, J. C., 405
Honegger, A. M., 225
Hong, Y.-M., 260
Hood, L. E., 18, 214, 237, 365
Hoogerbrugge, J. W., 518
Hoope, J., 401
Horio, T., 516
Horiuchi, T., 129, 130(23)

Hortsch, M., 203
Hosang, M., 61, 62, 69(4)
Hoshi, H., 149, 161
Hossler, P. A., 406, 407, 411
Hou, J., 149, 161
Houesche, F., 342
Housey, G. M., 62
Howard, B. H., 178
Howe, C. C., 161
Howell, K. E., 7
Howk, R., 171, 226, 392, 399(7)
Hsiung, N., 175
Hsu, C., 61, 62(2), 171
Hsu, Y.-M., 451, 455(3), 462(4), 464(3, 4)
Hsueh, A., 347
Hu, X., 97, 129
Huang, J. S., 401, 468, 469
Huang, K. P., 147
Huang, K.-S., 260, 262, 266(11), 267, 271
Huang, M. C., 273
Huang, S. S., 468, 469
Huber, R., 7
Huberman, E., 401
Hudson, L. G., 140, 236
Hudson, P. L., 359, 363(17), 364(17), 368
Hudson, P., 358
Hudziak, R. M., 276, 289
Hughes, R. C., 251, 255(5)
Hultberg, H., 5, 7(5), 10(5)
Humbel, R. E., 529
Humphrey, J. G., 189, 312
Hung, M. C., 274, 276(9, 10)
Hunkapillar, M. W., 214
Hunkapiller, M. G. V., 365
Hunkapiller, M. W., 18, 237
Hunter, T., 139, 141, 142(13), 233, 237, 494, 496, 499(1), 531
Hunter, W. M., 304, 307, 431, 467, 490
Hurni, W. M., 194
Hutchings, S., 530
Hutson, H. M., 360, 362(26)
Hutson, J. M., 359, 360, 362
Hutton, J. J., 504
Huynh, T., 21, 367

I

Ibbotson, K. J., 502
Ichihara, A., 332

Igarashi, K., 97, 98, 104(27), 109, 110(66), 112(66)
Igarashi, M., 348, 349(17)
Ihara, S., 494, 495(3)
Ihrie, E. J., 502
Iio, M., 374
Ikawa, H., 359, 362
Ikeda, T., 317, 322, 323(9), 327
Ikeda, Y., 462, 463(10)
Ikuta, S., 148, 150(5), 151(5), 152(5), 153(5)
Imamura, S., 260
Imamura, T., 134
Imanishi-Kari, T., 117, 119(5)
Imawari, M., 125
Inagami, T., 200, 214
Inenaga, D., 311
Ingber, D., 95
Ingraham, H. A., 241
Inooka, H., 203
Inoue, H., 416, 424
Inoué, S., 480
Isackson, P. J., 49, 56(14)
Isacson, O., 42, 46, 47
Ishai-Michaeli, R., 119, 123(7)
Ishibashi, S., 384
Ishii, S., 242
Ishikawa, F., 386, 387(7), 389(7), 391, 392(5), 399(5)
Itakura, K., 3, 436
Ito, K., 109, 110(66), 112(66), 358, 360, 361(32), 362(32)
Ives, R. A., 344, 345(16)
Iwamoto, M., 417, 424
Iwane, M., 97, 98, 104(27), 109, 110(66), 112(66)
Iwata, K. K., 303, 327

Jakewiew, S. B., 303
Jakobovits, A., 129, 135(20)
Jakowlew, S., 327
Jansson, B., 5, 8, 9(12)
Jarnagin, K., 20
Jarrett, J. A., 304, 317, 322, 327
Jaye, M., 19, 392, 399(7)
Jeanny, J. C., 469
Jeanteur, P., 137
Jendrisak, J., 399
Jinno, Y., 242
Johansson, A., 401
Johansson, B. G., 305
Johnson, A. C., 242
Johnson, C. M., 102
Johnson, D. E., 138, 171
Johnson, D., 61, 171
Johnson, E. M., 35, 69
Johnson, G. L., 69
Johnson, M. D., 62, 263
Johnson, M. S., 214
Johnson, P. F., 242
Johnsson, A., 334, 384, 401, 467
Jones, A. L., 480
Jones, K. A., 242
Jones, L. A., 274, 275(12), 289, 290, 291(1), 294(1), 297(1)
Jones, T. A., 7, 8
Jörnvall, H., 5, 7(5), 10, 11, 13(36), 14(36), 15(36)
Joseph-Silverstein, J., 96, 392, 482
Josephson, S., 5, 7(5, 6, 8), 8(12), 9, 10(5, 6, 8, 29), 12(6), 18(6)
Josso, N., 358, 359, 362
Joullie, M. M., 95
Jourdian, G. W., 529
Jullien, P., 304, 308(7), 328, 330(11, 13, 14), 332(13, 14), 336

J

Jackson, J. F., 387
Jacobs, J. W., 48
Jacobs, J., 97, 418
Jacobs, S. C., 102
Jacobs, S., 17, 18, 19, 20, 22, 25, 226, 228(4)
Jacobson, A., 50, 52(25), 398
Jaehnchen, L., 426
Jaffe, E. A., 394
Jagschies, G., 427

K

Kadonaga, J. T., 242
Kageyama, R., 242, 250(5, 6, 11)
Kaghad, M., 98, 104(29)
Kahn, C. R., 259
Kahn, P., 272
Kahn, R. C., 264, 267(17)
Kajiyama, G., 137

Kakinuma, M., 125, 136, 137(35)
Kakunaga, T., 260
Kamagata, S., 358, 359, 360, 362(9, 25)
Kamata, N., 273, 275(6)
Kamath, S. A., 251
Kamps, M. P., 494, 495, 496, 497(4, 5, 11), 499(11), 500, 501(4, 11)
Kan, M., 149, 161
Kan, Y. W., 20
Kanetly, H., 203
Kang, A. H., 530
Kannag, R., 260
Kannourakis, G., 521
Kapeller, R., 78, 79(2)
Kaplan, D. R., 78, 79(4), 83(9)
Karasik, A., 264, 267(17)
Karey, K. P., 10
Karin, M., 21, 50, 398
Karnovsky, M., 334
Kashles, O., 171
Kathuria, S., 18, 19, 20, 22, 25, 226, 228(4)
Kato, Y., 416, 417, 419(1, 6), 424
Katoh, O., 136
Kaufman, R. J., 368, 531
Kawai, T., 129, 130(23)
Kawaichi, M., 435
Kazakoff, P., 469
Kearney, J. F., 189, 311
Keck, P. J., 404
Keeble, W. W., 96, 97(7)
Keeler, M., 78, 79(10), 82(10)
Keith, D. E., 274, 275(12), 289, 290, 291(1), 294(1), 297(1)
Keller, T., 78, 79(10), 82(10)
Kelly, R. B., 62
Kelson, A., 521
Kennedy, B., 468
Kennett, R. H., 530
Kenney, J. S., 151
Kerlavage, A. R., 140
Keski-Oja, J., 329, 332(25), 333(25)
Keyt, B., 99, 101(43)
Khan, F. R., 147
Kiefer, D. M., 363
Kies, M. W., 321
Kiess, W., 26, 33(1), 35(1)
Kikkawa, U., 531
Kiliani, P. L., 511
Kim, J. W., 80
Kimata, K., 326

Kimura, J. H., 326
Kimura, T., 516
King, C. R., 171, 272, 273(5), 274, 275, 276(15), 277, 290, 293(4)
King, C. S., 200
King, L. E., 259
King, L., 256, 258(11)
Kingston, I. B., 19
Kingston, R. E., 228
Kinoshita, M., 416
Kipreos, D. T., 498
Kirsch, D. R., 95
Kirschmeier, P. T., 62
Kishimoto, T., 522
Kissane, J. M., 509
Kitano, C. T., 531
Kitazawa, T., 332
Klagsbrun, M., 91, 95, 96, 97, 98, 102, 104(29), 108, 117, 119, 123, 124(2, 3), 145, 148, 153, 391, 392(1), 395, 418, 446, 447(12), 448(13)
Kleinman, H. K., 446
Klenk, D. C., 255, 511
Klepper, R., 97, 99(14), 102(14), 104(14), 139, 145, 148
Klingbeil, C. K., 96
Klintworth, G. K., 440, 521
Knight, M. B., 323
Knof, P. D., 36
Knott, T. J., 401
Knowles, B. B., 161
Knusel, B., 35, 36(5), 37(5)
Knutson, V. P., 17
Koda, T., 125, 136, 137(35)
Koewn, W. A., 531
Koh, S., 46
Kohanski, R. A., 17
Kohler, G., 416, 417(9)
Kohli, V., 19
Kohr, W., 185
Kohzuki, H., 342
Koide, S. S., 359, 363(18)
Koike, T., 417
Kojima, I., 347, 348(15), 349(15)
Kokai, Y., 289
Komoriya, A., 203, 317
Kondaiah, P., 303, 327
Kondo, A., 129
Kondo, S., 349
Koprowski, H., 61, 62(3)

Kordower, J. H., 36, 46(13), 47(13)
Kornbluth, S., 498
Korsching, S., 37, 49
Korte, H., 240
Kovacina, K. S., 336
Kozak, M., 23
Kraft, R., 425, 427(4), 433(4), 436(4)
Kraus, M. H., 272, 273(5), 274, 275, 276(15), 290, 293(4)
Krause, D., 153
Krebs, E. G., 139
Kris, R., 192
Krivi, G., 404
Krohn, R. I., 255, 511
Kromer, L. F., 36, 38(18), 44(18)
Krycève-Martinerie, C., 328, 330(11)
Ku, L., 176
Kuang, W.-J., 73, 392, 399(13), 401(13), 404(13), 498
Kucherlapati, R. S., 531
Kuehn, W., 433
Kuenzel, W. J., 30
Kuhar, M. J., 31
Kull, F. C., 17
Kull, F. C., Jr., 17
Kung, H. F., 147
Kuntz, I. D., 7
Kuo, C. H., 74
Kurokawa, T., 97, 98, 104(27), 109, 110(66), 112(66)
Kurtz, A., 425, 426, 427(4), 430(5), 433(4, 5), 436
Kurz, J., 36
Kuszynski, C., 469
Kyhse-Anderson, J., 415

L

LaBella, L., 342, 343(10), 344(10)
Lacy, E., 21, 74, 519
Laemmli, U. K., 95, 112, 141, 145(12), 208, 221, 319, 375, 397, 454
Lake, M., 9, 10, 11, 13(36), 14(36), 15(36)
Lam, H.-Y. P., 318
Lamb, L. C., 328
Lambre, C. R., 304, 308(10)
Lams, B. E., 47
Lanahan, A. A., 36, 47(12)
Lanahan, A., 61, 62(3), 171
Lanahan, T., 61
Land, H., 122
Lane, M. D., 17
Lang, K., 339
Lang, R. A., 521
Lange, E., 405
Langen, P., 425, 427(2, 3, 4, 6), 432(6), 433, 436(4)
Langer, R., 443
Langone, J. J., 5
LaPolt, P., 347
Large, T. H., 49
Larhammar, D., 38, 49
Larkfors, L., 37
Lathe, R., 389
Lauer, J., 21
Laurent, M., 469
Laureys, G., 171
LaVeck, M. A., 331
Lavigne, M., 469
Lawn, R. M., 99, 101(43), 231
Lawrence, D. A., 304, 308(7), 327, 328, 330(11, 13, 14), 332(13, 14), 336
Lawrence, J. C., 69
Lawson, D. H., 380
Lawson, R. K., 102
Lax, I., 192, 226
Lazar, C. S., 233, 240, 241(25), 495
Lazarides, E., 337
Le Bon, T. R., 18
Le Bon, T., 19, 20, 22, 25
Le Vea, C. M., 289
LeBon, T., 226, 228(4)
Lechner, J. F., 331
Lecocq, J.-P., 19
Leder, P., 50, 52(24), 398, 434, 435(16)
Lee, A., 303, 327, 336(7)
Lee, B. A., 480
Lee, D. C., 185, 200
Lee, D., 98
Lee, F., 435
Lee, G. J., 498
Lee, J. M., 390
Lee, J., 19, 21
Lee, P. L., 171
Lee, P., 435
Lee, S. L., 325
Lee, W. H., 138
Lee-Ng, C. T., 98
Legai, L., 359

Lehmann, W., 425, 427(2, 3, 4, 6), 432(6), 433(4, 6), 436(4)
Leiberman, A. R., 42
Leiovich, S. J., 441
Leistner, I., 426
Lelias, J. M., 98, 104(29)
Lemas, V., 341, 342(4)
Leo, E. B., 328
Leof, E. B., 391, 392(3)
Lerah, H., 293
Lesis, G. D., 289
Lesniak, M. A., 26, 29(2), 33(1), 34(3), 35(1, 3)
Lessler, J. D., 334
Leung, D. W., 392, 399(13), 401(13), 404
Levey, A. I., 37
Levi-Montalcini, R., 35, 48
Levin, W. J., 274, 275, 289, 290, 291(1), 294(1), 297(1)
Levine, R. L., 531
Lewellen, T. K., 31
Lewis, M. E., 438
Lewis, P. R., 37
Lewis, R. B., 406
Li, C. H., 15
Liao, Y. C., 273
Liao, Y.-C., 20, 226
Liauzun, P., 98, 104(29)
Libby, P., 78, 79(1), 80, 83(1)
Libermann, T. A., 21, 192, 273
Liesegang, B., 189, 311
Liggitt, D., 522
Lim, G. J., 335
Lin, C. R., 240, 241(25)
Lin, P. H., 251, 255(1), 256, 257(1), 258(1, 12)
Lind, P., 401
Lindamood, T., 42
Lindberg, M., 5, 6(18), 7, 9(18)
Lindberg, R. A., 495, 498(6)
Lindburg, K., 337
Lindholm, D., 49
Lindmark, R., 5, 7(15)
Lindquist, P. B., 185, 187, 188(8), 192, 303, 327, 336(7)
Ling, N. A., 233
Ling, N., 97, 98, 99(14, 16), 102, 104(14), 138, 139, 145, 146, 148
Link, V. C., 278, 281, 282(6), 283(6), 285, 286(4)

Lioubin, M. N., 317, 322, 323(9), 327, 335, 336
Lipari, T., 185
Lippman, M. E., 290
Lippman, M. M., 334
Lips, D. L., 80
Little, P. F. R., 97, 124, 126, 127(3), 136
Littlefield, J. W., 188, 189(10), 311
Liu, W., 347
Livesey, S. A., 530
Lloyd, C. J., 214
Lloyd, C. W., 251, 255(3)
Lobb, R. R., 96, 97(3), 104(3), 108(3), 395
Lobb, R., 97, 418
Loh, D. Y., 117, 119(5)
Longo, F. M., 70
Loo, D. T., 337, 338(2)
Loo, D., 337, 338(3)
Loomis, J. L., 531
Loveless, J., 126
Löwenadler, B., 5, 8(12), 9, 10(29)
Lowenstein, J. M., 80
Lowry, O. H., 506
Lowry, O., 508
Lozzio, B. B., 342
Lozzio, C. B., 342
Lu, L., 341
Luben, R. A., 456, 457(7), 502
Lucas, C., 312
Lucas, R., 317, 327
Lujan, E., 18
Lukas, T. J., 263, 266(12)
Lusky, M., 176
Luzio, J. P., 511
Lyons, I., 521
Lyons, R. M., 329, 332(25), 333(25)
Lyser, K. M., 482

M

McCaffrey, T. A., 334
MacCarter, D. K., 411
McCarthy, T. L., 318
McCartney-Francis, N., 391, 392(2)
McClintock, R., 194
McClure, D. B., 530
McClure, D., 339
McCourt, D. W., 238
McCune, B. K., 185

Macdonald, R. J., 50, 292, 366, 398
McDonald, V. L., 200, 214, 218(1), 219(1), 220(1), 221(2)
McEver, R. P., 259
McFarlane, A. S., 467
McGann, L. E., 342
McGee, G. S., 96
McGrath, J., 273
McGray, P., 262, 266(11), 267, 271
McGuire, M. K. B., 502
McGuire, W. L., 275, 290, 291(3), 294(3), 297(3)
Maciag, T., 91, 96, 97(4), 102, 104(4), 108(4), 138, 158, 161(1)
McIlhinney, J., 192
McKeehan, W. L., 102, 149, 158, 159(2), 161, 162(2), 338, 530
McKnight, S. L., 242
MacLaughlin, D. T., 359, 360, 362(21, 22), 363(21, 22), 365(27), 367(24)
MacLeod, C. L., 234
McPherson, J. M., 317
McSheehy, P. M. J., 504
Madri, J. A., 446
Maeda, S., 125, 129, 130
Maguruma, Y., 512, 513(7, 8), 514(7, 8)
Maher, D. W., 480
Mahrenholz, A. M., 239
Maiag, T., 392, 399(7)
Maina, D. M., 510
Majerus, P. W., 80
Majeska, R. J., 508
Makino, K., 203
Malathi, P., 251
Maldonado, F., 317
Mallett, P., 251
Mallia, A. K., 255, 511
Manganaro, T. F., 359, 360, 363(19, 20), 364, 367(24), 369(36)
Maniatis, T., 21, 57, 59(32), 73, 74, 107, 121, 228, 230(6), 367, 519
Manley, J. L., 248
Mannoni, P., 342
Mansson, P. E., 149
Mansson, P.-E., 161
Manthorpe, M., 42, 47
Marchalonis, J. J., 467
Marcus, F., 112, 147
Marfatia, S., 360, 361(32), 362(32)
Marics, I., 97, 126, 127

Mariz, I. K., 529
Marks, B. J., 187
Markwell, M. A. K., 467
Marquardt, D. W., 307, 314(19)
Marquardt, H., 10, 200, 202(3), 214, 317, 322, 323(9), 327, 335, 336
Marsh, Y. V., 203
Marshak, D. R., 102
Marshall, J., 344
Marston, F. A. O., 3, 4(3)
Martensen, T. M., 531
Martial, J. A., 21, 50, 398
Martin, G. R., 129, 135(20), 446
Martin, P. L., 248
Martin, T. J., 502, 530
Martinez, D., 363
Marumoto, Y., 130
Mascarelli, F., 147
Mash, D. C., 36
Mashima, K., 516
Masiarz, F. R., 74, 98
Masiarz, F., 20
Mason, A. J., 303, 327, 336(7), 341, 521
Mason, A., 20, 226
Massagué, J., 17, 185, 192, 317, 327, 340, 529
Massoglia, S. L., 151
Massoglia, S., 110, 115(67), 394
Masson, M., 30
Masuda, Y., 512, 513(7, 8), 514(7, 8)
Masui, H., 234, 530
Masui, T., 304, 308(13), 328, 329(21), 331, 336(21)
Mather, J. P., 338, 530
Mather, J., 231, 347, 349, 350, 530
Mathews, M. B., 334
Mathews, M. C., 318
Matsuda, T., 522
Matsudaira, P., 379, 415
Matsui, S., 511, 513(1)
Matsukawa, S., 124, 125(1)
Matsumoto, K., 150, 514
Matsushima, S., 125
Matsuzaki, K., 150, 156(9), 157(9), 518
Mattaliano, R. J., 260, 261, 262, 266(11), 267, 271, 360, 364, 365, 366(39), 367(24), 369(36)
Mattei, M.-G., 97, 126, 127
Matuo, Y., 150, 156(9), 157(9), 511, 512, 513(1, 7, 8), 514(7, 8), 518

Maxam, A. M., 366, 367(40), 390
Mayo, K. E., 347
Mayo, K. H., 203
Mayo, K., 347
Maysinger, D., 36
Mazo, A. M., 368
Medina-Selby, A., 98
Mehlman, T., 102
Meier, R., 49
Meisenhelder, J., 233
Meister, A., 139
Mellor, A. L., 401
Melsert, R., 518
Melton, D. A., 79
Mendelsohn, J., 234
Mendez, B., 21, 50, 398
Mercer, E., 61, 171
Mergia, A., 97, 98, 99(25), 101(25), 102(25), 104(26), 117, 119(6), 392, 399(6)
Merler, E., 440
Merlino, G. T., 242, 250(5, 6, 11)
Merrifield, R. B., 188, 194, 203
Merril, C. R., 221, 455
Merryweather, J. P., 74
Messing, J., 9, 73, 99, 102, 106, 230, 390
Mesulam, M., 37
Metcalf, D., 521
Meunier, H., 341
Meunier, P. J., 509
Meyer, H. E., 240
Meyer, M., 49
Meyers, C. A., 317, 328
Meyers, C., 203
Mezzio, C., 446
Miajima, A., 129
Michael, H., 529
Mieth, M., 426
Miki, T., 97, 129
Miller, D. A., 303, 327, 336(7)
Miller, D. M., 317
Miller, E. J., 325
Miller, G. T., 260
Miller, J. W., 405, 409(3)
Milner, T. A., 37
Milstein, C., 151, 458
Mine, T., 347
Minick, C. R., 394
Mishell, B. B., 458
Misko, T. P., 49, 61, 62(2), 171
Misono, K. S., 263, 266(12)

Mitchell, D. C., 529
Mitchell, P. J., 242
Mitchell, R., 98, 393
Mitsui, Y., 125, 134
Miyada, C. G., 101
Miyagawa, K., 97, 124, 125, 127(3), 133, 135, 136, 137(35)
Miyajima, N., 137
Miyamoto, K., 348, 349(17)
Miyano, K., 167
Miyazaki, K., 516, 530
Miyazano, K., 332, 334(32)
Miyazono, K., 333, 334(32, 38), 336, 383, 384, 385(1, 5), 386, 387(7), 389(7), 391, 392(5), 399(5)
Miyozoki, K., 150
Moberly, L., 147
Moble, W. C., 70
Modi, W. E., 392, 399(7)
Mohler, M. A., 456, 457(7)
Moks, T., 4, 5, 7(5, 6, 7, 19), 8, 9, 10(5, 6), 12(6), 16(4)
Mole, J. E., 317, 327
Monaco, L., 72
Montelione, G. T., 203
Montesano, R., 446
Monu, C., 436
Moore, D. D., 228
Mora-Worms, M., 312
Moran, J., 481, 482(7)
Morén, A., 73, 74(4)
Morgan, C., 61, 171
Morgan, F. J., 405
Mori, M., 124, 125(1)
Morikawa, Y., 358
Morimoto, M., 203
Moritz, R. L., 214
Mormede, P., 138
Morris, N. R., 95
Morrison, D., 78, 79(3)
Morrison, J. R., 214
Morrissey, J. H., 319, 397, 414
Moscatelli, D., 96, 97, 104(5, 28), 146, 392, 404, 477
Moses, A. C., 18
Moses, H. L., 303, 316, 327, 328, 329, 331, 332(25), 333(25), 336(7), 391, 392(3)
Moss, B., 200
Mountjoy, K., 406
Mroczkowski, B., 263, 266(12)

Mudgett-Hunter, M., 358
Mueller, T., 425, 426, 427(4), 433(4), 436(4)
Mufson, E. J., 37
Muggia, V. A., 213
Müller, G., 447
Mulligan, R., 368
Mullis, K. B., 59
Mundy, G. R., 304, 308(9), 502, 503
Munoz, E. F., 318
Munson, P. J., 313
Muramatsu, H., 136
Muramatsu, M., 349
Muramatsu, T., 136
Murata, M., 349
Murby, M., 9
Murphy, J. B., 321
Murray, J., 91, 446, 448(13)
Murray, M., 272
Murrills, R. J., 304, 308(8)
Myers, J. N., 289
Myers, S. L., 405, 409(2), 411

N

Nachman, R. L., 394
Nagai, Y., 333
Nagao, M., 125
Nagashima, M., 185
Naito, K., 127
Najarian, R., 74
Nakabayashi, H., 167
Nakagama, H., 125
Nakagawa, S., 98
Nakakuki, M., 125
Nakamura, N., 514
Nakamura, T., 332, 348, 349(17)
Nakane, P. K., 312, 313(29)
Nakano, N., 522
Nakashima, K., 424
Nakatani, A., 342
Nakayama, D., 323
Nakazawa, T., 416
Napier, M., 185
Nasu, N., 416, 417(1), 419(1)
Nathan, R. M., 318
Nebelsick, J., 469
Negelein, E., 426
Nestor, J. J., 192
Nestor, J. J., Jr., 192, 202, 203

Neufeld, G., 96, 97(1), 98, 104(1), 108(1), 138, 148, 149, 391, 392, 393, 404(4), 469, 480
Nevo, Z., 416, 417(8)
Newby, A. C., 511
Newman, K. M., 97
Newman, S. R., 192, 202, 203
Ng, K. W., 502, 530
Niall, H. D., 48, 175
Niall, M., 502
Nice, E. C., 203, 214
Nicklen, S., 73, 390, 436
Nicoloso, M., 481
Niewiarowski, S., 405
Nikolics, K., 341, 521
Nilsson, B., 3, 5, 6(18), 7, 8, 9, 11, 12(1, 6, 39), 13(39), 16(39), 18(6)
Ninfa, E. G., 360, 365(27), 367(24)
Nishi, M., 203
Nishi, N., 150, 511, 512, 513(1, 7, 8), 514(7, 8)
Nishikawa, K., 148, 150, 151, 152(5), 153(5), 156(9), 157(9), 512, 513(7, 8), 514(7, 8), 518
Nishizuka, Y., 531
Nissley, S. P., 17
Nistér, M., 75
Niu, C. H., 204
Nixon, D. W., 380
Noguchi, S., 514
Noll, F., 425, 427(3)
Nomura, Y., 416, 417(1), 419(1)
Nonomura, N., 514
Nordfang, O., 74
Nowlan, P., 5
Nusbaum, H. R., 192

O

O'Brien, K. V., 80
O'Brien, S. J., 392, 399(7)
O'Brien, T. S., 46
O'Connell, C., 21
O'Connor-McCourt, M., 329, 334(24), 336(24)
O'Hare, M. J., 214
O'Keefe, E. J., 96
Oakley, B. R., 95
Obinata, M., 129, 130(23)

Odagiri, H., 97, 124, 125, 127(3)
Ogasawara, M., 10
Ogata, E., 347, 348(15), 349(15)
Ogawa, R., 494, 495(3)
Ogawa, Y., 317
Ohasa, S., 530
Ohlsson-Wilhelm, B. M., 342, 343(10), 344(10)
Ohmae, H., 416
Ohnishi, S., 125
Ohno, S., 522
Ohtsuka, M., 494, 495(3)
Oi, V., 311
Okabe, T., 384
Okamoto, A. K., 175, 185(4)
Okayama, H., 231, 367, 435
Old, L. J., 468
Oliff, A., 194, 196, 197, 363
Ollier-Hartmann, M. P., 469
Olson, B. J., 255, 511
Olson, L., 38
Olsson, A., 5, 7(5), 10(5)
Olwin, B. B., 469
Omann, G. M., 79
Oppenheim, J. J., 502
Oquendo, P., 373, 378(5)
Orci, L., 446
Oreffo, R. O. C., 304, 308(9)
Orita, M., 60
Oroszlan, S., 147
Orredson, S. U., 446
Orth, D. N., 531
Osborne, C. K., 502
Osterlof, B., 5, 7(6), 9(6), 10(6), 12(6), 18(6)
Östling, M., 5, 7(6), 9(6), 10(6), 12(6), 18(6)
Östman, A., 73, 74(4), 75, 467
Ostrander, F., 18
Otsu, K., 129
Otten, U., 36, 37(14), 49
Otto, A., 425, 427(4), 430(5), 433(4, 5), 436(4)
Ou, J., 20
Owada, M. K., 260
Owen, A. J., 468
Ozanne, B., 192

P

Pace, D. R., 311
Pachl, C., 74
Padhy, L. C., 272, 278
Paganelli, K. A., 511
Paik, S., 290
Paisley, T., 243
Paleus, S., 5, 8(12), 9(12)
Palisi, T. M., 109, 112(65)
Palladino, M. A., 312
Pallas, D. C., 78, 79(4)
Palm, G., 5, 8(12), 9, 10(29)
Palmieri, S., 344
Palmiter, R. D., 520
Paquette, T., 31
Parada, L. F., 122
Pardue, L. R., 213
Parham, P., 152
Parker, P. J., 236
Parker, P., 233
Parsons, L. M., 62
Partridge, N. C., 530
Pasquale, E. B., 495, 498(6)
Pastan, I., 242, 250(5, 6, 11)
Pastore, A., 203
Patchornik, A., 11, 13(37)
Patel, A. H., 5
Patterson, M. K., Jr., 530
Patzer, E. J., 312
Paulson, K. E., 498
Pavlakis, G. N., 176
Paxinos, G., 30, 39
Payne, L. S., 196
Payne, R. E., Jr., 96
Peck, W. A., 416, 417(9)
Pelicci, P. G., 342
Pellicer, A., 62, 63(9), 74
Penshow, J. D., 175
Pepinsky, B., 363
Pepinsky, R. B., 260, 261, 262, 263, 264, 265(9), 266(6, 9, 11), 267, 269, 271, 364, 365, 366(39), 369(36)
Pepper, D. S., 405
Pepperle, M., 425, 430(5), 433(5)
Perkins, A. S., 62
Perlman, D., 392, 399(8)
Perlmann, P., 312, 313(28)
Perlmutter, A. P., 368
Perruzzi, C. A., 404
Persson, H., 38, 49
Persson, I., 5, 7(6), 9(6), 10(6), 12(6), 18(6)
Pert, C. B., 26, 33(1), 34(3), 35(1)
Peterkofsky, B., 507, 508(11), 509, 510(11)
Peters, G., 97, 129, 137

Peterson, G. M., 36, 38(17), 44(17), 47
Petraglia, F., 347
Petrusz, P., 524
Petruzzelli, L. M., 20, 226
Pfeiffer, S. W., 347
Pfeilchifter, J. P., 503
Philipson, L., 5, 6(18), 8(12), 9(12, 18)
Phillips, H. S., 404
Phillips, L. S., 529
Philpott, K., 401
Picard, J. Y., 358
Picon, R., 360
Pieler, C., 303, 327
Pierce, G. B., 312, 313(29)
Pierce, G. F., 490
Pierce, J. H., 275, 276(15), 277
Piez, K. A., 317, 318, 324, 325, 424
Pilatte, Y., 304, 308(10)
Pilch, P. F., 17
Pircher, R., 304, 308(7), 328, 330(11, 13, 14), 332(13, 14)
Pittelkow, M. R., 96, 97(7), 512, 513(6)
Pittenger, M., 529
Pitts, S. L., 521, 522
Plackshear, P. J., 531
Placzek, M., 137
Plagemann, P. G. W., 531
Planche, J., 97, 126, 127
Plastuch, W., 77
Plotsky, P. M., 347
Plowman, G. D., 214, 221(2)
Poenie, M., 233
Pohl, G., 5, 7(5), 10(5)
Pollak, A., 342
Polverini, P. J., 98, 441, 448
Ponka, P., 344
Ponte, P., 98
Ponzetto, C., 501
Popescu, N. C., 274, 275(11), 290, 293(4)
Porter, L. P., 290
Poser, J. W., 509
Postlewaite, A. E., 530
Potter, A., 109, 112(65)
Potter, M., 189, 312
Potter, S. J., 74
Poulter, L., 98
Powell, A., 251, 255(3)
Powell, S. M., 358
Prats, A. C., 98, 104(29)
Prats, H., 98, 104(29), 480, 482(4), 484(4)

Pratt, D. J., 360, 365(27)
Pratt, D., 260
Preiser, H., 251
Press, M. F., 289, 290, 291(1), 294(1), 297(1)
Presta, M., 96, 97, 104(28), 392, 446
Primavera, M. V., 505
Proper, J. A., 316, 328, 329
Provenzano, M. D., 255, 511
Przybyla, A. E., 50, 292, 366, 398
Puma, J., 74
Puma, P., 69
Purchio, A. F., 335, 336

Q

Quiroga, M., 49, 74, 175
Quon, D., 21

R

Rabussay, D., 77
Rackoff, W. R., 235
Radbruch, A., 189, 311
Radeke, M. J., 49, 61, 62(2), 171
Raelson, M. E., 384
Ragin, R. C., 359, 360, 367(24)
Ragsdale, C. G., 406
Raines, E. W., 72, 529
Raisz, L. G., 416, 502, 510
Rajewski, K., 189, 311
Rall, L. B., 74, 175, 176
Ramachandran, J., 19, 20, 22, 25, 226, 228(4)
Ramachandran, K. L., 261, 262, 266(11), 360, 367(24)
Ramani, N., 206
Ranchalis, J. E., 335
Randall, R. J., 506
Randolph, A., 74
Rao, Ch. V., 206
Raptis, L., 78, 79(4)
Rastinejad, F., 448
Raulais, D., 147
Rave, N., 55
Ravera, M. W., 392, 399(7)
Rawson, C. L., 337, 338(2)
Rawson, C., 337, 338(3)
Ray, A., 399
Raybaud, F., 97, 126, 127

Razon, N., 192
Read, L. C., 10
Rechler, M. M., 17
Reddi, A. H., 328, 424
Reddy, M., 126
Reddy, V. B., 175
Redmond, C., 290
Reers, M., 427
Rees, D. A., 251, 255(3)
Rees-Jones, R. W., 17
Reeves, B., 192
Reichardt, L. F., 37, 49, 56(15)
Reichmann, G., 426
Reid, L. C. M., 180
Reidel, H., 225
Reilly, C. F., 334
Rennard, S. I., 326
Reyes, A. A., 436
Rhee, L., 303, 327, 336(7)
Rhee, S. G., 80
Riboloni, A., 365, 366(39)
Ricca, G. A., 392, 399(7)
Ricca, G., 171
Richardson, P. M., 35
Richmond, A., 373, 374, 376, 377(2), 378(1), 380, 382(1, 2, 14)
Riemen, M. W., 194, 196, 197, 363
Rifas, L., 416, 417(9)
Rifkin, D. B., 96, 97, 104(5, 28), 333, 392, 404
Riggs, A. D., 3
Riopelle, R. J., 35
Rios-Candelore, M., 97, 99(15)
Risau, W., 386, 387(7), 389(7), 391, 392(5), 399(5)
Ritchie, J. C., 405, 409(1, 2)
Rittenhouse, J., 112, 147
Rivier, C., 341
Rivier, J., 194, 341, 342(4), 347
Rizzino, A., 134, 469, 472, 530
Roach, P. J., 69, 239
Robel, P., 362
Roberts, A. B., 186, 303, 304, 309, 314(2, 20), 315(20), 316(1), 317, 318, 322, 323, 327, 328, 424, 447
Roberts, B., 214
Roberts, D. J., 407
Roberts, E., 153
Roberts, G. D., 102
Roberts, J. L., 524, 531

Roberts, T. M., 78, 79(4), 83(9)
Roberts, T., 78, 79
Robertson, D., 347
Robins, E., 509
Robinson, R. A., 328
Rochant, H., 342
Roche, N. S., 328, 424
Rodan, G. A., 508
Rodan, S. B., 508
Rodbard, D., 313, 530, 531
Rodkey, J., 97, 99(15)
Rodriguez, H., 398
Rodriguez-Boulan, E., 67
Roe, B. A., 99, 102(44), 242
Roeder, R. G., 248
Roeske, R. W., 239
Rogelj, S., 98, 117, 119(2), 123(2), 124(2, 3)
Rohde, K., 425, 427(4), 433(4), 436(4)
Rohde, R. F., 98, 109(39), 112(39)
Rojas, E., 260
Rojeski, M., 26, 33(1), 35(1)
Roller, P. P., 204
Rommerts, F. F. G., 518
Ron, D., 97, 129
Ronnett, G. V., 17
Rönnstrand, L., 75, 334, 384, 468
Roodman, G. D., 504
Rose, J. W., 109, 112(65)
Rose, T. M., 200
Rosebrough, N. J., 506
Rosen, D. M., 304, 317, 318, 324, 424
Rosen, L. M., 20
Rosen, O. M., 226
Rosenberg, R. D., 334
Rosenberg, R., 334
Rosenfeld, M. G., 233, 240, 241(25), 495
Rosenthal, A., 192
Ross, A. H., 61, 62, 68(7), 83, 494, 495
Ross, D., 342
Ross, R., 72, 468, 529
Roth, J., 26, 29(2), 33(1), 34(3), 35(1, 3), 531
Roth, R. A., 20, 336
Rougeon, F., 73
Rowland, A., 380
Rowley, P. T., 342, 343(10), 344(10)
Roy, R. G. B., 376, 380
Rubin, B. Y., 468
Rubin, E., 251
Rubin, J. B., 17
Rubin, J. S., 97, 98, 129

Rubin, R. A., 235
Ruderman, N. B., 78, 79(2)
Rudland, P. S., 214
Ruff, E., 469, 472
Russel, R. G. G., 502
Russell, R. G. G., 509
Ruta, M., 171
Rutter, W. J., 20, 49, 50, 175, 366, 522
Rutter, W., 387
Ryan, R. J., 331
Rybka, R., 108, 145, 153, 395
Rye, D. B., 37

S

Sacchi, N., 433
Saeki, Y., 129, 130
Saez, J., 350
Saez, S., 509
Sager, R., 373
Sahagian, G. G., 529
Saiki, R. K., 435
Saiki, R., 59
Saito, M., 333
Sakai, Y., 337
Sakamoto, H., 97, 124, 125, 127, 135, 136, 137
Sakamoto, S., 203
Sakano, K., 130
Sakata, S., 342
Sakuda, M., 424
Salmon, S. E., 502
Salvatierra, A., 36
Samanta, A. K., 380
Sambrook, J., 57, 59(32), 73, 107, 121, 228, 230(6), 367, 519
Samtleben, R., 425, 427(3)
Samuels, M., 248
Sanchez-Pescador, R., 175, 176
Sandberg, A. A., 511, 513(1)
Sandison, J. C., 440
Sanger, F., 73, 99, 102(44), 390, 436
Santon, J. B., 234
Sanzo, K., 404
Sara, V. R., 10
Sasada, R., 97, 98, 104(27), 109, 110(66), 112(66)
Sasak, H., 175
Sasaki, A., 125

Sasaki, H., 150, 156(9), 157(9), 518
Sasaki, N., 17
Sass, R. E., 290
Sasse, J., 95, 97, 102, 119, 123(7), 418
Sassone-Corsi, P., 243
Sato, B., 514
Sato, G. H., 338
Sato, G., 329, 339, 530
Sato, J. D., 167
Sato, K. Y., 109, 112(65)
Sato, K., 424
Sato, Y., 97, 129, 130, 333
Satoh, H., 136, 137(35)
Savage, C. R., Jr., 200, 214, 529
Savetnick, N., 522
Savouret, J. F., 50
Savouret, J.-F., 21, 398
Sawamura, S. J., 317
Sawchenko, P., 347
Sawyer, S. T., 261, 263(8), 264(8)
Scatchard, G., 165
Schaffhausen, B., 78, 83(9)
Schagger, H., 380
Schairer, H. U., 401
Schatteman, G. C., 36, 47(12)
Schatteman, G., 36, 46(13), 47(13)
Schechter, A. L., 63, 65, 71(12), 272, 274, 276(9)
Schechter, G. P., 502
Schechter, Y., 11, 13(37)
Schecter, A. L., 278
Scheidegger, D., 189
Scheraga, H. A., 203
Schien, P. H., 197
Schiffer, S. G., 98, 109(39), 112(39)
Schindler, D., 267, 269, 271(24)
Schlaepfer, D. D., 263, 265, 266(18), 272(15, 18)
Schleifer, L. S., 61, 62(5), 71(5), 72(5)
Schlessinger, J., 21, 170, 171, 192, 203, 225, 226, 273, 276, 277, 290
Schlusener, H., 303
Schmidt, H.-E., 425, 427(2, 3)
Schmidt, J. A., 344
Schneider, U., 426
Schofield, T. L., 197
Schreiber, A. B., 192, 202, 203
Schreurs, J., 129
Schubert, D., 146
Schultz, G. S., 206

Schultz, G., 201
Schuurs, A. H. W. M., 326
Schwab, M. E., 36, 37(14, 15)
Schwall, R. H., 341
Schwall, R., 349
Schwander, J., 17
Schwartz, B. D., 238
Schwartz, B. R., 360
Schwartz, N. B., 347
Schwarz, M. A., 49, 56(14)
Schweigerer, L., 96, 97(1), 104(1), 108(1), 138, 148, 391, 392(4), 393, 404(4), 480
Schweingruber, M. E., 105
Schweitzer, E. S., 62
Scott, J., 49, 175, 401
Scott, M. E., 405, 409(1, 2)
Scott, M. R. D., 64
Scott, R. E., 512, 513(6)
Scrace, G. T., 401
Sebald, W., 401
Sedlak, B. J., 384
Sedman, S. A., 221
Seeburg, P. H., 20, 21, 99, 101(43), 226, 231, 273, 401, 436, 521
Sefton, B. M., 141, 142(13), 237, 494, 495(4), 496(4), 497(4), 499(1), 500(4), 501(4), 531
Segarini, P. R., 304, 317, 318
Segarini, P., 323
Segatto, O., 275, 276(15), 277
Segui-Real, B., 446
Sehgal, A., 61, 62(3), 171
Seidman, J. G., 228
Seifert, J. M., 303
Seiger, Å., 38, 48
Seiler, M., 36, 37(15)
Sekiguchi, M., 124, 125(1)
Sekiya, T., 60
Selby, M., 49, 175
Selinfreund, R., 251, 255(1), 256, 257(1), 258(1, 12)
Semba, K., 273, 275(6)
Senger, D. R., 404
Seno, M., 98, 109, 110(66), 112(66)
Serrero, G., 530
Serunian, L. A., 78, 79, 80, 83(1)
Server, A. C., 48
Severinsson, L., 73, 74(4)
Seyedin, S. M., 304, 308(9), 317, 318, 324, 424

Seyedin, S., 503
Shackleford, G. M., 129, 135(20)
Shaffhusen, B., 78, 79(4)
Shamber, F., 19
Shao, L., 341, 342(4)
Sharp, P. A., 248
Sharpe, P. A., 368
Shastry, B. S., 248
Shaw, A., 49
Sheard, B., 203
Sheldon, D. L., 37
Shelton, D. L., 49
Shelton, D., 49, 56(15)
Shen, V., 416, 417(9)
Shenkman, J. B., 251
Shepard, H. M., 289
Sherwin, S. A., 192
Shia, M. A., 17
Shibai, H., 341, 347, 348, 349
Shih, C., 121, 272, 278
Shiigi, S. M., 458
Shimasaki, S., 102
Shimizu, K., 124, 125(1)
Shimomura, Y., 417, 418(13)
Shimosato, Y., 137
Shing, Y., 91, 92(5), 94(5), 95, 97, 102, 108, 145, 153, 395, 418, 446, 448(13)
Shipley, G. D., 96, 97(7), 328, 329(20), 331, 391, 392(3), 512, 513(6)
Shirahata, S., 337, 338(2)
Shmid, C., 17
Shoelson, S. E., 259, 264
Sholley, M., 444, 445(11)
Shooter, E. M., 48, 49, 61, 62, 69(4), 171, 529
Shows, T. B., 401
Shoyab, M., 200, 214, 218(1), 219(1), 220(1), 221(2)
Shutte, C. C. D., 37
Siddle, K., 511
Sidkey, Y. A., 441
Siegel, N. R., 317
Siepl, C., 317
Sigman, D. S., 365, 366(37)
Sigumura, T., 125
Silba, M., 98
Silber, S., 98
Silver, A., 37
Silver, I. A., 304, 308(8)
Silverstein, S., 62, 63(9), 74

Sim, G. K., 21, 74
Simon, M.-P., 137
Simonsen, C. C., 99, 101(43)
Simony-aontaine, J., 137
Simpson, R. J., 214
Simpson, T. L., 17
Sims, P. J., 259
Sinclair, L. K., 260, 261, 263(9), 264, 265(9), 266(6, 9), 267, 269, 271, 364, 365, 366(39), 369(36)
Sinclair, L., 262, 266(11)
Sirbasku, D. A., 10
Sirbasku, D., 374
Sisson, T. H., 410
Sjödahl, J., 5, 7
Sjogren, B., 10
Sjöquist, J., 5, 7
Sklar, L. A., 79
Skottner-Lundin, A., 9, 10(29)
Slack, J. M. W., 129
Slamon, D. J., 274, 275, 289, 290, 291(1), 294(1), 297(1)
Sloan, T. B., 405, 409(2)
Smart, J. E., 262, 266(11), 365, 366(39)
Smith, A. J. H., 99, 102(44)
Smith, D. H., 99, 101(43)
Smith, D. M., 304, 328, 329(21), 331, 334(31), 336(21, 31)
Smith, E. M., 406
Smith, J. A., 95, 98, 102, 104(29), 214, 228
Smith, J. M., 328
Smith, L. M., 304, 308(13)
Smith, M., 105, 203, 241
Smith, P. K., 255, 511
Smith, R., 129, 137
Smith, S., 102, 108, 145, 153, 395
Sodek, J., 530
Sofroniew, M. V., 42, 46, 47
Soltoff, S. P., 78, 79(1), 83(1)
Sommer, A., 96, 97, 98, 104(28, 35)
Sonnerfeld, E., 201
Sonoda, T., 514
Soreq, H., 192
Southern, E. M., 99
Southern, P. J., 122, 178
Sowder, R. C., 147
Spatala, A., 201
Speck, J. C., 467
Spener, F., 427
Spiess, J., 317, 373, 380(3)

Sporn, M. B., 186, 303, 304, 308(13), 309, 314(2, 20), 315(20), 316, 317, 318, 322, 323, 327, 328, 329(21), 331, 334(31), 336(21, 31)
Spudich, J. A., 324
Squires, C. H., 98
Squires, C., 105
St. John, T., 50, 53(26)
Stadig, B. K., 206
Stahlman, M. T., 263
Stanley, E., 521
Stanley, K., 12
Staros, J. V., 259
Steck, P. A., 451
Steele-Perkins, G., 336
Steimer, K. S., 98
Stein, G. H., 530
Stein, R. B., 194
Steinberg, S. M., 290
Steiner, D. J., 175, 185(4)
Steiner, M., 462, 463(10)
Stempien, S. A., 318, 324
Stenevi, U., 36, 37, 42(26, 27)
Stephens, L. A., 79
Sterky, C., 5, 7(5), 10(5)
Stern, A. S., 511
Stern, D. F., 119, 120(8), 121(8), 272, 274, 276(9), 278, 281, 496, 497(11, 12), 499(11, 12), 500, 501(11)
Steward, J. M., 194
Stewart, T. A., 521, 522
Stiles, C. D., 373, 378(5)
Stockdale, H., 130
Storm-Mathisen, J., 37
Story, M. T., 102
Stoscheck, C. M., 240
Stoscheck, C., 194
Stratowa, C., 387
Stratton, R. H., 242
Straus, J. W., 435
Stroobant, P., 401
Struhl, K., 228
Stuart, S. G., 274, 275(12), 289, 290, 291(1), 294(1), 297(1)
Su, F.-Y., 17
Subramani, S., 368
Sudo, K., 109, 110(66), 112(66), 125
Suematsu, S., 522
Sugano, K., 333
Sugimoto, Y., 78

Sugimura, T., 97, 124, 125, 127(3), 135, 136, 137, 274, 275(13), 290, 291(2)
Sugino, H., 348, 349(17)
Sullivan, R., 91, 95, 96, 97, 102, 108, 119, 123, 145, 153, 395, 418, 446, 448(13)
Sullivan, S. A., 468
Sun, Q., 425
Sundell, H., 263
Sutherland, J. A., 342
Suzuki, F., 150, 416, 417, 418(13), 419(1, 6), 424
Suzuki, Y., 60
Svendsen, C. N., 46
Svoboda, M. E., 17
Svobada, M., 17
Svoboda, I. M. E., 529
Swain, S. M., 290
Swann, D. A., 358, 362(3)
Sweet, R., 74
Swift, G. H., 50, 292, 398
Szonyi, E., 341, 521

T

Tabillo, A., 342
Tabuchi, M., 347, 348(15), 349(15)
Tahara, E., 137
Tai, J. Y., 196, 197
Taira, M., 124, 125(1, 2), 130(2)
Takahashi, M., 359, 360, 363(18)
Takaku, F., 125, 384, 386, 387(7), 389(7), 391, 392(5), 399(5)
Takano, K., 14, 15(42)
Takano, S., 341
Taketa, K., 167
Takezawa, M., 341
Takigawa, M., 416, 417(7)
Tally, M., 9, 15
Tam, A. W., 19, 20, 21, 22, 25, 226, 228(4)
Tam, J. P., 204
Tamm, J., 303, 327, 336(7)
Tamura, A., 203
Tandon, A. K., 290, 291(3), 294(3), 297(3)
Tang, J., 262, 266(11)
Taniuchi, M., 35, 69
Tappin, M. J., 203
Tashiro, K.-I., 446
Taube, C., 426
Tauber, J. P., 98, 104(29), 480, 482(4), 484(4)

Taylor, L. A., 360
Taylor, P., 79
Taylor, S. S., 140
Teissié, J., 480, 482(4), 484(4)
Teixido, J., 185
Telser, A., 414
Ten Dijke, P., 327
Teplitz, R. L., 436
Terada, M., 97, 124, 125, 127, 135, 136, 137, 138, 274, 275(13), 290, 291(2)
Terry, R. D., 38, 42(38), 43(38)
Teshima, S., 135
Testa, J. R., 342
Testa, U., 342
Teti, A., 505
Theillet, C., 137
Thode, K., 425
Thoenen, H., 35, 36, 37, 48, 49
Thom, D., 251, 255(3)
Thoma, R. S., 238
Thomas, H. G., 373, 374, 376, 377(2), 378(1), 380, 382(1, 2, 14)
Thomas, K. A., 133, 529
Thomas, K., 97, 99(15)
Thomas, P. S., 55, 56(31)
Thomas, T. C., 317
Thomas, T. P., 531
Thomopoulos, P., 342
Thompson, A. Y., 317, 318, 324
Thompson, B. L., 79
Thompson, N. L., 318
Thompson, R. C., 97, 104(28)
Thompson, S. A., 109, 112(65)
Thorell, J. I., 305
Thorén-Tolling, K., 5, 7(15)
Thotakura, N. T., 333
Thys, R., 175, 185(4)
Tiesman, J., 469
Timpl, R., 325
Tischer, E., 98
Titani, K., 348, 349(17)
Titeux, M., 342
Tizard, R., 260, 261, 262, 266(11), 360, 367(24)
Tjian, R., 242, 243
Todaro, G. J., 192, 200, 202, 203, 206, 208, 214, 218(1), 219(1), 220(1), 221(2), 331
Todaro, G. L., 263
Todaro, G., 214
Tolstoshev, P., 19

Tomita, S., 342
Torres, G., 360, 367(24)
Torvik, A., 42
Totsuka, Y., 347, 348(15), 349(15)
Townes, T. M., 520
Toyoshima, K., 137, 273, 274, 275(6), 275(13), 290, 291(2)
Tran, D., 358, 359
Traynor-Kaplan, A. E., 79
Trelstad, R. L., 360
Tremble, P. M., 73, 498
Truett, M. A., 74
Tsai, L. B., 98, 109(39), 112(39)
Tsang, M. L. S., 327
Tsang, M. L.-S., 317
Tsien, R. Y., 233
Tsubokawa, M., 19, 20, 22, 25, 226, 228(4)
Tsuda, T., 137
Tsugane, S., 137
Tsuji, M., 416, 417(1), 419(1)
Tsuji, T., 341
Tsutsumi, M., 135
Tucker, R. F., 316, 329, 331
Tuddenham, E. G. D., 99, 101(43)
Tumolo, A., 97, 98, 99(25), 101(25), 102(25), 104(26), 117, 119(6), 392, 399(6)
Ture, J.-M., 342
Turner, A. R., 342
Tuszynski, M. H., 46, 47
Twardzik, D. R., 192, 200, 202(3), 335
Twardzik, D., 201

U

Uchida, N., 514
Udove, J., 274, 275(12), 289
Udove, S., 290, 291(1), 294(1), 297(1)
Ueda, H., 243
Ueno, N., 97, 99(14, 16), 102, 104(14), 138, 145, 148
Ueno, S., 359, 360, 363(19)
Uher, L., 77
Uhl, G. K., 438
Uhlén, M., 4, 6(18), 7, 8, 9, 10(5, 6, 8), 12(6), 16(4)
Uitvlugt, N., 359, 362(22), 363(22)
Ullrich, A., 19, 20, 21, 22, 25, 49, 73, 175, 192, 225, 226, 228(4), 231, 274, 275, 276, 289, 290, 291(1, 3), 294(1, 3), 297(1, 3), 390, 401, 498

Unanue, E. M., 441
Unnerstall, 31
Unterberg, C., 427
Upton, F. M., 10
Urabe, A., 384
Urdea, J., 175
Urdea, M. S., 74, 176, 401
Urdea, M., 49
Urlaub, G., 62, 63(9)
Usher, P., 18
Ushiro, H., 194, 233
Usuki, K., 386, 387(7), 389(7), 391, 392(5), 399(5)

V

Vahlsing, H. L., 42, 47
Vaidyanathan, L., 272, 278
Vainchenker, W., 342
Vaitukaitis, J. L., 481
Vale, R. D., 529
Vale, W., 341, 342(4), 347, 348, 349(16)
Valentin, A., 509
Valenzuela, P., 74
van Boom, J. H., 436
van der Bosch, J., 530
van der Eb, A. J., 121, 436
van der Elst, H., 436
Van Driel, I. R., 401
van Dyke, P., 303
Van Weemen, B. K., 326
van Wyk, J. J., 17
Van Wyk, J. J., 529
Varmus, H. E., 129, 135(20)
Varon, S., 35, 36, 38, 42, 44(7, 17), 46(39), 47
Varticovski, L., 78, 79, 267, 269, 271(24)
Vassalli, P., 446
Vaughan, J., 341, 342(4), 347
Vaughn, K. M., 63, 65, 71(12)
Vedvick, T. S., 140, 236
Vehar, G. A., 99, 101(43), 231
Veira, J., 390
Veno, N., 139
Verge Isse, V. M. K., 35
Verlaan-deVries, M., 436
Verpault, C., 482
Vieira, J., 73, 102, 106
Viera, J., 9
Vigier, B., 359, 362
Vigier, P., 328

Vigny, M., 469
Vileval, J. L., 342
Vlodavsky, I., 119, 123(7)
Vogel, F., 426, 436
Von Hoff, D. D., 363
Von Huff, D. D., 213
Von Jagow, G., 380
Voss, P. G., 451
Vu, M. T., 441, 444(8)
Vuocolo, G. A., 196, 197

W

Wachter, L., 260
Wada, A., 136
Wada, F., 511, 512, 513(1, 7, 8), 514(7, 8)
Wada, M., 136, 137(35)
Wada, T., 289
Waddell, W. J., 321
Wadzinski, M., 95
Waegell, W., 311, 323
Wagner, R. M., 529
Wahl, G. M., 399
Wahl, M. I., 500
Wahl, S. M., 530
Wahl, S. W., 391, 392(2)
Wainer, B. H., 37
Wakefield, L. M., 186, 303, 304, 308(13), 314(2), 317, 323(11, 12), 327, 328, 329, 331, 334(24, 31), 336(21, 24)
Waker, B., 243
Wakshull, E. M., 256
Wakshull, E., 251, 255(1), 256(1, 2), 257(1), 258(1)
Walicke, P. A., 138, 140(7), 145, 146(18)
Wallace, D. M., 52
Wallace, J. C., 10
Wallace, R. B., 101, 436
Wallen, J. W., 363
Wallner, B. P., 260, 261, 262, 266(11), 360, 365, 366(39), 367(24)
Walsh, B. J., 241
Walter, J. L., 189, 312
Walton, G. M., 140, 236, 495
Walukat, G., 433
Walz, A., 406
Walz, D. A., 405, 406, 409(3)
Wang, J. L., 392, 399(6), 425, 451, 455(3), 462(4), 464(3, 4)
Wang, J. Y. J., 495, 498

Wang, Y., 239
Ward, D., 347
Warner, S. J. C., 80
Warran, T., 404
Warren, J., 153
Wasteson, Å., 334, 384, 401, 467
Watanabe, C., 398
Watanabe, R., 416
Watanabe, T., 494, 495(3)
Watanabe, Y., 494, 495(3)
Waterfield, M. D., 21, 192, 233, 236, 401
Watkins, P., 175
Watson, C. J., 387
Watson, C., 30, 39
Weatherbee, J. A., 317, 327
Weavers, E. D., 5
Webb, N. R., 200, 335
Weber, W., 234, 236, 240(9)
Wedegaertner, P. B., 240
Wegrzyn, R. J., 194, 196, 197
Wehrenberg, W., 138
Wei, C.-M., 175
Wei, S. J., 147
Weich, H. A., 401
Weinberg, R. A., 98, 117, 119, 120(8), 121, 122, 123(2), 124(2, 3), 272, 273, 274, 276(9, 10), 278, 281, 282(6), 283(6), 285, 496, 497(12), 499
Weinberg, W. A., 276
Weiner, R. I., 392
Weiner, R., 392, 393(14)
Weiner, W. J., 36
Weinstein, I. B., 62
Weisz, P. B., 95
Weksler, B. B., 334
Wells, A., 233
Wells, J. A., 11, 12(39), 13(39, 40), 16
Welt, F. G. P., 334
Wen, D., 303, 327, 336(7), 373, 374, 378(5), 380
Wennergren, S., 334, 384
Wernstedt, C., 332, 334(32), 386, 387(7), 389(7), 425, 427(4), 433(4), 436(4)
Wernsterdt, C., 391, 392(5), 399(5)
Weskamp, G., 49
Wessel, D., 18
West, B. L., 21, 50, 398
Westermark, B., 72, 75, 76, 334, 384, 401, 467
Whang, J. L., 97, 98, 99(25), 101(25), 102(25), 104(26), 117, 119(6)

Wharton, W., 251, 255(1), 256, 257(1), 258(1, 12)
White, B. A., 58
White, M. F., 78, 79(2), 259
White, M. R., 359, 362(22), 363(22)
White-Scharf, M. E., 117, 119(5)
Whitlock, C. A., 123
Whitman, M., 78, 79, 82(10), 83(9), 267, 269, 271(24)
Whitney, S. L., 405, 409(1, 2)
Whittaker, J., 175, 185(4)
Whittemore, S. R., 38, 48, 49
Whittle, N., 21, 192
Whittmore, S. R., 37
Wictorin, K., 36, 38, 46(39)
Widdicombe, J. H., 392, 393(15)
Widmmaier, R., 425, 427(6), 432(6), 433(6)
Wigler, M., 62, 63(9), 74
Wilcox, J. N., 521, 524
Wilcox, J., 531
Wilczek, J., 326
Wildeman, A. G., 243
Wiley, H. S., 241
Wilkinson, A. J., 203
Wille, J. J., Jr., 512, 513(6)
William, L. R., 47
Williams, C. H., 405, 409(2)
Williams, J., 178
Williams, L. R., 35, 36, 38, 42, 44(7, 17), 46(39)
Williams, L. T., 73, 79, 138, 171, 498
Williams, M. R., 508
Williams, R. D., 192
Williams, S. K., 446
Williams, S. R., 468
Williamson, B. D., 468
Williamson, D. J., 521
Wilson, S., 243
Winchell, L. F., 185
Winget, M., 289
Winkler, M. E., 187
Wion, K. L., 99, 101(43)
Wion, K., 231
Witkop, B., 10, 13(35), 14(35)
Witte, O. N., 123
Witters, L. A., 531
Wittwer, A. J., 323
Wohlhueter, R. M., 531
Wold, B., 74
Wolfe, R., 339, 530

Wollenberger, A., 433
Wong, H., 391, 392(2)
Wong, S. G., 274, 275, 289, 290, 291(1), 294(1), 297(1)
Wong, S. T., 185
Wong, W. L., 374
Woo, J. E., 70
Wood, F. T., 199
Wood, O. D., 502
Wood, W. I., 99, 101(42, 43), 231, 399
Woodgett, J. R., 139
Woodruff, T. K., 347
Woodruff, T., 347
Woods, S. G., 31
Woodward, S. C., 96
Woost, P. G., 206
Woost, P., 201
Wormsted, M. A., 176
Wozney, J. M., 293
Wrann, M., 303, 317
Wu, C., 243
Wu, J., 95
Wu, M. M., 199
Wu, R., 399, 530
Wu, T., 399
Wuthrich, K., 203
Wyatt, T., 446

X

Xu, Y.-h., 242

Y

Yamamoto, H., 137
Yamamoto, T., 137, 273, 274, 275(6, 13), 290, 291(2)
Yamamura, J., 522
Yamane, T., 167
Yamashita, N., 516
Yamashita, Y., 125
Yamuda, Y., 446
Yang-Feng, T. L., 73, 273, 498
Yang-Feng, T., 19, 20, 22, 25, 226, 228(4)
Yanisch-Perron, C., 9, 102
Yanish-Perron, C., 390
Yanishevsky, R., 530
Yannisch-Perron, C., 73

Yanosky, C., 105, 107
Yarden, Y., 21, 73, 170, 225, 498, 499
Yaron, M., 408
Yayon, A., 117
Yeh, H., 490
Ying, S. Y., 138
Ying, S.-Y., 340, 341(1)
Yokogawa, Y., 341
Yokota, J., 136, 137, 274, 275(13), 290, 291(2)
Yokota, T., 435
Yoneda, T., 417, 418(13)
Yoshida, H.-S. U. K., 47
Yoshida, M. C., 136, 137(35)
Yoshida, T., 97, 124, 125, 127(3), 135, 136, 137
Yoshitake, Y., 148, 150, 151, 152(5), 153(5), 156(9), 157(9), 512, 513(7, 8), 514(7, 8), 518
Young, A. T., 80
Young, J. D., 194
Young, R. A., 367, 399
Young, R., 21
Yu, A. L., 341, 342(4)
Yu, J., 341, 342(4)

Yuasa, Y., 125
Yutani, Y., 416, 417(7)

Z

Zabeau, M., 12
Zabelshansky, M., 171
Zachariae, C. O. C., 380
Zalta, J. P., 481
Zambonin-Zallone, A., 505
Zap, J., 17
Zenke, M., 243
Zentgraf, H., 344
Zetter, B. R., 93, 394, 404(18)
Zetter, B., 404
Zhan, X., 97, 126, 127, 129
Zhu, D., 95
Zick, Y., 17
Ziegler, S. F., 123
Zimarino, V., 243
Zokas, L. M., 495
Zokas, L., 260
Zoller, M. J., 105, 140, 241
Zumstein, P. P., 529

Subject Index

A

Activin, 360
 activity, 340–341, 347–348
 binding characteristics, 349
 erythroid differentiation bioassays for, 340–347
  cell culture for, 343
  general considerations, 342–343
  identification of hemoglobin-containing cells, 343–344
  measurement of hemoglobin accumulation, 344–346
 forms of, 341
 labeling, 349
 receptor, 348
 stimulation of erythroid differentiation, 341
 structure, 341
Adipocyte differentiation associated protein (P422), 425
African rat. See *Mastomys*
A431 human epidermoid carcinoma cells
 EGF-dependent phosphorylation of lipocortin-1 in, 263–265
 EGF receptor expression, 240–241
 EGF receptor purification from, 235
 growth, 234
 labeling with $^{32}PO_4$, 235
 nuclear extract
  fractionation of, 244–246
  preparation of, 243–244
 phospholipase C in, purification, using phosphotyrosine antibodies, 500
 purified cell membranes, EGF-dependent receptor autophosphorylation using, 256–259
Amphiregulin, 200
 activity, 214
 cell growth modulatory assay, 215–216
 forms of, 214
 purification, 213–224
  from TPA-induced cells, 214–224
   cell culture, 216
   collection of conditioned medium for, 216
   gel-permeation chromatography, 219–221
   reversed-phase liquid chromatography, 217–220
   SDS-polyacrylamide gel electrophoresis, 220–221
 sequence identity to EGF and transforming growth factor α, 214
Androgen-induced growth factors, purification, stabilization during, 514–516
Angiogenesis, 391
 in vitro assays of, 445–448
 in vivo assays of, 440–441
  in chick chorioallantoic membrane, 440–441
  in cornea, 441
   colloidal carbon-perfused whole mounts, preparation of, 444–445
   formation of corneal pocket for, 443–444
   preparation of conditioned media for, 441–443
   scoring of vascular responses, 444–446
 naturally occurring inhibitor of, 440–450
  purification, from baby hamster kidney fibroblast conditioned medium, 448–450
  purified, storage of, 449
 regulatory factors, 391
Angiogenin, 391
Autophosphorylation, 496
 of EGF receptor

EGF-stimulated, 496
  using purified cell membranes from A431 human epidermoid carcinoma cells, 256–259
  inhibition, 359
  growth factor, inhibition, 359–360

B

Bacterial expression systems
  for mammalian proteins, 3–4
  second generation, 4. See also Staphylococcal protein A, fusions
Baculovirus expression system, 129
  generation of recombinant virus, 130–131
Benzamidine–zinc biaffinity chromatography, of urokinase-related proteins, 95
Biaffinity chromatography
  applications of, 95
  of fibroblast growth factor, 91–95
Bombyx mori. See Silkworm expression system
Bone, organ culture, technique, 502, 510
Bone cells, human, culture, 509
Bone formation, assays for, 507–510
  alkaline phsophatase activity measurements, 507–508
  collagen synthesis measurements, 507–510
  mitogenesis assay, 507–508
  protein synthesis measurements, 507–510
Bone histomorphometry, quantitative, 506–507
Bone marrow culture system, 503–504
  autoradiography experiments using, 504
  morphologic studies using, 504
Bone morphogenesis factors, 360
Bone resorption, assays for, 502–507
  organ culture methods for, 502
Bovine aortic arch endothelial cells
  culture, 481
  DNA synthesis, measurement of, 481
  nuclear localization of basic fibroblast growth factor in, 480
Bovine mammary gland cDNA library
  screening of, 435–436
  synthesis of, 435

Bovine papillomavirus expression system, 175–176
  construction and preparation of vectors, 176–178

C

Capillary endothelial cells
  bovine adrenal cortex
    bioassay of vascular endothelial growth factor using, 394
    culture of, 394
  bovine brain cortex
    effect of basic fibroblast growth factor on, effect of monoclonal antibody on, 153–155
    growth, 153–154
    progeny of single cluster of, isolation of, 154–155
  DNA synthesis in, measurement of, 92–93
  proliferation, as bioassay of activity of recombinant basic fibroblast growth factor, 115–116
Cartilage
  growth factors from, purification of, 416–424
  source, 419
Cartilage-derived factors, 416–424
  assays, 417–418
  biological effects of, 416–417
  16-kDA, purification, 419–422
  molecular weight of, 423–424
  purification
    procedures, 419
    stabilization during, 518
  that stimulate chondrocyte colony formation in soft agar, purification, 422–423
Cartilage-inducing factors, 317
Cell membranes
  purification, 251–259
    procedure, 252–254
      large-scale, 251–252
      small-scale, 251–254
      speed of, 251–252
    purified
      characterization of, 255–256
      epidermal growth factor receptor profile of, 256–259

relative specific activity analysis, 255–256
resuspension, 254–255
storage, 255
transmission electron microscopy, 255–257
Cell surface receptors
  chimeric, 225–232
    characterization of, 231–232
    construction of, 225–231
    between EGF and insulin receptors, 225
    expression of, 231–232
    between insulin and insulin-like growth factor I receptors, 225–231
    of tyrosine kinase family, 225
Cellular retinoid-binding proteins, 425
Central nervous system, effects of nerve growth factor in, 35
CHAPS
  general properties of, 511–512
  noncytotoxic properties of, 512–514
  as noncytotoxic stabilizing agent for growth factors, 511–518
Chick brain, autoradiographic localization of IGF-I receptors in, 26–35
  binding studies, 28–31
    competition studies, 30–31
    incubation, 29
    materials, 28
    preincubation, 28–29
    reagents, 28
    rinsing, 29–30
    tissue fixation, 30
  tissue preparation, 27–28
Chick embryo, autoradiographic localization of IGF-I receptors in, 26–35
  binding studies, 28–31
    competition studies, 30–31, 34
    materials, 28
    preincubation, 28–29
    reagents, 28
    rinsing, 29–30
    tissue fixation, 30
  tissue preparation, 27–28
Chicken, osteoclasts, isolation of, 505
3-[(3-Cholamidopropyl)dimethylammonio]-1-propane sulfonate. *See* CHAPS

Chondrocyte colony-stimulating factor. *See also* Cartilage-derived factors
  purification, 422–423
Chondrocytes
  colony formation in soft agar, 417–418
  cartilage-derived factors that stimulate, purification, 422–423
  proteoglycan synthesis by, 416–417
Colony-stimulating factor-1, 78
Connective tissue activating peptide III, 405–416
  activity, 405
  analytic electrophoresis, 413–415
  analytic isoelectric focusing, 414
  assay methods, 406–415
    immunologic, 410–413
  bioassay, 406–410
    cell culture methods, 406–407
    [^{14}C]glucosamine incorporation assay, 407–408
    by [^{3}H]thymidine incorporation into fibroblast DNA, 408–409
  CTAP(des 1–4) form, 405, 407
  CTAP(des 1–13) form, 406
  CTAP(des 1–15) form, 406
  denaturing electrophoresis, 413–414
  double-antibody sandwich enzyme-linked immunosorbent assay, procedure, 412–413
  enzymatic cleavage, 415–416
  enzyme-linked immunosorbent assay, 411–412
  immunoaffinity chromatography, 410–411
  isoforms, 405–406
  microsequencing, 415
  optical density measurement, 413
  precursor, 405
  preparation of, 409
  purification, optimal conditions for, 415
  staining and blotting techniques for, 414–415
  structure, 407
CTAP-III. *See* Connective tissue activating peptide III
Cyclic AMP-dependent protein kinase
  phosphorylation of basic fibroblast growth factor, sites for, 138–139
  purified

phosphorylation of basic fibroblast
growth factor, 140–141
phosphorylation of epidermal growth
factor receptor *in vitro*, 235–236
source for, 140
β-Cyclodextrin–copper biaffinity column,
tetradecasulfated, for purification of
FGF from tumor, 95

D

Decapentaplegic complex of *Drosophila*,
348, 360
DNA, synthesis
in capillary endothelial cells, measurement of, 92–93
EGF-induced, inhibition, by synthetic
fragments of transforming growth
factor α, 207–208, 212
induction, by synthetic fragments of
transforming growth factor α, 207,
211

E

Endothelial cells. *See also* Bovine aortic
arch endothelial cells; Capillary endothelial cells; Human umbilical vein
endothelial cells
directional migration (chemotaxis),
assay, 446–447
proliferation assay, 446, 448
proteolytic enzyme activity, assay, 446
Epidermal growth factor, 200, 391
iodination of, 209
phosphorylation induced by, 208–209,
212
precursor
characteristics of, 175
as receptor, 175
recombinant
detection of, 181–185
in transfected cell lines
immunoblot analysis of, 181–183
immunoprecipitation of, 182–184
precursor cDNA, expression in animal
cells, 175–185
cell transfection, 178–180
construction and preparation of expression vector for, 176–178

mRNA blot hybridization analysis,
180–182
RNA isolation, 180–181
screening of transformants, 180–181
selection of transfected cells, by
neomycin resistance, 179–180
radioiodinated, binding to purified cell
membranes, 256, 258
receptor
activation, in Rat-1 or FR3T3 cells,
496–497
autophosphorylation, inhibition, 359
EGF-stimulated autophosphorylation,
496
gene
in vitro transcription assay, 247–250
in vitro transcription of, 242–250
sequence similarity to genes in
tumors, 272–273
transcription factors, 242
isolation of cell membranes for studies
of, 251–259
numbers, in various cell types, 496
phosphorylated residues, identification
of, 238–240
by inspection, 234, 238–239
by ^{32}P release during Edman degradation, 234, 239
split-filter technique for, 234, 239–240
phosphorylation, 233
in vitro, by purified protein kinases,
235–236
sites
characterization *in vivo* versus *in
vitro*, 241
identification of, 233–241
physiologic role of, 241
retention during purification, 240–241
physiological substrate of, 260–261
^{32}P-labeled
proteolytic digestion and peptide
purification, 236–238
purification of, 235
purification, 240
radiolabeling, *in vivo*, 234
tryptic phosphopeptide sequences and
elution behavior, 238
tyrosine kinase activity of, 233
structure, 203–204

Epidermal growth factor receptor transcription factor, 242
  purification of, 246–247
erbB-2/neu gene. See Oncogene, erbB-2/neu
Erythroid differentiation factor, 341
Erythropoietic factors, 360
Escherichia coli expression system, 129
  expression of recombinant human basic fibroblast growth factor in, 105–108
    growth and lysis of cells, 107–108
    purification of recombinant FGF, 108–109
    vector for, 105–107
  expression of recombinant insulin-like growth factors, 3–16
  preparation of recombinant transforming growth factor α as fusion protein in, 186–187
ETF. See Epidermal growth factor receptor transcription factor
Expression vector pBPVMTMhEGF, construction and preparation of, 176–178
Expression vector pEZZ
  construction of, 8–9
  structure, 8
Expression vector pMV7, 62, 64
Expression vector pTsF-9dH3, construction of, 105–107

F

Fatty acid-binding proteins, 425
FGF5. See also Oncogene, human, FGF-5
  homology with hst-1 protein, 128–129
FGF-6 gene, product, 97
FGR-s. See Fibroblast growth regulator, soluble form (FGR-s)
Fibroblast growth factor, 391–392. See also Heparin-binding growth factor; Heparin-binding growth factor family
  acidic, 96–97, 391
    activities of, 97
    and basic, sequence similarity, 95, 97
    homology with hst-1 protein, 128–129
    iodinated
      binding of, 478–479
      biological activity of, 479
      purification, stabilization during, 518

  iodination, 469
    lactoperoxidase method, 475–477
    purification, stabilization during, 518
    stabilization, during storage and dilution, 517–518
  assays, 91–93
  basic, 96–116, 391
    activity, 96, 138
    antibody
      affinity-purified, characterization of, 482–483
      enzyme-linked immunosorbent assay, 482
      preparation of, 481–482
      screening, 482
    bioavailability, 138, 140
    bovine, amino acid sequencing of, 102–103
    bovine brain
      cDNA, restriction map of, 118
      preparation of, 148–149
    cDNA clone encoding, isolation of, 98–102
    cloning, 97–98
    expression, analysis of, 123–124
    fusion with signal peptide sequence, 117–124
      materials for, 119
      method, 119–121
      overall strategy, 117–119
      plasmid resulting from (ppbFGF)
        cells expressing chimeric FGF from, 122–123
        construction of, 119–121
        structure of, 118
        transfection of NIH 3T3 cells with, 121–122
        transforming gene activity of, 117
        tumorigenicity of cells transfected with, 124
    homology-choice probe for, 98–101
    homology with hst-1 protein, 128–129
    human
      coding region, structure of, 103–105
      recombinant, 105–108
      sequencing of, 102–104
    iodinated
      binding of, 478–479
      biological activity of, 479
      intracellular distribution in growing and stimulated cells, 492–494

iodination, 469, 490–492
  lactoperoxidase method, 475–477
  localization of, 138
  localization of, in nucleus and nucleolus, 480
    cell system for, 481
    digitized video microscopy for, 484, 486
      digital image processing, 484–487
      fluorescence localization, 488
      fluorescence quantification, 488–490
      image analysis, 488–490
      mathematical image processing, 487–488
    indirect immunofluorescence, 481–485
    by ultrastructural immunocytochemistry, 490–494
  mitogenic activity, 96
  monoclonal antibody to, 148–157
    activity assay in plasma, 156–157
    applications of, 155–157
    determination of immunoglobulin classes, 152
    dissociation constant, determination of, 152–153
    effect on biological activities of basic FGF, 153–155
    enzyme-linked immunosorbent assay for, 151–152
    generation of athymic mice with high blood level of, 156
    production of, 150–152
      cell lines for, 150
      cloning, 151
      hybridoma generation, 151
      immunization of mice, 150
      media for, 150
    properties of, 152–155
    purification of, 152
    radioimmunoassay, 152–153
    specificity, determination of, 152–153
  phosphorylated
    biological assays of, 145–147
    detection of, 141–142
    heparin-binding assay, 146
    phosphoamino acids in, identification of, 141–142
    radioreceptor assay on BHK cells, 146–147
  phosphorylation
    by cells in culture, 144–145
    in vitro, by purified kinase, 140–141
    kinetic analyses of, 142–144
    by protein kinase C and cAMP-dependent protein kinase, sites for, 138–139
    posttranslational modification of, 140
    purification, stabilization during, 518
    radioiodination of, 149–150
  recombinant
    analysis of, 109–116
    bioassay, 115–116
    expression of, 98, 140
    heparin-TSK HPLC, 110–112
    purification of, 108–109
    reversed-phase HPLC, 110
    SDS–polyacrylamide gel electrophoresis, 112–115
  source, 140
  stabilization, during storage and dilution, 517–518
  structure, 148
  subcellular localization, by immunofluorescence microscopy, 483–485
  biaffinity chromatography of, 91–95
  in cartilage preparations, 424
  iodination of, chloramine-T method, 467
  isolation of, 93
  radioreceptor assay of, 418
  receptors, 158–174
Fibroblast growth factor family, 97
Fibroblast growth factor–like factor, in conditioned medium from SC-3 cells, purification, stabilization during, 514–516
Fibroblast growth regulator, soluble form (FGR-s), 425, 451
  fractionation of, 454–456
  growth-inhibitory activity of
    assays of, 452–454
    depletion of, on affinity column containing Antibody 2A4, 463–464
  effect of monoclonal antibody on, 463
  immunogen fraction, isolation of, 451–456
  in vitro immunization with, 456–457
  monoclonal antibody against, 451–464
    characterization of, 462
    derivation of

generation of hybridoma cells, 456–458
hybridoma screening assays, 458–462
immunogen for, 451–456
effect of, on growth-inhibitory activity of FGR-s, 463
isolation of, 462
properties of, 462–464
target of, identification of, 462
properties of, 451
Follicle-stimulating hormone, secretion, regulation of, 341, 347
Folliculostellate cells, primary cultures, 392–394

G

Gene transfer, 62–64
Glial fibrillary acidic protein
expression in mouse embryo cells, 337
immunoassay, in mouse embryo cells in culture, 337–340
Glycerophosphoinositides, novel, high-performance liquid chromatography of, 85–86
quantification of data, 86–87
gro gene, 373. *See also* Melanoma growth stimulatory activity
Growth factors. *See also* Peptide growth factor; *specific growth factor*
autophosphorylation, inhibition, 359–360
in vivo effects, transgenic mouse models for study of, 519–525
stabilization
CHAPS in, 511–518
during purification, 513–517
during storage and dilution, 517–518
Growth inhibitor. *See also* Mammary-derived growth inhibitor
from rabbit serum, purification, stabilization during, 516–517

H

H64A subtilisin
cleavage of insulin-like growth factor II on C-terminal side of PheAlaHisTyr, 11–12
cleavage of truncated insulin-like growth factor I on C-terminal side of PheAlaHisTyr, 11

cleavage of ZZ–IGF-II with, 13–14
cleavage of ZZ–tIGF-I with, 13–14
Heparin affinity column, preparation, 409–410
Heparin-binding growth factor, receptors for, 137–138
Heparin-binding growth factor-1, 97. *See also* Fibroblast growth factor
cell surface receptors
binding assay of, 161–165
binding to HepG2 cell surface receptors, 161–165
kinetics of, 165–168
iodinated
covalent cross-linking to HepG2 cells, 168–170
receptor complex with, cross-linking of, with disuccinimidyl suberate, 168–171
separation from free $Na^{125}I$, 160–161
iodination of, 158–160
preparation for iodination, 158–160
receptor sites on HepG2 cells, characterization of, 169–171
Heparin-binding growth factor-2, 97. *See also* Fibroblast growth factor
receptor gene, 171
Heparin-binding growth factor family, 128–129, 137–138, 158
mechanism of action of, 158
properties of, 158
Heparin–copper biaffinity chromatography, of fibroblast growth factor, 91–95
method, 94–95
principle of, 91–92
Heparin-TSK HPLC, of recombinant basic fibroblast growth factor, 110–112
High-performance liquid chromatography, of deacylated phospholipids, 85–86
HSTF1 gene, 137
hst-2/FGF6 protein, amino acid sequence, 128
hst-1 gene, 124–138. *See also* Oncogene, human, *hst*
cloning, from cosmid libraries, 126
expression of
in embryos and germ cell tumors, 135–136
in mouse teratocarcinoma cell line F9, compared to expression of *int-2*, 135–136

growth factor encoded by, 124–125
identification of, 124–125
  from normal genomic libraries, 126–128
  in nongastric cancers, 125–126
protein
  amino acid sequence, 125
  homology with fibroblast growth factors and related molecules, 128–129
  recombinant
    angiogenic activity of, 135
    growth factor activity of, 133–135
    purification of, 132–133
    synthesis of, 129–133
transcription, suppression in normal adult cells, 127–128
transforming potential of, 126–128
Human brain, nerve growth factor mRNA, 54
Human umbilical vein endothelial cells
  bioassay of vascular endothelial growth factor using, 394
  growth stimulation, by recombinant *hst-1* protein, 134
HUVE cells. *See* Human umbilical vein endothelial cells
Hybridoma
  generation of, 456–458
    fusion of immune spleen cells with myeloma cells, 458
    screening assay, 458–462

I

Immunoglobulin G, biotinylation of, 412
Inhibin, 360
  activity, 340–341, 347–348
  binding
    comparative analysis of techniques, 357–358
    to TM3 cells, 349–351
  fluorescein isothiocyanate conjugated, analysis of binding using fluorescence-activated cell sorting, 354–357
  fluorescein isothiocyanate conjugation of, 354–357
  forms of, 341, 348
  labeled, purification of, 352–353
  labeling, comparative analysis of techniques, 357–358
  receptor, 348
  recombinant
    iodination of, 354
    metabolic labeling of, 349
  structure, 341
  subunits, 341, 348
Insulin, 78
  receptor, 17. *See also* Cell surface receptors, chimeric
    autoradiographic localization in rat brain, 35
Insulin-like growth factor(s). *See also* ZZ-IGF fusion proteins
  affinity purification of, using IgG-binding ZZ fusion protein concept, 3–4
  I
    characterization, 14–15
    purification, 14–15
    radioreceptor assay of, 418
  receptor. *See also* Cell surface receptors, chimeric
    amino acid sequencing, 18–19, 23–25
    autoradiographic localization of, 26–35
    cDNA
      characterization, 22–26
      cloning, 21–22
    cDNA library, screening, 21–22
    functional and structural similarities to insulin receptor, 17–18
    functions, 17–18
    molecular weight, 17
    nucleotide sequence analysis, 23–25
    oligonucleotide probes for, 19–20
    precursor, 17
    purification, 18–19
    subunits, 17, 23–26
    tyrosine kinase enzymatic domains, characterization, 23–26
  release from ZZ fusion proteins, cleavage method for, 9–10
  truncated
    characterization, 14–15
    cleavage
      after methionine residue, with cyanogen bromide, 10–11
      on C-terminal side of PheAlaHis-Tyr with H64A subtilisin, 11
      on C-terminal side of PhePhe-ProArg sequence with thrombin, 11

on C-terminal side of tryptophan
with N-chlorosuccinimide, 11
purification, 14–15
release from ZZ fusion proteins,
cleavage method for, 9–11
structure, 10
II, 26
characterization, 14–15
cleavage
on C-terminal side of Met with
cyanogen bromide, 11
with H64A subtilisin on C-terminal
side of PheAlaHisTyr, 11–12
purification, 14–15
stabilization during, 518
receptor
autoradiographic localization in rat
brain, 31–35
in chick tissue, 31
release from ZZ fusion proteins,
cleavage method for, 9, 11–12
recombinant, expression and purification
of, 3–16
int-2 gene. *See also* Oncogene, murine,
int2
chromosomal localization of, 136–137
expression of, in mouse teratocarcinoma
cell line F9, compared to expression
of *hst-1*, 135–136
protein, homology with *hst-1* protein,
128–129

K

Keratinocyte growth factor, 97
homology with *hst-1* protein, 128–129
Keratinocytes, growth in culture, effects of
CHAPS on, 512–513

L

Ligand affinity chromatography, of fusion
proteins, 3–4
Lipids, thin-layer chromatography-purified,
deacylation of, 84
Lipocortin-1
EGF-dependent phosphorylation of, in
A431 cells, 263–265
phosphorylated
cyanogen bromide mapping of, 267–
270
tryptic peptide mapping of, 269–272

phosphorylation, by epidermal growth
factor receptor, 260–272
in cell-free system using A431 cell
membranes, 264–265
in vitro reactions, 262–263
labeling and immune precipitation
procedures, 261–262
methods, 261–263
phosphorylation site, localization of, by
peptide mapping, 267–272
purification of, 266–267
Lipocortins
calcium binding, 260
physiologic significance of, 260–261
properties of, 260
sequence similarities of, 260

M

Mammary-derived growth inhibitor, 425–
440
activity, 425
cDNA encoding
dot-blot analysis, 436–438
isolation of, 434–438
sequencing information, 436–438
expression, 425–426
coupling to differentiation in mammary gland, 426
gene transcription in mammary glands,
localization of, by *in situ* hybridization, 426, 438–440
growth inhibitor assay with Ehrlich
ascites mammary carcinoma cells,
425–427
immunoassay, 430–432
modulation of β_2-adrenergic receptor
function, assay of, 433–434
mRNA, preparation of, 434–435
physicochemical properties, 425
proliferation assay, using permanent
mammary carcinoma cell lines, 432–
433
proteins homologous to, 425
purification, 427–430
radioiodination, 430–431
structure, 425
Mastomys
kidney, nerve growth factor, mRNA, 54
submaxillary gland, nerve growth factor,
48
mRNA, 54

MCF-7 cells, TPA-induced
  cell culture, 216
  conditioned medium
    collection of, 216
    concentration and preparation of acid-soluble extract from, 216–217
    purification of amphiregulin from serum-free conditioned medium of, 214–224
Melanoma growth stimulatory activity, 373–383
  activity, 373
  assays, 374–375
  cell number assay, 375
  denaturing electrophoresis, 375
  gene, expression, 373
  [^{3}H]thymidine incorporation assay, 374–375
  natural and recombinant, similarities of, 382–383
  purification, 373
    materials, 374
    method, 375–377
  recombinant
    amino-terminal sequencing, 378–380
    bioassay, 382
    characterization of, 378–382
    immunoprecipitation, 380–382
    O-glycanase digestion, 382
    purification, method, 377–378
  structure, 373
Messenger RNA. *See also* Nerve growth factor, messenger RNA
  poly(A)$^+$, yield, from various species, 54
  purification, 49–55
    oligo(dT)-cellulose chromatography, 50, 52–53
    tissue homogenization in presence of guanidine thiocyanate, 49–52, 54
    tissue homogenization in presence of proteinase K, 53–54
MGSA. *See* Melanoma growth stimulatory activity
Mouse. *See also* Transgenic mouse models
  calvaria, assays of bone resorption using, 502–503
  embryonic cell cultures, 337
Müllerian inhibiting substance, 348, 358–369
  activity, 358–359
    augmentation, 359
  bioassay for, 360–362
  gene
    cloning, 365–367
    expression, 367–369
  inhibitors, 358–359
  physicochemical properties, 358
  production, 358–359
  purification, 362–365
    by immunoaffinity purification, 363–365
    by serial chromatography, 362–363
  structure, 358
Myelin P-2, 425

N

NAP-2, 406
Nerve growth factor
  delivery to brain. *See also* Rat brain, nerve growth factor infusion apparatus for
    means of, 48
  in vivo effects of, model systems for demonstration of, 35
  messenger RNA, 48–61
    autoradiography, 57–58
    denaturing agarose gel electrophoresis, 55
    dot blots, 58–59
    hybridization, 57–58
    Northern blotting, 55–56
    polymerase chain reaction, 59–60
    probe preparation, 56–57
    purification, 49–55
    quantitation, 57–58
    reprobing, 57–58
  neuron-specific effects of, *in vivo* assay of, 35–48
    effects on cholinergic neurons, 42–47
    effects on noncholinergic neurons, 47
    fimbria/fornix model for, 36–42
    infusions in nonhuman primates, 47
    model systems for, 36
  physiological role of, 48–49
  production in fibroblasts, 48
  radioiodination, 70
  receptors
    in adult mammals, 35–36
    affinity cross-linking of, 70–72
    affinity labeling, 69–72
    detection of, after gene transfer, 61–72

erythrocyte rosette assay, 65–67
gene transfer, 62–64
  by retroviral infection, 64
  indirect immunofluorescence, 62, 67–68
  in situ rosetting, 62
  monoclonal antibodies against, 61
species distribution, 48
tissue distribution, 48
NIH 3T3 cells
  anchorage-independent growth of, produced by *hst-1* protein, 134–135
  density-inhibited cultures, 451–452
    conditioned growth medium from, 452–453
      fractionation of components, 454–456
  DNA synthesis in, stimulation of, by recombinant *hst-1* protein, 133–134
  in vivo immunization of rats with, 457
  transfected with bFGF–signal peptide fusion construct, 117
    focus forming assay, 122
    selection of lines expressing chimeric FGF, 122–123
    tumorigenicity, 124
  transfection with bFGF–signal peptide fusion construct, method, 121–122
Nonhuman primates, neuron-specific effects of nerve growth factor in, *in vivo* assay of, 47

O

Oncogene
  erbB-2
    chromosomal localization of, 273
    product (gp185^{erbB-2}), 274
      antibodies, cytotoxic effects with tumor necrosis factor α, 289
    overexpression
      in model systems *in vitro*, 275–277
      in tumors, 289
  erbB-2/neu, 272–277
    choice of assay, 291
    DNA/RNA blot analysis, 291–294
    expression, abnormalities of, in tumors, 290
    gene amplification, 290–291, 500
    identification of, 272–274
    immunohistochemistry, 291, 294–300
      choice of antibodies against *erbB-2/neu* protein, 294–297
      controls, 299–300
      protocol for, 298–299
      tissue preparation, 297–298
      using phosphotyrosine antibodies, 500
    mRNA overexpression, 290–291
    nomenclature, 272–274
    normal and activated, transforming potential in model systems, 275–277
    Northern blot analysis, 293
    protein overexpression, 290–291
    Southern blot analysis, 291, 293–294
    tissue levels, quantification of, 290–300
    in tumors, role of, 274–275
    Western blot analysis, 291
  human
    FGF-5, product, 97
    hst, product, 97
  immunologic studies of, using phosphotyrosine antibodies, 500–501
  met, immunoblotting, using phosphotyrosine antibodies, 500–501
  murine, *int2*, product, 97
  neu, 272
    immunoprecipitation with phosphotyrosine antibodies, 499
    product (p185^{neu}), 277
      antibody-induced down-modulation of expression of, 281–282, 288
      assay for, 281–282
      phenotypic effects of, 282–285, 288
      monoclonal antibodies to, 278–281
        cytostatic and cytotoxic antitumor effects of, 287–288
        in vivo effects on tumorigenic growth of *neu*-transformed cells, 285–286, 288
        similarity to epidermal growth factor receptor, 277
  neuN, 274
  neuT, 274
  ras family, 272
Osteoclasts
  culture of, 504–505
  isolated
    activation, detection of, by increase in

tartrate-resistant acid phosphatase, 506
bone resorption assay using, 505–506
isolation, 504–505

P

Peptide growth factor
  immunoprecipitation, with phosphotyrosine antibodies, 498
  iodination of, 467–479
    with Bolton–Hunter reagent, 467–469
    chloramine-T method, 467–469
    choice of method, 467–468
    iodine monochloride method, 467–468
    Iodobeads method, 467–469
    Iodogen method, 467–469
    lactoperoxidase method, 467–469
    methods, 467
  localization of, in nucleus, 480–494
Peripheral nervous system, effects of nerve growth factor in, 35
Phosphatidylinositol 3,4-bisphosphate, 79
Phosphatidylinositol 3-phosphate, 78–79
  distribution, 79
Phosphatidylinositol 4-phosphate, 79
Phosphatidylinositol 3,4,5-trisphosphate, 79
Phosphoinositide kinase, in cell growth and transformation, 78
Phospholipase C, in A431 cells, purification, using phosphotyrosine antibodies, 500
Phospholipids
  deacylated, high-performance liquid chromatography of, 85–86
  thin-layer chromatography-purified, deacylation of, 84
Phosphorylation sites, identification of, 233–241
  in EGF receptor sequence
    materials for, 234–235
    principle of, 234
Phosphotyrosine
  antibodies, 494–501
    applications of, 494–495
    cross-reactivity, 495
    immunoblotting
      in pathological analysis of human tumors, 500–501
      in studies of human disease, 500
    immunoprecipitation with, 497–501
    bacterial colony and plaque screens, 498
    buffers for, 497
    growth factors without receptors, 498
    receptors without growth factors, 499
    substrate identification using, 499–500
    sources, 495
    in studies of human disease, 500
    substrate identification with, 499–500
  antisera, immunodetection methods for, 495
  monoclonal antibodies, sources, 495
  polyclonal antibodies, sources, 495
PI3-kinase, 78
  activation, 79
  in cellular proliferative response to PDGF, 79
  immunoprecipitation of, from PDGF-stimulated smooth muscle cells, 83–84
  novel phosphoinositides produced by, 78–80
    detection of, in vascular smooth muscle cells, 80–82
    high-performance liquid chromatography of, 85–87
    production of novel polyphosphoinositides, 83–84
Plasmid pspbFGF. See Fibroblast growth factor, basic, fusion with signal peptide sequence
Plasmid pTsF-9dH3, construction of, 105–107
Plasmid vector pEZZ
  construction of, 8–9
  structure, 8
Platelet-derived endothelial cell growth factor, 383–392
  activity, 383
  amino acid sequencing, 386–389
  assay for, 384
  cDNA cloning, 387–390
  expression of, 390
  purification, 384–386
  tissue distribution, 383
Platelet-derived growth factor, 383

A-type receptor. *See* Platelet-derived growth factor, α receptor
B-type receptor. *See* Platelet-derived growth factor, β receptor
human
  iodination, 467–469
    with Bolton–Hunter reagent, 473–475
    chloramine-T method, 469–471
  structure, 467
immunoprecipitation, with phosphotyrosine antibodies, 498
iodinated
  binding of, 472–473
  biological activity of, 473
iodination of, 467
isoforms, 72
nuclear targeting, 480
polyphosphoinositides produced in response to, 78–87
porcine
  iodination, 468–469
    with Bolton–Hunter reagent, 473–475
  structure, 467
receptor
  cDNA, 498
  immunoprecipitation, with phosphotyrosine antibodies, 498
  intrinsic kinase activity, 72
  α receptor, 72
    cloning and expression of, 75–77
    structure, 74
  β receptor, 72
    cloning and expression of, 72–74
    structure, 73–74
Polymerase chain reaction, for detection of nerve growth factor messenger RNA, 59–60
Protein
  expression, analysis in transgenic mouse models, 524
  stability, and fusion to staphylococcal protein A, 7
Protein kinase, purified
  growth factors as substrates for, 147
  phosphorylation of epidermal growth factor receptor *in vitro*, 235–236
Protein kinase C
  phosphorylation of basic fibroblast growth factor, sites for, 138–139

purified
  phosphorylation of basic fibroblast growth factor, 140–141
  phosphorylation of epidermal growth factor receptor *in vitro*, 235–236
  source for, 140
Protooncogene. *See* Oncogene

R

Rat. *See also* Mastomys
  fetal, assays of long bone resorption using, 502
  liver, nerve growth factor mRNA, 54
  osteoclasts, isolation of, 504–505
  sciatic nerve, nerve growth factor mRNA, 54
  submaxillary gland, nerve growth factor, 48
    mRNA, 54
Rat brain
  autoradiographic localization of IGF-I receptors in, 26–35
    binding studies, 28–31
    incubation, 28–29
    materials, 28
    preincubation, 28–29
    reagents, 28
    rinsing, 29–30
    tissue fixation, 30
    competition studies, 30–33
    tissue preparation, 27–28
  autoradiographic localization of IGF-II receptors in, 31–35
  autoradiographic localization of insulin receptors in, 35
  fimbria/fornix lesions
    distance of axotomy from cell body, 42
    surgery for, 39–41
    unilateral versus bilateral, 40–42
    for *in vivo* assay of neuron-specific effects of nerve growth factor, 36–42
  hippocampal source of nerve growth factor in, 37–38
  nerve growth factor infusion apparatus for, 39–41
  parenchymal damage from, 42
  nerve growth factor mRNA, 54

olfactory bulb, autoradiographic localization of IGF-I receptors in, 30–31
septohippocampal projection, neuron-specific effects of nerve growth factor in, 36–42
Receptor, immunoprecipitation, with phosphotyrosine antibodies, 499
Receptor tyrosine kinase, 17
  genes encoding, bacterial colony and plaque screens for, with phosphotyrosine antibodies, 498
  of growth factor receptors, assay, 496–497
  in human disease, studies of, 500
  mutations, bacterial colony and plaque screens for, with phosphotyrosine antibodies, 498
Recombinant DNA technology, 3
Reversed-phase HPLC, of recombinant basic fibroblast growth factor, 110
RNA. *See also* Messenger RNA
  analysis, in transgenic mouse models, 523–524

S

Serum-free mouse embryo cells, 337
  culture, 337–339
SFME. *See* Serum-free mouse embryo cells
Silkworm expression system, 129
Sp1 (epidermal growth factor receptor transcription factor), 242
Staphylococcal protein A
  fusions
    applications of, 5
    for expression and purification of insulin-like growth factor(s), 3, 5–8
  IgG-binding domains, 5–7
  proteolytic stability of, 7
  signal peptide, 5–7
  structure-function relationships in, 5–7

T

TC-10 medium, 130–131
12-*O*-Tetradecanoylphorbol 13-acetate, cell lines treated with, purification of amphiregulin from serum-free conditioned medium of, 214–224

β-TG. *See* Connective tissue activating peptide III, CTAP-III(des 1–4)
TPA. *See* 12-*O*-Tetradecanoylphorbol 13-acetate
Transforming gene. *See also* Fibroblast growth factor, basic, fusion with signal peptide sequence; *hst-1* gene
Transforming growth factor α, 185–191, 391
  activity, 185–186, 191
  amino acid sequence of, 193, 200
  antibody, 186
    purification of, 189–190
  antipeptide antibodies, generation of, 187–188
  assay of, 190–191
  characteristics of, 185
  enzyme-linked immunosorbent assay, 190
  expression as fusion protein in *Escherichia coli*, 186–187
  fragment sequences, selection of, 202–204
  hydrophilicity plot for, 202
  immunoassay, 186
  monoclonal antibodies, generation of, 188–189
  peptide–thyroglobulin conjugates
    antipeptide immunoglobulins, isolation of, 195
    immunization protocol, 195
    preparation of, 192, 194–195
  phosphorylation induced by, 208–209, 212
  polyclonal antibodies, generation of, 187
  precursor, 185, 200
  radioreceptor assay, 190–191
  receptor, 185, 192
  receptor binding activity
    analysis of, 192–193
    effect of antipeptide antibodies on, 192, 194–195
    synthetic peptide analysis, 192–194
  receptor-binding domain, 202, 213
  recombinant
    postsynthetic modification of, 192–193, 198–199
    preparation of, 186–187
    site-directed mutagenesis of, 192, 195–198
  structure, 185, 200, 203–204

structure–function analysis of, 191–200
synthetic fragments of
  biological activity of, 200–213
  chemical characterization of, 210
  competition with epidermal growth
    factor binding, 206–207, 210–211
  induction of DNA synthesis, 207, 211
  inhibition of EGF-induced DNA synthesis, 207–208, 212
  materials and methods for, 204–206
  phosphorylation induced by, 208–209, 212
  physicochemical properties of, 210
  preparation of, 204–206
  receptor binding, 202–203
  significance of findings with, 212–213
  stimulation of anchorage-independent cell growth, 208, 212
  tumorigenicity, 191–192
Transforming growth factor β, 348, 360, 391
  activity, 303, 327
  assays, 303–304
  astrocyte differentiation-inducing activity of, 337–340
  bioassay for, 314–315
  in cartilage extract, 424
  cell culture assay, with mink epithelial cells, 323–325
  chondrogenesis assay
    agarose gel, 324–326
    in presence of dihydrocytochalasin B, 326–327
  forms of, 303–304
  isoforms, 327
  latent, 327–336
    activation of, by heparin, 334
    identification of, sample preparation, 329–330
    from platelets, identification of, 330
  monoclonal antibody neutralization of, in vitro, 316
  polyclonal antibodies to, production of, 309–310
  purification, 328
  radioiodinated
    chromatographic purification of, 306
    Sephadex G-100 chromatography, 306
    storage, 306
    trichloroacetic acid precipitation, 307
  radioreceptor assay for, 308–309

receptor, 186, 303
receptor binding, 336
structure, 185–186, 303
Transforming growth factor $\beta 1$, 383
  activity, 317–318
  in bone formation, 318
  bovine and human, sequence similarity, 321–322
  latent
    activation of, 330–334
      by alkalinization, 332
      by boiling, 332
      by endoglycosidase F, 333
      enzymatic, 332–333
      by monosaccharides, 333–334
      by plasmin, 332–333
      by sialidase, 333
      by urea, 332
    from platelets, purification of, 334–336
  monoclonal antibodies to, production of, 311–312, 316
  physiologic role of, 317–318
  properties of, 317, 327–328
  purification, 318–321
  quantitation of, enzyme-linked immunosorbent assay for, 312–314
  radioimmunoassay for, 307–308
  radioiodination of, 304–306
    chloramine-T procedure, 305
    lactoperoxidase procedure, 305–306
  storage, 321–322
  structure, 317
Transforming growth factor $\beta 2$
  activity, 317–318
  in bone formation, 318
  bovine and human, sequence similarity, 321–322
  microheterogeneity of, 322–323
  properties of, 317
  purification, 318–323
  storage, 321–322
  structure, 317
  tissue distribution, 317
Transgenic mouse models
  growth factor expression in
    by generalized overexpression, 521–522
    site-directed, 522
    strategies for, 519
    by use of native gene, 521

for growth factor studies, 519–525
  analysis of offspring, 523–525
    assessment of levels of expression, 523–524
    pathological analysis, 525
    phenotypic characterization, 523
  transgene for
    construction of, 519–520
    testing, prior to injection into mouse, 520–521
Transphosphorylation, 496
Tumor cell growth inhibitor, from rabbit serum, purification, stabilization during, 516–517
Tumor growth, *in vivo*, effects of antireceptor antibodies on, 288–290
Tumor necrosis factor, 391
Tyrosine kinase. *See* Receptor tyrosine kinase

U

Urokinase-related proteins, benzamidine–zinc biaffinity chromatography of, 95

V

Vaccinia virus growth factor, 200
Vascular endothelial growth factor, 392–405
  angiogenic effect, 403–405
  bioassays, 394
  bovine, cDNA clones encoding
    isolation of, 398–401
    sequence information, 400
  characterization of, 402–404
  cloning, 398–402
  forms of, 402–403
  human, cDNA clones encoding, isolation of, 401–402
  in vivo functions, 404–405
  microsequencing, 398
  purification, 395–398
    concentration of conditioned medium, 395
    heparin-Sepharose affinity chromatography, 395–396
    high-performance liquid chromatography, 396–398
  transient expression of, in mammalian cells, 402
Vascular permeability factor, 404
Vascular smooth muscle cells, human
  cellular lipids
    deacylation of, 82–83
    extraction of, 81–82
    inorganic ^{32}P labeling, 80–81
    myo[^{3}H]inositol labeling, 80–81
  PDGF-stimulated, immunoprcipitation of PI3-kinase from, 83–84
  stimulation with PDGF, 81
Vgl, 360

Z

ZZ–IGF fusion proteins
  expression, 12
  purification, 12
  site-specific cleavage
    with HG64A subtilisin, 16
    methods, 16
ZZ–IGF-I
  cleavage, with hydroxylamine, 12–13
  structure, 6
ZZ–IGF-II
  cleavage
    with cyanogen bromide, 14
    with H64A subtilisin, 13–14
  structure, 6
ZZ–tIGF-I
  cleavage
    with H64A subtilisin, 13–14
    with N-chlorosuccinimide, 13
    with thrombin, 14
  partial cleavage, with cyanogen bromide, 13
  structure, 6

253738